FIFTH EDITION

Mathematics for the Trades

A GUIDED APPROACH

Robert A. Carman
Santa Barbara City College

Hal M. Saunders
Santa Barbara School District

PRENTICE HALL
Upper Saddle River, New Jersey   Columbus, Ohio

Library of Congress Cataloging-in-Publication Data

Carman, Robert A.
  Mathematics for the trades : a guided approach / Robert A. Carman,
Hal M. Saunders. — 5th ed.
     p.   cm.
  Includes indexes.
  ISBN 0-13-907783-9
  1. Mathematics.   I. Saunders, Hal M.   II. Title.
QA39.2.C35 1999
513′.14—dc21   98-18951
                                                            CIP

Cover art: "Tools of the Trades," © Cynthia Martin
Editor: Stephen Helba
Production Supervision: Clarinda Publication Services
Design Coordinator: Karrie M. Converse
Cover Design: Ceri Fitzgerald
Production Manager: Deidra M. Schwartz
Marketing Manager: Frank Mortimer, Jr.

This book was set in Century Schoolbook by The Clarinda
Company and was printed and bound by Courier—Kendallville, Inc.
The cover was printed by Phoenix Color Corp.

© 1999, 1996 by Prentice-Hall, Inc.
Simon & Schuster/A Viacom Company
Upper Saddle River, New Jersey 07458

Earlier editions © 1993, 1986, 1981 by Regents/Prentice Hall.

Printed in the United States of America

10 9 8 7 6 5 4 3 2

ISBN: 0-13-907783-9

Prentice-Hall International (UK) Limited, *London*
Prentice-Hall of Australia Pty. Limited, *Sydney*
Prentice-Hall of Canada, Inc., *Toronto*
Prentice-Hall Hispanoamericana, S.A., *Mexico*
Prentice-Hall of India Private Limited, *New Delhi*
Prentice-Hall of Japan, Inc., *Tokyo*
Simon & Schuster Asia Pte. Ltd., *Singapore*
Editora Prentice-Hall do Brasil, Ltda., *Rio de Janeiro*

Contents

CHAPTER 8 **Practical Plane Geometry**

CHAPTER 9 **Solid Figures and Geometric Constructions**

Contents

Preface

This book provides the practical mathematics skills needed in a wide variety of trade and technical areas, including electronics, auto mechanics, construction trades, air conditioning, machine technology, welding, drafting, and many other occupations. It is especially intended for students who have a poor math background and for adults who have been out of school for a time. Most of these students have had little success in mathematics, some openly fear it, and all need a direct, practical approach that emphasizes careful, complete explanations and actual on-the-job applications. This book is intended to provide practical help with real math, beginning at each student's own individual level of ability.

Features

Those who have difficulty with mathematics will find in this book several special features designed to make it most effective for them:

- Careful attention has been given to **readability.** Reading specialists have helped plan both the written text and the visual organization.
- A **diagnostic pretest** and performance **objectives** keyed to the text are included at the beginning of each unit. These clearly indicate the content of each unit and provide the student with a sense of direction.
- Each unit ends with a **problem set** covering the work of the unit.
- The **format** is clear and easy to follow. It respects the individual needs of each reader, providing immediate feedback at each step to ensure understanding and continued attention. The emphasis is on *explaining* concepts rather than simply *presenting* them. This is a practical presentation rather than a theoretical one.
- Special attention has been given to **on-the-job math skills,** using a wide variety of real problems and situations. Many problems parallel those that appear on professional and apprenticeship exams. The answers to all problems are given in the back of the book.
- A light, lively **conversational style** of writing and a pleasant, easy-to-understand visual approach are used. The use of humor is designed to appeal to students who have in the past found mathematics to be dry and uninteresting.

Four editions and a decade of experience with a wide variety of students indicate that this approach is successful—the book works and students learn, many of them experiencing success in mathematics for the first time.

Flexibility of use was a major criterion in the design of the book. Field testing and extensive experience with the first edition indicate that the book can be used successfully in a variety of course formats. It can be used as a textbook in traditional lecture-oriented courses. It is very effective in situations where an instructor wishes to modify a traditional course by devoting a portion of class time to independent study. The book is especially useful in programs of individualized or self-paced instruction, whether in a learning lab situation, with tutors, with audio tapes, or in totally independent study.

Calculators

Calculators are a necessary tool for workers in trade and technical areas, and we have recognized this by using calculators extensively in the text, both in finding numerical solutions to problems, including specific keystroke sequences, and in determining the values of transcendental functions. We have taken care to first explain all concepts and problem solving without the use of the calculator and to estimate and check answers. Many realistic problems included in the exercise sets involve large numbers, repeated calculations, and large quantities of information and are ideally suited to calculator use. They are representative of actual trades situations where a calculator is needed. Detailed instruction on the use of calculators is included in special sections at the end of appropriate chapters or is integrated into the text.

Supplements

An extensive package of supplementary materials is available for teachers and students using this textbook:

- An Instructor's Solutions Manual containing completely worked out solutions to all exercises and providing step-by-step procedures to help students learn problem-solving skills.
- Transparency Masters for over 130 illustrations, charts, and other visual materials.
- A Test Item File of approximately 1000 problem-type, multiple-choice, and fill-in questions for a full range of testing.
- The Prentice Hall Test Manager, a state-of-the art system that provides, on disk, the test item file.
- A text/workbook, *How to Study Technical Math,* to illustrate methods and techniques for solving a wide range of problems in technical math.

Fifth Edition

In response to suggestions from many instructors who have used earlier versions of the text, this fifth edition includes a number of substantial revisions and improvements:

- All problem sets and examples have been revised and brought up to date. A significant number of new applied problems have been added to the exercise sets, and additional worked examples have been added where needed throughout the text.
- Instructions on translating whole numbers and decimal numbers from numeral form to word form and the reverse have been added or expanded.
- The material on rounding has been revised and expanded.
- Instructions on the use of calculators with fractions have been revised to reflect the wide availability of fraction keys on scientific calculators. Instruction on determining square roots has been revised to reflect the use of calculators.
- Chapter 7 on basic algebra has been revised and the pace slowed. The section on solving simple equations has been made more accessible to students by splitting it into two sections, treating the solution of two-step equations separately.
- Explanations in the many step-by-step examples have been revised and simplified where needed.

Acknowledgments

It is a pleasure to acknowledge the help of many people who have contributed to the development of this book. 'Lyn Carman spent countless hours interviewing trades workers, union leaders, apprentices, teachers, and training program directors so that their experience could be used in developing realistic problems, an effective format, and appropriate content. Her contributions were invaluable to us.

The staff of Prentice Hall provided outstanding assistance at every step of the development and production of this fifth edition. We are especially grateful to editor Steve Helba for his guidance and assistance and to Louise Sette and Clarinda Publication Services for coordinating this edition through the production process.

We are indebted to the following teachers who read preliminary versions of the text and offered many helpful suggestions:

Robert Ahntholz, Coordinator—Learning Center, Central City Occupational Center, Los Angeles, California

Dean P. Athans, East Los Angeles College, Monterey Park, California

Frances L. Brewer, Vance-Granville Community College, Henderson, North Carolina

James W. Cox, Merced College, Merced, California

B. H. Dwiggins, Systems Development Engineer, Technovate, Inc., Tacoma, Washington

Kenneth R. Ebernard, Chabot College, Hayward, California

Hal Ehrenreich, Northcentral Technical College, Wausau, Wisconsin

Donald Fama, Chairperson, Mathematics-Engineering Science Department, Cayuga Community College, Auburn, New York

James Graham, Industrial Engineering College of Chicago, Chicago, Illinois

Mary Glenn Grimes, Bainbridge College, Bainbridge, Georgia

Ronald J. Gryglas, Tool and Die Institute, Chicago, Illinois

Vincent J. Hawkins, Chairperson, Department of Mathematics, Warwick Public Schools, Warwick, Rhode Island

Bernard Jenkins, Lansing Community College, Lansing, Michigan

Chris Johnson, Spokane Community College, Spokane, Washington

Judy Ann Jones, Madison Area Technical College, Madison, Wisconsin

Robert Kimball, Wake Technical College, Raleigh, North Carolina

J. Tad Martin, Technical College of Alamance, Haw River, North Carolina

Gregory B. McDaniel, Texas State Technical College, Texas

Sharon K. Miller, North Central Technical College, Mansfield, Ohio

David C. Mitchell, Seattle Central Community College, Seattle, Washington

Voya S. Moon, Western Iowa Technical Community College, Sioux City, Iowa

Robert E. Mullaney, Santa Barbara School District, Santa Barbara, California

R. O'Brien, SUNY-Canton, Canton, New York

Steven B. Ottmann, Southeast Community College, Lincoln, Nebraska

Emma M. Owens, Tri-County Technical College, Pendleton, South Carolina

David A. Palkovich, Spokane Community College, Spokane, Washington

William Poehler, Santa Barbara School District, Santa Barbara, California

Richard Powell, Fullerton College, Fullerton, California

Martin Prolo, San Jose City College, San Jose, California

Richard C. Spangler, Developmental Instruction Coordinator, Tacoma Community College Tacoma, Washington

Arthur Theobald, Bergen County Vocational School, Wayne, New Jersey

Cathy Vollstedt, North Central Technical College, Wausau, Wisconsin

Joseph Weaver, Associate Professor, State University of New York, Delhi, New York

Kay White, Walla Walla Community College, Walla Walla, Washington

Raymond E. Wilhite, Solano Community College, Vacaville, California

This book has benefited greatly from their excellence as teachers.

Finally, through every step of the seemingly endless sequence of researching, interviewing, testing, writing, and rewriting that makes a textbook, we have benefited from the patience, understanding, and concern of our wives, 'Lyn and Chris. They have made it a better book and a more pleasant experience than we would have otherwise had.

Robert A. Carman
Hal M. Saunders

Santa Barbara, California

How to Use This Book

In this book you will find many questions, not only at the end of each chapter or section, but on every page. This textbook is designed for those who need to learn the practical math used in the trades, and who want it explained carefully and completely at each step. The questions and explanations are designed so that you can:

- Start either at the beginning or where you need to start.
- Work on only what you need to know.
- Move as fast or as slowly as you wish.
- Skip material you already understand.
- Do as many practice problems as you need.
- Test yourself often to measure your progress.

In other words, if you find mathematics difficult and you want to be guided carefully through it, this book is designed for you.

This is no ordinary book. You cannot browse in it; you don't read it. You *work* your way through it. The ideas are arranged step by step in short portions or *frames*. Each frame contains information, careful explanations, examples, and questions to test your understanding. Read the material in each frame carefully, follow the examples, and answer the questions that lead to the next frame. Correct answers move you quickly through the book. Incorrect answers lead you to frames that provide further explanation. You move through this book frame by frame. Because we know that every person is different and has different needs, each major section of the book starts with a preview that will help you to determine the parts on which you need to work.

As you move through the book, you will notice that material not directly connected to the frames appears in special boxes. Read these at your leisure. They contain information that you may find interesting, helpful in your work, or even fun.

Most students hesitate to ask questions that nag at their understanding. They are fearful that "dumb questions" will humiliate them and reveal their lack of understanding. To relieve you of worry over dumb questions, we will ask and answer them for you. Thousands of students have taught us that "dumb questions" can produce smart students.

This textbook has been designed for students who not only will read it but will be working to achieve understanding and acquire skills.

Be alert for the following learning devices as you use this textbook.

Reference Head   Within each section of a chapter when new vocabulary is introduced or when new topics are presented, **reference headings** appear in the left margin as shown here. These signals show where the new information appears. Reference heads will be useful to you when you need to locate a word, phrase, equation, or topic, or when you need to review.

Note ▶ Every experienced teacher knows that certain mathematical concepts and procedures will present special difficulties for students. To help you with these, special notes are included in the text. A large triangle ▶ and a warning word appear in the left margin to indicate the start of the comment, and another triangle ◀ shows when it is completed. The word **NOTE**, as used at the start of this paragraph, calls your attention to conclusions or consequences that might be overlooked. ◀

Careful ▶ A **CAREFUL** comment points out a common mistake that you might make and shows you how to avoid it. ◀

Learning Help ▶ A **LEARNING HELP** gives you an alternative explanation or slightly different way of thinking about and working with the concepts being presented. ◀

An important feature of this textbook is the use of many worked examples.

Example: In each chapter carefully worked examples are included so that you can follow through the solution of all the various types of problems that are included in the problem sets. Examples cover both mathematical operations and applications.

Step 1 The solution of each worked example is usually organized in a step-by-step format.

Step 2 Explanations for each step are provided on the left, and corresponding mathematical operations are on the right.

Step 3 Color, ◁ boxed comments , and other graphical aids ↗ are used to highlight the important or tricky aspects of the solution.

 The calculator is an important tool for the modern trades worker or technician, and we assume in this textbook that once you have learned the basic operations of arithmetic you will use a calculator. Problems in the exercise sets or examples in the text that involve the use of a calculator are preceded by the calculator symbol shown here.

Solutions often include a display of the proper calculator key sequences. For example, the calculation

$$\frac{85.7 + (12.9)^2}{71.6}$$

would be shown as

85.7 ⊞ **12.9** $\boxed{x^2}$ ⊟ ÷ **71.6** ⊟ → 3.5210894

Exercises 0-1
Sample Problem Set At the conclusion of each section of each chapter you will find a set of problems covering the work of that section. These will include a number of routine or drill problems as well as applications or word problems. Each applied problem begins with an indication of the trades area from which it has been taken. Many of these applications have been obtained from trades workers in these areas.

Problem Set 0 Each chapter concludes with a set of problems reviewing all of the material covered in the chapter.

> Important rules, definitions, equations, or helpful hints are often placed in a box like this so that they will be easy to find.

If your approach to learning mathematics is to skim the text lightly on the way to puzzling through a homework assignment, you will have difficulty with this or any other textbook. If you are motivated to study mathematics so that you understand it and can use it correctly, this textbook is designed for you.

According to an old Spanish proverb, the world is an ocean and he who cannot swim will sink to the bottom. A study published by the U.S. Office of Education revealed that two-thirds of the skilled and semiskilled job opportunities on today's labor market are available only to those who have an understanding of the basic principles of arithmetic, algebra, and geometry. If the modern world of work is an ocean, the skill needed to keep afloat or even swim to the top is clearly mathematics. It is the purpose of this book to help you learn these basic skills.

Now, turn to page 1 and let's begin.

R. A. C.
H. M. S.

PREVIEW

1

Arithmetic of Whole Numbers

Objective	Sample Problems		Where to Go for Help	
			Page	Frame
When you finish this unit you will be able to:				
1. Add and subtract whole numbers.	(a) $67 + 58$		6	6
	(b) $7009 + 1598$			
	(c) $82 - 45$		16	12
	(d) $4035 - 1967$			
	(e) $14 + 31 + 59 - 67 + 22 + 37 - 19$			
2. Multiply and divide whole numbers	(a) 64×37		23	18
	(b) 305×243			
	(c) 908×705		33	27
	(d) $2006 \div 6$			
	(e) $7511 \div 37$			
3. Do word problems with whole numbers.	A metal casting weighs 680 lb; 235 lb of metal is removed during shaping. What is its finished weight?			

Name _____

Date _____

Course/Section _____

4. Use the correct order of operations with addition, subtraction, multiplication, and division.

(a) $6 + 9 \times 3$ _____ 41   **32**

(b) $35 - 14 \div 7$ _____

(c) $56 \div 4 \times 2 + 9 - 4$ _____

(d) $(23 - 7) \times 24 \div (12 - 4)$ _____

(Answers to these preview problems are given on page 577. Don't peek.)

If you are certain that you can work *all* these problems correctly, turn to page 47 for a set of practice problems. If you cannot work one or more of the preview problems, turn to the page indicated to the right of the problem. For those who wish to master this material with the greatest success, turn to frame **1** and begin to work there.

Arithmetic of Whole Numbers

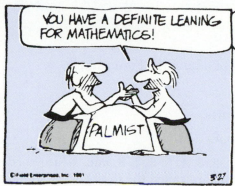

By permission of Johnny Hart and Creators Syndicate, Inc.

1 The average person a century ago used numbers to tell time, count, and keep track of money. Today, most people need to develop technical skills based on their ability to read, write, and work with numbers in order to earn a living. Although we live in an age of computers and calculators, much of the simple arithmetic used in industry, business, and the skilled trades is still done mentally or by hand. In fact, most trade and technical areas require you to *prove* that you can do the calculations by hand before you can get a job.

In the first part of this book we take a practical, how-to-do-it look at basic arithmetic: addition, subtraction, multiplication, and division, including fractions, decimal numbers, negative numbers, powers, and roots. Once we are past the basics, we will show you how to use a calculator to do such calculations. There are no quick and easy formulas here, but we do provide a lot of help for people who need to use mathematics in their daily work.

1-1 WORKING WITH WHOLE NUMBERS

The simplest numbers are the whole numbers—the numbers we use for counting the number of objects in a group. The whole numbers are 0, 1, 2, 3, . . . , and so on. For example, how many letters are in the collection shown in the margin?

Count them. Write your answer here _____, then turn to frame **3**.

---

2 Hi.

What are you doing here? Lost? Window shopping? Just passing through? Nowhere in this book are you directed to frame **2**. (Notice that the **2** to the left above? That's a frame number.) Remember, in this book you move from frame to frame as directed, but not necessarily in 1–2–3 order. Go slowly, follow directions, and you'll never get lost.

Now, return to **1** for a fresh start.

---

3 We counted 23. Notice that we can count the letters by grouping them into sets of ten:

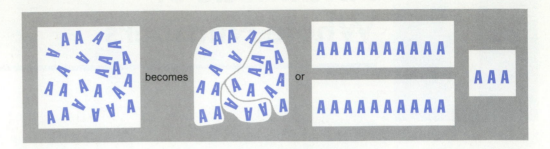

2 tens + 3 ones

20   + 3   or 23

Expanded Form   Mathematicians call this the *expanded form* of a number. For example,

46 = 40 + 6   = 4 tens + 6 ones

274 = 200 + 70 + 4 = 2 hundreds + 7 tens + 4 ones

305 = 300 + 5   = 3 hundreds + 0 tens + 5 ones

Only ten numerals or number symbols—0, 1, 2, 3, 4, 5, 6, 7, 8, and 9—are needed to write any number. These ten basic numerals are called the *digits* of the number. The digits 4 and 6 are used to write 46, the number 274 is a three-digit number, and so on.

Write out the following three-digit numbers in expanded form:

(a)   362 = _____ + _____ + _____ = _____ hundreds + _____ tens + _____ ones

(b)   425 = _____ + _____ + _____ = _____ hundreds + _____ tens + _____ ones

(c)   208 = _____ + _____ + _____ = _____ hundreds + _____ tens + _____ ones

Check your work in **4**.

---

4   (a)   *362 = 300 + 60 + 2 = 3 hundreds + 6 tens + 2 ones*

     (b)   *425 = 400 + 20 + 5 = 4 hundreds + 2 tens + 5 ones*

     (c)   *208 = 200 + 0 + 8 = 2 hundreds + 0 tens + 8 ones*

Notice that the 2 in 362 means something different from the 2 in 425 or 208. In 362 the 2 signifies two ones. In 425 the 2 signifies two tens. In 208 the 2 signifies two hundreds. Ours is a *place-value* system of naming numbers: the value of any digit depends on the place where it is located.

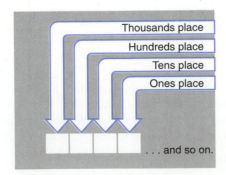

Being able to write a number in expanded form will help you to understand and remember the basic operations of arithmetic—even though you'll never find it on a blueprint or in a technical handbook.

This expanded-form idea is useful especially in naming very large numbers. Any large number given in numerical form may be translated to words by using the following diagram:

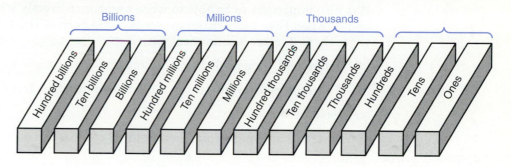

The number 14,237 can be placed in the diagram like this:

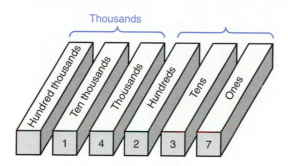

and read "fourteen thousand, two hundred thirty-seven."

The number 47,653,290,866 becomes

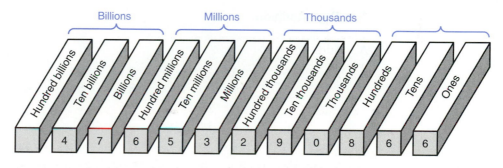

and is read "forty-seven billion, six hundred fifty-three million, two hundred ninety thousand, eight hundred sixty-six."

In each block of three digits read the digits in the normal way ("forty-seven," "six hundred fifty-three") and add the name of the block ("billion," "million"). Notice that the word "and" is not used in naming these numbers.

Use the diagram to name the following numbers.

(a) 4072   (b) 1,360,105

(c) 3,000,210   (d) 21,010,031,001

Check your answers in 5.

5 (a) Four thousand, seventy-two

(b) One million, three hundred sixty thousand, one hundred five

(c) Three million, two hundred ten

(d) Twenty-one billion, ten million, thirty-one thousand, one

It is also important to be able to write numbers correctly when you hear them spoken or when they are written in words. Read each of the following aloud and then write them in correct numerical form.

(a) Fifty-eight thousand, four hundred six

(b) Two hundred seventy-three million, five hundred forty thousand

(c) Seven thousand sixty

(d) Nine billion, six million, two hundred twenty-three thousand, fifty-eight

Check your answers in **6**.

---

6 (a) 58,406   (b) 273,540,000

(c) 7060   (d) 9,006,223,058

Addition of Whole Numbers

Adding whole numbers is fairly easy provided that you have stored in your memory a few simple addition facts. It is most important that you be able to add simple one-digit numbers mentally.

The following sets of problems in one-digit addition are designed to give you some practice. Work quickly. You should be able to answer all problems in a set in the time shown.

Practice   **One-Digit Addition**

A. Add.

7	5	2	5	8	2	3	8	9	7
3	6	9	7	8	5	6	7	3	6
6	8	9	3	7	2	9	9	7	4
4	5	6	5	7	7	4	9	2	7
9	2	5	8	4	9	6	4	8	8
7	6	5	9	5	5	6	3	2	3
5	6	7	7	5	6	2	3	6	9
8	7	5	9	4	5	8	7	8	8
7	5	9	4	3	8	4	8	5	7
4	9	2	6	8	6	9	4	8	8

Average time = 90 seconds
Record = 35 seconds

B. Add. Try to do all addition mentally.

2	7	3	4	2	6	3	5	9	5
5	3	6	5	7	7	4	7	6	2
4	2	5	8	9	8	4	8	3	8

6	5	4	8	6	9	7	4	8	1
2	4	2	1	8	3	1	9	4	8
7	5	9	9	8	5	6	1	6	7

1	9	3	1	7	2	9	9	8	5
9	9	1	6	9	9	8	5	3	4
2	1	4	3	6	1	2	1	3	7

Average time = 90 seconds
Record = 41 seconds

The answers are given on page 580

When you have had the practice you need, turn to **7** and continue.

Rounding Whole Numbers

7 In many situations a simplified approximation of a number is more useful than its exact value. For example, the accountant for a business may calculate its total monthly revenue as $247,563, but the owner of the business may find it easier to talk about the revenue as "about $250,000." The process of approximating a number is called *rounding*. Rounding numbers comes in handy when we need to make estimates or do "mental mathematics."

A number can be rounded to any desired place. For example, $247,563 is approximately

$247,560 rounded to the nearest ten,
$247,600 rounded to the nearest hundred,
$248,000 rounded to the nearest thousand,
$250,000 rounded to the nearest ten thousand, and
$200,000 rounded to the nearest hundred thousand.

To round a whole number, follow this step-by-step process:

Example: Round 247,563

		to the nearest hundred thousand	to the nearest ten thousand
Step 1	Determine the place to which the number is to be rounded. Mark it on the right with a ∧.	$2 ∧ 47,563	$24 ∧ 7,563
Step 2	If the digit to the right of the mark is less than 5, replace all digits to the right of the mark with zeros.	$200,000	
Step 3	If the digit to the right of the mark is equal to or larger than 5, increase the digit to the left by 1 and replace all digits to the right with zeros.		$250,000

Try these for practice. Round

(a) 73,856 to the nearest thousand

(b) 64 to the nearest ten

(c) 4852 to the nearest hundred

(d) 350,000 to the nearest hundred thousand

(e) 726 to the nearest hundred

Follow the three-step rule, then check your work in **8**.

---

8 (a) **Step 1** Place a mark to the right of the thousands place. The digit 3 is in the thousands place.

73 856
^

Step 2 Does not apply.

Step 3 The digit to the right of the mark, 8, is larger than 5. Increase the 3 to a 4 and replace all digits to the right with zeros.

74,000

(b) **Step 1** Place a mark to the right of the tens place. The digit 6 is in the tens place.

6 4
^

Step 2 The digit to the right of the mark, 4, is less than 5. Replace it with a zero.

60

(c) 4900   (d) 400,000   (e) 700

Doing arithmetic with one-digit numbers is very important. It is the key to any mathematical computation—even if you do the work on a calculator. Suppose that you need to find the total time spent on a job by two workers. You need to find the sum

31 hours + 48 hours = _____

Estimating What is the first step? Start adding digits? Punch in some numbers on your trusty calculator? Rattle your abacus? None of these. The first step is to *estimate* your answer. The most important rule in any mathematical calculation is:

> Know the approximate answer to any calculation before you calculate it.

Never do an arithmetic calculation until you know roughly what the answer is going to be. Always know where you are going.

Rounding to the nearest ten hours, the preceding sum can be estimated as 31 hours + 48 hours or approximately 30 hours + 50 hours or 80 hours, not 8 or 8000 or 800 hours. Having the estimate will keep you from making any major (and embarrassing) mistakes. Once you have a rough estimate of the answer, you are ready to do the arithmetic work.

Calculate 31 + 48 = _____

You don't really need an air-conditioned, solar-powered, talking calculator for that, do you? Work it out and check your answer in **9**.

---

9   You should have set it up like this:

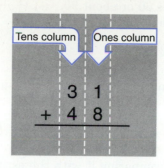

1. The numbers to be added are arranged vertically (up and down) in columns.

2. The right end or ones digits are placed in the ones column, the tens digits are placed in the tens column, and so on.

Avoid the confusion of   31   31
                        + 48   or   + 48

Careful ▶   Most often the cause of errors in arithmetic is carelessness, especially in simple tasks such as lining up the digits correctly. ◀

Once the digits are lined up the problem is easy.

$$\begin{array}{r} 31 \\ + 48 \\ \hline 79 \end{array}$$

Does the answer agree with your original estimate? Yes. The estimate, 80, is roughly equal to the actual sum, 79.

What we have just shown you is called the *guess n' check method* of doing mathematics.

1. **Estimate** the answer using rounded numbers.

2. **Work** the problem carefully.

3. **Check** your answer against the estimate. If they disagree, repeat both steps 1 and 2.

Most students worry about estimating, either because they won't take the time to do it or because they are afraid they might do it incorrectly. Relax. You are the only one who will know your estimate. Do it in your head, do it quickly, and make it reasonably accurate. Step 3 helps you to find incorrect answers before you finish the problem. The guess n' check method means that you never work in the dark; you always know where you are going.

Note ▶   Estimating is especially important in practical math, where a wrong answer is not just a mark on a piece of paper. An error may mean time and money lost. ◀

Here is a slightly more difficult problem:

27 lb + 58 lb = _____

Try it, then turn to **10**.

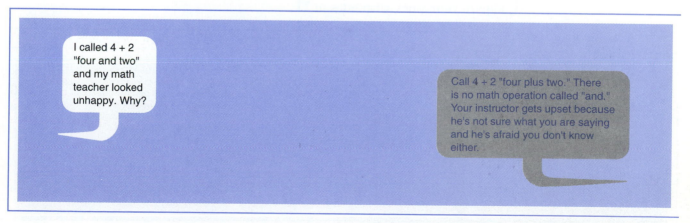

10   **First,** estimate the answer. 27 + 58 is roughly
30 + 60 or about 90.
The answer is about 90 lb.

Second, line up the digits in columns.

$$\begin{array}{r} 27 \\ + 58 \\ \hline \end{array}$$

The numbers to be added, 27 and 58 in this case, are usually called *addends*.

Third, add carefully.

$$\begin{array}{r} \overset{1}{}27 \\ + 58 \\ \hline 85 \end{array}$$

Finally, check your answer by comparing it with the estimate. The estimate 90 lb is roughly equal to the answer 85 lb—at least you have an answer in the right ballpark.

What does that little 1 above the tens digit mean? What really happens when you "carry" a digit. Let's look at it in detail. In expanded notation,

$$
\begin{aligned}
27 &\rightarrow 2 \text{ tens} + 7 \text{ ones} \\
+ 58 &\rightarrow \underline{5 \text{ tens} + 8 \text{ ones}} \\
&= 7 \text{ tens} + 15 \text{ ones} \\
&= 7 \text{ tens} + 1 \text{ ten} + 5 \text{ ones} \\
&= 8 \text{ tens} \qquad + 5 \text{ ones} \\
&= 85
\end{aligned}
$$

The 1 that is carried over to the tens column is really a ten.

Learning Help ▶ Trades people often must calculate exact answers mentally. To do a problem such as 27 + 58, a trick called "balancing" works nicely. Simply add 2 to the 58 to get a "round" number, 60, and subtract 2 from 27 to balance, keeping the total the same. Therefore, 27 + 58 is the same as 25 + 60, which is easy to add mentally to get 85. ◀

Following is a short set of problems. Add, and be sure to estimate your answers first.

(a) 429 + 738 = _____

(b) 446 + 867 = _____

(c) 2368 + 744 = _____

(d) 409 + 72 = _____

(e) Three bricklayers working together on a job each laid the following number of bricks in a day: 927, 1143, and 1065. How many bricks did all three lay that day?

Compare your work with ours in **11**.

Where do the + and – signs come from?

They were used in the fifteenth century to show that boxes of merchandise were overweight (+) or underweight (–). Within 40 years or so bookkeepers and mathematicians started using them. The + symbol came from the Latin word *et* meaning and.

11   (a)   **Estimate:**   Rounding each number to the nearest hundred, $400 + 700 = 1100$
Line up the digits: 429
             + 738

Calculate:

Step 1 $\overset{1}{4}29$
          + 738 $9 + 8 = 17$ Write 7; carry 1 ten.
             7

Step 2 $\overset{1}{4}29$
          + 738 $1 + 2 + 3 = 6$ Write 6
           67

Step 3 $\overset{1}{4}29$
          + 738 $4 + 7 = 11$ Write 11.
         1167

Check: The estimate 1100 and the answer 1167 are roughly equal.

(b)   **Estimate:** $400 + 900 = 1300$

Calculate:

Step 1 $\overset{1}{4}46$
          + 867 $6 + 7 = 13$ Write 3; carry 1 ten
            3

Step 2 $\overset{11}{4}46$ $1 + 4 + 6 = 11$ Write 1; carry 1 hundred.
          + 867
           13

Step 3 $\overset{11}{4}46$
          + 867 $1 + 4 + 8 = 13$ Write 13
         1313

Check: The estimate 1300 and the answer 1313 are roughly equal.

(c)   **Estimate:**   Rounding each number to the nearest hundred, $2400 + 700 = 3100$

Calculate: $\overset{111}{2}368$
          + 744
        3112

Check: The estimate 3100 and the answer 3112 are roughly equal.

(d)   **Estimate:** $400 + 100 = 500$

Calculate: $\overset{1}{4}09$
          + 72
        481

Check: The estimate 500 and the answer 481 are roughly equal.

(e)   **Estimate:** $900 + 1100 + 1100 = 3100$

Calculate: $\overset{111}{9}27$
          1143
         + 1065
         3135

Check: The estimate 3100 is roughly equal to the answer 3135.

Estimating answers is a very important part of any mathematics calculation, especially for the practical mathematics used in engineering, technology, and the trades. A successful builder, painter, or repairperson must make accurate estimates of job costs—business success depends on it. If you work in a technical trade, getting and keeping your job may depend on your ability to get the correct answer *every* time.

If you use a calculator to do the actual arithmetic, it is even more important to get a careful estimate of the answer first. If you plug a wrong number into the calculator, accidentally hit a wrong key, or unknowingly use a failing battery, the calculator may give you a wrong answer—lightning fast, but wrong. The estimate is your best insurance that a wrong answer will be caught immediately. Convinced?

Now, work the following problems for practice in adding whole numbers.

Exercises 1-1　**Working with Whole Numbers**

A.　Add.

1.　47 　　23	2.　27 　　38	3.　45 　　35	4.　38 　　65	5.　75 　　48
6.　26 　　98	7.　48 　　84	8.　67 　　69	9.　189 　　204	10.　508 　　495
11.　684 　　706	12.　432 　　399	13.　621 　　388	14.　747 　　59	15.　375 　　486
16.　4237 　　1288	17.　5076 　　4385	18.　7907 　　1395	19.　3785 　　7643	20.　6709 　　9006
21.　18745 　　6972	22.　40026 　　7085	23.　10674 　　397	24.　9876 　　4835	25.　78044 　　97684

B.　Arrange vertically and add.

1.　487 + 29 + 526 = _____

2.　715 + 4293 + 184 + 19 = _____

3.　1706 + 387 + 42 + 307 = _____

4.　456 + 978 + 1423 + 3584 = _____

5.　6284 + 28 + 674 + 97 = _____

6.　6842 + 9008 + 57 + 368 = _____

7.　322 + 46 + 5984 = _____

8.　7268 + 209 + 178 = _____

9.　5016 + 423 + 1075 = _____

10.　8764 + 85 + 983 + 19 = _____

11.　4 + 6 + 11 + 7 + 14 + 3 + 9 + 6 + 4 = _____

12.　12 + 7 + 15 + 16 + 21 + 8 + 10 + 5 + 30 + 17 = _____

13.　1 + 2 + 3 + 4 + 5 + 6 + 7 + 8 + 9 + 10 = _____

14.　22 + 31 + 43 + 11 + 9 + 1 + 19 + 12 = _____

15.　75 + 4 + 81 + 12 + 14 + 65 + 47 + 22 + 37 = _____

C. Writing and Rounding Whole Numbers

Write in words.

1. 357   2. 2304   3. 17,092   4. 207,630   5. 2,000,034

6. 10,007   7. 740,106   8. 5,055,550   9. 118,180,018   10. 6709

Write as numbers.

11. Three thousand six

12. Seventeen thousand twenty-four

13. Eleven thousand, one hundred

14. Three million, two thousand, seventeen

15. Four million, forty thousand, six

16. Seven hundred twenty million, ten

Round as indicated.

17. 357 to the nearest ten

18. 4386 to the nearest hundred

19. 4386 to the nearest thousand

20. 5386 to the nearest thousand

21. 225,799 to the nearest ten thousand

22. 225,799 to the nearest thousand

D. Applied Problems

1. **Electrical Technology**   In setting up his latest wiring job, an electrician cut the following lengths of wire: 387, 913, 76, 2640, and 845 ft. Find the total length of wire used.

2. **Building Construction**   The Acme Lumber Co. made four deliveries of 1-in. by 6-in. flooring: 3280, 2650, 2465, and 2970 fbm. What was the total number of board feet of flooring delivered? (The abbreviation for "board feet" is fbm, which is short for "feet board measure.")

3. **Machine Technology**   The stockroom has eight boxes of No. 10 hexhead cap screws. How many screws of this type are in stock if the boxes contain 346, 275, 84, 128, 325, 98, 260, and 120 screws, respectively?

4. **Office Services**   In calculating her weekly expenses, a contractor found that she had spent the following amounts: materials, $1386; labor, $2537; salaried help, $393; overhead expense, $832. What was her total expense for the week?

5. **Machine Technology**   The head machinist at Tiger Tool Co. is responsible for totaling time cards to determine job costs. He found that five different jobs this week took 78, 428, 143, 96, and 384 minutes each. What was the total time in minutes for the five jobs?

6. **Carpentry**   On a home construction job, a carpenter laid 1480 wood shingles the first day, 1240 the second, 1560 the third, 1320 the fourth, and 1070 the fifth day. How many shingles did he lay in five days?

7. **Industrial Technology**   Eight individually powered machines in a small production shop have motors using 420, 260, 875, 340, 558, 564, 280, and 310 watts each. What is the total wattage used when (a) the total shop is in operation? (b) the three largest motors are running? (c) the three smallest motors are running?

8. **Auto Mechanics**   A mechanic is taking inventory of oil in stock. He has 24 quarts of 10W-30, 8 quarts of 30W, 42 quarts of 20W-50, 16 quarts of 10W-40, and 21 quarts of 20W-40. How many total quarts of oil does he have in stock?

9. **Building Construction**   The Happy Helper building materials supplier has four piles of bricks containing 1250, 865, 742, 257 bricks. What is the total number of bricks they have on hand?

10. **Machine Technology** A machinist needs the following lengths of 1-in. diameter rod: 8 in., 14 in., 6 in., 27 in., and 42 in. How long a rod is required to supply all five pieces. (Ignore cutting waste.)

E. Calculator Problems

You probably own a calculator and, of course, you are eager to put it to work doing practical math calculations. In this book we include problem sets for calculator users. These problems are taken from real-life situations and, unlike most textbook problems, involve big numbers and lots of calculations. If you think that having an electronic brain-in-a-box means that you do not need to know basic arithmetic, you will be disappointed. The calculator helps you to work faster, but it will not tell you *what* to do or *how* to do it.

Detailed instruction on using a calculator with whole numbers appears on page 49.

Here are a few helpful hints for calculator users:

1. Always *estimate* your answer before doing a calculation.

2. *Check* your answer by comparing it with the estimate or by the other methods shown in this book Be certain that your answer makes sense.

3. If you doubt the calculator (they do break down, you know), put a problem in it whose answer you know, preferably a problem like the one you are solving.

1. **Manufacturing** The following table lists the number of widget fasteners made by each of the five machines at the Ace Widget Co. during the last ten working days.

| Day | Machine | | | | | Daily Totals |
	A	B	C	D	E	
1	347	402	406	527	237	
2	451	483	312	563	316	
3	406	511	171	581	289	
4	378	413	0	512	291	
5	399	395	452	604	342	
6	421	367	322	535	308	
7	467	409	256	578	264	
8	512	514	117	588	257	
9	302	478	37	581	269	
10	391	490	112	596	310	
Machine Total						

(a) Complete the table by finding the number of fasteners produced each day. Enter these totals under the column "Daily Totals" on the right.

(b) Find the number of fasteners produced by each machine during the ten-day period and enter these totals along the bottom row marked "Machine Totals."

(c) Does the sum of the daily totals equal the sum of the machine totals?

2. Add the following as shown.

(a) $ 67429
     6070
     4894
     137427
     91006
     399

(b) $216847
     9757
     86492
     4875
     386738
     28104

(c) $693884
     675489
     47039
     276921
     44682
     560487

(d) $4299 + $137 + $20 + $177 + $63 + $781 + $1008 + $671 =?

3. **Office Services** Joe's Air Conditioning Installation Co. has not been successful, and he is wondering if he should sell it and move to a better location. During the first six months of the year his expenses were:

Rent $1860
Supplies $2540
Part-time helper $2100
Transportation $948

Taxes $315
Advertising $250
Miscellaneous $187

His monthly income was:

January $609
February $1151
March $1269

April $1381
May $1687
June $1638

(a) What was his total expense for the six-month period?

(b) What was his total income for the six-month period?

(c) Rotate your calculator 180° to learn what Joe should do about this unhappy situation.

4. **Electrical Technology** A mapper is a person employed by an electrical utility company who has the job of reading diagrams of utility installations and listing the materials to be installed or removed by engineers. Part of a typical job list might look like this:

Installation (in Feet of Conductor)

Location Code	No. 12 BHD (bare, hard-drawn copper wire)	#TX (triplex)	410 AAC (all-aluminum conductor	110 ACSR (aluminum-core steel-reinforced conductor)	6B (No. 6, bare conductor)
A3	1740	40	1400		350
A4	1132		5090		2190
B1	500			3794	
B5		87	3995		1400
B6	4132	96	845		
C4		35		3258	2780
C5	3949		1385	1740	705

(a) How many total feet of each kind of conductor must the installer have to complete the job?

(b) How many feet of conductor are to be installed at each of the seven locations?

When you have completed these exercises, check your answers on page 580 and then continue in 12.

1-2 SUBTRACTION OF WHOLE NUMBERS

12 Subtraction is the reverse of addition.

Addition: $3 + 4 = \square$

Subtraction: $3 + \square = 7$

Written this way, a subtraction problem asks the question: How much must be added to a given number to produce a required amount?

Most often, however, the numbers in a subtraction problem are written using a minus sign $(-)$:

$17 - 8 = \square$ means that there is a number $\square$ such that $8 + \square = 17$

But we should remember that

$8 + 9 = 17$ or $17 - 8 = 9$ $\Longleftarrow$ Difference

When you set up the subtraction vertically, the *subtrahend* is the bottom number and the *minuend* is the top number.

The *difference* is the name given to the answer in a subtraction problem.

Solving simple subtraction problems depends on your knowledge of the addition of one-digit numbers.

For example, to solve the problem

$9 - 4 =$ _____

you probably go through a chain of thoughts something like this:

Nine minus four. Four added to what number gives nine? Five? Try it: four plus five equals nine. Right.

Subtraction problems with small whole numbers will be easy for you if you know your addition tables.

Here is a more difficult subtraction problem:

$47 - 23 =$ _____

What is the first step?

Work the problem and continue in 13.

Can you do a problem like 37 – 64?

In this section we are considering only subtraction of a smaller number from a larger number.. See Chapter 6 for information on that problem.

13   The **first** step is to estimate the answer—remember?

$47 - 23$ is roughly $50 - 20$ or 30.

The difference, your answer, will be about 30—not 3 or 10 or 300.

The **second** step is to write the numbers vertically as you did with addition. Be careful to keep the ones digits in line in one column, the tens digits in a second column, and so on.

$$\begin{array}{r} 4\ 7 \\ -2\ 3 \end{array}$$

Notice that the minuend is written above the subtrahend—larger number on top.

Once the numbers have been arranged in this way, the difference may be written by performing the following two steps:

Step 1

$$\begin{array}{r} 4\ 7 \\ -2\ 3 \\ \hline 4 \end{array}$$

ones digits: $7 - 3 = 4$

Step 2

$$\begin{array}{r} 4\ 7 \\ -2\ 3 \\ \hline 2\ 4 \end{array}$$

tens digits: $4 - 2 = 2$

The difference is 24, which agrees roughly with our estimate.

With some problems it is necessary to rewrite the larger number before the problem can be solved. For example, try this one:

$64 - 37 =$ _____

Check your work in **14**.

---

14   **First,** estimate the answer. Rounding to the nearest ten, $64 - 37$ is roughly $60 - 40$ or 20.

Second, arrange the numbers vertically in columns.
$$\begin{array}{r} 64 \\ -\ 37 \end{array}$$

Because 7 is larger than 4 we must "borrow" one ten from the 6 tens in 64. We are actually rewriting 64 (6 tens + 4 ones) as 5 tens + 14 ones. In actual practice our work would look like this:

Step 1   **Step 2**

$$\begin{array}{r} 64 \\ -\ 37 \end{array}\qquad \begin{array}{r} {}^{514}\\ 6\!\!\!/4 \\ -\ 37 \\ \hline 7 \end{array}$$

Borrow one ten, change the 6 in the tens place to 5, change 4 to 14, subtract $14 - 7 = 7$.

Step 3

$$\begin{array}{r} {}^{514}\\ 6\!\!\!/4 \\ -\ 37 \\ \hline 27 \end{array}$$

$14 - 7$

$5 - 3 = 2$

Double-check subtraction problems by adding the answer and the smaller number (subtrahend); their sum should equal the larger number.

Step 4   Check:
$$\begin{array}{r} 37 \\ +\ 27 \\ \hline 64 \end{array}$$

If you need to get an exact answer to a problem such as 64 − 37 mentally, add or subtract to make the subtrahend, 37, a "round" number. In this case, add 3 to make it 40. Since we're subtracting, we want the *difference,* not the *total,* to be the same or balance. Therefore, we also add 3 to the 64 to get 67. The problem becomes 64 − 37 = (64 + 3) − (37 + 3) = 67 − 40. Subtracting a round number is easy mentally: 67 − 40 = 27.* ◀

Try these problems for practice.

(a) 71
   − 39

(b) 263
   − 127

(c) 426
   −128

(d) 902
   −465

Solutions are given in **15**.

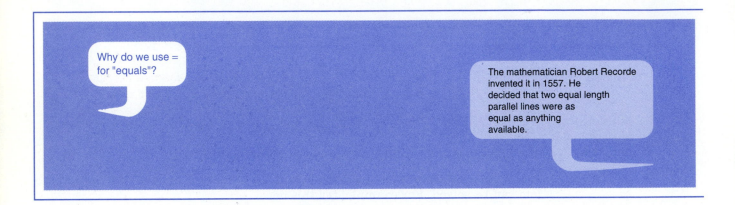

Why do we use = for "equals"?

The mathematician Robert Recorde invented it in 1557. He decided that two equal length parallel lines were as equal as anything available.

15 (a) **Estimate:** 70 − 40 = 30

Step 1   **Step 2**

   71          $\overset{6\,11}{\cancel{7}\cancel{1}}$ Borrow one ten from 70.
 − 39   − 39   change the 7 in the tens place to 6,
            ——   change the 1 in the ones place to 11.
             32

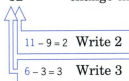

11 − 9 = 2 **Write 2**

6 − 3 = 3 **Write 3**

Check: The answer 32 is approximately equal to the estimate 30. Mentally, add 1 to each number,

71 − 39 = 72 − 40

then subtract.

72 − 40 = 32

(b) **Estimate:** 260 − 130 = 130.

*See Jack A. Hope, Barbara J. Reys, and Robert E. Reys, *Mental Math in Junior High* (Palo Alto, CA: Dale Seymour Publications, 1988).

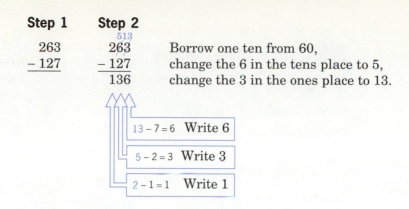

Step 1	Step 2	
263	$\overset{5\,13}{2\,6\,3}$	Borrow one ten from 60,
$-\,127$	$-\,127$	change the 6 in the tens place to 5,
	136	change the 3 in the ones place to 13.

$13 - 7 = 6$ Write 6

$5 - 2 = 3$ Write 3

$2 - 1 = 1$ Write 1

Check: The answer is approximately equal to the estimate.

(c) **Estimate:** $400 - 100 = 300$

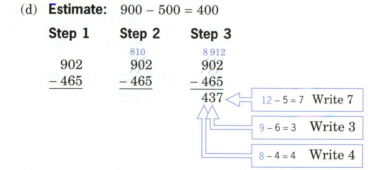

Step 1	Step 2	Step 3	
426	$\overset{1\,16}{4\,2\,6}$	$\overset{3\,11\,16}{4\,2\,6}$	Notice that in this case we borrow
$-\,128$	$-\,128$	$-\,128$	twice. Borrow one ten from the 20 in
	8	298	426 and make 16. Then borrow one
			hundred from the 400 in 426 to make 110.

$16 - 8 = 8$ Write 8

$11 - 2 = 9$ Write 9

$3 - 1 = 2$ Write 2

Check: The answer 298 is approximately equal to the estimate 300.

(d) **Estimate:** $900 - 500 = 400$

Step 1	Step 2	Step 3
902	$\overset{8\,10}{9\,0\,2}$	$\overset{8\,9\,12}{9\,0\,2}$
$-\,465$	$-\,465$	$-\,465$
		437

$12 - 5 = 7$ Write 7

$9 - 6 = 3$ Write 3

$8 - 4 = 4$ Write 4

Check: The answer 437 is roughly equal to the estimate 400.

In problem (d) we first borrow one hundred from 900 to get a 10 in the tens place. Then we borrow one 10 from the tens place to get a 12 in the ones place.

Now, do you want more worked examples of problems containing zeros, similar to the last one? If so, go to **17**.

Otherwise, go to **16** for a set of practice problems.

Exercises 1-2   **Subtraction of Whole Numbers**

A.   Subtract.

1. 13	2. 12	3. 8	4. 8	5. 11	6. 16
7	5	6	0	7	7

7. 10	8. 5	9. 12	10. 11	11. 10	12. 14
7	5	9	8	2	6

13. 12	14. 15	15. 9	16. 14	17. 9	18. 11
3	6	0	5	6	5

19. 15	20. 12	21. 13	22. 18	23. 16	24. 12
7	7	6	0	9	4

25. 0	26. 13	17. 17	28. 18	29. 14	30. 16
0	8	9	9	8	8

31. 15	32. 17	33. 14	34. 15	35. 13	36. 12
9	8	9	8	9	8

B.   Subtract.

1. 40	2. 78	3. 51	4. 36	5. 42	6. 52
27	49	39	17	27	16

7. 65	8. 46	9. 84	10. 70	11. 34	12. 56
27	17	38	48	9	18

13. 546	14. 409	15. 476	16. 330	17. 504	18. 747
357	324	195	76	96	593

19. 400	20. 803	21. 632	22. 438	23. 6218	24. 6084
127	88	58	409	3409	386

25. 13042	26. 57022	27. 5007	28. 10000	29. 48093	30. 27004
524	980	266	386	500	4582

C.   Applied Problems

1.   **Painting and Decorating**   In planning for a particular job, a painter buys $264 worth of materials. When the job is completed, she returns five empty paint drums for a credit of $17. What was the net amount of her bill?

2.   **Building Construction**   How many square feet of plywood remain from an original supply of 8000 sq ft after 5647 sq ft is used?

3.   **Welding**   A storage rack at the Tiger Tool Company contains 3540 ft of 1-in. stock. On a certain job 1782 ft is used. How much is left?

4.   **Welding**   Five pieces measuring 26, 47, 38, 27, and 32 cm are cut from a steel bar that was 200 cm long. Allowing for a total of 1 cm for waste in cutting, what is the length of the piece remaining?

5.   **Office Services**   Taxes on a group of factory buildings owned by the Ace Manufacturing Company amounted to $875,977 eight years ago. Taxes on

the same buildings last year amounted to $1,206,512. Find the increase in taxes.

6. **Office Services** In order to pay their bills, the owners of Edwards Plumbing Company made the following withdrawals from the their bank account: $42, $175, $24, $217, and $8. If the original balance was $3610, what was the amount of the new balance?

7. **Manufacturing** Which total volume is greater, four drums containing 72, 45, 39, and 86 liters, or three drums containing 97, 115, and 74 liters? By how much is it greater?

8. **Machine Technology** Determine the missing dimension (L) in the following drawings.

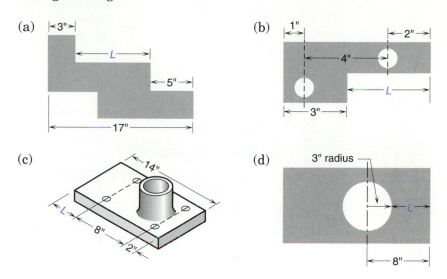

(a) (b) (c) (d)

9. **Auto Mechanics** A service department began the day with 238 gallons of coolant. During the day 64 gallons were used. How many gallons remained at the end of the day?

10. **Building Construction** A truck loaded with rocks weighs 14,260 lb. If the truck weighs 8420 lb, how much do the rocks weigh?

11. **Printing** A press operator has a total of 22,000 impressions to run for a job. If the operator runs 14,250 the first day, how many are left to run?

12. **Plumbing** In the following plumbing diagram, find pipe lengths A and B.

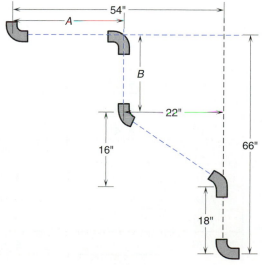

Problem 12

D. Calculator Problems

1. **Plumbing** The Karroll Plumbing Co. has 10 trucks and, for the month of April, the following mileage was recorded on each.

Truck No.	Mileage at Start	Mileage at End
1	58352	60027
2	42135	43302
3	76270	78007
4	40006	41322
5	08642	10002
6	35401	35700
7	79002	80101
8	39987	40122
9	10210	11671
10	71040	73121

Find the mileage traveled by each truck during the month of April and the total mileage of all vehicles.

2. Which sum is greater?

987654321		123456789
87654321		123456780
7654321		123456700
654321		123456000
54321	or	123450000
4321		123400000
321		123000000
21		120000000
1		100000000

3. If your income is $28245 per year and you pay $6959 in taxes, what is your take-home pay?

4. **Office Services** The income of the Smith Construction Company for the year is $437,672 and the total expenses are $320,867. Find the difference, Smith's profit, for that year.

5. Balance the following checking account record.

Date	Deposits	Withdrawals	Balance
7/1			$6375
7/3		$ 379	
7/4	$1683		
7/7	$ 474		
7/10	$ 487		
7/11		$2373	
7/15		$1990	
7/18		$ 308	
7/22		$1090	
7/26		$ 814	
8/1			A

(a) Find the new balance A.

(b) Keep a running balance by filling each blank in the balance column.

When you have completed these exercises, check your answers on page 581, then turn to **18** to study the multiplication of whole numbers.

---

17 Let's work through a few examples of subtraction problems involving zero digits.

Example 1: $400 - 167 = ?$

Step 1	Step 2	Step 3
	310	3 9 10
400	4̶0̶0̶	4̶0̶0̶
-167	-167	-167
		233

Do you see that we have rewritten 400 as $300 + 90 + 10$?

Check:
$$\begin{array}{r} 167 \\ +\,233 \\ \hline 400 \end{array}$$

Example 2: $5006 - 2487 = ?$

Step 1	Step 2	Step 3	Step 4	Check
	410	4 910	49 916	
5006	5̶0̶0̶6̶	5̶0̶0̶6̶	5̶0̶0̶6̶	2487
-2487	-2487	-2487	-2487	$+2519$
			2519	5006

Here is an example involving repeated borrowing.

Example 3: $24632 - 5718 = ?$

Step 1	Step 2	Step 3	Step 4	Check
	212	316212	11316212	
24632	24632	24632	24632	5718
-5718	-5718	-5718	-5718	$+18914$
	14	914	18914	24632

Any subtraction problem that involves borrowing should always be checked in this way. It is very easy to make a mistake in this process.

Now turn back to **16** for some practice problems on subtraction.

1-3 MULTIPLICATION OF WHOLE NUMBERS

18 In a certain football game, the West Newton Waterbugs scored five touchdowns at six points each. How many total points did they score through touchdowns? We can answer the question several ways:

1. Count points, ...

2. Add touchdowns, $6 + 6 + 6 + 6 + 6 = ?$

    or

3. Multiply $5 \times 6 = ?$

We're not sure about the mathematical ability of the West Newton scorekeeper, but most people would multiply. Multiplication is a shortcut method of performing repeated addition.

How many points did they score?

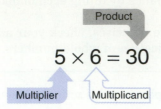

In a multiplication problem the *product* is the name given to the result of the multiplication. The numbers being multiplied are the *factors* of the product.

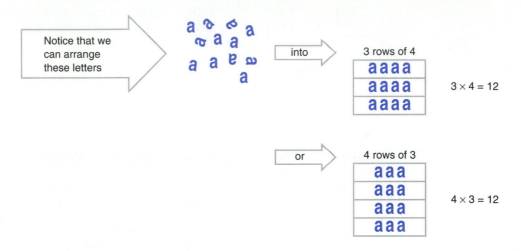

Changing the order of the factors does not change their product. This is called the *commutative* property of multiplication.

To become skillful at multiplication, you must know the one-digit multiplication table from memory. Even if you use a calculator for your work, you need to know the one-digit multiplication table to make estimates and to check your work.

Complete the table below by multiplying the number at the top by the number at the side and placing their product in the proper square. We have multiplied $3 \times 4 = 12$ and $2 \times 5 = 10$ for you.

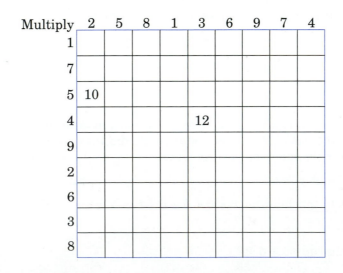

Multiply	2	5	8	1	3	6	9	7	4
1									
7									
5	10								
4					12				
9									
2									
6									
3									
8									

Check your work in **20**.

Chap. 1 Arithmetic of Whole Numbers

19 Practice   One-Digit Multiplication

Multiply as shown. Work quickly; you should be able to answer all problems in a set correctly in the time indicated. (These times are for community college students enrolled in a developmental math course.)

A. Multiply.

6	4	9	6	3	9	7	8	2	8
2	8	7	6	4	2	0	3	7	1

6	8	5	5	2	3	9	7	3	1
8	2	9	6	5	3	8	5	6	4

7	5	4	7	4	8	6	9	8	6
4	3	9	7	2	5	7	6	8	4

5	3	5	9	9	6	1	8	4	7
4	0	5	3	9	1	1	6	4	9

Average time = 100 sec
Record = 37 sec

B. Multiply.

2	6	3	5	6	4	4	8	2	7
8	5	3	7	3	5	7	6	6	9

8	0	2	3	1	5	6	9	5	8
4	6	9	8	9	5	4	5	2	9

3	7	5	6	9	2	7	8	9	2
5	7	8	9	4	4	6	8	0	2

5	9	1	8	6	4	9	0	2	7
5	3	7	7	6	3	9	4	1	8

Average time = 100 sec
Record = 36 sec

Check your answers on page 582.

Hop ahead to **21** to continue.

___

20   Here is the completed multiplication table:

Multiplication Table

×	2	5	8	1	3	6	9	7	4
1	2	5	8	1	3	6	9	7	4
7	14	35	56	7	21	42	63	49	28
5	10	25	40	5	15	30	45	35	20
4	8	20	32	4	12	24	36	28	16
9	18	45	72	9	27	54	81	63	36
2	4	10	16	2	6	12	18	14	8
6	12	30	48	6	18	36	54	42	24
3	6	15	24	3	9	18	27	21	12
8	16	40	64	8	24	48	72	56	32

If you are not able to perform these one-digit multiplications quickly from memory, you should practice until you can do so. A multiplication table is given at the back of the book. Use it if you need it.

Multiplying by Zero and One

Notice that the product of any number and 1 is that same number. For example,

$1 \times 2 = 2$

$1 \times 6 = 6$

or even

$1 \times 753 = 753$

Zero has been omitted from the table because the product of any number and zero is zero. For example,

$0 \times 2 = 0$

$0 \times 7 = 0$

$395 \times 0 = 0$

If you want some more practice in one-digit multiplication, turn back to **19**. Otherwise, go to **21**.

Multiplying by Larger Numbers

21   The multiplication of larger numbers is based on the one-digit number multiplication table. Find the product

$34 \times 2 = $ _____

Remember the procedure you followed for addition. What are the first few steps in this multiplication?

Try it, then go to **22**.

---

22   **First,** estimate the answer: $30 \times 2 = 60$. The actual product of the multiplication will be about 60.

Second, arrange the factors to be multiplied vertically, with ones digits in a single column, tens digits in a second column, and so on.

Finally, to make the process clear, let's write it in expanded form.

$$\begin{array}{r} 34 \\ \times 2 \end{array} \Rightarrow \begin{array}{r} 3 \text{ tens} + 4 \text{ ones} \\ \times 2 \\ \hline 6 \text{ tens} + 8 \text{ ones} = 60 + 8 = 68 \end{array}$$

Check: The guess 60 is roughly equal to the answer 68.

Now write the following multiplication in expanded form.

$$\begin{array}{r} 28 \\ \times 3 \end{array}$$

Check your work in **23**.

MULTIPLICATION SHORTCUTS

There are hundreds of quick ways to multiply various numbers. Most of them are quick only if you are already a math whiz. If you are not, the shortcuts will confuse more than help you. Here are a few that are easy to do and easy to remember.

1. To multiply by 10, attach a zero to the right end of the multiplicand. For example,

    $34 \times 10 = 340$

    $256 \times 10 = 2560$

    Multiplying by 100 or 1000 is similar.

    $34 \times 100 = 3400$

    $256 \times 1000 = 256000$

2. To multiply by a number ending in zeros, carry the zeros forward to the answer. For example,

    $$\begin{array}{r} 26 \\ \times 20 \\ \hline \end{array} \Rightarrow \begin{array}{r} 26 \\ \times 20 \\ \hline 520 \end{array}$$
    Multiply 26×2 and attach the zero on the right. The product is 520.

    $$\begin{array}{r} 34 \\ \times 2100 \\ \hline \end{array} \Rightarrow \begin{array}{r} 34 \\ \times 2100 \\ \hline 34\ \ \\ 68\ \ \ \\ \hline 71400 \end{array}$$

3. If both multiplier and multiplicand end in zeros, bring all zeros forward to the answer.

    $$\begin{array}{r} 230 \\ \times 200 \\ \hline \end{array} \Rightarrow \begin{array}{r} 230 \\ 200 \\ \hline 46000 \end{array}$$
    Attach three zeros to the product of 23×2.

    $$\begin{array}{r} 1000 \\ \times\ \ 100 \\ \hline 100{,}000 \end{array}$$
    This kind of multiplication is mostly a matter of counting zeros.

23 **Estimate:** $30 \times 3 = 90$ The answer is about 90.

$$\begin{array}{r} 28 \\ \times 3 \\ \hline \end{array} \Rightarrow \begin{array}{r} 2 \text{ tens} + 8 \text{ ones} \\ \times 3 \\ \hline 6 \text{ tens} + 24 \text{ ones} \end{array}$$

$= 6 \text{ tens} +\ \ 2 \text{ tens} + 4 \text{ ones}$

$= 8 \text{ tens} +\ \ 4 \text{ ones}$

$= 80 + 4$

$= 84$

Check: 90 is roughly equal to 84.

Of course, we do not normally use the expanded form; instead we simplify the work like this:

$$\begin{array}{r} \overset{2}{2}8 \\ \times\ \ 3 \\ \hline 84 \end{array}$$

$3 \times 8 = 24$ Write 4 and carry 2 tens.

$3 \times 2 \text{ tens} = 6 \text{ tens}$ $6 \text{ tens} + 2 \text{ tens} = 8 \text{ tens}$

Write 8.

Now try these problems to be certain you understand the process. Multiply as shown.

(a) 43
  $\times\ 5$

(b) 73
  $\times\ 4$

(c) 29
  $\times\ 6$

(d) 258
  $\times\ \ 7$

Check your answers in **24.**

---

24 (a) **Estimate:** $40 \times 5 = 200$ The answer is roughly 200.

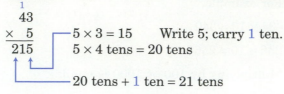

    $\overset{1}{43}$

  $\underline{\times\ 5}$ — $5 \times 3 = 15$ Write 5; carry 1 ten.

   215 — 5×4 tens = 20 tens

      — 20 tens + 1 ten = 21 tens

Check: The answer 215 is roughly equal to the estimate 200.

(b) **Estimate:** $70 \times 4 = 280$

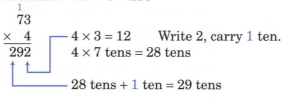

    $\overset{1}{73}$

  $\underline{\times\ \ 4}$ — $4 \times 3 = 12$ Write 2, carry 1 ten.

   292 — 4×7 tens = 28 tens

      — 28 tens + 1 ten = 29 tens

Check: The answer 292 is roughly equal to the estimate 280.

(c) **Estimate:** $30 \times 6 = 180$

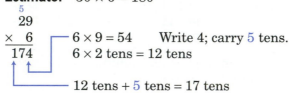

    $\overset{5}{29}$

  $\underline{\times\ \ 6}$ — $6 \times 9 = 54$ Write 4; carry 5 tens.

   174 — 6×2 tens = 12 tens

      — 12 tens + 5 tens = 17 tens

Check: The answer 174 is roughly equal to the estimate 180.

(d) **Estimate:** $300 \times 7 = 2100$

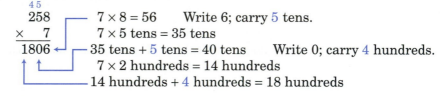

    $\overset{4\,5}{258}$

  $\underline{\times\ \ \ 7}$ — $7 \times 8 = 56$ Write 6; carry 5 tens.

  1806 — 7×5 tens = 35 tens

      — 35 tens + 5 tens = 40 tens   Write 0; carry 4 hundreds.

      — 7×2 hundreds = 14 hundreds

      — 14 hundreds + 4 hundreds = 18 hundreds

Check: The answer and estimate are roughly equal.

Learning Help ▶ Multiplying a two- or three-digit number by a one-digit number can be done mentally. For example, think of 43×5 as 40×5 plus 3×5, or $200 + 15 = 215$. For a problem such as 29×6, think 30×6 minus 1×6 or $180 - 6 = 174$. Tricks like this are very useful on the job when neither paper and pencil nor a calculator is at hand. ◀

Two-Digit Multiplications

Calculations involving two-digit multipliers are done in a similar way. For example, to multiply

  89
$\underline{\times 24}$

First, estimate the answer. Rounding each number to the nearest ten, $90 \times 20 = 1800$.

Second, multiply by the ones digit 4.

$$
\begin{array}{r}
\overset{3}{} \\
8\,9 \\
\times\ 2\,4 \\
\hline
3\,5\,6
\end{array}
$$

$4 \times 9 = 36$ Write 6; carry 3 tens.

4×8 tens $= 32$ tens

32 tens $+\ 3$ tens $= 35$ tens

Third, multiply by the tens digit 2.

$$
\begin{array}{r}
\overset{1}{} \\
\overset{3}{} \\
8\,9 \\
\times\ 2\,4 \\
\hline
3\,5\,6 \\
1\,7\,8\ \leftarrow
\end{array}
$$

$2 \times 9 = 18$ Write 8, carry 1.

$2 \times 8 = 16$

$16 + 1 = 17$

Leave a blank space here, since we are actually multiplying $89 \times 20 = 1780$.

Fourth, add the products obtained.

$$
\begin{array}{r}
3\,5\,6 \\
1\,7\,8 \\
\hline
2\,1\,3\,6
\end{array}
$$

Finally, check it. The estimate and the answer are roughly the same, or at least in the same ballpark.

Notice that the product in the third step, 178, is written one digit space over from the product from the second step. When we multiplied 2×9 to get 18 in step 3, we were actually multiplying $20 \times 9 = 180$, but the zero in 180 is usually omitted to save time.

Try these:

(a)
$$
\begin{array}{r}
64 \\
\times\ 37
\end{array}
$$

(b)
$$
\begin{array}{r}
327 \\
\times\ 145
\end{array}
$$

(c)
$$
\begin{array}{r}
342 \\
\times\ 102
\end{array}
$$

(d) The Good Value hardware store sold 168 tool boxes at \$47 each. What was their total income on this item?

Multiply as shown and check your work in **25**.

THE SEXY SIX

Here are the six most often missed one-digit multiplications:

Inside digits

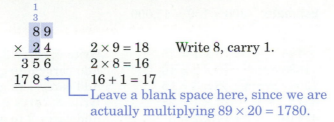

$9 \times 8 = 72$

$9 \times 7 = 63$

$9 \times 6 = 54$

$8 \times 7 = 56$

$8 \times 6 = 48$

$7 \times 6 = 42$

It may help you to notice that in these multiplications the "inside" digits, such as 8 and 7, are consecutive and the digits of the answer add to nine: $7 + 2 = 9$. This is true for *all* one-digit numbers multiplied by 9.

Be certain that you have these memorized.

(There is nothing very sexy about them, but we did get your attention, didn't we?)

25 (a) **Estimate:** $60 \times 40 = 2400$

```
    64
  × 37
   448  ←
   192  ←
  2368  ←
```

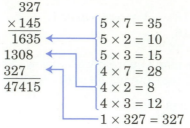

$7 \times 4 = 28$ Write 8; carry 2.
$7 \times 6 = 42$ Add carry 2 to get 44; write 44.
$3 \times 4 = 12$ Write 2; carry 1.
$3 \times 6 = 18$ Add carry 1 to get 19; write 19.
Add to obtain the answer.

(b) **Estimate:** $300 \times 150 = 45,000$

```
     327
   × 145
    1635  ←
    1308  ←
     327  ←
   47415
```

$5 \times 7 = 35$ Write 5; carry 3.
$5 \times 2 = 10$ Add carry 3 to get 13; write 3, carry 1.
$5 \times 3 = 15$ Add carry 1 to get 16; write 16.
$4 \times 7 = 28$ Write 8, carry 2.
$4 \times 2 = 8$ Add carry 2 to get 10; write 0, carry 1.
$4 \times 3 = 12$ Add carry 1 to get 13; write 13.
$1 \times 327 = 327$

The product is 47,415.

(c) **Estimate:** $300 \times 100 = 30,000$

```
     342
   × 102
     684  ←——— 2 × 342 = 684
     000  ←——— 0 × 342 = 000
     342  ←——— 1 × 342 = 342
   34884
```

The product is 34,884.

(d) To find the total, multiply 168 by 47.

```
     168
   ×  47
    1176
     672
    7896
```

The total income is $7896.

Be very careful when there are zeros in the multiplier, it is very easy to misplace one of those zeros. Do not skip any steps, and be sure to estimate your answer first.

Go to **26** for a set of practice problems on the multiplication of whole numbers.

26

Exercise 1-3   **Multiplication of Whole Numbers**

A. Multiply.

1. 7 <u>6</u>	2. 7 <u>8</u>	3. 6 <u>8</u>	4. 8 <u>9</u>	5. 9 <u>7</u>	6. 29 <u>3</u>
7. 9 <u>6</u>	8. 4 <u>7</u>	9. 5 <u>9</u>	10. 12 <u>7</u>	11. 37 <u>8</u>	12. 24 <u>7</u>
13. 72 <u>8</u>	14. 47 <u>9</u>	15. 64 <u>5</u>	16. 39 <u>4</u>	17. 58 <u>5</u>	18. 94 <u>6</u>

19. 32	20. 46	21. 72	22. 68	23. 54	24. 17
13	14	11	16	26	9

25. 47	26. 77	27. 48	28. 64	29. 90	30. 86
6	4	15	27	56	83

31. 34	32. 66	33. 59	34. 29	35. 78	36. 94
57	25	76	32	49	95

B. Multiply.

1. 305	2. 2006	3. 8043	4. 809	5. 3706
123	125	37	47	102

6. 708	7. 684	8. 2043	9. 2008	10. 563
58	45	670	198	107

11. 809	12. 609	13. 500	14. 542	15. 7009
9	7	50	600	504

16. 407	17. 316	18. 514	19. 807	20. 560
22	32	62	111	203

C. Practical Problems

1. **Plumbing** A plumber receives $46 per hour. How much is he paid for 40 hours of work?

2. **Electrical Technology** What is the total length of wire on 14 spools if each spool contains 150 ft?

3. **Building Construction** How many total lineal feet of redwood are there in 65 2-in. by 4-in. boards each 20 ft long?

4. **Automotive Services** An auto body shop does 17 paint jobs at $859 each and 43 paint jobs at $1165 each. How much money does the shop receive from these jobs?

5. **Office Services** Three different-sized boxes of envelopes contain 50, 100, and 500 envelopes, respectively. How many envelopes total are there in 18 boxes of the first size, 16 of the second size, and 11 of the third size?

6. **Machine Technology** A machinist needs 25 lengths of steel each 9 in. long. What is the total length of steel that he needs? No allowance is required for cutting.

7. **Machine Technology** The Ace Machine Company advertises that one of its machinists can produce 2 parts per hour. How many such parts can 27 machinists produce if they work 45 hours each?

8. **Building Construction** What is the horizontal distance in inches covered by 12 stair steps if each is 11 in. wide?

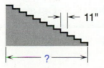

9. **Auto Mechanics** Five bolts are needed to mount a tire. How many bolts are needed to mount all four tires on 60 cars?

10. **Auto Mechanics**   Each of the 17 head bolts on a certain engine exerts a force of 17,364 lb. What is the total force that holds the head and block together?

11. **Printing**   A printer finds that nine 6-in. by 8-in. cards can be cut from a 19-in. by 25-in. sheet. How many such cards can be cut from 850 of the sheets?

12. **Industrial Technology**   The Hold Tite fastener plant produces 738 cotter pins per shift. If it operates two shifts per day, five days each week, how many cotter pins will the plant produce in six weeks?

13. **Machine Technology**   If a machine produces 16 screws per minute, how many screws will it produce in 24 hours?

D. Calculator Problems

1. **Office Services**   If an investment company started in business with $100,000 of capital on January 1 and, because of mismanagement, lost an average of $273 every day for a year, what would be its financial situation on the last day of the year?

2. Which of the following pay schemes gives you the most money over a one-year period?

   (a) $100 per day

   (b) $700 per week

   (c) $400 for the first month and a $400 raise each month

   (d) 1 cent for the first two-week pay period, 2 cents for the second two-week period, 4 cents for the third two-week period, and so on, the pay doubling each two weeks.

3. Multiply.

   (a) $12,345,679 \times 9 =$ (b) $15,873 \times 7 =$
   $12,345,679 \times 18 =$ $15,873 \times 14 =$
   $12,345,679 \times 27 =$ $15,873 \times 21 =$

   (c) $1 \times 1 =$ (d) $6 \times 7 =$
   $11 \times 11 =$ $66 \times 67 =$
   $111 \times 111 =$ $666 \times 667 =$
   $1111 \times 1111 =$ $6666 \times 6667 =$
   $11111 \times 11111 =$ $66666 \times 66667 =$

   Can you see the pattern in each of these?

4. **Machine Technology**   In the Aztec Machine Shop there are nine lathes each weighing 2285 lb, five milling machines each weighing 2570 lb, and three drill presses each weighing 395 lb. What is the total weight of these machines?

5. **Manufacturing**   The Omega Calculator Company makes five models of electronic calculators. The following table gives the daily production output.

Model	Alpha	Beta	Gamma	Delta	Tau
Cost of production of each model	$6	$17	$32	$49	$78
Number produced during typical day	117	67	29	37	18

Find the weekly (five typical day) production costs for each model.

Check your answers on page 582.

Turn to **27** to continue.

1-4 DIVISION OF WHOLE NUMBERS

27 Division is the reverse of multiplication. It enables us to separate a given quantity into equal parts. The mathematical phrase $12 \div 3$ is read "twelve divided by three," and it asks us to separate a collection of 12 objects into 3 equal parts. The mathematical phrases

$$12 \div 3 \qquad 3\overline{)12} \qquad \frac{12}{3} \qquad \text{and} \qquad 12/3$$

all represent division, and they are all read "twelve divided by three."

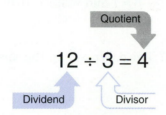

In this division problem, 12, the number being divided, is called the *dividend;* 3, the number used to divide is called the *divisor;* and 4, the result of the division, is called the *quotient,* from a Latin word meaning "how many times."

One way to perform division is to reverse the multiplication process.

$$24 \div 4 = \square \qquad \text{means that} \qquad 4 \times \square = 24$$

If the one-digit multiplication tables are firmly in your memory, you will recognize immediately that $\square = 6$.

Try these.

$35 \div 7 =$ _____ $\qquad$ $30 \div 5 =$ _____ $\qquad$ $18 \div 3 =$ _____

$28 \div 4 =$ _____ $\qquad$ $42 \div 6 =$ _____ $\qquad$ $63 \div 9 =$ _____

$45 \div 5 =$ _____ $\qquad$ $56 \div 7 =$ _____ $\qquad$ $72 \div 8 =$ _____

$70 \div 10 =$ _____

Check your answers in **28**.

---

28 $35 \div 7 = 5$ $\qquad$ $30 \div 5 = 6$ $\qquad$ $18 \div 3 = 6$

$28 \div 4 = 7$ $\qquad$ $42 \div 6 = 7$ $\qquad$ $63 \div 9 = 7$

$45 \div 5 = 9$ $\qquad$ $56 \div 7 = 8$ $\qquad$ $72 \div 8 = 9$

$70 \div 10 = 7$

You should be able to do all these quickly by working backward from the one-digit multiplication tables.

Careful ▶ Division by zero has no meaning in arithmetic. (Any number) $\div$ 0 or $\dfrac{(\text{any number})}{0}$ is not defined. But $0 \div (\text{any number}) = \dfrac{0}{(\text{any number})} = 0$ ◀

How do we divide dividends that are larger than 9×9 and therefore not in the multiplication table? Obviously, we need a better procedure.

Here is a step-by-step explanation of the division of whole numbers:

Example: Divide 96 ÷ 8 = _____.

First, *estimate* the answer: 100 ÷ 10 is about 10. The quotient or answer will be about 10.

Second, arrange the numbers in this way:

8)96

Third, divide using the following step-by-step procedure.

Step 1 $\begin{array}{r} 1 \\ 8\overline{)96} \end{array}$ 8 into 9? Once. Write 1 in the answer space above the 9.

Step 2 $\begin{array}{r} 1 \\ 8\overline{)96} \\ 8 \end{array}$ Multiply 8 × 1 = 8 and write the product 8 under the 9

Step 3 $\begin{array}{r} 1 \\ 8\overline{)96} \\ -8\downarrow \\ \hline 16 \end{array}$ Subtract 9 − 8 = 1 and write 1. Bring down the next digit, 6.

Step 4 $\begin{array}{r} 12 \\ 8\overline{)96} \\ -8 \\ \hline 16 \end{array}$ 8 into 16? Twice. Write 2 in the answer space above the 6.

Step 5 $\begin{array}{r} 12 \\ 8\overline{)96} \\ -8 \\ \hline 16 \\ -16 \end{array}$ Multiply 8 × 2 = 16 and write the product 16 under the 16.

Step 6 $\begin{array}{r} 12 \\ 8\overline{)96} \\ -8 \\ \hline 16 \\ -16 \\ \hline 0 \end{array}$ Subtract 16 − 16 = 0. Write 0.

0 ← The remainder is zero. 8 divides into 96 exactly 12 times.

Finally, *check* your answer. The answer 12 is roughly equal to the original estimate 10. As a second check, multiply the divisor 8 and the quotient 12. Their product should be the original dividend number, 8 × 12 = 96, which is correct.

Practice this step-by-step division process by finding 112 ÷ 7.

Our work is shown in **29**.

29 **Estimate:** 112 ÷ 7 is roughly 100 ÷ 5 or 20. The answer will be roughly 20.

$$
\begin{array}{r}
016 \\
7)\overline{112} \\
-7\downarrow \\
\hline
42 \\
-42 \\
\hline
0
\end{array}
$$

Step 1 7 into 1. Won't go, so put a zero in the answer place.

Step 2 7 into 11. Once. Write 1 in the answer space as shown.

Step 3 $7 \times 1 = 7$. Write 7 below 11.

Step 4 $11 - 7 = 4$. Write 4. Bring down the 2.

Step 5 7 into 42. Six. Write 6 in the answer space.

Step 6 $7 \times 6 = 42$. Subtract 42. The remainder is zero.

Check: The answer 16 is roughly equal to our original estimate of 20. Check again by multiplying. $7 \times 16 = 112$. Finding a quick estimate of the answer will help you avoid making mistakes.

Careful ▶ Notice that once the first digit of the answer has been obtained, there will be an answer digit for every digit of the dividend.

$$
\begin{array}{c}
16 \\
\Updownarrow \\
7)\overline{112} \quad \blacktriangleleft
\end{array}
$$

Example: 203 ÷ 7 = ?

Estimate: 203 ÷ 7 is roughly 210 ÷ 7 or 30. The answer will be roughly 30.

$$
\begin{array}{r}
029 \\
7)\overline{203} \\
-14\downarrow \\
\hline
63 \\
-63 \\
\hline
0
\end{array}
$$

Step 1 7 into 2. Won't go, so put a zero in the answer space.

Step 2 7 into 20. Twice. Write 2 in the answer space as shown.

Step 3 $7 \times 2 = 14$. Write 14 below 20.

Step 4 $20 - 14 = 6$. Write 6. Bring down the 3.

Step 5 7 into 63. Nine times. Subtract 63. The remainder is zero.

Check: The answer is 29, roughly equal to the original estimate of 30. Double-check by multiplying: $7 \times 29 = 203$.

The remainder is not always zero, of course. For example, to calculate 153 ÷ 4,

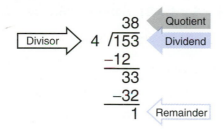

Write the answer as 38r1 to indicate that the quotient is 38 and the remainder is 1.

To check an answer with a remainder, first multiply the quotient by the divisor, then add the remainder. In this example,

$38 \times 4 = 152$

$152 + 1 = 153$

which checks.

Ready for some guided practice? Try these problems.

(a) 3174 ÷ 6   (b) 206 ÷ 6   (c) 59 ÷ 8

(d) 5084 ÷ 31   (e) 341 ÷ 43

(f)   Pete the Plumber billed $7548 for the labor on a large project. If his regular rate is $37 per hour, how many hours did he work?

Check your work in **30**.

---

30   (a) $3174 \div 6 = $ _____

Estimate: $6 \times 500 = 3000$ The answer will be roughly 500.

```
   0529
6) 3174
  −30
  ────
   17
  − 12
  ────
    54
  − 54
  ────
     0
```

Step 1   6 into 3? No. Write a zero. 6 into 31? 5 times. Write 5 above the 1.

Step 2 $6 \times 5 = 30$. Subtract $31 − 30 = 1$.

Step 3   Bring down 7. 6 into 17? Twice. Write 2 in the answer.

Step 4 $6 \times 2 = 12$. Subtract $17 − 12 = 5$.

Step 5   Bring down 4. 6 into 54? 9 times. Write 9 in the answer.

Step 6 $6 \times 9 = 54$. Subtract $54 − 54 = 0$.

Quotient = 529

The answer is 529.

Check: The estimate, 500, and the answer, 529, are roughly equal.

Double-check: $6 \times 529 = 3174$.

(b) $206 \div 6 = $ _____

Estimate: $6 \times 30 = 180$ The answer is about 30.

```
   034
6) 206
  −18
  ────
   26
  − 24
  ────
    2
```

Step 1   6 into 2? No. Write a zero. 6 into 20? Three times. Write 3 above the zero in the answer.

Step 2 $6 \times 3 = 18$. Subtract $20 − 18 = 2$.

Step 3   Bring down 6. 6 into 26? 4 times. Write 4 in the answer.

Step 4 $6 \times 4 = 24$. Subtract $26 − 24 = 2$. The remainder is 2.

Quotient = 34

The answer is 34 with a remainder of 2.

Check: The estimate 30 and the answer are roughly equal.

Double-check: $6 \times 34 = 204$ $204 + 2 = 206$.

(c) $59 \div 8 = $ _____

Estimate: $7 \times 8 = 56$. The answer will be about 7.

```
    7
8) 59
  −56
  ───
    3
```

8 into 5? No There is no need to write the zero.

The quotient is 7 with a remainder of 3.

Check it, and double-check by multiplying and adding the remainder.

(d) $5084 \div 31 = $ _____

Estimate: This is roughly about the same as $5000 \div 30$ or $500 \div 3$ or about 200. The quotient will be about 200.

$$\begin{array}{r} 164 \\ 31\overline{)\ 5084} \\ -31 \quad\ \\ \hline 198 \quad \\ -186 \quad \\ \hline 124 \\ -\ 124 \\ \hline 0 \end{array}$$

Step 1 31 into 5: No. 31 into 50? Yes, once. Write 1 above the zero.

Step 2 $31 \times 1 = 31$. Subtract $50 - 31 = 19$.

Step 3 Bring down 8. 31 into 198? (That is, about the same as 3 into 19.) Yes, 6 times; write 6 in the answer.

Step 4 $31 \times 6 = 186$. Subtract $198 - 186 = 12$.

Step 5 Bring down 4. 31 into 124? (That is, about the same as 3 into 12.) Yes, 4 times. Write 4 in the answer.

Step 6 $31 \times 4 = 124$. Subtract $124 - 124 = 0$.

The quotient is 164.

Check: The estimate is reasonably close to the answer.

Double-check: $31 \times 164 = 5084$.

Notice that in step 3 it is not at all obvious how many times 31 will go into 198. Again, you must make an educated guess and check your guess as you go along.

(e) $341 \div 43 = $ _____

Estimate: This is roughly $300 \div 40$ or $30 \div 4$ or about 7. The answer will be about 7 or 8.

$$\begin{array}{r} 7 \\ 43\overline{)\ 341} \\ -301 \\ \hline 40 \end{array}$$

Your first guess would probably be that 43 goes into 341 8 times (try 4 into 34), but $43 \times 8 = 344$, which is larger than 341. The quotient is 7 with a remainder of 40.

Check it.

(f) To determine the total bill, Pete multiplied his hourly rate by the number of hours. Therefore, if we know the total and the hourly rate, we reverse this process and divide the total by the hourly rate to find the number of hours.

Estimate: $7000 \div 30$ or $700 \div 3$ is about 200.

$$\begin{array}{r} 204 \\ 37\overline{)7548} \\ 74 \quad\ \\ \hline 148 \\ 148 \\ \hline \end{array}$$

On the second step notice that 37 does not go into 14. So we must write a zero in the tens place of the answer, above the 4, and bring down the 8.

The quotient is 204. Pete worked a total of 204 hours on the project.

Check: The answer 204 and the estimate 200 are roughly equal.

Double-check: $37 \times 204 = 7548$

Factors It is sometimes useful in mathematics to be able to write any whole number as a product of other numbers. If we write

$6 = 2 \times 3$ 2 and 3 are called **factors** of 6.

Of course, we could also write

$6 = 1 \times 6$ and see that 1 and 6 are also factors of 6.

The factors of 6 are 1, 2, 3, and 6.

The factors of 12 are 1, 2, 3, 4, 6, and 12.

The factors of 30 are 1, 2, 3, 5, 6, 10, 15, and 30.

Any number is exactly divisible by its factors; that is, every factor divides the number with zero remainder.

Primes

For some numbers the only factors are 1 and the number itself. For example, the factors of 7 are 1 and 7 because $7 = 1 \times 7$. There are no other numbers that divide 7 with remainder zero. Such numbers are known as prime numbers. A **prime number** is one for which there are no factors other than 1 and the prime itself.

Here is a list of the first few prime numbers.

2, 3, 5, 7, 11, 13, 17, 19, 23, 29, 31, 37, 41, 43, 47

(Notice that 1 is not listed. All prime numbers have two different factors: 1 and the number itself. The number 1 has only 1 factor—itself.)

Prime Factors

The **prime factors** of a number are those factors that are prime numbers. For example, the prime factors of 6 are 2 and 3. The prime factors of 30 are 2, 3, and 5. You will see later that the concept of prime factors is useful when working with fractions.

To find the prime factors of a number a **factor-tree** is often helpful.

Example: Find the prime factors of 132.

First, write the number 132 and draw two branches below it.

Second, beginning with the smallest prime in our list, 2, test for divisibility. $132 = 2 \times 66$. Write the factors below the branches.

Next, repeat this procedure on the nonprime factor, 66. $66 = 2 \times 33$

Finally, continue dividing until all branches end with a prime.

Therefore, $132 = 2 \times 2 \times 3 \times 11$

Check by multiplying.

The following problem set provides practice with factors and primes.

Factors and Primes

A. List all the factors of each of the following whole numbers. If the whole number given is a prime, label it as a prime.

1. 4	2. 10	3. 9	4. 11	5. 24
6. 18	7. 15	8. 3	9. 14	10. 17
11. 21	12. 32	13. 20	14. 23	15. 29
16. 26	17. 44	18. 90	19. 39	20. 13

B. List the prime factors of each of the whole numbers given in part A. Check your answers on page 583.

Now hop to **31** for a set of problems on the division of whole numbers.

SOME DIVISION "TRICKS OF THE TRADE"

1. If a number is exactly divisible by 2, that is, divisible with zero remainder, it will end in an even digit, 0, 2, 4, 6, or 8.

    Example: We know that 374 is exactly divisible by 2 since it ends in the even digit 4.

    $$\begin{array}{r} 187 \\ 2\overline{)374} \end{array}$$

2. If a number is exactly divisible by 3, the sum of its digits is exactly divisible by 3.

    Example: The number 2784 has the sum of digits $2 + 7 + 8 + 4 = 21$. Since 21 is exactly divisible by 3, we know that 2784 is also exactly divisible by 3.

    $$\begin{array}{r} 928 \\ 3\overline{)2784} \end{array}$$

3. A number is exactly divisible by 4 if its last two digits are exactly divisible by 4 or are both zero.

    Example: We know that the number 3716 is exactly divisible by 4 since 16 is exactly divisible by 4.

    $$\begin{array}{r} 929 \\ 4\overline{)3716} \end{array}$$

4. A number is exactly divisible by 5 if its last digit is either 5 or 0.

    Example: The numbers 875, 310, and 33,195 are all exactly divisible by 5.

5. A number is exactly divisible by 8 if its last three digits are divisible by 8.

    Example: The number 35120 is exactly divisible by 8 since the number 120 is exactly divisible by 8.

6. A number is exactly divisible by 9 if the sum of its digits is exactly divisible by 9.

    Example: The number 434,673 has the sum of digits 27. Since 27 is exactly divisible by 9, we know that 434,673 is also exactly divisible by 9.

    $$\begin{array}{r} 48297 \\ 9\overline{)434673} \end{array}$$

7. A number is exactly divisible by 10 if it ends with a zero.

    Example: The numbers 230, 77,380, and 100,200 are all exactly divisible by 10.

Can you combine the tests for divisibility by 2 and 3 to get a rule for divisibility by 6?

Exercises 1-4   **Division of Whole Numbers**

A. Divide.

1. $63 \div 7$	2. $92 \div 8$	3. $72 \div 0$
4. $37 \div 5$	5. $71 \div 7$	6. $6 \div 6$
7. $\dfrac{32}{4}$	8. $\dfrac{28}{7}$	9. $\dfrac{54}{9}$
10. $245 \div 7$	11. $167 \div 7$	12. $228 \div 4$
13. $310 \div 6$	14. $3310 \div 3$	15. $\dfrac{147}{7}$
16. $7\overline{)364}$	17. $6\overline{)222}$	18. $4\overline{)201}$
19. $322 \div 14$	20 $382 \div 19$	21. $936 \div 24$
22. $700 \div 28$	23. $730 \div 81$	24. $\dfrac{901}{17}$
25. $31\overline{)682}$	26. $27\overline{)1724}$	27. $42\overline{)371}$

B. Divide.

1. $61\overline{)7320}$	2. $33\overline{)303}$	3. $16\overline{)904}$
4. $2001 \div 21$	5. $2016 \div 21$	6. $1000 \div 7$
7. $2000 \div 9$	8. $2400 \div 75$	9. $14\overline{)4275}$
10. $71\overline{)6005}$	11. $53\overline{)6307}$	12. $3\overline{)9003}$
13. $7\overline{)3507}$	14. $6\overline{)48009}$	15. $6\overline{)3624}$
16. $3\overline{)62160}$	17. $15\overline{)3000}$	18. $67\overline{)3354}$
19. $24\overline{)2596}$	20. $47\overline{)94425}$	21. $38\overline{)22800}$
22. $231\overline{)14091}$	23. $411\overline{)42020}$	24. $603\overline{)48843}$
25. $111\overline{)11111}$	26. $102\overline{)2004}$	27. $405\overline{)7008}$

C. Applied Problems

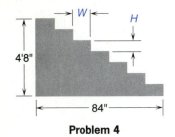

Problem 4

1. **Machine Technology**   A machinist has a piece of bar stock 243 in. long. If she must cut nine equal pieces, how long will each piece be? (Assume no waste to get a first approximation.)

2. **Carpentry**   How many rafters 32 in. long can be cut from a piece of lumber 192 in. long?

3. **Building Construction**   If subflooring is laid at the rate of 85 sq ft per hour, how many hours will be needed to lay 1105 sq ft?

4. **Building Construction**   The illustration shows a stringer for a short flight of stairs. Determine the missing dimensions H and W.

5. **Building Construction**   How many joists spaced 16 in. o.c. (on center) are required for a floor 36 ft long? (Add one joist for a starter.)

6. **Carpentry**   If rafters are placed 24 in. o.c., how many are required for both sides of a gable roof that is 48 ft long? Add one rafter on each side for a starter.

7. **Carpentry**   A stairway in a house is to be built with 18 risers. If the distance from the top of the first floor to the top of the second floor is 10 ft 6 in., what is the height of each step?

8. **Manufacturing**   What is the average horsepower of six engines having the following horsepower ratings: 385, 426, 278, 434, 323, and 392? (*Hint:* To find the average, add the six numbers and divide their sum by 6.)

9. **Plumbing**   Mr. Martinez, owner of the Maya Restaurant, hired a plumber who worked for 18 hours installing some new equipment. The total bill for wages and supplies used was $932. If the cost for materials was $176, calculate the plumber's hourly wage.

10. A truck traveling at an average rate of 54 miles per hour covered a distance of 486 miles in one run. Allowing 2 hours for refueling and meals, how long did it take to complete the trip?

11. **Auto Mechanics**   A mechanic needs to order 480 spark plugs. If there are eight plugs in a box, how many boxes does he need?

12. **Metalworking**   One cubic foot of a certain alloy weighs 375 lb. How many cubic feet does a 4875-lb block occupy?

13. **Printing**   A printer needs 13,500 sheets of paper for a job. If there are 500 sheets in a ream, how many reams does the printer need to get from the stockroom?

14. **Wastewater Technology**   A water tank contains 18,000 gallons. If 500 gallons per day are used from this tank, how long will it last?

D. Calculator Problems

1. Calculate the following.

   (a) $\dfrac{1347 \times 46819}{3}$

   (b) $\dfrac{76459 + 93008 + 255}{378}$

   (c) $\dfrac{4008 + 408 + 48}{48}$

   (d) $\dfrac{9909 \times 9090}{3303}$

2. **Masonry**   If 6-in. Norwegian brick is on sale at $67 per hundred, calculate the cost of the 14,000 bricks needed on a construction project.

3. **Manufacturing**   The specified shear strength of a $\frac{1}{8}$-in. 2117-T3 (AD) rivet is 344 lb. If it is necessary to assure a strength of 6587 lb to a riveted joint in an aircraft structure, how many rivets must be used?

4. A train of 74 railway cars weighed a total of 9,108,734 lb. What was the average weight of a car?

5. **Manufacturing**   A high-speed stamping machine can produce small flat parts at the rate of 96 per minute. If an order calls for 297,600 parts to be stamped, how many hours will it take to complete the job? (60 minutes = 1 hour.)

6. **Painting and Decorating**   Following are four problems from the national apprentice examination for painting and decorating contractors.

   (a) $11,877,372 \div 738$

   (b) $87,445,005 \div 435$

   (c) $1,735,080 \div 760$

   (d) $206,703 \div 579$

Check your answers on page 583, then turn to **32** to study another important topic in the arithmetic of whole numbers.

1-5   ORDER OF OPERATIONS

32   What is the value of the arithmetic expression

$3 \times 4 + 2 = ?$

If we multiply first, $(3 \times 4) + 2 = 12 + 2 = 14$

but if we add first, $3 \times (4 + 2) = 3 \times 6 = 18$

To avoid any possible confusion in situations like this, mathematicians have adopted a standard order of operations for arithmetic calculations. When two or more of the four basic arithmetic operations (addition, subtraction, multiplication, and division) are combined in the same calculation, follow these rules.

Rule 1   Perform any calculations shown inside parentheses first.

Rule 2   Perform all multiplications and divisions next, working from left to right.

Rule 3   Perform additions and subtractions last, working from left to right.

For example, $3 \times 4 + 2 = 12 + 2$ Multiply first, Rule 2. $3 \times 4 = 12$

$ = 14$ Add last.

Perform the following calculations, using the order of operation rules.

(a) $5 + 8 \times 7$ (b) $12 \div 6 - 2$

(c) $(26 - 14) \times 2$ (d) $12 \times (6 + 2) \div 6 - 7$

Check your work in **33**.

---

33 (a) $5 + 8 \times 7 = 5 + 56$ Do the multiplication first. Rule 2.
 $(8 \times 7 = 56)$

$ = 61$ Then add, Rule 3.

(b) $12 \div 6 - 2 = 2 - 2$ First divide. $(12 \div 6 = 2)$
$ = 0$ Then subtract, Rule 3.

(c) $(26 - 14) \times 2 = 12 \times 2$ Perform the operation in parentheses first, Rule 1.
 $(26 - 14 = 12)$

$ = 24$ Then multiply, Rule 2.

(d) $12 \times (6 + 2) \div 6 - 7 = 12 \times 8 \div 6 - 7$ Work inside parentheses first, Rule 1.
 $(6 + 2 = 8)$

$ = 96 \div 6 - 7$ Multiply, Rule 2. $(12 \times 8 = 96)$
$ = 16 - 7$ Divide, Rule 2. $(96 \div 6 = 16)$
$ = 9$ Subtract, Rule 3.

Notice in Exercise (d) that the multiplication and division without parentheses are performed working left to right.

Using parentheses helps guard against confusion. For example, the calculation $12 - 4 \div 2$ should be written $12 - (4 \div 2)$ to avoid making a mistake.

Perform the following calculations, using the three rules of order of operations.

(a) $240 \div (18 + 6 \times 2) - 2$ (b) $6 \times 5 + 14 \div (6 + 8) - 3$

(c) $6 + 2 \times 3 \div (7 - 4) + 5$

Check your work in **34**.

---

34 (a) $240 \div (18 + 6 \times 2) - 2 = 240 \div (18 + 12) - 2$ Perform the multiplication inside the parentheses first. $(6 \times 2 = 12)$

$$= 240 \div 30 - 2$$

Perform the sum inside the parentheses next. $(18 + 12 = 30)$

$$= 8 - 2$$

Next divide, Rule 2. $(240 \div 30 = 8)$

$$= 6$$

Finally, subtract, Rule 3.

(b) $6 \times 5 + 14 \div (6 + 8) - 3 = 6 \times 5 + 14 \div 14 - 3$ Perform the operations inside parentheses first.

$$= 30 + 14 \div 14 - 3$$

Next, multiply, then divide, going left to right $(14 \div 14 = 1)$

$$= 30 + 1 - 3$$

Finally, do the addition and subtraction left to right,

$$= 28$$

Rule 3.

(c) $6 + 2 \times 3 \div (7 - 4) + 5 = 6 + 2 \times 3 \div 3 + 5$ Parentheses first, Rule 1.

$$= 6 + 6 \div 3 + 5$$

Multiply, then divide, left to right,

$$= 6 + 2 + 5$$

Rule 2. $(6 \div 3 = 2)$

$$= 13$$

Add last, Rule 3.

When division is written with a fraction bar or *vinculum*, all calculations above the bar or below the bar should be done before the division. For example,

$$\frac{24 + 12}{6 - 2} \qquad \text{should be thought of as} \qquad \frac{(24 + 12)}{(6 - 2)}$$

First, simplify the top and bottom to get $\frac{36}{4}$.

Then, divide to obtain the answer $\frac{36}{4} = 9$.

Note ▶ This is very different from the calculation

$$\frac{24}{6} + \frac{12}{2} = 4 + 6 = 10$$

where we use Rules 2 and 3, doing the divisions first and the additions last. ◀

Now turn to **35** for a set of exercises involving the order of operations.

35

Exercises 1-5   **Order of Operations**

A. Perform all operations in the correct order.

1.	$2 + 8 \times 6$	2.	$20 - 3 \times 2$
3.	$40 - 20 \div 5$	4.	$16 + 32 \div 4$
5.	$16 \times 3 + 9\grave{\ }$	6.	$2 \times 9 - 4$
7.	$48 \div 8 - 2$	8.	$64 \div 16 + 8$
9.	$(5 + 9) \times 3$	10.	$(18 - 12) \div 6$
11.	$24 \div (6 - 2)$	12.	$9 \times (8 + 3)$

13. $16 + 5 \times (3 + 6)$

14. $8 + 3 \times (9 - 4)$

15. $(23 + 5) \times (12 - 8)$

16. $(17 - 9) \div (6 - 2)$

17. $6 + 4 \times 7 - 3$

18. $24 - 8 \div 2 + 6$

19. $5 \times 8 + 6 \div 6 - 12 \times 2$

20. $24 \div 8 - 14 \div 7 + 8 \times 6$

21. $2 \times (6 + 4 \times 9)$

22. $54 \div (8 - 3 \times 2)$

23. $(4 \times 3 + 8) \div 5$

24. $(26 \div 2 - 5) \times 4$

25. $8 - 4 + 2$

26. $24 \div 6 \times 2$

27. $18 \times 10 \div 5$

28. $22 + 11 - 7$

29. $12 - 7 - 3$

30. $48 \div 6 \div 2$

31. $12 - (7 - 3)$

32. $18 \div (3 \times 2)$

33. $\dfrac{36}{9} + \dfrac{27}{3}$

34. $\dfrac{36 - 27}{9 - 6}$

35. $\dfrac{44 + 12}{11 - 3}$

36. $\dfrac{44}{11} + \dfrac{12}{3}$

37. $\dfrac{6 + 12 \times 4}{15 - 3 \times 2}$

38. $\dfrac{36 - (7 - 4)}{5 + 3 \times 2}$

39. $\dfrac{12 + 6}{3 + 6} + \dfrac{24}{6} - 6 \div 6$

40. $8 \times 5 - \dfrac{2 + 4 \times 12}{18 - 4 \times 2} + 72 \div 9$

B. Applied Problems

1. **Painting and Decorating** A painter ordered 3 gallons of antique white latex paint for $8 a gallon and 5 gallons of white enamel for $10 a gallon. Write out a mathematical statement giving the total cost of the paint. Calculate the cost.

2. **Landscape Architecture** On a certain landscaping job, Steve charged a customer $168 for labor and $7 each for eight flats of plants. Write a single mathematical statement giving the total cost of this job, then calculate the cost.

3. **Electrical Technology** An electrician purchased 12 wall switches at $8 each and received a $7 credit for each of the five power outlet panels he returned. Write out a mathematical statement that gives the amount of money he spent, then calculate this total.

4. **Auto Mechanics** At the beginning of the day on Monday, the parts department has on hand 520 spark plugs. Mechanics in the service department estimate they will need about 48 plugs per day. A new shipment of 300 will arrive on Thursday. Write out a mathematical statement that gives the number of spark plugs on hand at the end of the day on Friday. Calculate this total.

C. Calculator Problems

1. $462 + 83 \times 95$

2. $425 \div 25 + 386$

3. $7482 - 1152 \div 12$

4. $1496 - 18 \times 13$

5. $(268 + 527) \div 159$

6. $2472 \times (1169 - 763)$

7. $612 + 86 \times 9 - 1026 \div 38$

8. $12 \times 38 + 46 \times 19 - 1560 \div 24$

9. $3579 - 16 \times (72 + 46)$

10. $273 + 25 \times (362 + 147)$

11. $864 \div 16 \times 27$

12. $973 - (481 + 327)$

13. $(296 + 18 \times 48) \times 12$ 14. $(27 \times 18 - 66) \div 14$

15. $\dfrac{3297 + 1858 - 493}{48 \times 16 - 694}$ 16. $\dfrac{391}{17} + \dfrac{4984}{89} - \dfrac{1645}{47}$

Check your answers on page 583, then turn to **36** for a set of problems on the arithmetic of whole numbers, with many practical applications of this mathematics.

Arithmetic of Whole Numbers

36 Answers are given on page 584.

A. Perform the arithmetic as shown.

1. $24 + 69$

2. $38 + 45$

3. $456 + 72$

4. $43 + 817$

5. $396 + 538$

6. $2074 + 906$

7. $43 - 28$

8. $93 - 67$

9. $734 - 85$

10. $315 - 119$

11. $543 - 348$

12. $3401 - 786$

13. 376×4

14. 489×7

15. 67×21

16. 45×82

17. 207×63

18. 314×926

19. 5236×44

20. 4018×392

21. $259 \div 7$

22. $1704 \div 8$

23. $42\overline{)2394}$

24. $34\overline{)2108}$

25. $1440 \div 160$

26. $11309 \div 263$

27. $\dfrac{1314}{73}$

28. $\dfrac{23 \times 51}{17}$

29. $\dfrac{36 \times 91}{13 \times 42}$

30. $(18 + 5 \times 9) \div 7$

31. $120 - 40 \div 8$

32. $32 \div 4 + 16 \div 2 \times 4$

33. $3 \times 4 - 15 \div 3$

34.   139
      407
    $+ \ \ 81$

35.   308
      793
    $+ 144$

36.   194
      271
    $+ 368$

B. Practical Problems

1. **Electrical Technology**   From a roll of No. 12 wire, an electrician cut the following lengths: 6, 8, 20, and 9 ft. How many feet of wire did he use?

2. **Machine Technology**   A machine shop bought 14 steel rods of $\frac{7}{8}$-in.-diameter steel, 23 rods of $\frac{1}{2}$-in. diameter, 8 rods of $\frac{1}{4}$-in. diameter, and 19 rods of 1-in. diameter. How many rods were purchased?

3. **Interior Design**   In estimating the floor area of a house, Jean listed the rooms as follows: living room, 346 sq ft; dining room, 210 sq ft; four bedrooms, 164 sq ft each; two bathrooms, 96 sq ft each; kitchen, 208 sq ft; and family room, 280 sq ft. What is the total floor area of the house?

4. **Metalworking**   A metal casting weighs 680 lb, and 235 lb of metal is removed during shaping. What is its finished weight?

5. **Electrical Technology**   Roberto has 210 ft of No. 14 wire left on a roll. If he cuts it into pieces 35 ft long, how many pieces will he get?

6. **Building Construction**   How many hours will be required to install the flooring of a 2160-sq ft house if 90 sq ft can be installed in 1 hour?

Name

Date

Course/Section

7. **Construction** The weights of seven cement platforms are 210, 215, 245, 217, 220, 227, and 115 lb. What is the average weight per platform? (Add up and divide by 7.)

8. **Interior Design** A room has 26 sq yd of floor space. If carpeting costs $19 per sq yd, what will it cost to carpet the room?

9. A portable TV set can be bought for $95 down and 12 payments of $37 each. What is its total cost?

10. **Electrical Technology** A 500-ft spool of No. 14-2TF wire is being used on an electrical wiring job. If 248 ft of wire is used on the job, how much wire is left on the spool?

11. Texas has more miles of highway than any other state. In a recent survey, Texas had a paved road mileage of 294,491; California was next with 169,047 miles; Illinois was third with 137,149 miles; and Kansas was fourth with 133,280. What was the total paved road mileage for these four states?

12. **Office Services** During the first three months of the year, the Print Rite Company reported the following sales of printers:

January $35,724
February $27,162
March $42,473

What is their sales total for this quarter of the year?

13. **Fire Protection** A fire control pumping truck can move 156,000 gallons of water in 4 hours of continuous pumping. What is the flow rate (gallons per minute) for this truck?

14. **Building Construction** A pile of lumber contains 170 boards 10 ft long, 118 boards 12 ft long, 206 boards 8 ft long, and 19 boards 16 ft long.
(a) How many boards are in the pile?
(b) Calculate the total length, or linear feet, of lumber in the pile.

15. **Marine Technology** If the liquid pressure on a surface is 17 psi (pounds per square inch), what is the total force on a surface area of 167 sq in.?

16. **Auto Mechanics** A rule of thumb useful to auto mechanics is that a car with a pressure radiator cap will boil at a temperature of $(3 \times \text{Cap Rating}) + 212$ degrees. What temperature will a Dino V6 reach with a cap rating of 17?

17. **Machine Technology** If it takes 45 minutes to set up a job and cut the teeth on a gear blank, how many hours will be needed for a job that requires cutting 32 such gear blanks?

18. **Plumbing** To construct a certain pipe system, the following material is needed:
6 90° elbows at 85 cents each
5 tees at 91 cents each
3 couplings at 68 cents each
4 pieces of pipe each 76 in. long
6 pieces of pipe each 24 in. long
4 pieces of pipe each 17 in. long
The pipe is schedule 40 1-in. PVC pipe, costing 54 cents per linear foot. Find the total cost of the material for the system.

19. **Manufacturing** A gallon is a volume of 231 cu in. If a storage tank holds 380 gallons of oil, what is the volume of the tank in cubic inches?

20. **Machine Technology** The Ace Machine Tool Co. received an order for 15,500 flanges. If two dozen flanges are packed in a box, how many boxes are needed to ship the order?

21. **Manufacturing** At 8 A.M. the revolution counter on a diesel engine reads 460089. At noon it reads 506409. What is the average rate, in revolutions per minute, for the machine?

22. **Fire Protection** The pressure reading coming out of a pump is 164 lb. It is estimated that every 50-ft section of hose reduces the pressure by 7 lb. What will the estimated pressure be at the nozzle end of nine 50-ft sections?

23. **Carpentry** A fireplace mantel 6 ft long is to be centered along a wall 18 ft long. How far from each end of the wall should the ends of the mantel be?

37 Since its introduction in the early 1970s, the handheld electronic calculator has quickly become an indispensable tool in modern society. From clerks, carpenters, and shoppers to technicians, engineers, and scientists, people in virtually every occupation now use calculators to perform mathematical tasks. With this in mind, we have included in this book special instructions, examples, and exercises demonstrating the use of a calculator wherever it is appropriate.

To use a calculator intelligently and effectively in your work, you should remember the following:

1. Whenever possible, make an estimate of your answer before entering a calculation. Then check this estimate against your calculator result. You may get an incorrect answer by accidentally pressing the wrong keys or by entering numbers in an incorrect sequence.

2. Always try to check your answers to equations and word problems by substituting them back into the original problem statement.

3. Organize your work on paper before using a calculator, and record any intermediate results that you may need later.

4. Round your answer whenever necessary.

This is the first of several sections, spread throughout the text, designed to teach you how to use a calculator with the mathematics taught in this course. In addition to these special sections, the solutions to many worked examples include the appropriate calculator key sequences and displays. Most exercise sets contain a special section giving calculator problems.

How to Select a Calculator There are many types of calculators available. They differ in size, shape, color, cost, and most important, in how they work and what they can do. To use a calculator efficiently with the mathematical operations covered in this book, you will need a scientific calculator with the following characteristics.

- A display of at least eight digits.
- At least one memory (look for keys marked $\boxed{\text{X→M}}$, $\boxed{\text{STO}}$, or $\boxed{\text{M+}}$).
- Keys for calculating square root $\boxed{\sqrt{}}$, powers $\boxed{x^2}$ $\boxed{y^x}$, trigonometric functions $\boxed{\sin}$, $\boxed{\cos}$, $\boxed{\tan}$, inverse trigonometric functions $\boxed{\sin^{-1}}$, $\boxed{\cos^{-1}}$, $\boxed{\tan^{-1}}$, as well as the four basic arithmetic operations $\boxed{+}$ $\boxed{-}$ $\boxed{\times}$ $\boxed{\div}$. Calculators that can perform all these functions usually include additional functions that may be useful in more advanced courses.
- Algebraic logic.

Becoming Familiar with Your Calculator First, check out the machine. You'll find an on–off switch somewhere (unless it is solar powered), a display, and a keyboard with both *numerical* (0, 1, 2, 3, . . ., 9) and basic *function* (+, −, ×, ÷, =, .) keys usually arranged like this:

7	8	9	÷
4	5	6	×
1	2	3	−
0	.	=	+

Next, locate the *clear* $\boxed{\text{C}}$, *clear entry* $\boxed{\text{CE}}$, and memory keys. There are usually three memory keys, one for storing the displayed information $\boxed{\text{X→M}}$ or $\boxed{\text{STO}}$, one for retrieving information from memory $\boxed{\text{RM}}$ or $\boxed{\text{RCL}}$, and one for adding the number in the display to the number in memory $\boxed{\text{M+}}$ or $\boxed{\text{SUM}}$. Finally, look for the parentheses keys $\boxed{(}$ and $\boxed{)}$.

The *clear* key $\boxed{\text{C}}$ "clears" all entries from the work space of the calculator, and causes the number zero to appear in the display. Pressing the clear key means that you want to start the entire calculation over or begin a new calculation.

Keyboard	**Display**
$\boxed{\text{C}}$	*0.*

The *clear entry* key $\boxed{\text{CE}}$ "clears" the display only, leaves the work space unchanged,

and causes a zero to appear in the display. Pressing the clear entry key $\boxed{\text{CE}}$ means that you wish to remove the last entry only.

Keyboard	**Display**
$\boxed{\text{CE}}$	*0.*

Basic Operations Every number is entered into the calculator digit by digit, left to right. For example, to enter the number 438, simply press the 4, 3, and 8 keys and the display will read *438.* . If you make an error in entering the number, press either the clear $\boxed{\text{C}}$ or clear entry $\boxed{\text{CE}}$ key and begin again. Note that the calculator does not know if the number you are entering is 4, 43, 438, or something even larger, until you stop entering numerical digits and press a function key such as $\boxed{+}$, $\boxed{-}$, $\boxed{\times}$, $\boxed{\div}$ or $\boxed{=}$. For example, to add 438 + 266

first, estimate the answer: 400 + 300 = 700

then, enter this sequence of keys,

4 3 8 $\boxed{+}$ **2 6 6** $\boxed{=}$ → *704.*

Notice that the calculator replaced the display of 438 with the display of 266 after the $\boxed{+}$ key was pressed, and that the sum was displayed immediately after the $\boxed{=}$ key was pressed.

All four basic operations work the same way. Try the following problems for practice.

(a) 789 − 375   (b) 246 × 97

(c) 2314 ÷ 26   (d) 46129 + 8596

Check your key sequences in **38**.

---

38   (a) **7 8 9** $\boxed{-}$ **3 7 5** $\boxed{=}$ → *414.*

(b) **2 4 6** $\boxed{\times}$ **9 7** $\boxed{=}$ → *23862.*

(c) **2 3 1 4** $\boxed{\div}$ **2 6** $\boxed{=}$ → *89.*

(d) **4 6 1 2 9** $\boxed{+}$ **8 5 9 6** $\boxed{=}$ → *54725.*

Don't forget to estimate the answer before using the calculator.

Combined Operations A calculator can operate on only two numbers at a time, but very often you need to use the result of one calculation in a second calculation or in a long string of calculations. Because the answer to any calculation remains in the work space of the calculator, you can carry out a very complex chain of calculations without ever stopping the calculator to write down an intermediate step. For example, to add

$$387 + 942 + 876 + 88$$

first, estimate: $400 + 900 + 900 + 100 = 2300$

then, use the calculator as follows:

3 8 7 ⊞ **9 4 2** ⊞ **8 7 6** ⊞ **8 8** ⊟ →   *2293.*

Notice that it was not necessary to press the ⊟ key until the very end, and that the display showed the running total every time you hit the ⊟ key.

In Section 1-5 you saw that when two or more operations are combined in the same calculation a standard order of operations must be followed. Fortunately, if your calculator uses algebraic logic, you do not need to worry about entering such a calculation in the correct order. Scientific calculators with algebraic logic automatically use the correct order of operations. To test your calculator, enter this sequence:

2 ⊞ **3** ⊠ **4** ⊟ →   *14.*

If your calculator displays this answer, it is using algebraic logic.

Notice that when you pressed the ⊠ key the calculator did not display the sum of 2 and 3. Because a multiplication must be done first, the calculator waited to perform the multiplication before doing the addition.

Enter each of the following calculations as written. Compare your answers with ours in **39**.

(a) $480 - 1431 \div 53$ (b) $72 \times 38 + 86526 \div 69$

(c) $2478 - 726 + 598 \times 12$ (d) $271440 \div 48 \times 65$

---

39   (a)   **4 8 0** ⊟ **1 4 3 1** ⊡ **5 3** ⊟ →   *453.*

      (b)   **7 2** ⊠ **3 8** ⊞ **8 6 5 2 6** ⊡ **6 9** ⊟ →   *3990.*

      (c)   **2 4 7 8** ⊟ **7 2 6** ⊞ **5 9 8** ⊠ **1 2** ⊟ →   *8928.*

      (d)   **2 7 1 4 4 0** ⊡ **4 8** ⊠ **6 5** ⊟ →   *367575.*

Parentheses and Memory As you have already learned, parentheses, brackets, and other grouping symbols are used in a written calculation to signal a departure from the standard order of operations. Fortunately, every scientific calculator has parentheses keys. To perform a calculation with grouping symbols, either enter the calculation inside the parentheses first and press ⊟, or enter the calculation as it is written but use the parentheses keys. For example, to enter $12 \times (28 + 15)$ use either the sequence

2 8 ⊞ **1 5** ⊟ ⊠ **1 2** ⊟ →   *516.*

or

1 2 ⊠ ⦅ **2 8** ⊞ **1 5** ⦆ ⊟ →   *516.*

Learning Help ▶ In complex problems it is useful to do the calculation both ways as a way of checking your answer. ◀

When division problems are given in terms of fractions, parentheses are implied but not shown. For example, the problem

$\frac{48 + 704}{117 - 23}$ is equal to $(48 + 704) \div (117 - 23)$

The fraction bar acts as a grouping symbol. The addition and subtraction in the top and bottom of the fraction must be performed before the division. Again you may use either parentheses or the $\boxed{=}$ key to simplify the top half of the fraction, but after pressing $\boxed{=}$ you must use parentheses to enter the bottom half. Here are the two possible sequences:

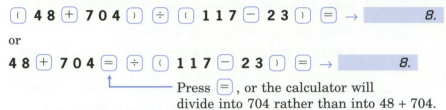

or

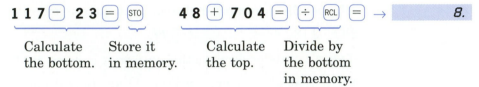

Press $\boxed{=}$, or the calculator will divide into 704 rather than into 48 + 704.

The memory keys provide a third option for this problem. Memory is used to store a result for later use. In this case, we can calculate the value of the bottom first, store it in memory, then calculate the top and divide by the contents of memory. The entire sequence looks like this:

$\boxed{1}\ \boxed{1}\ \boxed{7}\ \boxed{-}\ \boxed{2}\ \boxed{3}\ \boxed{=}\ \boxed{\text{STO}}$ $\qquad$ $\boxed{4}\ \boxed{8}\ \boxed{+}\ \boxed{7}\ \boxed{0}\ \boxed{4}\ \boxed{=}\ \boxed{\div}\ \boxed{\text{RCL}}\ \boxed{=}\ \to$ _8._

Calculate the bottom.   Store it in memory.   Calculate the top.   Divide by the bottom in memory.

Here are a few practice problems involving grouping symbols. Use the method indicated.

(a) $(4961 - 437) \div 52$ Do not use the parentheses keys.

(b) $56 \times (38 + 12 \times 17)$ Use the parentheses keys.

(c) $2873 - (56 + 83) \times 16$ Use the parentheses keys.

(d) $\frac{263 \times 18 - 41 \times 12}{18 \times 16 - 17 \times 11}$ Use memory.

(e) $12 \times 16 - \frac{7 + 9 \times 17}{81 - 49} + 12 \times 19$ Use parentheses.

(f) Repeat problem (e) using the memory to store the value of the fraction, then doing the calculation left to right.

Check your answers in **40**.

40

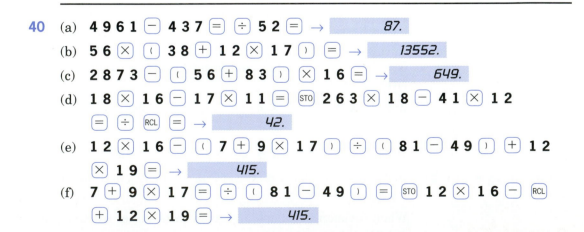

(a) $\boxed{4}\boxed{9}\boxed{6}\boxed{1}\ \boxed{-}\ \boxed{4}\boxed{3}\boxed{7}\ \boxed{=}\ \boxed{\div}\ \boxed{5}\boxed{2}\ \boxed{=}\ \to$ _87._

(b) $\boxed{5}\boxed{6}\ \boxed{\times}\ \boxed{(}\ \boxed{3}\boxed{8}\ \boxed{+}\ \boxed{1}\boxed{2}\ \boxed{\times}\ \boxed{1}\boxed{7}\ \boxed{)}\ \boxed{=}\ \to$ _13552._

(c) $\boxed{2}\boxed{8}\boxed{7}\boxed{3}\ \boxed{-}\ \boxed{(}\ \boxed{5}\boxed{6}\ \boxed{+}\ \boxed{8}\boxed{3}\ \boxed{)}\ \boxed{\times}\ \boxed{1}\boxed{6}\ \boxed{=}\ \to$ _649._

(d) $\boxed{1}\boxed{8}\ \boxed{\times}\ \boxed{1}\boxed{6}\ \boxed{-}\ \boxed{1}\boxed{7}\ \boxed{\times}\ \boxed{1}\boxed{1}\ \boxed{=}\ \boxed{\text{STO}}\ \boxed{2}\boxed{6}\boxed{3}\ \boxed{\times}\ \boxed{1}\boxed{8}\ \boxed{-}\ \boxed{4}\boxed{1}\ \boxed{\times}\ \boxed{1}\boxed{2}$
$\boxed{=}\ \boxed{\div}\ \boxed{\text{RCL}}\ \boxed{=}\ \to$ _42._

(e) $\boxed{1}\boxed{2}\ \boxed{\times}\ \boxed{1}\boxed{6}\ \boxed{-}\ \boxed{(}\ \boxed{7}\ \boxed{+}\ \boxed{9}\ \boxed{\times}\ \boxed{1}\boxed{7}\ \boxed{)}\ \boxed{\div}\ \boxed{(}\ \boxed{8}\boxed{1}\ \boxed{-}\ \boxed{4}\boxed{9}\ \boxed{)}\ \boxed{+}\ \boxed{1}\boxed{2}$
$\boxed{\times}\ \boxed{1}\boxed{9}\ \boxed{=}\ \to$ _415._

(f) $\boxed{7}\ \boxed{+}\ \boxed{9}\ \boxed{\times}\ \boxed{1}\boxed{7}\ \boxed{=}\ \boxed{\div}\ \boxed{(}\ \boxed{8}\boxed{1}\ \boxed{-}\ \boxed{4}\boxed{9}\ \boxed{)}\ \boxed{=}\ \boxed{\text{STO}}\ \boxed{1}\boxed{2}\ \boxed{\times}\ \boxed{1}\boxed{6}\ \boxed{-}\ \boxed{\text{RCL}}$
$\boxed{+}\ \boxed{1}\boxed{2}\ \boxed{\times}\ \boxed{1}\boxed{9}\ \boxed{=}\ \to$ _415._

Fractions

	Objective	Sample Problems		Where to Go for Help	
				Page	Frame

When you finish this unit you will be able to:

1. Work with fractions.

    (a) Write as a mixed number $\dfrac{31}{4}$.   _____   57   **7**

    (b) Write as an improper fraction $3\dfrac{7}{8}$.   _____   58   **9**

    Write as an equivalent fraction.

    (c) $\dfrac{5}{16} = \dfrac{?}{64}$ _____   59   **10**

    (d) $1\dfrac{3}{4} = \dfrac{?}{32}$ _____

    (e) Write in lowest terms $\dfrac{10}{64}$.   _____   61   **12**

    (f) Which is larger, $1\dfrac{7}{8}$ or $\dfrac{5}{3}$?   _____   62   **14**

2. Multiply and divide fractions.

    (a) $\dfrac{7}{8} \times \dfrac{5}{32}$ _____   65   **18**

    (b) $4\dfrac{1}{2} \times \dfrac{2}{3}$ _____

    (c) $\dfrac{3}{5}$ of $1\dfrac{1}{2}$ _____

    (d) $\dfrac{3}{4} \div \dfrac{1}{2}$ _____   70   **24**

    (e) $2\dfrac{7}{8} \div 1\dfrac{1}{4}$ _____

    (f) $4 \div \dfrac{1}{2}$ _____

Name

Date

Course/Section

3. Add and subtract fractions.

(a) $\dfrac{7}{16} + \dfrac{3}{16}$ ————   75   **29**

(b) $1\dfrac{3}{16} + \dfrac{3}{4}$ ————

(c) $\dfrac{3}{4} - \dfrac{1}{5}$ ————

(d) $4 - 1\dfrac{5}{16}$ ————

(Answers to these preview problems are given on page 577.)

If you are certain that you can work *all* these problems correctly, turn to page 87 for a set of practice problems. If you cannot work one or more of the preview problems, turn to the page indicated to the right of the problem. For those who wish to master this material with the greatest success, turn to frame **1** and begin work there.

2 Fractions

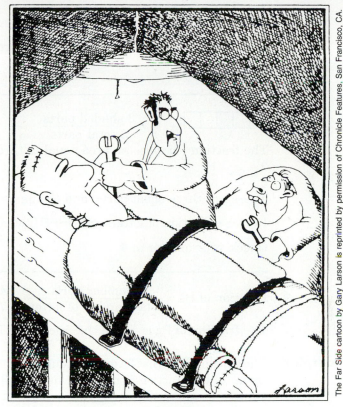

The Far Side cartoon by Gary Larson is reprinted by permission of Chronicle Features, San Francisco, CA.

"Fool! This is an eleven-sixteenths...I asked for a five-eighths!"

2-1 WORKING WITH FRACTIONS

1 The word *fraction* comes from a Latin word meaning "to break," and fraction numbers are used when we need to break down standard measuring units into smaller parts. Anyone doing practical mathematics problems will find that fractions are used in many different trade and technical areas. Unfortunately, fraction arithmetic usually *must* be done with pencil and paper, since not all calculators handle fractions directly.

Look at the following piece of lumber. What happens when we break it into equal parts?

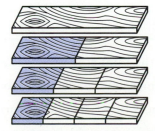

2 parts, each one-half of the whole

3 parts, each one-third of the whole

4 parts, each one-fourth of the whole

Divide this area [　　　　　] into fifths by drawing vertical lines.

Try it, then go to **2**.

2 Notice that the five parts or "fifths" are equal in area. ▭▭▭▭▭

A fraction is normally written as the division of two whole numbers: $\frac{2}{3}$, $\frac{3}{4}$, or $\frac{9}{16}$. One of the five equal areas above would be "one-fifth" or $\frac{1}{5}$ of the entire area.

▧▭▭▭▭

How would you label this portion of the area? ▧▧▧▭▭ = ?

Continue in **3**.

---

3 ▧▧▧▭▭ $\frac{3}{5} = \frac{3 \text{ shaded parts}}{5 \text{ total parts}}$

The fraction $\frac{3}{5}$ implies an area equal to three of the original parts.

$$\frac{3}{5} = 3 \times \left(\frac{1}{5}\right)$$

There are three equal parts, and the name of each part is $\frac{1}{5}$ or one-fifth.

In this collection of letters, HHHHPPT, what fraction are Hs?

Check your answer in **4**.

---

4 Fraction of Hs = $\dfrac{\text{number of Hs}}{\text{total number of letters}} = \dfrac{4}{7}$ (read it "four sevenths")

The fraction of Ps is $\frac{2}{7}$, and the fraction of Ts is $\frac{1}{7}$.

Numerator The two numbers that form a fraction are given special names to simplify talking about them. In the fraction $\frac{3}{5}$ the upper number 3 is called the *numerator* from the Latin *numero* meaning "number." It is a count of the number of parts.

Denominator The lower number 5 is called the *denominator* from the Latin *nomen* or "name." It tells us the name of the part being counted. The numerator and denominator are called the *terms* of the fraction.

$$\frac{3}{5}$$

◁ Numerator, the number of parts

◁ Denominator, the name of part, "fifths"

Learning Help ▶ A handy memory aid is to remember that the denominator is the "down part"—D for down. ◀

A paperback book costs $6 and I have $5. What fraction of its cost do I have? Write the answer as a fraction.

numerator = _____ , denominator = _____

Check your answer in **5**.

---

5 $ $ $ $ $ $5 is $\frac{5}{6}$ of the total cost.
  ⏟
  5   ↑ numerator = 5, denominator = 6

Complete these sentences by writing in the correct fraction.

(a) If we divide a length into eight equal parts, each part will be _____ of the total length.

(b) Then three of these parts will represent _____ of the total length.

(c)   Eight of these parts will be _____ of the total length.

(d)   Ten of these parts will be _____ of the total length.

Check your answers in **6**.

---

6   (a) $\frac{1}{8}$ (b) $\frac{3}{8}$ (c) $\frac{8}{8}$ (d) $\frac{10}{8}$

Proper Fraction   The original length is used as a standard, and any other length—smaller or larger—can be expressed as a fraction of it. A *proper fraction* is a number less than 1, as you would suppose a fraction should be. It represents a quantity less than the standard. For example, $\frac{1}{2}$, $\frac{2}{3}$, and $\frac{17}{20}$ are all proper fractions. Notice that for a proper fraction, the numerator is less than the denominator—the top number is less than the bottom number in the fraction.

Improper Fraction   An *improper fraction* is a number greater than 1 and represents a quantity greater than the standard. If a standard length is 8 in., a length of 11 in. will be $\frac{11}{8}$ of the standard. Notice that for an improper fraction the numerator is greater than the denominator—top number greater than the bottom number in the fraction.

Circle the proper fractions in the following list.

$\frac{3}{2}$ $\frac{3}{4}$ $\frac{7}{8}$ $\frac{5}{4}$ $\frac{15}{12}$ $\frac{1}{16}$ $\frac{35}{32}$ $\frac{7}{50}$ $\frac{65}{64}$ $\frac{105}{100}$

Go to **7** when you have finished.

---

7   You should have circled the following proper fractions: $\frac{3}{4}$, $\frac{7}{8}$, $\frac{1}{16}$, $\frac{7}{50}$. In each fraction the numerator (top number) is less than the denominator (bottom number). Each of these fractions represents a number less than 1.

The improper fraction $\frac{7}{3}$ can be shown graphically as follows:

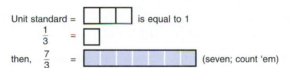

We can rename this number by regrouping.

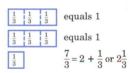

Mixed Numbers   A *mixed number* is an improper fraction written as the sum of a whole number and a proper fraction.

$\frac{7}{3} = 2 + \frac{1}{3}$ or $2\frac{1}{3}$ We usually omit the + sign and write $2 + \frac{1}{3}$ as $2\frac{1}{3}$, and read it as "two and one-third." The numbers $1\frac{1}{2}$, $2\frac{2}{5}$, and $16\frac{2}{3}$ are all written as mixed numbers.

To write an improper fraction as a mixed number, divide numerator by denominator and form a new fraction as shown:

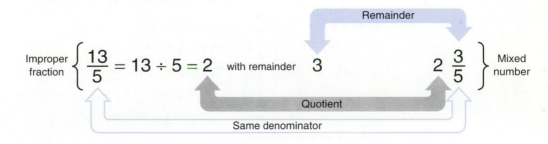

$$\frac{13}{5} = 5\overline{)13} = 2\frac{3}{5}$$

Quotient

10

3 ← Remainder

Now you try it. Rename $\frac{23}{4}$ as a mixed number. $\frac{23}{4} =$ _____

Follow the procedure shown above, then hop to **8**.

---

8 $\frac{23}{4} = 23 \div 4 = 5$ with remainder $3 \longrightarrow 5\frac{3}{4}$

If in doubt, check your work with a diagram like this:

If in doubt, check your work with a diagram like this:

■■■■■■■■■■■■■■■■■■■■■■■

23

= (grid) } 5 rows of 4

■■■   3 remaining

Now try these for practice. Write each improper fraction as a mixed number.

(a) $\frac{9}{5}$ (b) $\frac{13}{4}$ (c) $\frac{27}{8}$ (d) $\frac{31}{4}$ (e) $\frac{41}{12}$ (f) $\frac{17}{2}$

The answers are given in **9**.

---

9   (a) $\frac{9}{5} = 1\frac{4}{5}$ (b) $\frac{13}{4} = 3\frac{1}{4}$ (c) $\frac{27}{8} = 3\frac{3}{8}$

     (d) $\frac{31}{4} = 7\frac{3}{4}$ (e) $\frac{41}{12} = 3\frac{5}{12}$ (f) $\frac{17}{2} = 8\frac{1}{2}$

The reverse process, rewriting a mixed number as an improper fraction, is equally simple.

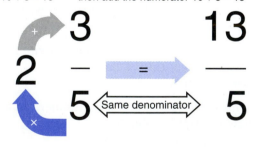

Work in a clockwise direction, first multiply, $5 \times 2 = 10$
then add the numerator $10 + 3 = 13$

$10 + 3 = 13$

$2\frac{3}{5} - = \frac{13}{5}$ Same denominator

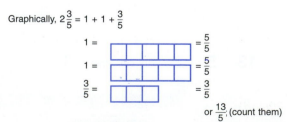

Graphically, $2\frac{3}{5} = 1 + 1 + \frac{3}{5}$

$1 =$ (grid) $= \frac{5}{5}$

$1 =$ (grid) $= \frac{5}{5}$

$\frac{3}{5} =$ (grid) $= \frac{3}{5}$

or $\frac{13}{5}$ (count them)

Now you try it. Rewrite these mixed numbers as improper fractions.

(a) $3\frac{1}{5}$ (b) $4\frac{3}{8}$ (c) $1\frac{1}{16}$ (d) $5\frac{1}{2}$ (e) $15\frac{3}{8}$ (f) $9\frac{3}{4}$

Check your answers in **10**.

---

10 (a) **Step 1** $3 \times 5 = 15$

Step 2 $15 + 1 = 16 \leftarrow$ The new numerator

Step 3 $3\frac{1}{5} = \frac{16}{5} \swarrow$ The original denominator

(b) $4\frac{3}{8} = \frac{35}{8}$ (c) $1\frac{1}{16} = \frac{17}{16}$ (d) $5\frac{1}{2} = \frac{11}{2}$

(e) $15\frac{3}{8} = \frac{123}{8}$ (f) $9\frac{3}{4} = \frac{39}{4}$

Equivalent Fractions

$\frac{1}{2}$

$\frac{2}{4}$

Two fractions are said to be *equivalent* if they represent the same number. For example, $\frac{1}{2} = \frac{2}{4}$ since both fractions represent the same portion of some standard amount.

There is a very large set of fractions equivalent to $\frac{1}{2}$.

$$\frac{1}{2} = \frac{2}{4} = \frac{3}{6} = \frac{4}{8} = \frac{5}{10} = \cdots = \frac{46}{92} = \frac{61}{122} = \frac{1437}{2874} \quad \text{and so on}$$

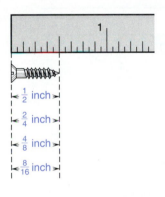

Each fraction is the same number, and we can use these fractions interchangeably in any mathematics problem.

To obtain a fraction equivalent to any given fraction, multiply the original numerator and denominator by the same nonzero number. For example,

Multiply top and bottom by 3

$$\frac{1}{2} = \frac{1 \times \boxed{3}}{2 \times \boxed{3}} = \frac{3}{6} \qquad \frac{2}{3} = \frac{2 \times \boxed{5}}{3 \times \boxed{5}} = \frac{10}{15}$$

Multiply top and bottom by 5

Rewrite the fraction $\frac{3}{4}$ as an equivalent fraction with denominator equal to 20.

$$\frac{3}{4} = \frac{?}{20}$$

Check your work in **11**.

---

11 $\quad \dfrac{3}{4} = \dfrac{3 \times \square}{4 \times \square} = \dfrac{?}{20} \qquad 4 \times \square = 20$, so $\square$ must be 5.

$\qquad = \dfrac{3 \times \boxed{5}}{4 \times \boxed{5}} = \dfrac{15}{20}$

The number value of the fraction has not changed; we have simply renamed it.

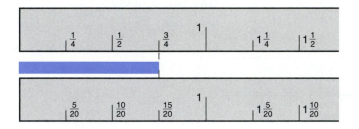

Practice with these.

(a) $\quad \dfrac{5}{8} = \dfrac{?}{32}$ $\qquad$ (b) $\quad \dfrac{7}{16} = \dfrac{?}{48}$ $\qquad$ (c) $\quad 1\dfrac{2}{3} = \dfrac{?}{12}$

Look in **12** for the answers.

12 $\quad$ (a) $\quad \dfrac{5}{8} = \dfrac{5 \times 4}{8 \times 4} = \dfrac{20}{32}$ $\qquad$ (b) $\quad \dfrac{7}{16} = \dfrac{7 \times 3}{16 \times 3} = \dfrac{21}{48}$

$\qquad$ (c) $\quad 1\dfrac{2}{3} = \dfrac{5}{3} = \dfrac{5 \times 4}{3 \times 4} = \dfrac{20}{12}$

Writing in Lowest Terms

Very often in working with fractions you will be asked to *write a fraction in lowest terms*. This means to replace it with the most simple fraction in its set of equivalent fractions. To write $\frac{15}{30}$ in its lowest terms means to replace it by $\frac{1}{2}$. They are equivalent.

$$\frac{15}{30} = \frac{15 \div 15}{30 \div 15} = \frac{1}{2}$$

We have divided both the top and bottom of the fraction by 15.

In general, you would write a fraction in lowest terms this way:

Example: Write $\frac{30}{48}$ in lowest terms.

First, find the largest number that divides both top and bottom of the fraction exactly.

$$30 = 5 \times 6$$

$$48 = 8 \times 6$$

In this case, the factor 6 is the largest number that divides both parts of the fraction exactly.

Second, eliminate this common factor by dividing.

$$\frac{30}{48} = \frac{30 \div 6}{48 \div 6} = \frac{5}{8}$$

The fraction $\frac{5}{8}$ is the simplest fraction equivalent to $\frac{30}{48}$. No whole number greater than 1 divides both 5 and 8 exactly.

Example: Write $\frac{90}{105}$ in lowest terms.

$$\frac{90}{105} = \frac{90 \div 15}{105 \div 15} = \frac{6}{7}$$

This process of eliminating a common factor is usually called *canceling*. When you cancel a factor, you divide both top and bottom of the fraction by that number. We would write $\frac{90}{105}$ as $\frac{6 \times \cancel{15}}{7 \times \cancel{15}} = \frac{6}{7}$. When you "cancel," you must divide by a *pair* of common factors, one factor in the numerator and the same factor in the denominator.

Example: In roof construction, the *pitch* or steepness of a roof is defined as the fraction

$$\text{Pitch} = \frac{\text{rise}}{\text{span}}$$

Rise 8 ft

Span 24 ft

The *rise* is the vertical distance between the ridge beam and the supporting plate. The *span* is the horizontal distance between the outer faces of the supporting walls. For the roof shown,

$$\text{Pitch} = \frac{8 \text{ ft rise}}{24 \text{ ft span}} = \frac{8}{24} = \frac{8 \div 8}{24 \div 8} = \frac{1}{3}$$

Pitch is usually written as a fraction in lowest terms.

Write the following fractions in lowest terms.

(a) $\frac{6}{8}$ (b) $\frac{12}{16}$ (c) $\frac{2}{4}$ (d) $\frac{4}{12}$

(e) $\dfrac{15}{84}$ (f) $\dfrac{21}{35}$ (g) $\dfrac{12}{32}$

(h) Calculate the pitch of a storage shed roof having a rise of 6 ft and a span of 64 ft.

The answers are given in **13**.

In $\dfrac{2 \times 3}{3 \times 5}$ I can cancel the 3s, right? Like $\dfrac{2 \times \cancel{3}}{\cancel{3} \times 5}$

Right. Just so you know that cancelling the 3s means dividing top and bottom by 3.

---

13 (a) $\dfrac{6}{8} = \dfrac{6 \div 2}{8 \div 2} = \dfrac{3}{4}$ (b) $\dfrac{12}{16} = \dfrac{12 \div 4}{16 \div 4} = \dfrac{3}{4}$

(c) $\dfrac{2}{4} = \dfrac{2 \div 2}{4 \div 2} = \dfrac{1}{2}$ (d) $\dfrac{4}{12} = \dfrac{4 \div 4}{12 \div 4} = \dfrac{1}{3}$

(e) $\dfrac{15}{84} = \dfrac{15 \div 3}{84 \div 3} = \dfrac{5}{28}$ (f) $\dfrac{21}{35} = \dfrac{21 \div 7}{35 \div 7} = \dfrac{3}{5}$

(g) $\dfrac{12}{32} = \dfrac{12 \div 4}{32 \div 4} = \dfrac{3}{8}$ (h) Pitch $= \dfrac{6 \text{ ft}}{64 \text{ ft}} = \dfrac{6 \div 2}{64 \div 2} = \dfrac{3}{32}$

This one is a little tricky. Write $\dfrac{6}{3}$ in lowest terms. Check your answer in **14**.

---

14 $\dfrac{6}{3} = \dfrac{6 \div 3}{3 \div 3} = \dfrac{2}{1}$ or simply 2

Any whole number may be written as a fraction by using a denominator equal to 1.

$3 = \dfrac{3}{1}$ $4 = \dfrac{4}{1}$ and so on.

Writing numbers in this way will be helpful when you learn to do arithmetic with fractions.

Comparing Fractions If you were offered your choice between $\frac{2}{3}$ of a certain amount of money and $\frac{5}{8}$ of it, which would you choose? Which is the larger fraction, $\frac{2}{3}$ or $\frac{5}{8}$?

Can you decide? Try. Rewriting the fractions as equivalent fractions will help. The answer is given in **15**.

---

15 To compare two fractions, rename each by changing them to equivalent fractions with the same denominator.

$\dfrac{2}{3} = \dfrac{2 \times 8}{3 \times 8} = \dfrac{16}{24}$ and $\dfrac{5}{8} = \dfrac{5 \times 3}{8 \times 3} = \dfrac{15}{24}$

Now compare the new fractions: $\frac{16}{24}$ is greater than $\frac{15}{24}$.

Learning Help ▶   1. The new denominator is the product of the original ones ($24 = 8 \times 3$).

2. Once both fractions are written with the same denominator, the one with the larger numerator is the larger fraction. (16 of the fractional parts is more than 15 of them.) ◀

Which of the following quantities is the larger?

(a) $\frac{3}{4}$ in. or $\frac{5}{7}$ in.

(b) $\frac{7}{8}$ or $\frac{19}{21}$

(c) 3 or $\frac{40}{13}$

(d) $1\frac{7}{8}$ lb or $\frac{5}{3}$ lb

(e) $2\frac{1}{4}$ ft or $\frac{11}{6}$ ft

(f) $\frac{5}{16}$ or $\frac{11}{36}$

Check your answer in **16**.

---

16   (a) $\frac{3}{4} = \frac{21}{28}, \frac{5}{7} = \frac{20}{28}; \frac{21}{28}$ is larger than $\frac{20}{28}$, so $\frac{3}{4}$ in. is larger than $\frac{5}{7}$ in.

(b) $\frac{7}{8} = \frac{147}{168}, \frac{19}{21} = \frac{152}{168}; \frac{152}{168}$ is larger than $\frac{147}{168}$, so $\frac{19}{21}$ is larger than $\frac{7}{8}$.

(c) $3 = \frac{39}{13}; \frac{40}{13}$ is larger than $\frac{39}{13}$, so $\frac{40}{13}$ is larger than 3.

(d) $1\frac{7}{8} = \frac{15}{8} = \frac{45}{24}, \frac{5}{3} = \frac{40}{24}; \frac{45}{24}$ is larger than $\frac{40}{24}$, so $1\frac{7}{8}$ lb is larger than $\frac{5}{3}$ lb.

(e) $2\frac{1}{4} = \frac{9}{4} = \frac{54}{24}, \frac{11}{6} = \frac{44}{24}; \frac{54}{24}$ is larger than $\frac{44}{24}$, so $2\frac{1}{4}$ ft is larger than $\frac{11}{6}$ ft.

(f) $\frac{5}{16} = \frac{180}{576}, \frac{11}{36} = \frac{176}{576}; \frac{180}{576}$ is larger than $\frac{176}{576}$, so $\frac{5}{16}$ is larger than $\frac{11}{36}$.

Now turn to **17** for some practice in working with fractions.

17

Exercises 2-1   **Working with Fractions**

A. Write as an improper fraction.

1. $2\frac{1}{3}$
2. $7\frac{1}{2}$
3. $8\frac{3}{8}$
4. $1\frac{1}{16}$
5. $2\frac{7}{8}$

6. 2
7. $2\frac{2}{3}$
8. $4\frac{3}{64}$
9. $4\frac{5}{6}$
10. $1\frac{13}{16}$

B. Write as a mixed number.

1. $\frac{17}{2}$
2. $\frac{8}{5}$
3. $\frac{11}{8}$
4. $\frac{40}{16}$
5. $\frac{3}{2}$

6. $\frac{11}{3}$
7. $\frac{100}{6}$
8. $\frac{4}{3}$
9. $\frac{80}{32}$
10. $\frac{5}{2}$

C. Write in lowest terms.

1. $\frac{12}{16}$
2. $\frac{4}{6}$
3. $\frac{6}{16}$
4. $\frac{18}{4}$
5. $\frac{4}{10}$

6. $\frac{35}{30}$
7. $\frac{24}{30}$
8. $\frac{10}{4}$
9. $4\frac{3}{12}$
10. $\frac{34}{32}$

11. $\dfrac{42}{64}$ 12. $\dfrac{10}{35}$ 13. $\dfrac{15}{36}$ 14. $\dfrac{45}{18}$ 15. $\dfrac{38}{24}$

D. Complete.

1. $\dfrac{7}{8} = \dfrac{?}{16}$ 2. $\dfrac{3}{4} = \dfrac{?}{16}$ 3. $\dfrac{1}{8} = \dfrac{?}{64}$

4. $\dfrac{3}{8} = \dfrac{?}{64}$ 5. $1\dfrac{1}{4} = \dfrac{?}{16}$ 6. $2\dfrac{7}{8} = \dfrac{?}{32}$

7. $3\dfrac{3}{5} = \dfrac{?}{10}$ 8. $1\dfrac{1}{16} = \dfrac{?}{32}$ 9. $1\dfrac{40}{60} = \dfrac{?}{3}$

10. $4 = \dfrac{?}{6}$ 11. $2\dfrac{5}{8} = \dfrac{?}{16}$ 12. $2\dfrac{5}{6} = \dfrac{?}{12}$

E. Which is larger?

1. $\dfrac{3}{5}$ or $\dfrac{4}{7}$ 2. $\dfrac{3}{2}$ or $\dfrac{13}{8}$ 3. $1\dfrac{1}{2}$ or $1\dfrac{3}{7}$

4. $\dfrac{3}{4}$ or $\dfrac{13}{16}$ 5. $\dfrac{7}{8}$ or $\dfrac{5}{6}$ 6. $2\dfrac{1}{2}$ or $1\dfrac{11}{8}$

7. $1\dfrac{2}{5}$ or $\dfrac{6}{4}$ 8. $\dfrac{3}{16}$ or $\dfrac{25}{60}$ 9. $\dfrac{13}{5}$ or $\dfrac{5}{2}$

10. $3\dfrac{1}{2}$ or $2\dfrac{7}{4}$ 11. $\dfrac{3}{8}$ or $\dfrac{5}{12}$ 12. $1\dfrac{1}{5}$ or $\dfrac{8}{7}$

F. Practical Problems

1. **Carpentry**   Maria, an apprentice carpenter, measured the length of a 2 by 4 as $15\dfrac{6}{8}$ in. Express this measurement in lowest terms.

2. **Electrical Technology**   An electrical light circuit in John's welding shop had a load of 2800 watts. He changed the circuit to ten 150-watt bulbs and six 100-watt bulbs. What fraction represents a comparison of the new load with the old load?

3. The numbers $\dfrac{22}{7}$, $\dfrac{19}{6}$, $\dfrac{47}{15}$, $\dfrac{25}{8}$, and $\dfrac{41}{13}$ are all reasonable approximations to the number π. Which is the largest approximation? Which is the smallest approximation?

4. **Sheet Metal Technology**   Which is thicker, a $\dfrac{3}{16}$-in. sheet of metal or a $\dfrac{13}{64}$-in. fastener?

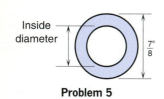

Inside diameter

$\dfrac{7}{8}$"

Problem 5

5. **Plumbing**   Is it possible to have a $\dfrac{7}{8}$-in pipe with an inside diameter of $\dfrac{29}{32}$-in.?

6. **Sheet Metal Technology**   Fasteners are equally spaced on a metal vent cover, with 9 spaces between fasteners covering 24 in. Write the distance between spaces as a mixed number.

$$\dfrac{24}{9} = ?$$

7. **Printing**   A printer has 15 rolls of newsprint in the warehouse. What fraction of this total will remain if six rolls are used?

8. **Machine Technology**   A machinist who had been producing 40 parts per day increased the output to 60 parts per day by going to a faster machine. How much faster is the new machine? Express your answer as a mixed number.

9. **Landscaping**   Before it can be used, a 12-ounce container of liquid fertilizer must be mixed with 48 ounces of water. What fraction of fertilizer is in the final mixture?

When you have had the practice you need, check your answers on page 584 then turn to **18**.

2-2 MULTIPLICATION OF FRACTIONS

18 The simplest arithmetic operation with fractions is multiplication and, happily, it is easy to show graphically. The multiplication of a whole number and a fraction may be illustrated this way.

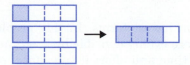

$$3 \times \frac{1}{4} = \frac{1}{4} + \frac{1}{4} + \frac{1}{4} = \frac{3}{4} \qquad \text{three segments each } \frac{1}{4} \text{ unit long.}$$

Any fraction such as $\frac{3}{4}$ can be thought of as a product: $3 \times \frac{1}{4}$

The product of two fractions can also be shown graphically.

$$\frac{1}{2} \times \frac{1}{3} \qquad \text{means} \qquad \frac{1}{2} \text{ of } \frac{1}{3}$$

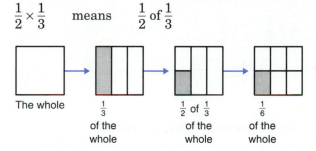

The whole | $\frac{1}{3}$ of the whole | $\frac{1}{2}$ of $\frac{1}{3}$ of the whole | $\frac{1}{6}$ of the whole

The product $\frac{1}{2} \times \frac{1}{3}$ is

$$\frac{1}{2} \times \frac{1}{3} = \frac{1}{6} = \frac{1 \text{ shaded area}}{6 \text{ equal areas in the square}}$$

In general, we calculate this product as

$$\frac{1}{2} \times \frac{1}{3} = \frac{1 \times 1}{2 \times 3} \qquad \overleftarrow{\text{Multiply the numerators (top)}} \qquad \frac{1}{2} \times \frac{1}{3} = \frac{1}{6}$$
$$\underleftarrow{\text{Multiply the denominators (bottom)}}$$

The product of two fractions is a fraction whose numerator is the **product of their numerators** and whose denominator is the **product of their denominators**.

Multiply $\frac{5}{6} \times \frac{2}{3}$.

(a) $\frac{5}{6} \times \frac{2}{3} = \frac{10}{18}$ If you think this is correct, go to **19**.

(b) $\frac{5}{6} \times \frac{2}{3} = \frac{5}{9}$ If you think this is correct, go to **20**.

(c) I don't know how to do it and I can't figure out how to draw the little boxes.   Go to **21**.

19 Good.

$$\frac{5}{6} \times \frac{2}{3} = \frac{5 \times 2}{6 \times 3} = \frac{10}{18}$$

Now write this answer in lowest terms and turn to **20**.

---

20 Excellent.

$$\frac{5}{6} \times \frac{2}{3} = \frac{5 \times 2}{6 \times 3} = \frac{10}{18} = \frac{5 \times 2}{9 \times 2} = \frac{5}{9}$$

Note ▶ Always write your answer in lowest terms. In this problem you probably recognized that both 10 and 18 were evenly divisible by 2, so you canceled out that common factor. It will save you time and effort if you cancel common factors, such as the 2 above, *before* you multiply this way:

$$\frac{5 \times \overset{1}{\cancel{2}}}{\underset{3}{\cancel{6}} \times 3} = \frac{5}{9} \quad \blacktriangleleft$$

Example: Whole Number Times a Fraction

$$6 \times \frac{1}{8} = \frac{6}{1} \times \frac{1}{8}$$

$$= \frac{\overset{3}{\cancel{6}} \times 1}{1 \times \underset{4}{\cancel{8}}} = \frac{3}{4}$$

Example: Mixed Number Times a Fraction

$$2\frac{1}{2} \times \frac{3}{4} = \frac{5}{2} \times \frac{3}{4} \qquad \text{Write the mixed number } 2\frac{1}{2}$$

$$\text{as the improper fraction } \frac{5}{2}.$$

$$= \frac{5 \times 3}{2 \times 4} = \frac{15}{8}$$

$$= 1\frac{7}{8} \qquad \text{Multiply, then write the product as a mixed number.}$$

Example: Mixed Number Times a Mixed Number

$$3\frac{1}{3} \times 2\frac{1}{4} = \frac{10}{3} \times \frac{9}{4}$$

$$= \frac{\overset{5}{\cancel{10}} \times \overset{3}{\cancel{9}}}{\underset{1}{\cancel{3}} \times \underset{2}{\cancel{4}}}$$

$$= \frac{15}{2} = 7\frac{1}{2}$$

In many word problems the words "of" or "product of" appear as signals that you are to multiply. For example, the phrase "one-half of 16" means

$$\frac{1}{2} \times 16 = \frac{1}{2} \times \frac{16}{1} = \frac{16}{2} = 8, \text{ and the phrase "the product of } \frac{2}{3} \text{ and } \frac{1}{4}\text{"}$$

should be translated as $\frac{2}{3} \times \frac{1}{4} = \frac{2}{12} = \frac{1}{6}$.

Now test your understanding with these problems. Multiply as shown. Change any mixed numbers to improper fractions *before* you multiply.

(a) $\dfrac{7}{8}$ of $\dfrac{2}{3} =$ _____

(b) $\dfrac{8}{12} \times \dfrac{3}{16} =$ _____

(c) $\dfrac{3}{32}$ of $\dfrac{4}{15} =$ _____

(d) $\dfrac{15}{4} \times \dfrac{9}{10} =$ _____

(e) $\dfrac{3}{2}$ of $\dfrac{2}{3} =$ _____

(f) $1\dfrac{1}{2} \times \dfrac{2}{5} =$ _____

(g) $4 \times \dfrac{7}{8} =$ _____

(h) $3\dfrac{5}{6} \times \dfrac{3}{10} =$ _____

(i) $1\dfrac{4}{5} \times 1\dfrac{3}{4} =$ _____

(j) Polly the Plumber needs six lengths of PVC pipe each $26\dfrac{3}{4}$ in. long. What total length of pipe will she need?

Remember to write your answer in lowest terms. The step-by-step answers are given in **22**.

---

21 Now, now, don't panic. You needn't draw the little boxes to do the calculation. Try it this way:

$$\dfrac{5}{6} \times \dfrac{2}{3} = \dfrac{5 \times 2}{6 \times 3}$$

Multiply the numerators

Multiply the denominators

Finish the calculation and then return to **18** and choose an answer.

---

22 (*Hint:* "Of" means multiply.)

(a) $\dfrac{7}{8} \times \dfrac{2}{3} = \dfrac{7 \times 2}{(4 \times 2) \times 3} = \dfrac{7}{4 \times 3} = \dfrac{7}{12}$

Eliminate common factors before you multiply.

Your work will look like this when you learn to do these operations mentally:

$$\dfrac{7}{\underset{4}{8}} \times \dfrac{\overset{1}{2}}{3} = \dfrac{7}{12}$$

(b) $\dfrac{8}{12} \times \dfrac{3}{16} = \dfrac{\overset{1}{8} \times \overset{1}{3}}{(4 \times 3) \times (8 \times 2)} = \dfrac{1}{4 \times 2} = \dfrac{1}{8}$

or $\dfrac{\overset{1}{8}}{\underset{4}{12}} \times \dfrac{\overset{1}{3}}{\underset{2}{16}} = \dfrac{1}{8}$

(c) $\dfrac{\overset{1}{3}}{\underset{8}{32}} \times \dfrac{\overset{1}{4}}{\underset{5}{15}} = \dfrac{1}{40}$

(d) $\dfrac{15}{4} \times \dfrac{9}{\underset{2}{10}} \overset{3}{} = \dfrac{27}{8} = 3\dfrac{3}{8}$

(e) $\dfrac{\overset{1}{3}}{\underset{1}{2}} \times \dfrac{\overset{1}{2}}{\underset{1}{3}} = 1$

(f) $1\dfrac{1}{2} \times \dfrac{2}{5} = \dfrac{3}{2} \times \dfrac{\overset{1}{2}}{5} = \dfrac{3}{5}$

If you don't remember how to change a mixed number to an improper fraction see frame **9**.

(g) $4 \times \dfrac{7}{8} = \dfrac{4}{1} \times \dfrac{7}{8} = \dfrac{7}{2} = 3\dfrac{1}{2}$

(h) $3\dfrac{5}{6} \times \dfrac{3}{10} = \dfrac{23}{6} \times \dfrac{3}{10} = \dfrac{23}{20} = 1\dfrac{3}{20}$

(i) $1\dfrac{4}{5} \times 1\dfrac{3}{4} = \dfrac{9}{5} \times \dfrac{7}{4} = \dfrac{63}{20} = 3\dfrac{3}{20}$

(j) Total length of pipe = number of pieces × length of each piece

$$= 6 \times 26\dfrac{3}{4}$$

$$= 6 \times \dfrac{107}{4}$$

$$= \dfrac{6}{1} \times \dfrac{107}{4}$$

$$= \dfrac{321}{2} = 160\dfrac{1}{2} \text{ in.}$$

Now turn to **23** for a set of practice problems.

___

23

Exercises 2-2   **Multiplication of Fractions**

A. Multiply and write the answer in lowest terms.

1. $\dfrac{1}{2} \times \dfrac{1}{4}$
2. $\dfrac{2}{5} \times \dfrac{2}{3}$
3. $\dfrac{4}{5} \times \dfrac{1}{6}$
4. $6 \times \dfrac{1}{2}$

5. $\dfrac{8}{9} \times 3$
6. $\dfrac{11}{12} \times \dfrac{4}{15}$
7. $\dfrac{8}{3} \times \dfrac{5}{12}$
8. $\dfrac{7}{8} \times \dfrac{13}{14}$

9. $\dfrac{12}{8} \times \dfrac{15}{9}$
10. $\dfrac{4}{7} \times \dfrac{49}{2}$
11. $4\dfrac{1}{2} \times \dfrac{2}{3}$
12. $6 \times 1\dfrac{1}{3}$

13. $2\dfrac{1}{6} \times 1\dfrac{1}{2}$
14. $\dfrac{5}{7} \times 1\dfrac{7}{15}$
15. $4\dfrac{3}{5} \times 15$
16. $10\dfrac{5}{6} \times 3\dfrac{3}{10}$

17. $34 \times 2\dfrac{3}{17}$
18. $7\dfrac{9}{10} \times 1\dfrac{1}{4}$
19. $11\dfrac{6}{7} \times \dfrac{7}{8}$

20. $18 \times 1\dfrac{5}{27}$
21. $\dfrac{1}{2} \times \dfrac{1}{2} \times \dfrac{1}{2}$
22. $1\dfrac{4}{5} \times \dfrac{2}{3} \times \dfrac{1}{4}$

23. $\dfrac{1}{4} \times \dfrac{2}{3} \times \dfrac{2}{5}$
24. $2\dfrac{1}{2} \times \dfrac{3}{5} \times \dfrac{8}{9}$
25. $\dfrac{2}{3} \times \dfrac{3}{2} \times 2$

B. Find.

1. $\dfrac{1}{2}$ of $\dfrac{1}{3}$
2. $\dfrac{1}{4}$ of $\dfrac{3}{8}$
3. $\dfrac{2}{3}$ of $\dfrac{3}{4}$
4. $\dfrac{7}{8}$ of $\dfrac{1}{2}$

5. $\dfrac{1}{2}$ of $1\dfrac{1}{2}$
6. $\dfrac{3}{4}$ of $1\dfrac{1}{4}$
7. $\dfrac{5}{8}$ of $2\dfrac{1}{10}$
8. $\dfrac{5}{3}$ of $1\dfrac{2}{3}$

9. $\dfrac{4}{3}$ of $\dfrac{3}{4}$
10. $\dfrac{3}{5}$ of $1\dfrac{1}{6}$
11. $\dfrac{7}{8}$ of $1\dfrac{1}{5}$
12. $\dfrac{3}{5}$ of 4

13. $\frac{7}{16}$ of 6   14. $\frac{5}{16}$ of $1\frac{1}{7}$ 15. $\frac{3}{8}$ of $2\frac{2}{3}$ 16. $\frac{15}{16}$ of $1\frac{3}{5}$

C. Practical Problems

1. **Building Construction** Find the width of floor space covered by 38 boards with $3\frac{5}{8}$-in. exposed surface each.

2. **Building Construction** There are 14 risers in the stairs from the basement to the first floor of a house. Find the total height of the stairs if the risers are $7\frac{1}{8}$ in. high.

3. **Roofing** Shingles are laid so that 5 in. or $\frac{5}{12}$ ft is exposed in each layer. How many feet of roof will be covered by 28 courses?

4. **Carpentry** A board $5\frac{3}{4}$ in. wide is cut to three-fourths of its original width. Find the new width.

5. **Carpentry** What length of 2-in. by 4-in. material will be required to make 6 bench legs each $28\frac{1}{4}$ in. long?

6. **Electrical Technology** Find the total length of 12 pieces of wire each $9\frac{3}{16}$ in. long.

7. **Auto Services** If a car averages $22\frac{4}{5}$ miles to a gallon of gas, how many miles can it travel on 14 gallons of gas?

8. **Machine Technology** What is the shortest bar that can be used for making six chisels each $6\frac{1}{8}$ in. in length?

9. **Manufacturing** How many pounds of grease are contained in a barrel if a barrel holds $46\frac{1}{2}$ gallons, and a gallon of grease weighs $7\frac{2}{3}$ lb?

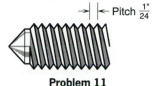
Pitch $\frac{1"}{24}$

Problem 11

10. **Electrical Technology** An electrician cut eight pieces of copper tubing from a coil. Each piece is $14\frac{3}{4}$ ft long. What is the total length used?

11. **Machine Technology** How far will a nut advance if it is given 18 turns on a $\frac{1}{4}$-in. 24-NF (National Fine thread) bolt? (*Hint:* The designation 24-NF means that the nut advances $\frac{1}{24}$ in. for each complete turn.)

12. **Building Construction** What width of floor space can be covered by 48 boards each with $4\frac{3}{8}$ in. of exposed surface?

13. **Drafting** If $\frac{3}{8}$ in. on a drawing represents 1 ft, how many inches on the drawing will represent 26 ft?

14. **Manufacturing** What is the volume of a rectangular box with interior dimensions $12\frac{1}{2}$ in. long, $8\frac{1}{8}$ in. wide, and $4\frac{1}{4}$ in. deep? (*Hint:* Volume = length × width × height.)

15. **Machine Technology** How long will it take to machine 45 pins if each pin requires $6\frac{3}{4}$ minutes? Allow 1 minute per pin for placing stock in the lathe.

16. **Manufacturing** There are 231 cu in. in a gallon. How many cubic inches are needed to fill a container with a rated capacity of $4\frac{1}{3}$ gallons?

17. **Printing** In a print shop, 1 unit of labor is equal to $\frac{1}{6}$ hour. How many hours are involved in 64 units of work?

18. **Carpentry** Find the total width of 36 2-by-4s if the finished width of each board is actually $3\frac{1}{2}$ in.

19. **Photography** A photograph must be reduced to four-fifths of its original size to fit the space available in a newspaper. Find the length of the reduced photograph if the original was $6\frac{3}{4}$ in. long.

20. **Printing** A bound book weighs $1\frac{5}{8}$ lb. How many pounds will 20 cartons of 12 books each weigh?

21. **Printing** There are 6 picas in 1 in. If a line of type is $3\frac{3}{4}$ in. long, what is this length in picas?

22. **Wastewater Technology** The normal daily flow of raw sewage into a treatment plant is 32 MGD (million gallons per day). Due to technical problems one day, the plant had to cut back to three-fourths of its normal intake. What was this reduced flow?

23. **Masonry** What is the height of 12 courses of $2\frac{1}{4}$-in. bricks with $\frac{3}{8}$-in. mortar joints? (12 rows of brick and 11 mortar joints)

24. **Plumbing** To find the degree measure of the bend of a pipe fitting, multiply the fraction of bend by 360 degrees. What is the degree measure of a $\frac{1}{8}$ bend? A $\frac{1}{5}$ bend? A $\frac{1}{6}$ bend?

25. **Plumbing** A drain must be installed with a grade of $\frac{1}{8}$ in. of vertical drop per foot of horizontal run. How much drop will there be for 26 ft of run?

26. **Machine Technology** The center-to-center distance between consecutive holes in a strip of metal is $\frac{5}{16}$ in. What is the total distance x between the first and last centers as shown in the figure?

27. **Sheet Metal Technology** The allowance for a wired edge on fabricated metal is $2\frac{1}{2}$ times the diameter of the wire. Calculate the allowance for a wired edge if the diameter of the wire is $\frac{3}{16}$ in.

Check your answers on page 584, then continue in **24**.

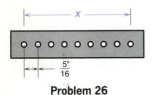

Problem 26

2-3 DIVISION OF FRACTIONS

24 Addition and multiplication are both reversible arithmetic operations. For example,

2×3 and 3×2 both equal 6

$4 + 5$ and $5 + 4$ both equal 9

The order in which you add or multiply is not important. This reversibility is called the *commutative* property of addition and multiplication.

In division this type of exchange is not allowed, and because it is not allowed many people find division very troublesome. In the division of fractions it is very important that you set up the problem correctly.

The phrase "8 divided by 4" can be written as $8 \div 4$ or $\frac{8}{4}$

4 is the divisor

In this problem you are being asked to divide a set of eight objects into sets of four objects.

In the division $5 \div \frac{1}{2}$, which number is the divisor?

Check your answer in **25**.

25 The divisor is $\frac{1}{2}$.

The division $5 \div \frac{1}{2}$, read "5 divided by $\frac{1}{2}$," asks how many $\frac{1}{2}$-unit lengths are included in a length of 5 units.

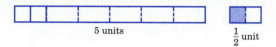

5 units $\frac{1}{2}$ unit

Division answers the question: How many of the divisor are in the dividend?

$8 \div 4 = \square$ asks you to find how many 4s are in 8.
It is easy to see that $\square = 2$.

$5 \div \frac{1}{2} = \square$ asks you to find how many $\frac{1}{2}$s are in 5. Do you see that $\square = 10$?

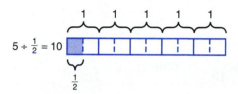

$5 \div \frac{1}{2} = 10$

There are ten $\frac{1}{2}$-unit lengths contained in the 5-unit length.

Using a drawing of this sort to solve a division problem is difficult and clumsy. We need a simple rule. Here it is:

> To divide by a fraction, invert the divisor and multiply.

Inverting a fraction means to switch the numerator and denominator.

Example: $5 \div \frac{1}{2} = ?$

$$5 \div \frac{1}{2} \quad = \quad 5 \times \frac{2}{1} = \frac{10}{1} = 10 \qquad \text{as shown graphically}$$

Invert — Multiply

Example: $\frac{2}{5} \div \frac{1}{2} = ?$

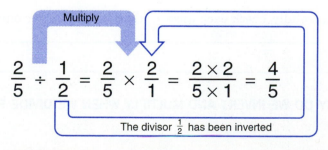

$$\frac{2}{5} \div \frac{1}{2} = \frac{2}{5} \times \frac{2}{1} = \frac{2 \times 2}{5 \times 1} = \frac{4}{5}$$

The divisor $\frac{1}{2}$ has been inverted

We have converted a division problem that is difficult to picture into simple multiplication.

The final, and very important, step in every division is checking the answer. To check, multiply the divisor and the quotient and compare this answer with the original fraction or dividend.

If $\frac{3}{5} \div \frac{2}{3} = \frac{9}{10}$ then $\frac{2}{3} \times \frac{9}{10}$ should equal $\frac{3}{5}$.

$$\overset{1}{\cancel{2}} \times \frac{\overset{3}{\cancel{9}}}{\underset{5}{\cancel{10}}} \Big/ \underset{1}{} = \frac{1 \times 3}{1 \times 5} = \frac{3}{5}$$

You try it:

$$\frac{7}{8} \div \frac{3}{2} = \underline{\hspace{3cm}}$$

Solve by inverting the divisor and multiplying.
Check your work in **26**.

26

$$\frac{7}{8} \div \frac{3}{2} = \frac{7}{8} \times \frac{2}{3} = \frac{7 \times \overset{1}{\cancel{2}}}{\underset{4}{} } = \frac{7}{12} \qquad \textbf{Check:} \quad \frac{\overset{1}{\cancel{3}}}{} \times \frac{7}{\underset{4}{\cancel{12}}} = \frac{7}{8}$$

Learning Help ▶ The chief source of confusion for many people in dividing fractions is deciding which fraction to invert. It will help if you:

1. Put every division problem in the form

    (dividend) ÷ (divisor)

    then invert the divisor, and finally, multiply to obtain the quotient.

2. Check your answer by multiplying. The product

    (divisor) × (quotient or answer)

    should equal the dividend. ◀

Here are a few problems to test your understanding.

(a) $\frac{2}{5} \div \frac{3}{8}$ (b) $\frac{7}{40} \div \frac{21}{25}$ (c) $3\frac{3}{4} \div \frac{5}{2}$

(d) $4\frac{1}{5} \div 1\frac{4}{10}$ (e) $3\frac{2}{3} \div 3$ (f) Divide $\frac{3}{4}$ by $2\frac{5}{8}$.

(g) Divide $1\frac{1}{4}$ by $1\frac{7}{8}$.   (h) How many sheets of plywood, each $\frac{3}{4}$ in. thick, are in a stack 18 in. high?

Work carefully, check each answer, then turn to **27** for our worked solutions.

WHY DO WE INVERT AND MULTIPLY WHEN WE DIVIDE FRACTIONS?

The division $8 \div 4$ can be written $\frac{8}{4}$. Similarly, $\frac{1}{2} \div \frac{2}{3}$ can be written $\dfrac{\frac{1}{2}}{\frac{2}{3}}$.

To simplify this fraction, multiply by $\dfrac{\frac{3}{2}}{\frac{3}{2}}$ (which is equal to 1).

$$\frac{\frac{1}{2}}{\frac{2}{3}} = \frac{\frac{1}{2} \times \frac{3}{2}}{\frac{2}{3} \times \frac{3}{2}} = \frac{\frac{1}{2} \times \frac{3}{2}}{1} = \frac{1}{2} \times \frac{3}{2}$$

$$\frac{2}{3} \times \frac{3}{2} = \frac{2 \times 3}{3 \times 2} = \frac{6}{6} = 1$$

Therefore, $\dfrac{1}{2} \div \dfrac{2}{3} = \dfrac{1}{2} \times \dfrac{3}{2}$. We have inverted the fraction $\dfrac{2}{3}$ and multiplied by it.

27 (a) $\dfrac{2}{5} \div \dfrac{3}{8} = \dfrac{2}{5} \times \dfrac{8}{3} = \dfrac{16}{15} = 1\dfrac{1}{15}$

The answer is $1\dfrac{1}{15}$. **Check:** $\dfrac{\overset{1}{\cancel{3}}}{\cancel{8}} \times \dfrac{\overset{2}{\cancel{16}}}{\cancel{15}} = \dfrac{2}{5}$

(b) $\dfrac{7}{40} \div \dfrac{21}{25} = \dfrac{7}{\underset{8}{\cancel{40}}} \times \dfrac{\overset{5}{\cancel{25}}}{\underset{3}{\cancel{21}}} = \dfrac{5}{24}$ **Check:** $\dfrac{\overset{7}{\cancel{21}}}{\underset{5}{\cancel{25}}} \times \dfrac{\overset{1}{\cancel{5}}}{\underset{8}{\cancel{24}}} = \dfrac{7}{40}$

(c) $3\dfrac{3}{4} \div \dfrac{5}{2} = \dfrac{15}{4} \div \dfrac{5}{2} = \dfrac{\overset{3}{\cancel{15}}}{\underset{2}{\cancel{4}}} \times \dfrac{\overset{1}{\cancel{2}}}{\cancel{5}} = \dfrac{3}{2} = 1\dfrac{1}{2}$ **Check:** $\dfrac{5}{2} \times \dfrac{3}{2} = \dfrac{15}{4} = 3\dfrac{3}{4}$

(d) $4\dfrac{1}{5} \div 1\dfrac{4}{10} = \dfrac{21}{5} \div \dfrac{14}{10} = \dfrac{\overset{3}{\cancel{21}}}{\underset{1}{\cancel{5}}} \times \dfrac{\overset{2}{\cancel{10}}}{\underset{2}{\cancel{14}}} = \dfrac{6}{2} = 3$

Check: $1\dfrac{4}{10} \times 3 = \dfrac{\overset{7}{\cancel{14}}}{\underset{5}{\cancel{10}}} \times \dfrac{3}{1} = \dfrac{21}{5} = 4\dfrac{1}{5}$

(e) $3\dfrac{2}{3} \div 3 = \dfrac{11}{3} \div \dfrac{3}{1} = \dfrac{11}{3} \times \dfrac{1}{3} = \dfrac{11}{9} = 1\dfrac{2}{9}$

Check: $3 \times \dfrac{11}{9} = \dfrac{3}{1} \times \dfrac{11}{\underset{3}{\cancel{9}}} = \dfrac{11}{3} = 3\dfrac{2}{3}$

(f) $\dfrac{3}{4} \div 2\dfrac{5}{8} = \dfrac{3}{4} \div \dfrac{21}{8} = \dfrac{\overset{1}{\cancel{3}}}{\underset{1}{\cancel{4}}} \times \dfrac{\overset{2}{\cancel{8}}}{\underset{7}{\cancel{21}}} = \dfrac{2}{7}$ **Check:** $2\dfrac{5}{8} \times \dfrac{2}{7} = \dfrac{\overset{3}{\cancel{21}}}{\underset{4}{\cancel{8}}} \times \dfrac{\overset{1}{\cancel{2}}}{\underset{1}{\cancel{7}}} = \dfrac{3}{4}$

(g) $1\dfrac{1}{4} \div 1\dfrac{7}{8} = \dfrac{5}{4} \div \dfrac{15}{8} = \dfrac{\overset{1}{\cancel{5}}}{\underset{1}{\cancel{4}}} \times \dfrac{\overset{2}{\cancel{8}}}{\underset{3}{\cancel{15}}} = \dfrac{2}{3}$

Check: $1\dfrac{7}{8} \times \dfrac{2}{3} = \dfrac{\overset{5}{\cancel{15}}}{\underset{4}{\cancel{8}}} \times \dfrac{\overset{1}{\cancel{2}}}{\cancel{3}} = \dfrac{5}{4} = 1\dfrac{1}{4}$

(h) A question of the form "How many X are in Y?" tells us we must divide Y by X. In this problem, divide 18 in., the total height, by $\frac{3}{4}$ in., the thickness of each sheet.

$$18 \div \frac{3}{4} = \overset{6}{\cancel{18}} \times \frac{4}{\underset{1}{\cancel{3}}} = 24 \qquad \text{There are 24 sheets of plywood in the stack.}$$

Turn to 28 for a set of practice problems on dividing fractions.

Exercises 2-3   **Division of Fractions**

A. Divide and write the answer in lowest terms.

1. $\frac{5}{6} \div \frac{1}{2}$

2. $6 \div \frac{2}{3}$

3. $\frac{5}{12} \div \frac{4}{3}$

4. $8 \div \frac{1}{4}$

5. $\frac{6}{16} \div \frac{3}{4}$

6. $\frac{1}{2} \div \frac{1}{2}$

7. $\frac{3}{16} \div \frac{6}{8}$

8. $\frac{3}{4} \div \frac{5}{16}$

9. $1\frac{1}{2} \div \frac{1}{6}$

10. $6 \div 1\frac{1}{2}$

11. $3\frac{1}{7} \div 2\frac{5}{14}$

12. $3\frac{1}{2} \div 2$

13. $6\frac{2}{5} \div 5\frac{1}{3}$

14. $10 \div 1\frac{1}{5}$

15. $8 \div \frac{1}{2}$

16. $\frac{2}{3} \div 6$

17. $\dfrac{12}{\dfrac{2}{3}}$

18. $\dfrac{\dfrac{3}{4}}{\dfrac{7}{8}}$

19. $\dfrac{\dfrac{5}{2}}{\dfrac{2}{3}}$

20. $\dfrac{1\frac{1}{2}}{2\frac{1}{2}}$

21. $\frac{5}{16} \div \frac{3}{8}$

22. $\frac{7}{12} \div \frac{2}{3}$

23. $\frac{7}{32} \div 1\frac{3}{4}$

24. $1\frac{2}{3} \div 1\frac{1}{4}$

B. Practical Problems

1. **Drafting**   How many feet are represented by a 4-in. line if it is drawn to a scale of $\frac{1}{2}$ in. = 1 ft?

2. **Drafting**   If $\frac{1}{4}$ in. on a drawing represents 1 ft 0 in., then $3\frac{1}{2}$ in. on the drawing will represent how many feet?

3. **Carpentry**   How many boards $4\frac{5}{8}$ in. wide will it take to cover a floor 222 in. wide?

4. **Building Construction**   How many supporting columns $88\frac{1}{2}$ in. long can be cut from six pieces each 22 ft long? (*Hint:* Be careful of units.)

5. **Building Construction**   How many pieces of $\frac{1}{2}$-in. plywood are there in a stack 42 in. high?

6. **Drafting**   If $\frac{1}{4}$ in. represents 1 ft 0 in. on a drawing, how many feet will be represented by a line $10\frac{1}{8}$ in. long?

7. **Masonry**   If we allow $2\frac{5}{8}$ in. for the thickness of a course of brick, including mortar joints, how many courses of brick will there be in a wall $47\frac{1}{4}$ in. high?

8. **Plumbing**   How many lengths of pipe $2\frac{5}{8}$ ft long can be cut from a pipe 21 ft long?

9. **Machine Technology**   How many pieces $6\frac{1}{4}$ in. long can be cut from 35 metal rods each 40 in. long? Disregard waste.

10. **Machine Technology**   The architectural drawing for a room measures $3\frac{5}{8}$ in. by $4\frac{1}{4}$ in. If $\frac{1}{8}$ in. is equal to 1 ft on the drawing, what are the actual dimensions of the room?

11. **Printing**   How many full $3\frac{1}{2}$-in. sheets can be cut from $24\frac{3}{4}$-in. stock?

12. **Machine Technology**   The feed on a boring mill is set for $\frac{1}{32}$ in. How many revolutions are needed to advance the tool $3\frac{3}{8}$ in.?

13. **Machine Technology**   If the pitch of a thread is $\frac{1}{18}$ in., how many threads are needed for the threaded section of a pipe to be $2\frac{1}{2}$ in. long?

14. **Building Construction**   The floor area of a room on a house plan measures $3\frac{1}{2}$ in. by $4\frac{5}{8}$ in. If the drawing scale is $\frac{1}{4}$ in. represents 1 ft, what is the actual size of the room?

Check your answers on page 585, then continue in **29**.

2-4   ADDITION AND SUBTRACTION OF FRACTIONS

Addition

29   At heart, adding fractions is a matter of counting:

$$\frac{1}{5} + \frac{3}{5} = \frac{1+3}{5} = \frac{4}{5}$$

$\frac{1}{5}$ [   ]   1 fifth
       +
$\frac{3}{5}$ [   ]   + 3 fifths
       =
$\frac{4}{5}$ [   ]   = 4 fifths, count them.

Add $\dfrac{1}{8} + \dfrac{3}{8} =$ _____

This is easy to see with measurements:

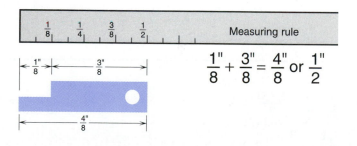

$$\frac{1''}{8} + \frac{3''}{8} = \frac{4''}{8} \text{ or } \frac{1''}{2}$$

Add $\dfrac{2}{7} + \dfrac{3}{7} =$ _____

Check your answer in **30**.

30 $\dfrac{2}{7} + \dfrac{3}{7} = \dfrac{2+3}{7} = \dfrac{5}{7}$

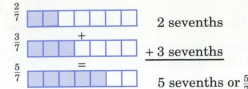

$\dfrac{2}{7}$ 2 sevenths

$\dfrac{3}{7}$ + 3 sevenths

$\dfrac{5}{7}$ 5 sevenths or $\dfrac{5}{7}$

Like Fractions   Fractions having the same denominator are called *like* fractions. In the preceding problem, $\dfrac{2}{7}$ and $\dfrac{3}{7}$ both have denominator 7 and are like fractions. Adding like fractions is easy: *first,* add the numerators to find the numerator of the sum and *second,* use the denominator the fractions have in common as the denominator of the sum.

$\dfrac{2}{9} + \dfrac{5}{9} = \dfrac{2+5}{9}$ ← Add numerators

← Same denominator

Adding three or more like fractions is easy: $\dfrac{3}{12} + \dfrac{1}{12} + \dfrac{5}{12} = ?$

Add these fractions, then turn to **31**.

---

31 $\dfrac{3}{12} + \dfrac{1}{12} + \dfrac{5}{12} = \dfrac{3+1+5}{12} = \dfrac{9}{12} = \dfrac{3}{4}$

Notice that we write the sum in lowest terms.

Try these problems for exercise.

(a) $\dfrac{1}{8} + \dfrac{3}{8}$

(b) $\dfrac{7}{9} + \dfrac{5}{9}$

(c) $2\dfrac{1}{5} + 3\dfrac{3}{5}$

(d) $\dfrac{1}{7} + \dfrac{4}{7} + \dfrac{5}{7} + 1\dfrac{2}{7} + \dfrac{8}{7}$

(e) $2 + 3\dfrac{1}{2}$

(f) $3\dfrac{1}{8} + 2\dfrac{3}{8}$

Go to **32** to check your work.

---

32 (a) $\dfrac{1}{8} + \dfrac{3}{8} = \dfrac{1+3}{8} = \dfrac{4}{8} = \dfrac{1}{2}$

(b) $\dfrac{7}{9} + \dfrac{5}{9} = \dfrac{7+5}{9} = \dfrac{12}{9} = \dfrac{4}{3} = 1\dfrac{1}{3}$

(c) $2\dfrac{1}{5} + 3\dfrac{3}{5} = 2 + 3 + \dfrac{1}{5} + \dfrac{3}{5} = 5 + \dfrac{4}{5} = 5\dfrac{4}{5}$

(d) $\dfrac{1}{7} + \dfrac{4}{7} + \dfrac{5}{7} + 1\dfrac{2}{7} + \dfrac{8}{7} = \dfrac{1}{7} + \dfrac{4}{7} + \dfrac{5}{7} + \dfrac{9}{7} + \dfrac{8}{7} = \dfrac{1+4+5+9+8}{7} = \dfrac{27}{7} = 3\dfrac{6}{7}$

(e) $2 + 3\dfrac{1}{2} = 2 + 3 + \dfrac{1}{2} = 5\dfrac{1}{2}$. Remember: $3\dfrac{1}{2}$ means $3 + \dfrac{1}{2}$.

(f) $3\dfrac{1}{8} + 2\dfrac{3}{8} = (3 + 2) + \left(\dfrac{1}{8} + \dfrac{3}{8}\right) = 5 + \dfrac{4}{8} = 5 + \dfrac{1}{2} = 5\dfrac{1}{2}$

This can also be done by rewriting each of the mixed numbers as an improper fraction before adding.

$$3\frac{1}{8} + 2\frac{3}{8} = \frac{25}{8} + \frac{19}{8} = \frac{44}{8} = \frac{11}{2} \qquad \text{Write in lowest terms.}$$

$$= 5\frac{1}{2} \qquad \text{Rewrite as a mixed number.}$$

Note ▶ If the addition is done using improper fractions, large and unwieldy numerators may result. Be careful. ◀

Unlike Fractions How do we add fractions whose denominators are not the same? For example, how do we add $\frac{2}{3} + \frac{3}{4}$?

The problem is to find a name for this new number. One way to find it is to change these fractions to equivalent fractions with the same denominator.

$$\frac{2}{3} = \frac{2 \times 4}{3 \times 4} = \frac{8}{12}$$

$$\frac{3}{4} = \frac{3 \times 3}{4 \times 3} = \frac{9}{12}$$

Note ▶ You should remember that we discussed equivalent fractions in **10**. Return for a quick review if you need it. ◀

Now add $\frac{2}{3} + \frac{3}{4}$ using the equivalent fractions, then continue in **33**.

---

33 $\qquad \dfrac{2}{3} + \dfrac{3}{4} = \dfrac{8}{12} + \dfrac{9}{12} = \dfrac{17}{12} = 1\dfrac{5}{12}$

We change the original fractions to equivalent fractions with the same denominator and then add as before.

Least Common Denominator How do you know what number to use as the new denominator? In general, you cannot simply guess at the best new denominator. We need a method for finding it from the denominators of the fractions to be added. The new denominator is called the *least common denominator*, abbreviated LCD.

Example: Suppose that we want to add the fractions $\frac{1}{8} + \frac{5}{12}$.

The first step is to find the LCD of the denominators 8 and 12. To find the LCD, follow this procedure.

Step 1 Write the denominators.

Step 2 Test each denominator for division by the first prime number, 2. If 2 divides one of the denominators exactly, write 2 on the left, write the quotient below, and bring down the other denominator. If 2 divides both denominators exactly, bring down both quotients. In this example, the number 2 will divide both denominators.

$$\begin{array}{r} 8 \quad 12 \\ 2 \;\big|\; \overline{8 \quad 12} \\ \overline{4 \quad 6} \end{array}$$

Step 3 Repeat the process, dividing by 2 if either number is exactly divisible by 2. If neither number is divisible by 2, test for division by the next prime number, 3.

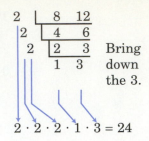

$2 \cdot 2 \cdot 2 \cdot 1 \cdot 3 = 24$

Step 4 Continue the process until all the numbers on the left and below the lines are either 1 or prime. The product of these numbers is the LCD.

LCD = 24

The least common denominator of 8 and 12 is 24. It is the smallest number that both 8 and 12 divide exactly.

Another Example: Find the LCD of 12 and 45.

```
2 | 12   45
 2 |  6   45
  3 |  3   45
   3 |  1   15
       1    5  → 2 · 2 · 3 · 3 · 1 · 5 = 180
              The LCD is 180.
```

Once you have found the LCD, add the original fractions by rewriting them with this new denominator.

$$\frac{1}{8} = \frac{?}{24} \qquad \frac{1}{8} = \frac{1 \times 3}{8 \times 3} = \frac{3}{24}$$

$$\frac{5}{12} = \frac{?}{24} \qquad \frac{5}{12} = \frac{5 \times 2}{12 \times 2} = \frac{10}{24}$$

Now we can add the fractions:

$$\frac{1}{8} + \frac{5}{12} = \frac{3}{24} + \frac{10}{24}$$

$$= \frac{13}{24}$$

Use the method described above to find the LCD of the numbers 12 and 15. Check your work in **34**.

A CALCULATOR METHOD FOR FINDING THE LCD

Here is an alternative method for finding the LCD that you might find easier than the method described in **33**. Follow these two steps.

Step 1 Choose the larger denominator and write down a few multiples of it.

Example: To find the LCD of 12 and 15, first write down a few of the multiples of the larger number, 15. The multiples are 15, 30, 45, 60, 75, and so on.

Step 2 Test each multiple until you find one that is exactly divisible by the smaller denominator.

Example: 15 is not exactly divisible by 12. 30 is not exactly divisible by 12. 45 is not exactly divisible by 12, 60 *is* exactly divisible by 12. The LCD is 60.

This method of finding the LCD has the advantage that you can use it with an electronic calculator. For this example the calculator steps would look like this:

15 ⌹ **12** ⌹ → **1.25** *Not* a whole number; therefore, *not* exactly divisible by 12.

15 ⌧ **2** ⌹ **12** ⌹ → **2.5** Second multiple: answer is *not* a whole number.

15 ⌧ **3** ⌹ **12** ⌹ → **3.75** Third multiple: answer is *not* a whole number.

15 ⌧ **4** ⌹ **12** ⌹ → **5.** Fourth multiple: answer *is* a whole number; therefore, the LCD is 4 × 15 or 60.

---

34 **Step 1** Write the denominators: 12   15.

Step 2 Divide by 2. Repeat until no number is divisible by 2.

```
2 | 12  15      ←——  Divide 12 by 2.
2 | 6   15      ←——  15 is not exactly
    3   15            divisible by 2, so
                      bring it down
                      unchanged.
```

Step 3 Divide by the next prime, 3.

Step 4 Multiply. LCD = 60

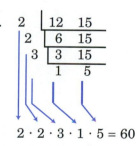

$$2 \cdot 2 \cdot 3 \cdot 1 \cdot 5 = 60$$

The process is exactly the same no matter how many numbers you begin with. To find the LCD of 4, 6, and 8:

```
2 | 4  6  8
  2 | 2  3  4
    2 | 1  3  2    Bring down the 3.
        1  3  1
```

LCD = 2 · 2 · 2 · 1 · 3 · 1 = 24

Ready for a bit of practice in finding LCDs?

Find the LCD of 12 and 18. Check your work in **35**.

---

35   **Step 1**   12   18

Step 2   Divide by 2.

$$\begin{array}{c|cc} 2 & 12 & 18 \\ 2 & 6 & 9 \leftarrow \text{9 is not divisible by 2.} \\ & 3 & 9 \leftarrow \text{Bring down the 9.} \end{array}$$

Step 3   Divide by 3.

$$\begin{array}{c|cc} 2 & 12 & 18 \\ 2 & 6 & 9 \\ 3 & 3 & 9 \\ & 1 & 3 \end{array}$$

Step 4   Multiply. $\text{LCD} = 2 \cdot 2 \cdot 3 \cdot 1 \cdot 3 = 36$

Notice that we start the dividing with 2 and continue until the remaining numbers cannot be divided further.

Practice by finding the LCD for each of the following sets of numbers.

(a)   2 and 4   (b)   8 and 4   (c)   6 and 3

(d)   5 and 4   (e)   9 and 15   (f)   15 and 24

(g)   4, 5, and 6   (h)   4, 8, and 12   (i)   12, 15, and 21

Check your answers in **36**.

---

36   (a)   4   (b)   8   (c)   6   (d)   20   (e)   45

(f)   120   (g)   60   (h)   24   (i)   420

In order to use the LCD to add fractions, rewrite the fractions with the LCD as the new denominator, then add the new equivalent fractions.

For example, add $\frac{1}{6} + \frac{5}{8}$.

First, find the LCD. The LCD of 6 and 8 is 24.

Next, rewrite the two fractions with denominator 24.

$$\frac{1}{6} = \frac{1 \times 4}{6 \times 4} = \frac{4}{24}$$

$$\frac{5}{8} = \frac{5 \times 3}{8 \times 3} = \frac{15}{24}$$

Finally, add the new equivalent fractions.

$$\frac{1}{6} + \frac{5}{8} = \frac{4}{24} + \frac{15}{24} = \frac{19}{24}$$

Try it: Add $\frac{3}{8} + \frac{1}{10}$.

Check your work in **37**.

---

37   The LCD of 8 and 10 is 40.

$$\frac{3}{8} = \frac{3 \times 5}{8 \times 5} = \frac{15}{40}$$

$$\frac{1}{10} = \frac{1 \times 4}{10 \times 4} = \frac{4}{40}$$

$$\frac{3}{8} + \frac{1}{10} = \frac{15}{40} + \frac{4}{40} = \frac{19}{40}$$

Ready for some practice at this?

Find the LCD, rewrite the fractions, and add.

(a) $\dfrac{1}{2} + \dfrac{1}{4}$ (b) $\dfrac{3}{8} + \dfrac{1}{4}$ (c) $\dfrac{1}{6} + \dfrac{2}{3} + \dfrac{5}{9}$

(d) $\dfrac{3}{4} + \dfrac{1}{6}$ (e) $\dfrac{2}{5} + \dfrac{1}{4}$

(f) The Monterey Canning Co. packs $8\frac{1}{2}$ ounces of tuna into each can. If the can itself weighs $1\frac{7}{8}$ ounces what is the total weight of a can of Monterey tuna?

Check your work in **38**.

---

38 (a) The LCD of 2 and 4 is 4.

$$\frac{1}{2} = \frac{?}{4} = \frac{1 \times \boxed{2}}{2 \times \boxed{2}} = \frac{2}{4}$$

$$\frac{1}{2} + \frac{1}{4} = \frac{2}{4} + \frac{1}{4} = \frac{3}{4}$$

(b) The LCD of 8 and 4 is 8.

$$\frac{1}{4} + \frac{?}{8} = \frac{1 \times \boxed{2}}{4 \times \boxed{2}} = \frac{2}{8}$$

$$\frac{3}{8} + \frac{1}{4} = \frac{3}{8} + \frac{2}{8} = \frac{5}{8}$$

(c) The LCD of 6, 3, and 9 is 18.

$$\frac{1}{6} = \frac{?}{18} = \frac{1 \times \boxed{3}}{6 \times \boxed{3}} = \frac{3}{18} \qquad \frac{2}{3} = \frac{?}{18} = \frac{2 \times \boxed{6}}{3 \times \boxed{6}} = \frac{12}{18}$$

$$\frac{5}{9} = \frac{?}{18} = \frac{5 \times \boxed{2}}{9 \times \boxed{2}} = \frac{10}{18}$$

$$\frac{1}{6} + \frac{2}{3} + \frac{5}{9} = \frac{3}{18} + \frac{12}{18} + \frac{10}{18} = \frac{25}{18} = 1\frac{7}{18}$$

(d) The LCD of 4 and 6 is 12.

$$\frac{3}{4} = \frac{?}{12} = \frac{3 \times \boxed{3}}{4 \times \boxed{3}} = \frac{9}{12} \qquad \frac{1}{6} = \frac{?}{12} = \frac{1 \times \boxed{2}}{6 \times \boxed{2}} = \frac{2}{12}$$

$$\frac{3}{4} + \frac{1}{6} = \frac{9}{12} + \frac{2}{12} = \frac{11}{12}$$

(e) The LCD of 5 and 4 is 20.

$$\frac{2}{5} = \frac{?}{20} = \frac{2 \times \boxed{4}}{5 \times \boxed{4}} = \frac{8}{20} \qquad \frac{1}{4} = \frac{?}{20} = \frac{1 \times \boxed{5}}{4 \times \boxed{5}} = \frac{5}{20}$$

$$\frac{2}{5} + \frac{1}{4} = \frac{8}{20} + \frac{5}{20} = \frac{13}{20}$$

(f) $8\dfrac{1}{2} + 1\dfrac{7}{8} = (8 + 1) + \left(\dfrac{1}{2} + \dfrac{7}{8}\right)$

$\qquad = 9 + \left(\dfrac{4}{8} + \dfrac{7}{8}\right)$ since $\dfrac{1}{2} = \dfrac{4}{8}$

$\qquad = 9 + \dfrac{11}{8} = 9 + 1 + \dfrac{3}{8}$

$\qquad = 10\dfrac{3}{8}$ oz

Subtraction   Once you have mastered the process of adding fractions, subtraction is very simple indeed. To find $\frac{3}{8} - \frac{1}{8}$, notice that the denominators are the same. We can subtract the numerators and write this difference over the common denominator.

$$\frac{3}{8} - \frac{1}{8} = \frac{3 - 1}{8}$$

     | Subtract numerators |

     | Same denominator |

$$= \frac{2}{8} \quad \text{or} \quad \frac{1}{4}$$

To find $\frac{3}{4} - \frac{1}{5}$, first find the LCD of 4 and 5.

The LCD of 4 and 5 is 20.

Find equivalent fractions with a denominator of 20:

$$\frac{3}{4} = \frac{?}{20} = \frac{3 \times 5}{4 \times 5} = \frac{15}{20}$$

$$\frac{1}{5} = \frac{?}{20} = \frac{1 \times 4}{5 \times 4} = \frac{4}{20}$$

Subtract the equivalent fractions:

$$\frac{3}{4} - \frac{1}{5} = \frac{15}{20} - \frac{4}{20} = \frac{15 - 4}{20} = \frac{11}{20}$$

The procedure is exactly the same as for addition.

If the fractions to be subtracted are given as mixed numbers, it is usually simplest to work with them as improper fractions.

Example: $3\frac{1}{2} - 1\frac{1}{3} = \frac{7}{2} - \frac{4}{3}$ The LCD of 2 and 3 is 6.

$$\frac{7}{2} = \frac{7 \times 3}{2 \times 3} = \frac{21}{6} \qquad \frac{4}{3} = \frac{4 \times 2}{3 \times 2} = \frac{8}{6}$$

Then, $3\frac{1}{2} - 1\frac{1}{3} = \frac{21}{6} - \frac{8}{6} = \frac{13}{6} = 2\frac{1}{6}$

Because $\frac{1}{2}$ is greater than $\frac{1}{3}$, we could have worked with the whole-number parts separately.

$$3\frac{1}{2} - 1\frac{1}{3} = (3 - 1) + \left(\frac{1}{2} - \frac{1}{3}\right)$$

$$= 2 + \left(\frac{3}{6} - \frac{2}{6}\right)$$

$$= 2 + \frac{1}{6} = 2\frac{1}{6}$$

Arranged vertically:

$$3\frac{1}{2} \rightarrow \quad 3 + \frac{1}{2} \rightarrow \quad 3 + \frac{3}{6}$$

$$-1\frac{1}{3} \rightarrow -\left(1 + \frac{1}{3}\right) \rightarrow \quad \underline{-1 - \frac{2}{6}}$$

$$2 \ \frac{1}{6} \quad \text{or} \quad 2\frac{1}{6}$$

Example: $5\frac{1}{4} - 2\frac{3}{8} = \frac{21}{4} - \frac{19}{8}$

$$= \frac{42}{8} - \frac{19}{8} = \frac{23}{8} = 2\frac{7}{8}$$

If one of the fractions is a whole number, write it as a fraction first, then add or subtract as with any other fraction.

Example: $3 - \dfrac{2}{5} = \dfrac{3}{1} - \dfrac{2}{5}$

$$= \dfrac{15}{5} - \dfrac{2}{5}$$

$$= \dfrac{13}{5} \quad \text{or} \quad 2\dfrac{3}{5}$$

Try these problems for practice in subtracting fractions. Solutions are in **39**.

(a) $\dfrac{7}{8} - \dfrac{5}{8}$ (b) $9\dfrac{13}{16} - 3\dfrac{1}{4}$ (c) $\dfrac{4}{5} - \dfrac{1}{6}$

(d) $6 - 2\dfrac{3}{4}$ (e) $7\dfrac{1}{6} - 2\dfrac{5}{8}$

(f) A bar $6\dfrac{3}{16}$ in. long is cut from a piece $24\dfrac{3}{8}$ in. long. If $\dfrac{5}{32}$ in. is wasted in cutting, what length remains?

---

39 (a) $\dfrac{7}{8} - \dfrac{5}{8} = \dfrac{7-5}{8} = \dfrac{2}{8} = \dfrac{1}{4}$ (b) $9\dfrac{13}{16} - 3\dfrac{1}{4} = (9-3) + \left(\dfrac{13}{16} - \dfrac{1}{4}\right)$

$$= 6 + \left(\dfrac{13}{16} - \dfrac{4}{16}\right)$$

$$= 6\dfrac{9}{16}$$

(c) The LCD of 5 and 6 is 30.

$$\dfrac{4}{5} = \dfrac{?}{30} = \dfrac{4 \times 6}{5 \times 6} = \dfrac{24}{30}$$

$$\dfrac{1}{6} = \dfrac{?}{30} = \dfrac{1 \times 5}{6 \times 5} = \dfrac{5}{30}$$

$$\text{so} \ \dfrac{4}{5} - \dfrac{1}{6} = \dfrac{24}{30} - \dfrac{5}{30}$$

$$= \dfrac{19}{30}$$

(d) $6 - 2\dfrac{3}{4} = \dfrac{24}{4} - \dfrac{11}{4} = \dfrac{24-11}{4} = \dfrac{13}{4} = 3\dfrac{1}{4}$

$6 = \dfrac{6}{1} = \dfrac{6 \times 4}{1 \times 4}$ $(2 \times 4) + 3$

(e) The LCD of 6 and 8 is 24. $\dfrac{1}{6} = \dfrac{4}{24}$ $\dfrac{5}{8} = \dfrac{15}{24}$

so $7\dfrac{1}{6} = 7\dfrac{4}{24} = \dfrac{172}{24}$ and $2\dfrac{5}{8} = 2\dfrac{15}{24} = \dfrac{63}{24}$

$$7\dfrac{1}{6} - 2\dfrac{5}{8} = \dfrac{172}{24} - \dfrac{63}{24} = \dfrac{109}{24} = 4\dfrac{13}{24}$$

(f) $24\dfrac{3}{8} - \dfrac{5}{32} - 6\dfrac{3}{16} = 24\dfrac{12}{32} - \dfrac{5}{32} - 6\dfrac{6}{32}$

$$= (24 - 6) + \left(\dfrac{12}{32} - \dfrac{5}{32} - \dfrac{6}{32}\right)$$

$$= 18 + \dfrac{1}{32} = 18\dfrac{1}{32} \ \text{in.}$$

Now turn to **40** for a set of problems on adding and subtracting fractions.

Addition and Subtraction of Fractions

A. Add or subtract as shown.

1. $\dfrac{1}{16} + \dfrac{3}{16}$ 2. $\dfrac{5}{12} + \dfrac{11}{12}$ 3. $\dfrac{5}{16} + \dfrac{7}{16}$

4. $\dfrac{2}{6} + \dfrac{3}{6}$ 5. $\dfrac{3}{4} - \dfrac{1}{4}$ 6. $\dfrac{13}{16} - \dfrac{3}{16}$

7. $\dfrac{3}{5} - \dfrac{1}{5}$ 8. $\dfrac{5}{12} - \dfrac{2}{12}$ 9. $\dfrac{5}{16} + \dfrac{3}{16} + \dfrac{7}{16}$

10. $\dfrac{1}{8} + \dfrac{3}{8} + \dfrac{7}{8}$ 11. $1\dfrac{7}{8} - \dfrac{3}{8}$ 12. $3\dfrac{9}{16} - 1\dfrac{5}{16}$

13. $\dfrac{1}{4} + \dfrac{1}{2}$ 14. $\dfrac{7}{16} + \dfrac{3}{8}$ 15. $\dfrac{5}{8} + \dfrac{1}{12}$

16. $\dfrac{5}{12} + \dfrac{3}{16}$ 17. $\dfrac{1}{2} - \dfrac{3}{8}$ 18. $\dfrac{5}{16} - \dfrac{3}{32}$

19. $\dfrac{15}{16} - \dfrac{1}{2}$ 20. $\dfrac{7}{16} - \dfrac{1}{32}$ 21. $\dfrac{3}{5} + \dfrac{1}{8}$

22. $\dfrac{2}{3} + \dfrac{4}{5}$ 23. $\dfrac{7}{8} - \dfrac{2}{5}$ 24. $\dfrac{4}{9} - \dfrac{1}{4}$

25. $\dfrac{1}{2} + \dfrac{1}{4} - \dfrac{1}{8}$ 26. $\dfrac{11}{16} - \dfrac{1}{8} - \dfrac{1}{3}$ 27. $1\dfrac{1}{2} + \dfrac{1}{4}$

28. $2\dfrac{7}{16} + \dfrac{3}{4}$ 29. $2\dfrac{1}{2} + 1\dfrac{5}{8}$ 30. $2\dfrac{8}{32} + 1\dfrac{1}{10}$

31. $2\dfrac{1}{3} + 1\dfrac{1}{5}$ 32. $1\dfrac{7}{8} + \dfrac{1}{4}$ 33. $4\dfrac{1}{8} - 1\dfrac{3}{4}$

34. $5\dfrac{3}{4} - 2\dfrac{1}{12}$ 35. $3\dfrac{1}{5} - 2\dfrac{1}{12}$ 36. $5\dfrac{1}{3} - 2\dfrac{2}{5}$

B. Add or subtract as shown.

1. $8 - 2\dfrac{7}{8}$ 2. $3 - 1\dfrac{3}{16}$ 3. $3\dfrac{5}{8} - \dfrac{13}{16}$

4. $\dfrac{1}{2} + \dfrac{1}{3} + \dfrac{1}{4} + \dfrac{1}{5}$ 5. $\dfrac{1}{2} + \dfrac{1}{4} + \dfrac{1}{8}$ 6. $6\dfrac{1}{2} + 5\dfrac{3}{4} + 8\dfrac{1}{8}$

7. $\dfrac{7}{8} - 1\dfrac{1}{4} + 2\dfrac{1}{2}$ 8. $1\dfrac{3}{8}$ subtracted from $4\dfrac{3}{4}$

9. $2\dfrac{3}{16}$ less than $4\dfrac{7}{8}$ 10. $6\dfrac{2}{3}$ reduced by $1\dfrac{1}{4}$

11. $2\dfrac{3}{5}$ less than $6\dfrac{1}{2}$ 12. By how much is $1\dfrac{8}{7}$ larger than $1\dfrac{7}{8}$?

C. Practical Problems

1. **Machine Technology** The time sheet for operations on a machine part listed the following (in minutes): chucking, $\frac{3}{4}$; spotting and drilling, $3\frac{1}{3}$; facing, $1\frac{2}{3}$; grinding, $4\frac{1}{2}$; reaming, $\frac{2}{5}$. What was the total time for the operations?

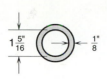

Problem 5

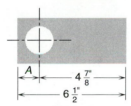

Problem 6

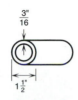

Problem 9

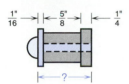

Problem 12

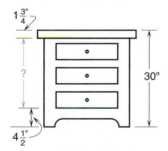

Problem 16

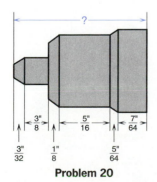

Problem 20

2. **Carpentry**   A countertop is made of $\frac{5}{8}$-in. particleboard and is covered with $\frac{3}{16}$-in. laminated plastic. What width of metal edging is needed to finish off the edge?

3. **Machine Technology**   The diameter of a steel shaft is reduced $\frac{7}{1000}$ in. The original diameter of the shaft was $\frac{850}{1000}$ in. What is the new diameter of the shaft?

4. **Plumbing**   If a piece of $\frac{3}{8}$-in.-i.d. (inside diameter) copper tubing measures $\frac{9}{16}$ in. o.d. (outside diameter), what is the wall thickness?

5. **Manufacturing**   What is the outside diameter of tubing whose inside diameter is $1\frac{5}{16}$ in. and whose wall thickness is $\frac{1}{8}$ in.?

6. **Machine Technology**   How long a bolt is needed to go through a piece of tubing $\frac{5}{8}$ in. long, a washer $\frac{1}{16}$ in. thick, and a nut $\frac{1}{4}$ in. thick?

7. **Office Services**   Newspaper ads are sold by the column inch (c.i.). What is the total number of column inches for a month in which a plumbing contractor has had ads of $6\frac{1}{2}$, $5\frac{3}{4}$, $3\frac{1}{4}$, $4\frac{3}{4}$, and 5 c.i.?

8. **Plumbing**   While installing water pipes, a plumber used pieces of pipe measuring $2\frac{3}{4}$, $4\frac{1}{3}$, $3\frac{1}{2}$, and $1\frac{1}{4}$ ft. How much pipe would remain if these pieces were cut from a 14-ft length of pipe? (Ignore waste in cutting.)

9. **Electrical Technology**   A piece of electrical pipe conduit has a diameter of $1\frac{1}{2}$ in. and a wall thickness of $\frac{3}{16}$ in. What is its inside diameter?

10. **Metalworking**   What is the total length of a certain machine part that is made by joining four pieces that measure $3\frac{1}{8}$, $1\frac{5}{32}$, $2\frac{7}{16}$, and $1\frac{1}{4}$ in.?

11. **Carpentry**   A blueprint requires four separate pieces of wood measuring $5\frac{3}{8}$, $8\frac{1}{4}$, $6\frac{9}{16}$, and $2\frac{5}{8}$ in. How long a piece of wood is needed to cut these pieces if we allow $\frac{1}{2}$ in. for waste?

12. **Drafting**   Find the missing dimension A in the drawing shown.

13. **Machine Technology**   Two splice plates are cut from a piece of sheet steel that has an overall length of $18\frac{5}{8}$ in. The plates are $9\frac{1}{4}$ in. and $6\frac{7}{16}$ in. long. How much material remains from the original piece if each saw cut removes $\frac{1}{16}$ in.?

14. **Printing**   A printer has $2\frac{3}{4}$ rolls of a certain kind of paper in stock. He must do three jobs that require $\frac{5}{8}$, $1\frac{1}{2}$, and $\frac{3}{4}$ roll, respectively. Does he have enough?

15. **Auto Mechanics**   The front and rear axles of a car need to be aligned. If the length measurement on one side of the car is 8 ft $5\frac{1}{4}$ in. and 8 ft $4\frac{7}{8}$ in. on the other side, how much shifting must be done to bring the axles into alignment?

16. **Woodworking**   A cabinet 30 in. high must have a $4\frac{1}{2}$-in. base and a $1\frac{3}{4}$-in. top. How much space is left for drawers?

17. **Printing**   Before it was trimmed, a booklet measured $8\frac{1}{4}$ in. high by $6\frac{3}{4}$ in. wide. If each edge of the height and one edge of the width were trimmed $\frac{1}{4}$ in., what is the finished size?

18. **Carpentry**   A wall has $\frac{1}{2}$-in. paneling covering $\frac{3}{4}$-in. drywall attached to a $3\frac{3}{4}$-in. stud. What is the total thickness of the three components?

19. **Machine Technology**   The large end of a tapered pin is $2\frac{15}{16}$ in. in diameter, while the small end is $2\frac{3}{8}$ in. in diameter. Calculate the difference to get the amount of taper.

20. **Machine Technology**   Find the total length of the metal casting shown.

Check your answers on page 585.

Turn to **41** for a set of problems on the arithmetic of fractions, with many practical applications.

Fractions

41. Answers are given on page 585.

A. Write as an improper fraction.

1. $1\frac{1}{8}$ 2. $4\frac{1}{5}$ 3. $1\frac{2}{3}$ 4. $2\frac{3}{16}$

5. $3\frac{3}{32}$ 6. $2\frac{1}{16}$ 7. $1\frac{5}{8}$ 8. $3\frac{7}{16}$

Write as a mixed number.

9. $\frac{10}{4}$ 10. $\frac{19}{2}$ 11. $\frac{25}{3}$ 12. $\frac{9}{8}$

13. $\frac{25}{16}$ 14. $\frac{21}{16}$ 15. $\frac{35}{4}$ 16. $\frac{7}{3}$

Write in lowest terms.

17. $\frac{6}{32}$ 18. $\frac{8}{32}$ 19. $\frac{12}{32}$ 20. $\frac{18}{24}$

21. $\frac{5}{30}$ 22. $1\frac{12}{21}$ 23. $1\frac{16}{20}$ 24. $3\frac{10}{25}$

Complete these.

25. $\frac{3}{4}=\frac{?}{12}$ 26. $\frac{7}{16}=\frac{?}{64}$ 27. $2\frac{3}{4}=\frac{?}{16}$ 28. $1\frac{3}{8}=\frac{?}{32}$

29. $5\frac{2}{3}=\frac{?}{12}$ 30. $1\frac{4}{5}=\frac{?}{10}$ 31. $1\frac{1}{4}=\frac{?}{12}$ 32. $2\frac{3}{5}=\frac{?}{10}$

Circle the larger number.

33. $\frac{7}{16}$ or $\frac{2}{15}$ 34. $\frac{2}{3}$ or $\frac{4}{7}$ 35. $\frac{13}{16}$ or $\frac{7}{8}$ 36. $1\frac{1}{4}$ or $\frac{7}{6}$

37. $\frac{13}{32}$ or $\frac{3}{5}$ 38. $\frac{2}{10}$ or $\frac{3}{16}$ 39. $1\frac{7}{16}$ or $\frac{7}{4}$ 40. $\frac{3}{32}$ or $\frac{1}{9}$

Name _____

Date _____

B. Multiply or divide as shown.

Course/Section _____

1. $\frac{1}{2}\times\frac{3}{16}$ 2. $\frac{3}{4}\times\frac{2}{3}$ 3. $\frac{7}{16}\times\frac{4}{3}$ 4. $\frac{15}{64}\times\frac{1}{12}$

5. $1\frac{1}{2} \times \frac{5}{6}$ **6.** $3\frac{1}{16} \times \frac{1}{5}$ **7.** $\frac{3}{16} \times \frac{5}{12}$ **8.** $14 \times \frac{3}{8}$

9. $\frac{3}{4} \times 10$ **10.** $\frac{1}{2} \times 1\frac{1}{3}$ **11.** $18 \times 1\frac{1}{2}$ **12.** $16 \times 2\frac{1}{8}$

13. $2\frac{2}{3} \times 4\frac{3}{8}$ **14.** $3\frac{1}{8} \times 2\frac{2}{5}$ **15.** $\frac{1}{2} \div \frac{1}{4}$ **16.** $\frac{2}{5} \div \frac{1}{2}$

17. $4 \div \frac{1}{8}$ **18.** $8 \div \frac{3}{4}$ **19.** $\frac{2}{3} \div 4$ **20.** $1\frac{1}{2} \div 2$

21. $3\frac{1}{2} \div 5$ **22.** $1\frac{1}{4} \div 1\frac{1}{2}$ **23.** $2\frac{3}{4} \div 1\frac{1}{8}$ **24.** $3\frac{1}{5} \div 1\frac{5}{7}$

C. Add or subtract as shown.

1. $\frac{3}{8} + \frac{7}{8}$ **2.** $\frac{1}{2} + \frac{3}{4}$ **3.** $\frac{3}{32} + \frac{1}{8}$ **4.** $\frac{3}{8} + 1\frac{1}{4}$

5. $\frac{3}{5} + \frac{5}{6}$ **6.** $\frac{5}{8} + \frac{1}{10}$ **7.** $\frac{9}{16} - \frac{3}{16}$ **8.** $\frac{7}{8} - \frac{1}{2}$

9. $\frac{11}{16} - \frac{1}{4}$ **10.** $\frac{5}{6} - \frac{1}{5}$ **11.** $\frac{7}{8} - \frac{3}{10}$ **12.** $1\frac{1}{2} - \frac{3}{32}$

13. $2\frac{1}{8} + 1\frac{1}{4}$ **14.** $1\frac{5}{8} + \frac{13}{16}$ **15.** $6 - 1\frac{1}{2}$ **16.** $3 - 1\frac{7}{8}$

17. $3\frac{2}{3} - 1\frac{7}{8}$ **18.** $2\frac{1}{4} - \frac{5}{6}$ **19.** $\frac{1}{2} + \frac{1}{3} + \frac{1}{5}$ **20.** $1\frac{1}{2} + 1\frac{1}{4} + 1\frac{1}{5}$

21. $3\frac{1}{2} - 2\frac{1}{3}$ **22.** $2\frac{3}{5} - 1\frac{4}{15}$ **23.** $2 - 1\frac{3}{5}$ **24.** $4\frac{5}{6} - 1\frac{1}{2}$

D. Practical Problems

1. **Welding**   In a welding job three pieces of 2-in. I-beam with lengths $5\frac{7}{8}$, $8\frac{1}{2}$, and $22\frac{3}{4}$ in. are needed. What is the total length of I-beam needed? (Do not worry about the waste in cutting.)

2. **Machine Technology**   How many pieces of $10\frac{5}{16}$-in. bar can be cut from a stock 20-ft bar? The metal is torch cut and allowance of $\frac{3}{16}$ in. kerf (waste) should be made for each piece. (*Hint:* 20 ft = 240 in.)

3. **Building Construction**   If an I-beam is to be $24\frac{3}{8}$ in. long with a tolerance of $\pm 1\frac{1}{4}$ in., find the longest and shortest acceptable lengths.

4. **Auto Mechanics**   In squaring a damaged auto frame, the mechanic measured the diagonals as $77\frac{3}{16}$ and $69\frac{5}{8}$ in. How much is the difference to be equalized?

5. **Machine Technology**   A shaft $1\frac{7}{8}$ in. in diameter is turned down on a lathe to a diameter of $1\frac{3}{32}$ in. What is the difference in diameters?

6. **Machine Technology**   A bar $14\frac{5}{16}$ in. long is cut from a piece $25\frac{1}{4}$ in. long. If $\frac{3}{32}$ in. is wasted in cutting, will there be enough left to make another bar $10\frac{3}{8}$ in. long?

7. **Manufacturing**   A cubic foot contains roughly $7\frac{1}{2}$ gallons. How many cubic feet are there in a tank containing $34\frac{1}{2}$ gallons?

8. **Manufacturing**   Find the total width of the three pieces of steel plate shown.

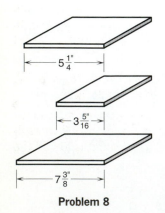

$5\frac{1}{4}''$

$3\frac{5}{16}''$

$7\frac{3}{8}''$

Problem 8

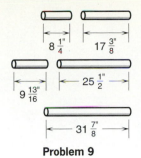

Problem 9

9. **Machine Technology** What would be the total length of the bar formed by welding together the five pieces of bar stock shown?

10. **Machine Technology** The Ace Machine Shop has the job of producing 32 zinger bars. Each zinger bar must be turned on a lathe from a piece of stock $4\frac{7}{8}$ in. long. How many feet of stock will they need?

11. **Carpentry** What is the thickness of a tabletop made of $\frac{3}{4}$-in. plywood and covered with a $\frac{3}{16}$-in. sheet of glass?

12. **Building Construction** For the wooden form shown, find the lengths A, B, C, and D.

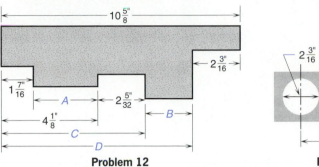

Problem 12

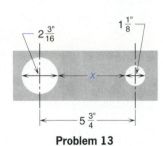

Problem 13

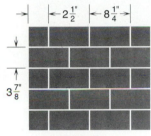

Problem 14

13. **Carpentry** Find the spacing x between the holes.

14. **Masonry** Find the height of the five-course (five-bricks-high) brick wall shown if each brick is $2\frac{1}{2}$ in. by $3\frac{7}{8}$ in. by $8\frac{1}{4}$ in. and all mortar joints are $\frac{1}{2}$ in.

15. **Masonry** If the wall in problem 14 has 28 stretchers (bricks laid lengthwise), what is its length?

16. **Electrical Technology** An electrical wiring job requires the following lengths of 14/2 BX cable: seven pieces each $6\frac{1}{2}$ ft long, four pieces each $34\frac{3}{4}$ in. long, and nine pieces $19\frac{3}{8}$ in. long. What is the total length of cable needed?

17. **Printing** An invitation must be printed on card stock measuring $4\frac{1}{4}$ in. wide by $5\frac{1}{2}$ in. long. The printed material covers a space measuring $2\frac{1}{8}$ in. wide by $4\frac{1}{8}$ in. long. If the printed material is centered in both directions, what are the margins?

18. **Printing** As a rule of thumb, the top margin of a page of a book should be $\frac{2}{5}$ of the total margin, and the bottom margin should be $\frac{3}{5}$ of the total margin. If the print takes up $9\frac{1}{2}$ in. of an 11-in.-long page, what should the top and bottom margins be? (*Hint:* The total margin = 11 in. − $9\frac{1}{2}$ in. = $1\frac{1}{2}$ in.)

19. **Welding** A 46-in. bar must have 9 equally spaced holes drilled through the centerline. If the centers of the two end holes are each $2\frac{1}{4}$ in. in from their respective ends, what should the center-to-center distance of the holes be? (*Hint:* There are 8 spaces between holes.)

20. **Welding** A piece of metal must be cut to a length of $22\frac{3}{8}$ in. ± $\frac{1}{16}$ in. What are the longest and shortest acceptable lengths? (*Hint:* The symbol ± means to add $\frac{1}{16}$ in. to get the longest length and subtract $\frac{1}{16}$ in. to get the shortest length. Longest = $22\frac{3}{8}$ in. + $\frac{1}{16}$ in. = ? Shortest = $22\frac{3}{8}$ in. − $\frac{1}{16}$ in. = ?)

21. **Machine Technology** If a shaft turns at 18 revolutions per minute, and the tool feed is $\frac{1}{64}$ in. per revolution, how long will it take to advance $7\frac{1}{2}$ in.?

22. **Sheet Metal Technology** The total allowance for both edges of a grooved seam is three times the width of the seam. Half of this total is added to each edge of the seam. Find the allowance for each edge of a grooved seam if the width of the seam is $\frac{5}{16}$ in.

PREVIEW
3

Decimal Numbers

Objective	Sample Problems		Where to Go for Help	
			Page	Frame
When you finish this unit you will be able to:				
1. Add, subtract, multiply, and divide decimal numbers.	(a) $5.82 + 0.096$	_____	97	**6**
	(b) $3.78 - 0.989$	_____	99	**7**
	(c) $27 - 4.03$	_____		
	(d) 7.25×0.301	_____	104	**10**
	(e) $104.2 \div 0.032$	_____	107	**13**
	(f) $0.09 \div 0.0004$	_____		
	(g) $20.4 \div 6.7$ (round to three decimal places)	_____		
2. Find averages.	Find the average of 4.2, 4.8, 5.7, 2.5, 3.6, 5.0	_____	114	**20**
3. Work with decimal fractions.	(a) Write as a decimal number $\frac{3}{16}$.	_____	117	**23**
	(b) Write as a decimal number: One hundred six and twenty-seven thousandths	_____	95	**2**
	(c) Write in words 26.035	_____	93	**1**
	(d) $1\frac{2}{3} + 1.785$	_____		
	(e) $4.1 \times 2\frac{1}{4}$	_____		
	(f) $1\frac{5}{16} \div 4.3$ (round to three decimal places)	_____		

Name _____

Date _____

Course/Section _____

4. Solve practical problems involving decimal numbers.

(a) **Electrician** Six recessed lights must be installed in a strip of ceiling. The cans cost $16.45 each, the fixtures run $8.75 each, the bulbs are $5.20 each, miscellaneous hardware and wiring amounts to $23.45, and labor is estimated at 3.25 hours at $32.50 per hour. What will the cost of the job be?

104   10

_____

(b) **Machinist** A container of 175 bolts weighs 61.3 lb. If the container itself weighs 1.8 lb, how much does each bolt weigh?

107   13

_____

(Answers to these preview problems are given on page 577.)

If you are certain that you can work *all* these problems correctly, turn to page 127 for a set of practice problems. If you cannot work one or more of the preview problems, turn to the page indicated to the right of the problem. Those who wish to master this material with the greatest success should turn to frame 1 and begin work there.

3

Decimal Numbers

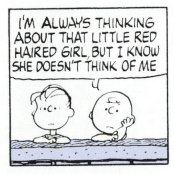

PEANUTS reprinted by permission of UFS, Inc.

3-1 ADDITION AND SUBTRACTION OF DECIMALS

Place Value of Decimal Numbers

1   By now you know that whole numbers are written in a form based on powers of ten. A number such as

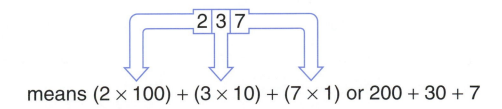

means $(2 \times 100) + (3 \times 10) + (7 \times 1)$ or $200 + 30 + 7$

This way of writing numbers can be extended to fractions. A *decimal* number is a fraction whose denominator is 10 or some multiple of 10.

A decimal number may have both a whole-number part and a fraction part. For example, the number 324.576 means

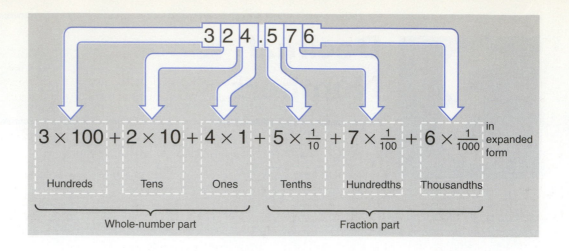

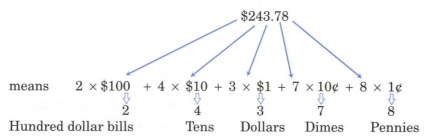

You are already familiar with this way of interpreting decimal numbers from working with money.

$$\$243.78$$

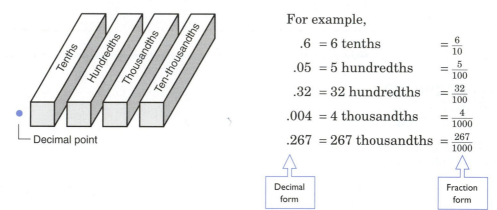

means   2 × $100   + 4 × $10 + 3 × $1 + 7 × 10¢ + 8 × 1¢

   2   4   3   7   8

Hundred dollar bills   Tens   Dollars   Dimes   Pennies

To write a decimal number in words, remember this diagram:

For example,

$$.6 \ = 6 \text{ tenths} \qquad = \frac{6}{10}$$

$$.05 \ = 5 \text{ hundredths} \qquad = \frac{5}{100}$$

$$.32 \ = 32 \text{ hundredths} \qquad = \frac{32}{100}$$

$$.004 \ = 4 \text{ thousandths} \qquad = \frac{4}{1000}$$

$$.267 \ = 267 \text{ thousandths} = \frac{267}{1000}$$

Decimal form   Fraction form

Write the following decimal numbers in words.

(a) 0.56   (b) 19.278   (c) 6.4

(d) 0.07   (e) 5.064   (f) 0.0018

Check your answers in **2**.

---

2 (a) Fifty-six hundredths

(b) Nineteen and two hundred seventy-eight thousandths

(c) Six and four tenths

(d) Seven hundredths

(e) Five and sixty-four thousandths

(f) Eighteen ten-thousandths

Learning Help ▶ The word **and** represents the decimal point. Everything preceding **and** is the whole-number part, and everything after **and** is the decimal part of the number. ◀

It is also important to be able to perform the reverse process—that is, write a decimal in numerical form if it is given in words. For example, to write "six and twenty-four thousandths" as a numeral,

First, write the whole-number part (six). This is the part before the "and." Follow this by a decimal point—the "and."

6.

Next, draw as many blanks as the decimal part indicates. In this case, the decimal part is "thousandths," so we allow for three decimal places. Draw three blanks.

6.__ __ __

Finally, write a number giving the decimal (twenty-four). Write it so it *ends* on the far right blank. Fill in any blank decimal places with zeros.

6.__ 2 4

6.024

Now you try it. Write each of the following as decimal numbers.

(a) Five thousandths

(b) One hundred and six tenths

(c) Two and twenty-eight hundredths

(d) Seventy-one and sixty-two thousandths

(e) Three and five hundred eighty-nine ten-thousandths

Check your work in **3**.

---

3 (a) 0.005   (b) 100.6   (c) 2.28   (d) 71.062   (e) 3.0589

Expanded Form   The decimal number .267 can be written in expanded form as

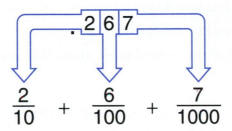

$$\frac{2}{10} \; + \; \frac{6}{100} \; + \; \frac{7}{1000}$$

Write the decimal number .526 in expanded form. Check your answer in **4**.

---

4 $.526 = \dfrac{5}{10} + \dfrac{2}{100} + \dfrac{6}{1000}$

We usually write a decimal number less than 1 with a zero to the left of the decimal point.

.526 would be written 0.526

.4   would be written 0.4

.001 would be written 0.001

It is easy to mistake .4 for 4, but the decimal point in 0.4 cannot be overlooked. That zero out front will help you remember where the decimal point is located.

To help get these ideas clear in your mind, write the following in expanded form.

(a) 86.42   (b) 43.607   (c) 14.5060   (d) 235.22267

Compare your answers with ours in **5**.

---

5 (a) 86.42 $= 8 \times 10 + 6 \times 1 + 4 \times \frac{1}{10} + 2 \times \frac{1}{100}$

                   $= 80 \quad + \quad 6 \quad + \quad \frac{4}{10} \quad + \quad \frac{2}{100}$

(b) 43.607 $= 4 \times 10 + 3 \times 1 + 6 \times \frac{1}{10} + 0 \times \frac{1}{100} + 7 \times \frac{1}{1000}$

                   $= 40 \quad + \quad 3 \quad + \quad \frac{6}{10} \quad + \quad \frac{0}{100} \quad + \quad \frac{7}{1000}$

(c) 14.5060 $= 10 + 4 + \frac{5}{10} + \frac{0}{100} + \frac{6}{1000} + \frac{0}{10000}$

(d) 235.22267 $= 200 + 30 + 5 + \frac{2}{10} + \frac{2}{100} + \frac{2}{1000} + \frac{6}{10000} + \frac{7}{100000}$

Learning Help ▶ Notice that the denominators in the decimal fractions change by a factor of 10. For example,

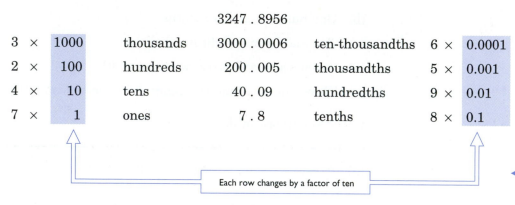

Decimal Digits   In the decimal number 86.423 the digits 4, 2, and 3 are called *decimal digits*.

The number 43.6708 has four decimal digits: 6, 7, 0, and 8.

The number 5376.2 has one decimal digit: 2.

All digits to the right of the decimal point, those that name the fractional part of the number, are decimal digits.

How many decimal digits are included in each of these numbers?

(a) 1.4   (b) 315.7   (c) 0.425   (d) 324.0075

Count them, then turn to **6**.

---

6 (a) one   (b) one   (c) three   (d) four

We will use the idea of decimal digits often in doing arithmetic with decimal numbers.

The decimal point is simply a way of separating the whole-number part from the fraction part. It is a place marker. In whole numbers the decimal point usually is not written, but it is understood to be there.

The whole number 2 is written 2. as a decimal.

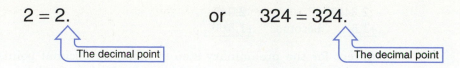

$$2 = 2. \qquad \text{or} \qquad 324 = 324.$$

The decimal point   The decimal point

This is very important. Many people make big mistakes in arithmetic because they do not know where that decimal point should go.

Very often, additional zeros are attached to the decimal number without changing its value. For example,

$$8.5 = 8.50 = 8.5000 \qquad \text{and so on}$$

$$6 = 6. = 6.0 = 6.000 \qquad \text{and so on}$$

The value of the number is not changed, but the additional zeros may be useful, as we shall see.

Addition   Because decimal numbers represent fractions with denominators equal to multiples of ten, addition is very simple.

$$2.34 = 2 + \frac{3}{10} + \frac{4}{100}$$
$$+5.23 = 5 + \frac{2}{10} + \frac{3}{100}$$
$$\overline{7 + \frac{5}{10} + \frac{7}{100}} = 7.57$$

Adding like fractions

Of course, we do not need this clumsy business in order to **add decimal numbers.** As with whole numbers, we may arrange the digits in vertical columns and add directly. Let's add 1.45 + 3.42

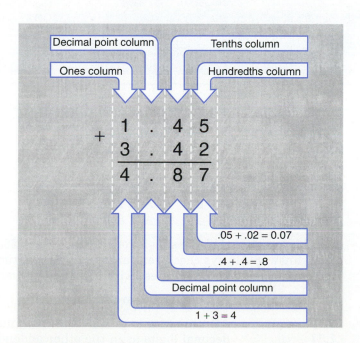

Decimal point column   Tenths column

Ones column   Hundredths column

$$
\begin{array}{r}
1 \,.\, 4 \;\; 5 \\
+\; 3 \,.\, 4 \;\; 2 \\
\hline
4 \,.\, 8 \;\; 7
\end{array}
$$

.05 + .02 = 0.07

.4 + .4 = .8

Decimal point column

1 + 3 = 4

Digits of the same power of ten are placed in the same vertical column. Decimal points are always lined up vertically.

If one of the numbers is written with fewer decimal digits than the other, attach as many zeros as needed so that both have the same number of decimal digits.

$$\begin{array}{r} 2.345 \\ +1.5 \\ \hline \end{array} \quad \text{becomes} \quad \begin{array}{r} 2.345 \\ +1.500 \\ \hline \end{array}$$

Except for the preliminary step of lining up decimal points, addition of decimal numbers is exactly the same process as addition of whole numbers.

Add the following decimal numbers.

(a) $4.02 + $3.67   = _____

(b) 13.2 + 1.57   = _____

(c) 23.007 + 1.12   = _____

(d) 14.6 + 1.2 + 3.15   = _____

(e) 5.7 + 3.4   = _____

(f) $9 + $0.72 + $6.09 = _____

(g) 0.07 + 6.79 + 0.3 + 3 = _____

(h) A sheet metal worker uses 2.36, 7, 3.9, and 0.6 ounces of cleaning concentrate on successive jobs. What total amount of concentrate has he used during this time?

Arrange each sum vertically, placing the decimal points in the same column, then add as with whole numbers.

Check your work in **7**.

What do you mean by "one decimal place"?

One decimal digit or one digit to the right of the decimal point. 4.2 has one decimal digit, 4.23 has two, 4.234 has three, . . . Got it?

7

┌──── Decimal points are lined up vertically.

(a) $4.02
     $3.67
     $7.69 ┌─0.02 + 0.07 = 0.09   Add cents.
            └─0.0 + 0.6 = 0.6   Add 10-cent units.
              ─ 4 + 3   = 7   Add dollars.

As a check, notice that the sum is roughly $4 + $4 or $8, which agrees with the actual answer. Always check your answer by first estimating it, then comparing your estimate or rough guess with the final answer.

┌──────── Decimal points are in line.

(b) 13.20 ◄──────── Attach a zero to provide the same number of
      1.57   decimal digits as in the other addend.
     14.77
       │
       └──────── Place answer decimal point in the same vertical line.

Check: $13 + 2 = 15$, which agrees roughly with the answer.

(c)
$$\begin{array}{r} 23.007 \\ +\ 1.120 \\ \hline 24.127 \end{array}$$
← Attach extra zero.

(d)
$$\begin{array}{r} 14.60 \\ 1.20 \\ +\ 3.15 \\ \hline 18.95 \end{array}$$
Attach extra zeros.

(e)
$$\begin{array}{r} \overset{1}{5.7} \\ +\ 3.4 \\ \hline 9.1 \end{array}$$
$0.7 + 0.4 = 1.1$ Write 0.1.
Carry 1.
Add: $1 + 5 + 3 = 9$

(f)
$$\begin{array}{r} \overset{1}{\$\ 9.00} \\ 0.72 \\ 6.09 \\ \hline \$15.81 \end{array}$$

(g)
$$\begin{array}{r} \overset{1\ 1}{0.07} \\ 6.79 \\ 0.30 \\ 3.00 \\ \hline 10.16 \end{array}$$
Attach extra zeros.

(h)
$$\begin{array}{r} \overset{1}{2.36} \\ 7.00 \\ 3.90 \\ 0.60 \\ \hline 13.86 \end{array}$$
Attach extra zeros.

Careful ▶ You must line up the decimal points vertically to be certain of getting a correct answer. ◀

Subtraction Subtraction is equally simple if you line up the decimal points carefully and attach any needed zeros before you begin work. As in the subtraction of whole numbers, you may need to "borrow" to complete the calculation.

For example, $\$437.56 - \$41 = $ _____ is

$$\begin{array}{r} \overset{313}{\$4\cancel{3}7.56} \\ -\$\ 41.00 \\ \hline \$396.56 \end{array}$$
Decimal points are in a vertical line.
Attach zeros (remember that $41 is $41. or $41.00).

or again $19.452 - 7.3617 = $ _____

$$\begin{array}{r} \overset{315110}{19.\cancel{4}520} \\ -\ 7.3617 \\ \hline 12.0903 \end{array}$$
Decimal points are in a vertical line.
Attach zero.

Answer decimal point is in same vertical line.

Try these problems to test yourself on the subtraction of decimal numbers.

(a) $\$37.66 - \$14.57 = $ _____   (b) $248.3 - 135.921 = $ _____

(c) $6.4701 - 3.2 = $ _____   (d) $7.304 - 2.59 = $ _____

(e) $\$20 - \$7.74 = $ _____   (f) $36 - 11.132 = $ _____

(g) If 0.037 in. of metal is machined from a rod exactly 10 in. long, what is the new length of the rod?

Work carefully. The answers are given in **8**.

8

(a)
$$\begin{array}{r} \overset{516}{\$37.66} \\ -\$14.57 \\ \hline \$23.09 \end{array}$$
Line up decimal points.

Check: $\$14.57 + \$23.09 = \$37.66$

(b)
$$\overset{7\ 12910}{\cancel{2}\cancel{4}\cancel{8}.\cancel{3}00}$$
    248.300 ◄─┐ Line up decimal points.
   −135.921   └─ Attach zeros.
    112.379   **Check:** 135.921 + 112.379 = 248.300
        ↑_____ Answer decimal point is in the same vertical line.

(c)
    6.4701
   −3.2000 ◄─── Attach zeros.
    3.2701   **Check:** 3.2000 + 3.2701 = 6.4701

(d)
$$\overset{6\ 1210}{7.\cancel{3}\cancel{0}\cancel{4}}$$
    7.304
   −2.590 ◄─── Attach zero.
    4.714   **Check:** 2.590 + 4.714 = 7.304

(e)
$$\overset{19\ 910}{\$\cancel{2}\cancel{0}.\cancel{0}\cancel{0}}$$
    $20.00 ◄─── Attach zeros.
   −$ 7.74
    $12.26   **Check:** $7.74 + $12.26 = $20.00

(f)
$$\overset{5\ 9\ 910}{\cancel{3}\cancel{6}.\cancel{0}\cancel{0}0}$$
    36.000 ◄─── Attach zeros.
   −11.132
    24.868   **Check:** 11.132 + 24.868 = 36.000

(g)
$$\overset{9\ 9\ 910}{\cancel{1}\cancel{0}.\cancel{0}\cancel{0}0}$$
    10.000 ◄─── Attach zeros.
   − 0.037
    9.963   **Check:** 0.037 + 9.963 = 10.000

Notice that each problem is checked by comparing the sum of the answer and the number subtracted with the first number. You should also start by estimating the answer. Whether you use a calculator or work it out with pencil and paper, checking your answer is important if you are to avoid careless mistakes.

Learning Help ▶

The balancing method used to add and subtract whole numbers mentally can also be used with decimal numbers.

For example, to add 12.8 and 6.3 mentally, first increase 12.8 by 0.2 to get a whole number 13. To keep the total the same, balance by subtracting 0.2 from 6.3 to get 6.1. Now 12.8 + 6.3 = 13.0 + 6.1, which is easy to do mentally. The answer is 19.1.

To subtract 34.6 − 18.7, add 0.3 to 18.7 to get the whole number 19. Balance this by adding 0.3 to 34.6 to get 34.9. Now 34.6 − 18.7 = 34.9 − 19.0. If this is still too tough to do mentally, balance again to make it 35.9 − 20.0 or 15.9. ◄

Now, for a set of practice problems on addition and subtraction of decimal numbers, turn to **9**.

9
Exercises 3-1   **Addition and Subtraction of Decimals**

A. Write in words.

1. 0.72	2. 8.7	3. 12.36	4. 0.05
5. 3.072	6. 14.091	7. 3.0024	8. 6.0083

Write as a number.

9. Four thousandths
10. Three and four tenths
11. Six and seven tenths
12. Five thousandths
13. Twelve and eight tenths
14. Three and twenty-one thousandths
15. Ten and thirty-two thousandths
16. Forty and seven tenths
17. One hundred sixteen ten-thousandths
18. Forty-seven ten-thousandths
19. Two and three hundred seventy-four ten-thousandths
20. Ten and two hundred twenty-two ten-thousandths

B. Add or subtract as shown.

1. $14.21 + 6.8$
2. $75.6 + 2.57$
3. $\$2.83 + \$12.19$
4. $\$52.37 + \$98.74$
5. $0.687 + 0.93$
6. $0.096 + 5.82$
7. $507.18 + 321.42$
8. $212.7 + 25.46$
9. $45.6725 + 18.058$
10. $390 + 72.04$
11. $19 - 12.03$
12. $7.83 - 6.79$
13. $\$33.40 - \$18.04$
14. $\$20.00 - \$13.48$
15. $75.08 - 32.75$
16. $40 - 3.82$
17. $\$30 - \$7.98$
18. $\$25 - \$0.61$
19. $130 - 16.04$
20. $19 - 5.78$
21. $37 + 0.09 + 3.5 + 4.605$
22. $183 + 3.91 + 45 + 13.2$
23. $\$14.75 + \$9 + \$3.76$
24. $148.002 + 3.4$
25. $68.708 + 27.18$
26. $35.36 + 4.347$
27. $47.04 - 31.88$
28. $180.76 - 94.69$
29. $26.45 - 17.832$
30. $92.302 - 73.6647$
31. $6.4 + 17.05 + 7.78$
32. $212.4 + 76 + 3.79$
33. $26.008 - 8.4$
34. $36.4 - 7.005$
35. $0.0046 + 0.073$
36. $0.038 + 0.00462$
37. $28.7 - 7.38 + 2.9$
38. $0.932 + 0.08 - 0.4$
39. $6.01 - 3.55 - 0.712$
40. $2.92 - 1.007 - 0.08$

C. Practical Problems

1. **Machine Technology**   What is the combined thickness of these five shims: 0.008, 0.125, 0.150, 0.185, and 0.005 in.?

2. **Electrical Technology**   The combined weight of a spool and the wire it carries is 13.6 lb. If the weight of the spool is 1.75 lb, what is the weight of the wire?

3. **Electrical Technology**   The following are diameters of some common household wires: No. 10 is 0.102 in., No. 11 is 0.090 in., No. 12 is 0.081 in., No. 14 is 0.064 in., and No. 16 is 0.051 in.

(a) The diameter of No. 16 wire is how much smaller than the diameter of No. 14 wire?

(b) Is No. 12 wire larger or smaller than No. 10 wire? What is the difference in their diameters?

(c) John measured the thickness of a wire with a micrometer as 0.066 in. Assuming that the manufacturer was slightly off, what wire size did John have?

4. **Painting and Decorating** In estimating a finishing job, a painter included the following items:

Material $377.85
Trucking $ 15.65
Profit $250
Labor $745.50
Overhead $ 37.75

What was his total estimate for the job?

5. **Metalworking** Find A, B, and C.

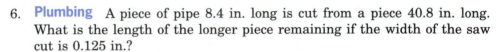

Problem 5

6. **Plumbing** A piece of pipe 8.4 in. long is cut from a piece 40.8 in. long. What is the length of the longer piece remaining if the width of the saw cut is 0.125 in.?

7. **Machine Technology** A certain machine part is 2.345 in. thick. What is its thickness after 0.078 in. is ground off?

8. **Masonry** Find the total cost of the materials for a certain masonry job if sand cost $43.65, cement cost $214.47, a steel lintel cost $32.65, and brick cost $476.28.

9. **Masonry** The specifications for a reinforced masonry wall called for 1.5 sq in. of reinforcing steel per square foot of cross-sectional area. If the three pieces of steel being used had cross sections of 0.125, 0.2, and 1.017 sq in., did they meet the specifications for a 1-sq-ft area?

10. **Building Construction** A plot plan of a building site showed that the east side of the house was 46.35 ft from the east lot line, and the west side of the house was 41.65 ft from the west lot line. If the lot was 156 ft wide along the front, how wide is the house?

11. **Auto Mechanics** A mechanic must estimate the total time for a particular servicing. He figures 0.3 hour for an oil change, 1.5 hours for a tune-up, 0.4 hour for a brake check, and 1.2 hours for air-conditioning service. What is the total number of hours of his estimate?

12. **Auto Mechanics** A heated piston measures 8.586 cm in diameter. When cold it measures 8.573 cm in diameter. How much does it expand when heated?

13. **Auto Mechanics** A piston must fit in a bore with a diameter of 3.375 in. What must be the diameter of the piston given the clearance shown in the diagram?

14. **Auto Mechanics** In squaring a damaged car frame, an autobody worker measured the diagonals between the two front cross-members. One diag-

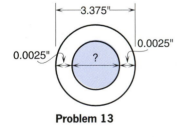

Problem 13

onal was 196.1 cm and the other was 176.8 cm. What is the difference that must be adjusted?

D. Calculator Problems (If you need help using a calculator with decimals, see **30** on page 131.)

1. Balance this checkbook by finding the closing balance as of November 4.

Date	Balance	Withdrawals	Deposits
Oct. 1	$367.21		
Oct. 3		$167.05	
Oct. 4		104.97	
Oct. 8			$357.41
Oct. 16		87.50	
Oct. 18		9.43	
Oct. 20		30.09	
Oct. 22			364.85
Oct. 27		259.47	
Oct. 30		100.84	
Nov. 2		21.88	
Nov. 4	?		

2. What is the actual cost of the following car?

"Sticker price"	$17,745.00
Dealer preparation	206.50
Leather interior	875.40
Antilock brakes	939.00
CD player	419.95
Moonroof	735.50
Tax and license	618.62
Less trade-in	$ 1780.00

3. Add as shown.

(a)
```
0.0067
0.032
0.0012
0.0179
0.045
0.5
0.05
0.0831
0.004
```

(b)
```
1379.4
204.5
16.75
300.04
2070.08
167.99
43.255
38.81
19.95
```

(c)
```
14.07
67.81
132.99
225.04
38.02
4
16.899
7.007
4.6
```

(d) $0.002 + 17.1 + 4.806 + 9.9981 - 3.1 + 0.701 - 1.001 - 14 - 8.09 + 1.0101$

Check your answers in page 586, then continue in **10**.

10 A decimal number is really a fraction with a multiple of 10 as denominator. For example,

$$0.5 = \frac{5}{10} \qquad 0.3 = \frac{3}{10} \qquad \text{and} \qquad 0.85 = \frac{85}{100}$$

Multiplication of decimals is easy to understand if we think of it in this way:

$$0.5 \times 0.3 = \frac{5}{10} \times \frac{3}{10} = \frac{15}{100} = 0.15$$

Learning Help ▶ To estimate the product of 0.5 and 0.3, remember that if two numbers are both less than 1, their product must be less than 1. ◀

Multiplication Of course it would be very, very clumsy and time consuming to calculate every decimal multiplication this way. We need a simpler method. Here is the procedure most often used:

Step 1 Multiply the two decimal numbers as if they were whole numbers. Pay no attention to the decimal points.

Step 2 The sum of the decimal digits in the two numbers being multiplied will give you the number of decimal digits in the answer.

Example: Multiply 3.2 by 0.41.

Step 1 Multiply, ignoring the decimal points.

$$\begin{array}{r} 32 \\ \times\ 41 \\ \hline 1312 \end{array}$$

Step 2 Count decimal digits in each number: 3.2 has *one* decimal digit (the 2), and 0.41 has *two* decimal digits (the 4 and the 1). The total number of decimal digits in the two factors is three. The answer will have *three* decimal digits. Count over *three* digits from right to left in the answer.

1.312   three decimal digits

Check: 3.2×0.41 is roughly $3 \times \frac{1}{2}$ or about $1\frac{1}{2}$. The answer 1.312 agrees with our rough guess. Remember, even if you use a calculator to do the actual work of arithmetic, you must *always* estimate your answer first and check it afterward.

Try these simple decimal multiplications.

(a) $2.5 \times 0.5 = $ _____

(b) $0.1 \times 0.1 = $ _____

(c) $10 \times 0.6 = $ _____

(d) $2 \times 0.4 = $ _____

(e) $2 \times 0.003 = $ _____

(f) $0.01 \times 0.02 = $ _____

(g) $0.04 \times 0.005 = $ _____

(h) If asphalt tile weighs 1.1 lb per square foot, what is the weight of tile covering 8 sq ft?

Check your answers in **11**.

11 (a) 2.5 × 0.5 = _____

First multiply 25 × 5 = 125. Second, count decimal digits.
2.5 × 0.5

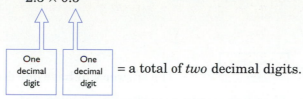

= a total of *two* decimal digits.

Count over *two* decimal digits from the right: 1.25 two decimal digits.

The product is 1.25.

Check: $2 \times \frac{1}{2}$ is about 1, so the answer seems reasonable.

(b) 0.1 × 0.1   1 × 1 = 1

Count over *two* decimal digits from the right. Since there are not two decimal digits in the product, attach a zero on the left: 0.01

So 0.1 × 0.1 = 0.01.

Check: $\frac{1}{10} \times \frac{1}{10} = \frac{1}{100}$.

Two decimal digits

(c) 10 × 0.6   10 × 6 = 60

Count over *one* decimal digit from the right: 6.0
So 10 × 0.6 = 6.0.

Learning Help ▶ Notice that multiplication by 10 simply shifts the decimal place one digit to the right.

10 × 6.2   = 62
10 × 0.075 = 0.75
10 × 8.123 = 81.23   and so on ◀

(d) 2 × 0.4   2 × 4 = 8

Count over *one* decimal digit: .8

2 × 0.4 = 0.8

(e) 2 × 0.003   2 × 3 = 6

Three decimal digits

Count over three decimal digits. Attach two zeros as place holders.

0.006   2 × 0.003 = 0.006

(f) 0.01 × 0.02   1 × 2 = 2

Two decimal digits   Two decimal digits

Total of four decimal digits

Count over four decimal digits. Attach three zeros.

0.0002   0.01 × 0.02 = 0.0002

Four decimal digits

(g) 0.04 × 0.005   4 × 5 = 20

Two decimal digits   Three decimal digits

Total of five decimal digits

Count over five decimal digits. Attach three zeros.

0.00020   0.04 × 0.005 = 0.00020 = 0.0002

Five decimal digits

(h) We are given the weight of 1 sq ft and we need to calculate the weight of 8 sq ft. Multiply

1.1×8 $11 \times 8 = 88$

One decimal digit

The answer will have *one* decimal digit: $1.1 \times 8 = 8.8$

The weight of 8 sq ft of tile is 8.8 lb.

Learning Help ▶

1. Do not try to do this entire process mentally until you are certain you will not misplace zeros.

2. Always estimate before you begin the arithmetic, and check your answer against your estimate. ◀

Multiplication of larger decimal numbers is performed in exactly the same manner. Try these:

(a) $4.302 \times 12.05 =$ _____   (b) $6.715 \times 2.002 =$ _____

(c) $3.144 \times 0.00125 =$ _____

(d) How much does 25.8 sq ft of heavy plastic sheeting weigh if 1 sq ft weighs 2.37 lb?

Look in **12** for the answers.

---

12 (a) **Estimate:** $4 \times 12 = 48$ The answer will be about 48.

Multiply: 4302
           $\times\, 1205$
         _____
           5183910

(If you cannot do this multiplication correctly, turn to page 26 in Section 1-3 for help with the multiplication of whole numbers.)

The two factors being multiplied have a total of five decimal digits (three in 4.302 and two in 12.05). Count over five decimal digits from the right in the answer.

51.83910

So that $4.302 \times 12.05 = 51.83910 = 51.8391$

Check: The answer 51.8391 is approximately equal to the estimate of 48.

(b) **Estimate:** 6.7×2 is about 7×2 or 14.

Multiply: 6.715   6.715 has *three* decimal digits
            $\times\, 2.002$ 2.002 has *three* decimal digits
          _____   a total of *six* decimal digits
           13.443430

six decimal digits

$6.715 \times 2.002 = 13.44343$

Check: The answer and estimate are roughly equal.

(c) **Estimate:** 3×0.001 is about 0.003

Multiply:

$$\begin{array}{r} 3.144 \\ \times\, 0.00125 \\ \hline .00393000 \end{array}$$

3.144 has *three* decimal digits
0.00125 has *five* decimal digits
a total of *eight* decimal digits

eight decimal digits   In order to count over 8 digits we had to attach 2 zeros on the left.

$$3.144 \times 0.00125 = 0.00393$$

Check: The answer and estimate are roughly equal.

(d) **Estimate:** 25.8×2.37 is about 30×2 or 60 lb.

Multiply:

$$\begin{array}{r} 25.8 \\ \times\, 2.37 \\ \hline 61.146 \end{array}$$

one decimal digit
two decimal digits
three decimal digits

Check: The answer and estimate are roughly equal.

Note ▶   When multiplying, you need not line up the decimal points. ◀

One nice thing about a calculator is that it counts decimal digits and automatically adjusts the answer in any calculation. But to get an estimate of the answer, you must still understand the process.

Now turn to **13** for a look at the division of decimal numbers.

Division

13   Division of decimal numbers is very similar to the division of whole numbers. For example,

$6.8 \div 1.7$ can be written $\dfrac{6.8}{1.7}$

and if we multiply both top and bottom of the fraction by 10 we have

$$\frac{6.8}{1.7} = \frac{6.8 \times \boxed{10}}{1.7 \times \boxed{10}} = \frac{68}{17}$$

But you should know how to divide these whole numbers.

$68 \div 17 = 4$

Therefore, $6.8 \div 1.7 = 4$.

To divide decimal numbers, use the following procedure.

Step 1   Write the divisor and dividend in standard long-division form.

Example:
$6.8 \div 1.7$
$1.7\overline{)6.8}$

Step 2   Shift the decimal point in the divisor to the right so as to make the divisor a whole number.

$1.7.\overline{)}$

Step 3   Shift the decimal point in the dividend the same amount. (Add zeros if necessary.)

$1.7.\overline{)6.8.}$

Step 4   Place the decimal point in the answer space directly above the new decimal position in the dividend.

$17.\overline{)68.}$

Step 5   Now divide exactly as you would with whole numbers. The decimal points in divisor and dividend may now be ignored.

$$\begin{array}{r} 4 \\ 17.\overline{)68.} \\ -68 \\ \hline 0 \end{array}$$

$6.8 \div 1.7 = 4$

Notice in steps 2 and 3 that we have simply multiplied both divisor and dividend by 10.

If there is a remainder in step 5, we must add additional zeros to complete the division. Keep this in mind as you try the next example.

Divide: 1.38 ÷ 2.4

Work carefully, then compare your work with ours in **14**.

---

14 Let's do it step by step.

$$2.4\overline{)1.38}$$

$$24\overline{)13.8}$$ Shift both decimal points one digit to the right to make the divisor (2.4) a whole number (24).

$$24\overline{)13.8}$$ Place the decimal point in the answer space.

$$\begin{array}{r} .5 \\ 24\overline{)13.8} \\ 12\,0 \\ \hline 1\,8 \end{array}$$ Divide as usual. 24 goes into 138 five times.

There is a remainder (18), so we must keep going.

$$\begin{array}{r} .57 \\ 24\overline{)13.80} \\ 12\,0 \\ \hline 1\,80 \\ 1\,68 \\ \hline 12 \end{array}$$ Attach a zero to the dividend and bring it down.

Divide 180 by 24. 24 goes into 180 seven times.

Now the remainder is 12.

$$\begin{array}{r} .575 \\ 24\overline{)13.800} \\ 12\,0 \\ \hline 1\,80 \\ 1\,68 \\ \hline 120 \\ 120 \\ \hline 0 \end{array}$$ Attach another zero and bring it down.

24 goes into 120 exactly five times.

The remainder is zero. 1.38 ÷ 2.4 = 0.575

Check: 1.38 ÷ 2.4 is roughly 1 ÷ 2 or 0.5

Double-Check: 2.4 × 0.575 = 1.38. Always multiply the answer by the divisor to double-check the work. Do this even if you are using a calculator.

Note ▶ In some problems, we may continue to attach zeros and divide yet never get a remainder of zero. We will examine problems like this in **19**.

How would you do this one?

2.6 ÷ 0.052 = ?

Look in **15** for the solution after you have tried it.

---

15 $$0.052.\overline{)2.6}$$

To shift the decimal place three digits in the dividend, we must attach two zeros to its right.

$$0.052.\overline{)2.600.}$$ Now place the decimal point in the answer space above that in the dividend.

$$\begin{array}{r} 50. \\ 52\overline{)2600.} \\ \underline{260} \quad \longleftarrow \quad 5 \times 52 = 260 \\ 0 \\ \underline{0} \end{array}$$

$2.6 \div 0.052 = 50$ **Check:** $0.052 \times 50 = 2.6$

Shifting the decimal point three digits and attaching zeros to the right of the decimal point in this way is equivalent to multiplying both divisor and dividend by 1000.

Note ▶ In this last example inexperienced students think the problem is finished after the first division because they get a remainder of zero. You must keep dividing at least until all the places up to the decimal point in the answer are filled. The answer here is 50, not 5. ◀

Try these problems.

(a) $3.5 \div 0.001$ = _____

(b) $9 \div 0.02$ = _____

(c) $0.365 \div 18.25$ = _____

(d) $8.8 \div 3.2$ = _____

(e) $7.230 \div 6$ = _____

(f) $3 \div 4$ = _____

(g) $30.24 \div 0.42$ = _____

(h) $273.6 \div 0.057$ = _____

(i)   How many pieces of plywood 0.375 in. thick are in a stack 30 in. high?

The answers are given in **16**.

ARITHMETIC "TRICKS OF THE TRADE"

1. To divide a number by 5, use the fact that 5 is one-half of 10. First, multiply the number by 2, then divide by 10 by shifting the decimal point one place to the left.

    Example: $64 \div 5$ $64 \times 2 \div 10 = 128 \div 10 = 12.8$

2. To divide a number by 25, first multiply by 4, then divide by 100, shifting the decimal point two places left.

    Example: $304 \div 25 = 304 \times 4 \div 100 = 1216 \div 100 = 12.16$

3. To divide a number by 20, first divide by 2, then divide by 10.

    Example: $86 \div 20 = 86 \div 2 \div 10 = 43 \div 10 = 4.3$

4. To multiply a number by 20, first multiply by 2, then multiply by 10, shifting the decimal point one place to the right.

    Example: $73 \times 20 = 73 \times 2 \times 10 = 146 \times 10 = 1460$

When doing "mental arithmetic" like this, it is important that you start with a rough estimate of the answer.

16 (a) $0.001.\overline{)3.500.}$

$$3500.$$
$$1.\overline{)3500.}$$

$3.5 \div 0.001 = 3500$

Check: $0.001 \times 3500 = 3.5$

(b) $0.02\overline{)9.00.}$ Shift the decimal point two places to the right. Divide 900 by 2.

$$450.$$
$$2.\overline{)900.}$$

$9 \div 0.02 = 450$

Check: $0.02 \times 450 = 9$

Dividing a whole number by a decimal is very troublesome for most people.

(c) $18.25.\overline{)0.36.5}$

$$.02$$
$$1825.\overline{)36.50}$$
$$\underline{36.50}$$
$$0$$

1825 does not go into 365, so place a zero above the 5. Attach a zero after the 5.

1825 goes into 3650 twice. Place a 2 in the answer space above the zero.

$0.365 \div 18.25 = 0.02$

Check: $18.25 \times 0.02 = 0.365$

(d) $3.2.\overline{)8.8.}$

$$2.75$$
$$32.\overline{)88.00}$$
$$\underline{64}\downarrow$$
$$240$$
$$\underline{224}\downarrow$$
$$160$$
$$\underline{160}$$
$$0$$

$2 \times 32 = 64$ Subtract, attach a zero, and bring it down.

$7 \times 32 = 224$ Subtract, attach another zero, and bring it down.

$5 \times 32 = 160$ $8.8 \div 3.2 = 2.75$

Check: The estimated answer is $9 \div 3$ or 3.

Double-check: $3.2 \times 2.75 = 8.8$

(e) $6\overline{)7.230}$

$$1.205$$
$$6\overline{)7.230}$$
$$\underline{-6}\downarrow$$
$$1\,2$$
$$\underline{-1\,2}\downarrow$$
$$03$$
$$\underline{0}\downarrow$$
$$30$$
$$\underline{-30}$$
$$0$$

The divisor 6 is a whole number, so we can bring the decimal point in 7.23 up to the answer space.

Check: $1.205 \times 6 = 7.230$

$7.230 \div 6 = 1.205$

(f) $4\overline{)3.00}$

$$.75$$
$$4\overline{)3.00}$$
$$\underline{2\,8}$$
$$20$$
$$\underline{20}$$

Check: $0.75 \times 4 = 3.00$

$3 \div 4 = 0.75$

(g) $0.42.\overline{)30.24}$

$$72$$
$$0.42.\overline{)30.24}$$
$$\underline{29\,4}$$
$$84$$
$$\underline{84}$$

Check: $72 \times 0.42 = 30.24$

110

$$
\begin{array}{r}
4\ 800 \\
\text{(h)} \quad 0.057.\overline{)273.600} \\
228 \\
\overline{45\ 6} \\
45\ 6
\end{array}
$$

Check $4800 \times 0.057 = 273.6$

$$
\begin{array}{r}
80 \\
\text{(i)} \quad 0.375.\overline{)30.000} \\
30\ 00 \\
\overline{0}
\end{array}
$$

Check: $80 \times 0.375 = 30$

There are 80 pieces of plywood in the stack.

If the dividend is not exactly divisible by the divisor, either stop the process after some preset number of decimal places in the answer, or round the answer. We do not generally indicate a remainder in decimal division.

Turn to **17** for some rules for rounding.

Rounding Decimal Numbers

17 The process of rounding a decimal number is very similar to the procedure for rounding a whole number. The only difference is that after rounding to a given decimal place, all digits to the right of that place are dropped. The following examples illustrate this difference.

Round 35782.462 to the	nearest thousand	nearest hundredth
Step 1 Place a ∧ mark to the right of the place to which the number is to be rounded.	35∧782.462	35782.46∧2
Step 2 If the digit to the right of the mark is less than 5, replace all digits to the right of the mark with zeros.		35782.460 → Drop this decimal digit zero.
If these zeros are decimal digits, discard them.		The rounded number is 35782.46
Step 3 If the digit to the right of the mark is equal to or larger than 5, increase the digit to the left by 1 and replace all digits to the right with zeros.	36 000.000 → Drop these right-end decimal zeros	
Drop all right-end decimal zeros, but keep zero placeholders.	36000 ↗ Keep these zeros as placeholders.	
	The rounded number is 36000.	

Careful ▶ Drop only the decimal zeros that are to the right of the ∧ mark. For example, to round 6.4086 to three decimal digits, we write 6.408∧6 which becomes 6.4090 or 6.409. We dropped the end zero because it was a decimal zero to the *right* of the mark. But we retained the other zero because it is needed as a placeholder.

HOW TO NAME DECIMAL NUMBERS

The decimal number 3,254,935.4728 should be interpreted as

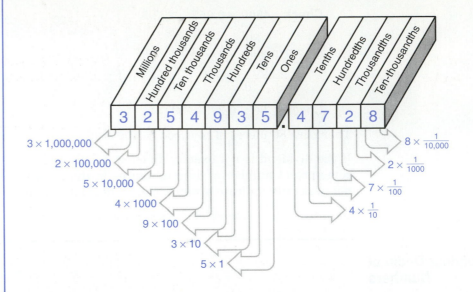

It may be read "three million, two hundred fifty-four thousand, nine hundred thirty-five, and four thousand seven hundred twenty-eight ten-thousandths."

Notice that the decimal point is read "and."

It is useful to recognize that, for example, the digit 8 represents 8 ten-thousandths or $\frac{8}{10,000}$, and the digit 7 represents 7 hundredths or $\frac{7}{100}$. Most often, however, this number is read more simply as "three million, two hundred fifty-four thousand, nine hundred thirty-five, *point* four, seven, two, eight." This way of reading the number is easiest to write, to say, and to understand.

Try it with these.

(a) Round 74.238 to two decimal places.

(b) Round 8.043 to two decimal places.

(c) Round 0.07354 to three decimal places.

(d) Round 7.98 to the nearest tenth.

Follow the rules, then check your work in **18**.

---

18 (a) 74.238 is 74.24 to two decimal places.
(Write 74.23̬8 and note that 8 is larger than 5, so increase the 3 to 4 and drop the last digit because it is a decimal digit.)

(b) 8.043 is 8.04 to two decimal places.
(Write 8.04̬3 and note that 3 is less than 5 so drop it.)

(c)   0.07354 is 0.074 to three decimal places.
(Write 0.073ˇ54 and note that the digit to the right of the mark is 5; therefore, change the 3 to a 4 to get 0.074ˇ00. Finally, drop the digits on the right to get 0.074.)

(d)   7.98 is 8.0 to the nearest tenth.
(Write 7.9ˇ8 and note that 8 is greater than 5, so increase the 9 to 0 and the 7 to 8. Drop the digit in the hundredths place.)

There are a few very specialized situations where this rounding rule is not used:

1.   Some engineers use a more complex rule when rounding a number that ends in 5.

2.   In business, fractions of a cent are usually rounded up to determine selling price. Three items for 25 cents or $8\frac{1}{3}$ cents each is rounded to 9 cents each.

Our rule will be quite satisfactory for most of your work in arithmetic.

Rounding During Division

In some division problems this process of dividing will never result in a zero remainder. At some point the answer must be rounded.

To round answers in a division problem, first continue the division so that your answer has one place more than the rounded answer will have, then round it. For example, in the division problem

4.7 ÷ 1.8 = ?

to get an answer rounded to one decimal place, first, divide to two decimal places:

4.7 ÷ 1.8 = 2.61 . . .

then round back to one decimal place:

4.7 ÷ 1.8 = 2.6 rounded

For the following problem, divide and round your answer to two decimal places.

6.84 ÷ 32.7 = _____

Careful now.

Check your work in **19**.

---

19   32.7.)6.8.4

```
        .209        Carry the answer to three decimal places.
327.)68.400         Notice that two zeros must be attached to the dividend.
      65 4↓↓         2 × 327 = 654
      3 000
      2 943          9 × 327 = 2943
```

0.209 rounded to two decimal places is 0.21.

6.84 ÷ 32.7 = 0.21   rounded to two decimal places

Check: 32.7 × 0.21 = 6.867, which is approximately equal to 6.84. (The check will not be exact because we have rounded.)

The ability to round numbers is especially important for people who work in the practical, trade, or technical areas. You will need to round answers to practical problems if they are obtained "by hand" or with a calculator. Rounding is discussed in more detail in Chapter 5.

Now turn to **20** to continue.

Averages

20 Suppose that you needed to know the diameter of a steel connecting pin. As a careful and conscientious worker, you would probably measure its diameter several times with a micrometer, and you might come up with a sequence of numbers like this (in inches):

1.3731, 1.3728, 1.3736, 1.3749, 1.3724, 1.3750

What is the actual diameter of the pin? The best answer is to find the **average** value or **arithmetic mean** of these measurements.

$$\text{Average} = \frac{\text{sum of the measurements}}{\text{number of measurements}}$$

For the preceding problem,

$$\text{Average} = \frac{1.3731 + 1.3728 + 1.3736 + 1.3749 + 1.3724 + 1.3750}{6}$$

$$= \frac{8.2418}{6} = 1.3736333\ldots$$

$$= 1.3736 \quad \text{rounded to four decimal places}$$

When you calculate the average of a set of numbers, the usual rule is to round the answer to the same number of decimal places as the least precise number in the set—that is, the one with the fewest decimal digits. If the numbers to be averaged are all whole numbers, the average will be a whole number.

Example: The average of 4, 6, 4, and 5 is $\frac{19}{4} = 4.75$ or 5 when rounded.

If one number of the set has fewer decimal places than the rest, round off to agree with the least precise number.

Example: The average of 2.41, 3.32, 5.23, 3.51, 4.1, and 4.12 is

$$\frac{22.69}{6} = 3.78166\ldots \text{ and this answer should be rounded to 3.8 to}$$

agree in precision with the least precise number, 4.1.

To avoid confusion, we will usually give directions for rounding.

Try it. Find the average for each of these sets of numbers.

(a) 8, 9, 11, 7, 5   (b) 0.4, 0.5, 0.63, 0.2

(c) 2.35, 2.26, 2.74, 2.55, 2.6, 2.31

Check your work in **21**.

---

21 (a) Average $= \dfrac{8 + 9 + 11 + 7 + 5}{5} = \dfrac{40}{5} = 8$

(b) Average $= \dfrac{0.4 + 0.5 + 0.63 + 0.2}{4} = \dfrac{1.73}{4} = 0.4324 \approx 0.4$ rounded

(c) Average $= \dfrac{2.35 + 2.26 + 2.74 + 2.55 + 2.6 + 2.31}{6}$

$$= \frac{14.81}{6} = 2.468333 \ldots$$

≈ 2.5 rounded to agree in precision with 2.6, the least precise number in the set (The symbol $\approx$ means "approximately equal to.")

Now turn to **22** for a set of exercises on the multiplication and division of decimal numbers.

Exercises 3-2   **Multiplication and Division of Decimals**

22   A.   Multiply or divide as shown.

1. 0.01×0.001	2. 10×2.15	3. 0.04×100
4. 0.3×0.3	5. 0.7×1.2	6. 0.005×0.012
7. 0.003×0.01	8. 7.25×0.301	9. 2×0.035
10. $0.2 \times 0.3 \times 0.5$	11. $0.6 \times 0.6 \times 6.0$	12. $2.3 \times 1.5 \times 1.05$
13. $3.618 \div 0.6$	14. $3.60 \div 0.03$	15. $4.40 \div 0.22$
16. $6.5 \div 0.05$	17. $0.0405 \div 0.9$	18. $0.378 \div 0.003$
19. $3 \div 0.05$	20. $10 \div 0.001$	21. $4 \div 0.01$
22. $2.59 \div 70$	23. $44.22 \div 6.7$	24. $104.2 \div 0.0320$
25. $484 \div 0.8$	26. 6.05×2.3	27. 0.0027×1.4
28. $0.0783 \div 0.27$	29. $0.00456 \div 0.095$	30. $800 \div 0.25$
31. $324 \div 0.0072$	32. $0.08322 \div 228$	33. $0.0092 \div 115$
34. 0.047×0.024	35. 0.0056×0.065	36. $0.02 \times 0.06 \times 0.04$
37. $0.008 \times 0.4 \times 0.03$	38. 123.4×0.45	39. 0.062×27.5

B.   Divide and round as indicated.
Round to two decimal digits.

1. $10 \div 3$	2. $5 \div 6$	3. $2.0 \div 0.19$
4. $3 \div 0.081$	5. $0.023 \div 0.19$	6. $12.3 \div 4.7$
7. $2.37 \div 0.07$	8. $6.5 \div 1.31$	

Round to the nearest tenth.

9. $100 \div 3$	10. $21.23 \div 98.7$	11. $1 \div 4$
12. $100 \div 9$	13. $0.006 \div 0.04$	14. $1.008 \div 3$

Round to three decimal places.

15. $10 \div 70$	16. $0.09 \div 0.402$	17. $0.091 \div 0.0014$
18. $3.41 \div 0.257$	19. $6.001 \div 2.001$	20. $123.21 \div 0.1111$

C.   Word Problems

1.   **Painting and Decorating**   A spray painting outfit is advertised for $420. It can also be bought "on time" for 24 payments of $22.75 each. How much extra do you pay by purchasing it on the installment plan?

2.   If the telephone rates between Zanzibar and Timbuctou are $3.10 for the first minute and $1.25 for each additional minute, what will be the cost of an 11-minute telephone call?

3.   **Metalworking**   Find the average weight of five castings that weigh 17, 21, 12, 20.6, and 23.4 lb.

4. **Office Services** If you work $8\frac{1}{4}$ hours on Monday, 10.1 hours on Tuesday, 8.5 hours on Wednesday, 9.4 hours on Thursday, $6\frac{1}{2}$ hours on Friday, and 4.2 hours on Saturday, what is the average number of hours worked per day?

5. **Machine Technology** For the following four machine parts, find W, the number of pounds per part; C, the cost of the metal per part; and T, the total cost.

Metal Parts	Number of Feet Needed	Number of Pounds per Foot	Cost per Pound	Pounds (W)	Cost per Part (C)
A	3.7	4.5	$0.98		
B	10.2	2.3	$0.89		
C	9	0.95	$1.05		
D	0.75	3.7	$2.15		

6. **Machine Technology** How much does 15.7 sq ft of No. 16 gauge steel weigh if 1 sq ft weighs 2.55 lb?

7. **Building Construction** A 4-ft by 8-ft sheet of $\frac{1}{4}$-in. plywood has an area of 32 sq ft. If the weight of $\frac{1}{4}$-in. plywood is 1.5 lb/sq ft, what is the weight of the sheet?

8. **Manufacturing** Each inch of 1-in.-diameter cold-rolled steel weighs 0.22 lb. How much would a piece weigh that was 38 in. long?

9. **Welding** A truck can carry a load of 5000 lb. Assuming that it could be cut to fit, how many feet of steel beam weighing 32.6 lb/ft can the truck carry?

10. **Fire Protection** One gallon of water weighs 8.34 lb. How much weight is added to a fire truck when its tank is filled with 750 gallons of water?

11. **Electrician** Voltage values are often stated in millivolts [1 millivolt (mV) equals 0.001 volt] and must be converted to volts to be used in Ohm's law. If a voltage is given to be 75 mV, how many volts is this?

12. **Industrial Technology** An industrial engineer must estimate the cost of building a storage tank. The tank requires 208 sq ft of material at $7.29 per square foot. It also requires 4.5 hours of labor at $22.40 per hour plus 1.6 hours of labor at $15.60 per hour. Find the total cost of the tank.

13. **Sheet Metal Technology** How many sheets of metal are in a stack 5 in. high if each sheet is 0.0149 in. thick?

14. **Metalworking** A casting weighs 3.68 lb. How many castings are contained in a load weighing 5888 lb?

15. A barrel partially filled with liquid fertilizer weighs 267.75 lb. The empty barrel weighs 18 lb, and the fertilizer weighs 9.25 lb per gallon. How many gallons of fertilizer are in the barrel?

D. Calculator Problems

1. Divide. $9.87654321 \div 1.23456789$.

Notice anything interesting? (Divide it out to eight decimal digits.) You should be able to get the correct answer even if your calculator will not accept a nine-digit number.

2. Divide. (a) $\frac{1}{81}$ (b) $\frac{1}{891}$ (c) $\frac{1}{8991}$

Notice a pattern? (Divide them out to about eight decimal places.)

3. **Metalworking**   The outside diameter of a steel casting is measured six times at different positions with a vernier caliper. The measurements are 4.2435, 4.2426, 4.2441, 4.2436, 4.2438, and 4.2432 in. Find the average diameter of the casting.

4. **Printing**   To determine the thickness of a sheet of paper, five batches of 12 sheets each are measured with a micrometer. The thickness of each of the five batches of 12 is 0.7907, 0.7914, 0.7919, 0.7912, and 0.7917 mm. Find the average thickness of a single sheet of paper.

5. **Building Construction**   The John Hancock Towers office building in Boston had a serious problem when it was built: When the wind blew hard, the pressure caused its windows to pop out! The only reasonable solution was to replace all 10,344 windows in the building at a cost of $6,000,000. What was the replacement cost per window? Round to the nearest dollar.

6. **Food Services**   In purchasing food for his chain of burger shops, José pays $3.25 per pound for cheese to be used in cheeseburgers. He calculates that an order of 5180 lb will be enough for 165,000 cheeseburgers.

    (a)   How much cheese will he use on each cheeseburger? (Round to three decimal places.)

    (b)   What is the cost, to the nearest cent, of the cheese used on a cheeseburger?

7. **Plumbing**   The pressure in psi (pounds per square inch) in a water system at a point 140.8 ft below the water level in a storage tank is given by the formula

$$P = \frac{62.4 \times 140.8 \times 0.43}{144}$$

Find P to the nearest tenth.

When you have finished these exercises, check your answers on page 587, then continue in **23**.

3-3 DECIMAL FRACTIONS

23   Since decimal numbers are fractions, they may be used, as fractions are used, to represent a part of some quantity. For example, recall that

"$\frac{1}{2}$ of 8 equals 4"   means $\frac{1}{2} \times 8 = 4$

and therefore,

"0.5 of 8 equals 4"   means $0.5 \times 8 = 4$

The word *of* used in this way indicates multiplication, and decimal numbers are often called "decimal fractions."

It is very useful to be able to convert any number from fraction form to decimal form. Simply divide the top by the bottom of the fraction. If the division has no remainder, the decimal number is called a *terminating* decimal. For example,

$$\frac{5}{8} = \underline{\qquad ? \qquad}$$

```
     .625
8 )5.000   ← Attach as many zeros as needed.
  4 8
    20
    16
    40
    40
     0   ← Zero remainder; therefore, the decimal terminates or ends.
```

$$\frac{5}{8} = 0.625$$

If the decimal does not terminate, you may round it to any desired number of decimal digits. For example,

$$\frac{2}{13} = \underline{\qquad ? \qquad}$$

```
        .1538
13 )2.0000   ← Attach zeros.
    1 3
      70
      65
      50
      39
     110
     104
       6   ← Remainder is not equal to zero.
```

$$\frac{2}{13} = 0.154 \quad \text{rounded to three decimal places}$$

Convert the following fractions to decimal form and round if necessary to two decimal places.

(a) $\frac{4}{5}$ (b) $\frac{2}{3}$ (c) $\frac{17}{7}$ (d) $\frac{5}{6}$ (e) $\frac{7}{16}$ (f) $\frac{5}{9}$

Our work is shown in **24**.

24

(a) $\frac{4}{5} = \underline{\qquad ? \qquad}$

```
    .8
5 )4.0
```

$$\frac{4}{5} = 0.8$$

(b) $\frac{2}{3} = \underline{\qquad ? \qquad}$

```
    .666 . . .
3 )2.000
   1 8
     20
     18
     20
     18
      2
```

$$\frac{2}{3} = 0.67 \quad \text{rounded to two decimal places}$$

Notice that in order to round to *two* decimal places, we must carry the division out to at least *three* decimal digits.

(c) $\dfrac{17}{7} =$ _____?_____

$$
\begin{array}{r}
2.428 \\
7\overline{)17.000} \\
14 \\
\hline
3\,0 \\
2\,8 \\
\hline
20 \\
14 \\
\hline
60 \\
56 \\
\hline
4
\end{array}
$$

$\dfrac{17}{7} = 2.43$ rounded to two decimal digits

(d) $\dfrac{5}{6} =$ _____?_____

$$
\begin{array}{r}
.833\ldots \\
6\overline{)5.000} \\
4\,8 \\
\hline
20 \\
18 \\
\hline
20 \\
18 \\
\hline
2
\end{array}
$$

$\dfrac{5}{6} = 0.83$ rounded to two decimal places

(e) $\dfrac{7}{16} =$ _____?_____

$$
\begin{array}{r}
.4375 \\
16\overline{)7.0000} \\
6\,4 \\
\hline
60 \\
48 \\
\hline
120 \\
112 \\
\hline
80 \\
80 \\
\hline
0
\end{array}
$$

$\dfrac{7}{16} = 0.4375$ or 0.44 rounded to two decimal digits

(f) $\dfrac{5}{9} =$ _____?_____

$$
\begin{array}{r}
.55 \\
9\overline{)5.00} \\
4\,5 \\
\hline
50 \\
45 \\
\hline
5
\end{array}
$$

$\dfrac{5}{9} = 0.555\ldots$ or 0.56 rounded

Repeating Decimal   Decimal numbers that do not terminate will repeat a sequence of digits. This type of decimal number is called a *repeating* decimal. For example,

$$\frac{1}{3} = 0.3333\ldots$$

where the three dots are read "and so on," and they tell us that the digit 3 continues without end.

Similarly, $\frac{2}{3} = 0.6666\ldots$ and $\frac{3}{11}$ is

```
      .2727
11)3.0000
    2 2
    ‾‾‾
     80
     77
     ‾‾
      30
      22
      ‾‾
       80
       77  ← The reminder 3 is equal to the original dividend.
       ‾‾    This tells us that the decimal quotient repeats itself.
        3
```

$\frac{3}{11} = 0.272727\ldots$

If you did the division in problem (c) with a calculator, you may have noticed that

$\frac{17}{7} = 2.\overset{\frown}{428571}\overset{\frown}{428571}\ldots$ The digits 428571 repeat endlessly.

Mathematicians often use a shorthand notation to show that a decimal repeats.

Write $\frac{1}{3} = 0.\overline{3}$ or $\frac{2}{3} = 0.\overline{6}$

where the bar means that the digits under the bar repeat endlessly.

$\frac{3}{11} = 0.\overline{27}$ means $0.272727\ldots$ and $\frac{17}{7} = 2.\overline{428571}$

Write $\frac{41}{33}$ as a repeating decimal using the "bar" notation.

Check your answer in **25**.

_____

25

```
       1.24
33)41.00
   33
   ‾‾
    8 0 ⎤
    6 6 ⎥
    ‾‾‾ ⎥  These remainders are the same, so we know that further
    1 40⎥  division will produce a repeat of the digits 24 in the answer.
    1 32⎥
    ‾‾‾‾⎥
       8 ⎦
```

$\frac{41}{33} = 1.242424\ldots = 1.\overline{24}$

Converting Fractions to Decimals

Some fractions are used so often in practical work that it is important for you to know their decimal equivalents as shown on p. 121.

Quick now, without looking at the table, fill in the blanks in the problems shown.

$\frac{1}{2} =$ _____ $\frac{3}{4} =$ _____ $\frac{3}{8} =$ _____ $\frac{7}{8} =$ _____

$\frac{1}{4} =$ _____ $\frac{5}{8} =$ _____ $\frac{3}{16} =$ _____ $\frac{7}{16} =$ _____

$\frac{1}{8} =$ _____ $\frac{5}{16} =$ _____ $\frac{9}{16} =$ _____ $\frac{13}{16} =$ _____

$\frac{1}{16} =$ _____ $\frac{11}{16} =$ _____ $\frac{15}{16} =$ _____

DECIMAL–FRACTION EQUIVALENTS

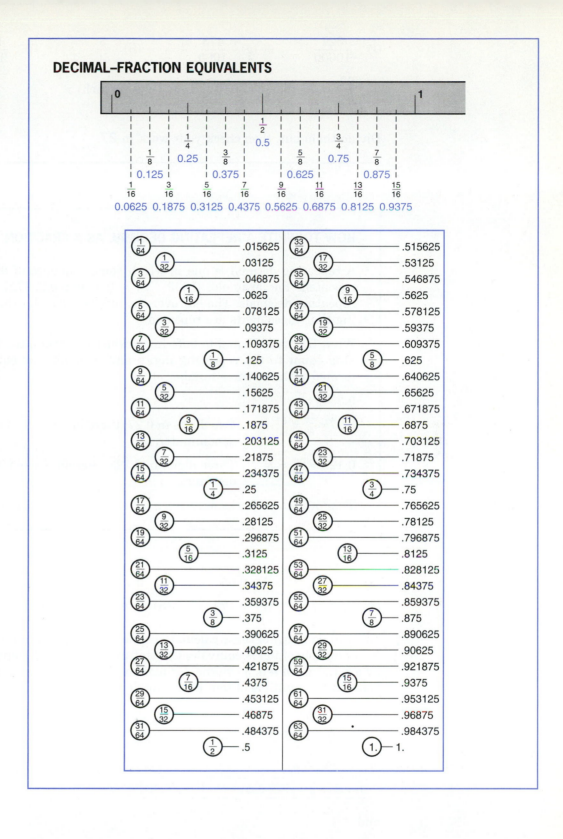

Check your work against the table when you are finished, then turn to **26** and continue.

---

26 Convert each of the following fractions into decimal form.

(a) $\dfrac{95}{100}$ (b) $\dfrac{1}{20}$ (c) $\dfrac{7}{10}$ (d) $\dfrac{4}{1000}$ (e) $\dfrac{11}{1000}$

(f) $\dfrac{327}{10000}$ (g) $\dfrac{473}{1000}$ (h) $\dfrac{3}{50}$ (i) $\dfrac{1}{25}$

(j) $\dfrac{27}{64}$ in. (to the nearest thousandth of an inch)

Divide them out, then check your work in **27**.

HOW TO WRITE A REPEATING DECIMAL AS A FRACTION

A repeating decimal is one in which some sequence of digits is endlessly repeated. For example, $0.333\ldots = 0.\overline{3}$ and $0.272727\ldots = 0.\overline{27}$ are repeating decimals. The bar over the number is a shorthand way of showing that those digits are repeated.

What fraction is equal to $0.\overline{3}$? To answer this, form a fraction with numerator equal to the repeating digits and denominator equal to a number formed with the same number of 9s.

$0.\overline{3} = \dfrac{3}{9} = \dfrac{1}{3}$

$0.\overline{27} = \dfrac{27}{99} = \dfrac{3}{11}$ Two digits in $0.\overline{27}$; therefore, use 99 as the denominator.

$0.\overline{123} = \dfrac{123}{999} = \dfrac{41}{333}$ Three digits in $0.\overline{123}$; therefore, use 999 as the denominator.

This procedure works only when *all* of the decimal part repeats.

---

27   (a) 0.95   (b) 0.05   (c) 0.7   (d) 0.004   (e) 0.011

      (f) 0.0327   (g) 0.473   (h) 0.06   (i) 0.04   (j) 0.422 in.

You will do many of these calculations more quickly if you remember that dividing by a multiple of ten is equivalent to shifting the decimal point to the left. To divide by a multiple of ten, move the decimal point to the left as many digits as there are zeros in the multiple of ten. For example,

$$\dfrac{95}{100} = 0.95$$

Two zeros | Move the decimal point two digits to the left

and

$$\dfrac{11}{1000} = 0.011$$

3 zeros | Move the decimal point three digits to the left

Now turn to **28** for a set of exercises on decimal fractions.

Exercises 3-3 Decimal Fractions

28 A. Write as decimal numbers. (Round to two decimal digits if necessary.)

1. $\frac{1}{4}$ 2. $\frac{2}{3}$ 3. $\frac{3}{4}$ 4. $\frac{2}{5}$

5. $\frac{4}{5}$ 6. $\frac{5}{6}$ 7. $\frac{2}{7}$ 8. $\frac{4}{7}$

9. $\frac{6}{7}$ 10. $\frac{3}{8}$ 11. $\frac{6}{8}$ 12. $\frac{1}{10}$

13. $\frac{3}{10}$ 14. $\frac{2}{12}$ 15. $\frac{5}{12}$ 16. $\frac{3}{16}$

17. $\frac{6}{16}$ 18. $\frac{9}{16}$ 19. $\frac{13}{16}$ 20. $\frac{3}{20}$

21. $\frac{7}{32}$ 22. $\frac{13}{20}$ 23. $\frac{11}{24}$ 24. $\frac{39}{64}$

25. $\frac{3}{100}$ 26. $\frac{213}{1000}$ 27. $\frac{19}{1000}$ 28. $\frac{17}{50}$

B. Calculate in decimal form. (Round to agree with the number of decimal digits in the decimal number.)

1. $2\frac{3}{5} + 1.785$ 2. $\frac{1}{5} + 1.57$ 3. $3\frac{7}{8} - 2.4$

4. $1\frac{3}{16} - 0.4194$ 5. $2\frac{1}{2} \times 3.15$ 6. $1\frac{3}{25} \times 2.08$

7. $3\frac{4}{5} \div 2.65$ 8. $3.72 \div 1\frac{1}{4}$ 9. $2.76 + \frac{7}{8}$

10. $16\frac{3}{4} - 5.842$ 11. $3.14 \times 2\frac{7}{16}$ 12. $1.17 \div 1\frac{3}{32}$

C. Practical Problems

1. **Pharmacy**   If one tablet of calcium pantothenate contains 0.5 gram, how much is contained in $2\frac{3}{4}$ tablets? How many tablets are needed to make up 2.6 grams?

2. **Building Construction**   Estimates of matched or tongue-and-groove (T & G) flooring—stock that has a tongue on one edge—must allow for the waste from milling. To allow for this waste, $\frac{1}{4}$ of the area to be covered must be added to the estimate when 1-in. by 4-in. flooring is used. If 1-in. by 6-in. flooring is used, $\frac{1}{6}$ of the area must be added to the estimate. (*Hint:* The floor is a rectangle. Area of a rectangle = length × width.)

(a) A house has a floor size 28 ft long by 12 ft wide. How many square feet of 1-in. by 4-in. T & G flooring will be required to lay the floor?

(b) A motel contains 12 units or rooms, and each room is 22 ft by 27 ft or 594 sq ft. In five rooms the builder is going to use 4-in. stock and in the other seven rooms 6-in.. stock—both being T & G. How many square feet of each size will be used?

(c) A contractor is building a motel with 24 rooms. Eight rooms will be 16 ft by 23 ft, nine rooms 18 ft by 26 ft, and the rest of the rooms 14 ft by 20 ft. How much did she pay for flooring, using 1-in. by 4-in. T & G at $2820 per 1000 sq ft?

3. **Carpentry**   If a carpenter lays $10\frac{1}{2}$ squares of shingles in $4\frac{1}{2}$ days, how many squares does he do in one day?

4. **Machine Technology**   A bearing journal measures 1.996 in. If the standard readings are in $\frac{1}{32}$-in. units, what is its probable standard size as a fraction?

5. **Upholstering**   Complete the following invoice for upholstery fabric.

(a) $5\frac{1}{4}$ yd @ $12.37 = \underline{\hspace{1.5cm}}$

(b) $23\frac{3}{4}$ yd @ $16.25 = \underline{\hspace{1.5cm}}$

(c) $31\frac{5}{6}$ yd @ $9.95 = \underline{\hspace{1.5cm}}$

(d) $16\frac{2}{3}$ yd @ $17.75 = \underline{\hspace{1.5cm}}$

Total $\underline{\hspace{1.5cm}}$

6. **Building Construction**   A land developer purchased a piece of land containing 437.49 acres. He plans to divide it into a 45-acre recreational area and lots of $\frac{3}{4}$ acre each. How many lots will he be able to form from this piece of land?

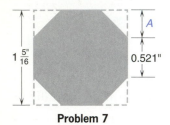

Problem 7

7. **Machine Technology**   Find the corner measurement A needed to make an octagonal end on a square bar as shown in the diagram.

8. **Auto Mechanics**   A cylinder is normally $3\frac{7}{16}$ in. in diameter. It is rebored 0.040 in. larger. What is the size of the rebored cylinder? (Give your answer in decimal form.)

9. Plumbers deal with measurements in inches and common fractions of an inch, while surveyors use feet and decimal fractions of a foot. Often one trade needs to interpret the measurements of the other.

(a)   **Surveying**   A drain has a run of 52 ft at a grade of $\frac{1}{8}$ in./ft. The high end of the drain has an elevation of 126.70 ft. What is the elevation at the low end?

(b)   **Plumbing**   The elevation at one end of a lot is 84.25 ft, and the elevation at the other end is 71.70 ft. Express the difference in elevation in feet and inches and round to the nearest $\frac{1}{8}$ in.

10. **Plumbing**   To fit a 45° connection, a plumber can approximate the diagonal length by multiplying the offset length by 1.414. Find the diagonal length of a 45° fitting if the offset length is $13\frac{11}{16}$ in. Express the answer as a fraction to the nearest $\frac{1}{16}$ in.

D.   Calculator Problems

1. **Sheet Metal Technology**   In the table on page 125 the first column lists the U.S. Standard sheet metal gauge number. The second column gives the equivalent thickness as a fraction in inches. Complete the third column, giving the thickness as a decimal number rounded to the nearest thousandth of an inch.

2. **Building Construction**   Calculate the total cost of the following building materials.

8024 ft of flooring	@ $855.75 per 1000 ft	\underline{\hspace{1cm}}
10,227 ft of siding	@ $162.50 per 100 ft	\underline{\hspace{1cm}}
24,390 bricks	@ $447 per 1000	\underline{\hspace{1cm}}
19,200 sq ft of shingles	@ $76.95 per 100 sq ft	\underline{\hspace{1cm}}
	Total	\underline{\hspace{1cm}}

3. **Office Services** Machine helpers at the CALMAC Tool Co. earn $6.72 per hour. The company pays time and one-half for hours over 40 per week. Sunday hours are paid at double time. How much would you earn if you worked the following hours: Monday, 9; Tuesday, 8; Wednesday, $9\frac{3}{4}$; Thursday, 8; Friday, $10\frac{1}{2}$; Saturday, $5\frac{1}{4}$; Sunday, $4\frac{3}{4}$?

When you have finished these exercises, check your answers on page 587, and then go to **29** for a set of practice problems on decimal numbers.

Gauge No.	Fraction Thickness (in.)	Decimal Thickness (in.)	Gauge No.	Fraction Thickness (in.)	Decimal Thickness (in.)
7-0	$\frac{1}{2}$		14	$\frac{5}{64}$	
6-0	$\frac{15}{32}$		15	$\frac{9}{128}$	
5-0	$\frac{7}{16}$		16	$\frac{1}{16}$	
4-0	$\frac{13}{32}$		17	$\frac{9}{160}$	
3-0	$\frac{3}{8}$		18	$\frac{1}{20}$	
2-0	$\frac{11}{32}$		19	$\frac{7}{160}$	
0	$\frac{5}{16}$		20	$\frac{3}{80}$	
1	$\frac{9}{32}$		21	$\frac{11}{320}$	
2	$\frac{17}{64}$		22	$\frac{1}{32}$	
3	$\frac{1}{4}$		23	$\frac{9}{320}$	
4	$\frac{15}{64}$		24	$\frac{1}{40}$	
5	$\frac{7}{32}$		25	$\frac{7}{320}$	
6	$\frac{13}{64}$		26	$\frac{3}{160}$	
7	$\frac{3}{16}$		27	$\frac{11}{640}$	
8	$\frac{11}{64}$		28	$\frac{1}{64}$	
9	$\frac{5}{32}$		29	$\frac{9}{640}$	
10	$\frac{9}{64}$		30	$\frac{1}{80}$	
11	$\frac{1}{8}$		31	$\frac{7}{640}$	
12	$\frac{7}{64}$		32	$\frac{13}{1280}$	
13	$\frac{3}{32}$		—	—	—

Decimal Numbers

29   Answers are on page 588.

A. Write in words.

1.	0.91	**2.**	0.84	**3.**	23.164	**4.**	63.219
5.	9.3	**6.**	3.45	**7.**	10.06	**8.**	15.037

Write as decimal numbers.

9.   Seven hundredths

10.   Eighteen thousandths

11.   Two hundred and eight tenths

12.   Sixteen and seventeen hundredths

13.   Sixty-three and sixty-three thousandths

14.   One hundred ten and twenty-one thousandths

15.   Five and sixty-three ten-thousandths

16.   Eleven and two hundred eighteen ten-thousandths

B. Add or subtract as shown.

1.	$4.39 + 18.8$	**2.**	$18.8 + 156.16$
3.	$\$7.52 + \$11.77$	**4.**	$26 + 0.06$
5.	$3.68 - 1.74$	**6.**	$\$12.46 - \$8.51$
7.	$104.06 - 15.80$	**8.**	$16 - 3.45$
9.	$264.3 + 12.804$	**10.**	$0.232 + 5.079$
11.	$165.4 + 73.61$	**12.**	$245.94 + 7.07$
13.	$116.7 - 32.82$	**14.**	$4.07 - 0.085$
15.	$0.42 + 1.452 + 31.8$	**16.**	$\frac{1}{2} + 4.21$
17.	$3\frac{1}{5} + 1.08$	**18.**	$1\frac{1}{4} - 0.91$
19.	$3.045 - 1\frac{1}{8}$	**20.**	$8.1 + 0.47 - 1\frac{4}{5}$

Name

Date

Course/Section

C. Calculate as shown.

1. 0.004×0.02
2. 0.06×0.05
3. 1.4×0.6
4. 3.14×12
5. $0.2 \times 0.6 \times 0.9$
6. 5.3×0.4
7. 6.02×3.3
8. $3.224 \div 2.6$
9. $187.568 \div 3.04$
10. $0.078 \div 0.3$
11. $0.6 \times 3.15 \times 2.04$
12. $3.78 + 4.1 \times 6.05$
13. $0.008 - 0.001 \div 0.5$
14. $3.1 \times 4.6 - 2.7 \div 0.3$
15. $\dfrac{4.2 + 4.6 \times 1.2}{2.73 \div 21 + 0.41}$
16. $\dfrac{7.2 - 3.25 \div 1.3}{3.5 + 5.7 \times 4.0 - 2.8}$
17. $1.2 + 0.7 \times 2.2 + 1.6$
18. $1.2 \times 0.7 + 2.2 \times 1.6$

Round to two decimal digits.

19. $0.007 \div 0.03$
20. $3.005 \div 2.01$
21. $17.8 \div 6.4$
22. $0.0041 \div 0.019$

Round to three decimal places.

23. $0.04 \div 0.076$
24. $234.1 \div 465.8$
25. $17.6 \div 0.082$
26. $0.051 \div 1.83$

Round to the nearest tenth.

27. $0.08 \div 0.053$
28. $3.05 \div 0.13$
29. $18.76 \div 4.05$
30. $0.91 \div 0.97$

D. Write as a decimal number.

1. $\dfrac{1}{16}$
2. $\dfrac{7}{8}$
3. $\dfrac{5}{32}$
4. $\dfrac{7}{16}$
5. $\dfrac{11}{8}$
6. $1\dfrac{3}{16}$
7. $2\dfrac{11}{32}$
8. $1\dfrac{1}{6}$
9. $2\dfrac{2}{3}$
10. $1\dfrac{1}{32}$
11. $2\dfrac{5}{16}$

Write in decimal form, calculate, round to two decimal digits.

12. $4.82 \div \dfrac{1}{4}$
13. $11.5 \div \dfrac{3}{8}$
14. $1\dfrac{3}{16} \div 0.62$
15. $2\dfrac{3}{4} \div 0.035$
16. $0.45 \times 2\dfrac{1}{8}$
17. $0.068 \times 1\dfrac{7}{8}$

E. Practical Problems

1. **Machine Technology** A $\frac{5}{16}$-in. bolt weighs 0.43 lb. How many bolts are there in a 125-lb keg?

2. **Metalworking** Twelve equally spaced holes are to be drilled in a metal strip $34\frac{1}{4}$-in. long, with 2 in. remaining on each end. What is the distance, to the nearest hundredth of an inch, from center to center of two consecutive holes?

3. **Plumbing** A plumber finds the length of pipe needed to complete a bend by performing the following multiplication:

   Length needed = $0.01745 \times$ (radius of the bend) $\times$ (angle of bend)

   What length of pipe is needed to complete a 35° bend with a radius of 16 in.?

4. **Printing** During the first six months of the year the Busy Bee Printing Co. spent the following amounts on paper:

January $470.16	February $390.12	March $176.77
April $200.09	May $506.45	June $189.21

   What is the average monthly expenditure for paper?

5. **Machine Technology** The *feed* of a drill is the distance it advances with each revolution. If a drill makes 310 rpm and drills a hole 2.125 in. deep in 0.800 minute, what is the feed?

   $$\left(Hint:\ \text{Feed} = \frac{\text{total depth}}{(\text{rpm}) \times (\text{time of drilling in minutes})}.\right)$$

6. **Electrical Technology** The resistance of an armature while it is cold is 0.208 ohm. After running for several minutes, the resistance increases to 1.340 ohms. Find the increase in resistance of the armature.

7. **Painting and Decorating** The Ace Place Paint Co. sells paint in steel drums each weighing 36.4 lb empty. If 1 gallon of paint weighs 9.06 lb, what is the total weight of a 50-gallon drum of paint?

8. **Electrical Technology** The diameter of No. 12 bare copper wire is 0.08081 in., and the diameter of No. 15 bare copper wire is 0.05707 in. How much larger is No. 12 wire than No. 15 wire?

9. **Machine Technology** What is the decimal equivalent of each of the following drill bit diameters?

   (a) $\frac{3}{16}$ in.   (b) $\frac{5}{32}$ in.   (c) $\frac{3}{8}$ in.   (d) $\frac{13}{64}$ in.

10. What is the fractional equivalent in 64ths of an inch of the following dimensions? (*Hint:* Use the table on page 121.)

    (a) 0.135 in.   (b) 0.091 in.   (c) 0.295 in.   (d) 0.185 in.

11. **Sheet Metal Technology** The following table lists the thickness in inches of several sizes of sheet steel:

U.S. Gauge	Thickness (in.)
35	0.0075
30	0.0120
25	0.0209
20	0.0359
15	0.0673
10	0.1345
5	0.2092

    (a) What is the difference in thickness between 30 gauge and 25 gauge sheet?

(b) What is the difference in thickness between seven sheets of 25 gauge and four sheets of 20 gauge sheet?

(c) What length of $\frac{3}{16}$-in.-diameter rivet is needed to join one thickness of 25 gauge sheet to a strip of $\frac{1}{4}$-in. stock? Add $1\frac{1}{2}$ times the diameter of the rivet to the length of the rivet to assure that the rivet is long enough that a proper rivet head can be formed.

(d) What length of $\frac{5}{32}$-in. rivet is needed to join two sheets of 20 gauge sheet steel? (Don't forget to add the $1\frac{1}{2}$ times rivet diameter.)

12. **Building Construction** When the owners of the Better Builder Co. completed a small construction job, they found that the following expenses had been incurred: labor, $672.25; gravel, $86.77; sand, $39.41; cement, $180.96; and bricks, $204.35. What total bill should they give the customer if they want to make $225 profit on the job?

13. **Building Construction** One lineal foot of 12-in. I-beam weighs 25.4 lb. What is the length of a beam that weighs 444.5 lb?

14. **Sheet Metal Technology** To determine the average thickness of a metal sheet, a sheet metal worker measures it at five different locations. His measurements are 0.0401, 0.0417, 0.0462, 0.0407, and 0.0428 in. What is the average thickness of the sheet?

15. **Drafting** A dimension in a technical drawing is given as $2\frac{1}{2}$ in. with a tolerance of $\pm\frac{1}{32}$ in. What are the maximum and minimum permissible dimensions written in decimal form?

16. **Auto Mechanics** The four employees of the Busted Body Shop earned the following amounts last week: $411.76, $396.21, $408.18, and $476.35. What is the average weekly pay for the employees of the shop?

17. **Electrical Technology** If 14-gauge Romex cable sells for 19 cents per foot, what is the cost of 1210 ft of this cable?

18. **Electrical Technology** In Lucy's first week as an electrician, she earned $894.38 for 37.5 hours of work. What is her hourly rate of pay?

19. **Welding** A welder finds that 2.083 cu ft of acetylene gas is needed to make one bracket. How much gas will be needed to make 27 brackets?

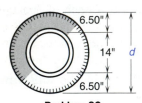

6.50"

14" d

6.50"

Problem 20

20. **Auto Mechanics** What is the outside diameter of a 6.50 × 14 tire? (See the figure.)

21. **Metalworking** How many 2.34-in. spacer blocks can be cut from a 2-in. by 2-in. square bar 48 in. long? Allow $\frac{1}{8}$ in. waste for each saw cut.

22. **Auto Mechanics** The Fixum Auto Shop charges customers a flat hourly rate of $62.50 for labor on all jobs. They pay their motor overhaul mechanic $34.25 per hour, brake mechanic $32.20 per hour, and general mechanic helpers $18.65 per hour. A motor overhaul takes 23.5 hours of labor, and a brake relining takes 6.25 hours of labor.

(a) What is the total cost to the customer for these two jobs?

(b) What does the shop pay the brake mechanic?

(c) What does the shop pay the motor overhaul mechanic?

(d) What does the shop pay the helper for the two jobs?

(e) What amount of money does the shop keep to cover overhead and profit?

23. **Machine Technology** A machinist estimates the following times for fabricating a certain part: 0.6 hour for setup, 2.4 hours of turning, 5.2 hours of milling, 1.4 hours of grinding, and 1.8 hours of drilling. What is the total time needed to make the part?

24. **Printing** A 614-page book was printed on paper 0.00175 in. thick and finished with a cover 0.165 in. thick. What was the total thickness of the bound book? (Be sure to count the cover twice and remember that there are two pages of the book for every one sheet of paper.)

25. **Printing** A certain type sets at an average of 14 characters to the inch. How many lines will it take to set 4325 characters to a width of 6 in.?

26. **Fire Protection** One gallon of water weighs approximately 8.34 lb. When sent to a particularly rough terrain, a Forestry Service truck is allowed to carry only 1.5 tons of weight. How many gallons of water can it carry? (1 ton = 2000 pounds.)

Using a Calculator, II: Decimals and Fractions

30 To perform addition, subtraction, multiplication, and division of decimal numbers on a calculator, use the same procedures outlined in Chapter 1, and press the $\boxed{\cdot}$ key to enter the decimal point. For example, to calculate $243.78 + 196.1 \times 2.75$ use the keystroke sequence

243 $\boxed{\cdot}$ **78** $\boxed{+}$ **196** $\boxed{\cdot}$ **1** $\boxed{\times}$ **2** $\boxed{\cdot}$ **75** $\boxed{=}$ → 　783.055

The decimal point key was shown as a special symbol here for emphasis, but in future calculator sequences in this book we will save space by not showing the decimal point as a separate key.

Notice that in the preceding calculation, the numbers and operations were entered, and the calculator followed the standard order of operations.

When you work with decimal numbers, remember that for a calculator with an eight-digit display if an answer is a nonterminating decimal or a terminating decimal with more than eight digits, the calculator will display a number rounded to eight digits.* Knowledge of rounding is especially important in such cases because the result must often be rounded further. Work the following problems for practice and round as indicated.

(a) $28.75 + 161.49 - 37.60$

(b) 2.8×0.85 (Round to the nearest tenth.)

(c) $2347.68 \div 12.9$ (Round to the nearest whole number.)

(d) $46.8 - 27.3 \times 0.49$ (Round to the nearest hundredth.)

(e) $\dfrac{16500 + 3700}{12 \times 68}$ (Round to the nearest tenth.)

(f) $(3247.9 + 868.7) \div 0.816$ (Round to the nearest ten.)

(g) $6 \div 0.07 \times 0.8 + 900$ (Round to the nearest hundred.)

Check your answers in 31.

---

*Some calculators will truncate an answer instead of rounding it. This means that they will show the first eight digits of the result. Be sure you know which type of calculator you have.

31 (a) **28.75** $\boxed{+}$ **161.49** $\boxed{-}$ **37.6** $\boxed{=}$ → ⎡ *152.64* ⎤

Final zeros to the right of the
decimal point may be omitted.

(b) **2.8** $\boxed{\times}$ **.85** $\boxed{=}$ ⎡ *2.38* ⎤ or 2.4 rounded

We need not enter the
leading zero in 0.85

(c) **2347.68** $\boxed{\div}$ **12.9** $\boxed{=}$ → ⎡ *181.99070* ⎤ or 182 rounded

(d) **46.8** $\boxed{-}$ **27.3** $\boxed{\times}$ **.49** $\boxed{=}$ → ⎡ *33.423* ⎤ or 33.42 rounded

(e) **16500** $\boxed{+}$ **3700** $\boxed{=}$ $\boxed{\div}$ $\boxed{(}$ **12** $\boxed{\times}$ **68** $\boxed{)}$ → ⎡ *24.754902* ⎤ or 24.8 rounded

(f) **3247.9** $\boxed{+}$ **868.7** $\boxed{=}$ $\boxed{\div}$ **.816** $\boxed{=}$ → ⎡ *5044.8529* ⎤ or 5040 rounded

(g) **6** $\boxed{\div}$ **.07** $\boxed{\times}$ **.8** $\boxed{+}$ **900** $\boxed{=}$ → ⎡ *968.57143* ⎤ or 1000 rounded

Calculating with Fractions Fractions can be entered directly on most calculators, and the results of arithmetic calculations with fractions can be displayed as fractions or decimals. If your calculator has a $\boxed{a\frac{b}{c}}$ key, you may enter fractions or mixed numbers directly into your machine without using the division key.

The calculator display will indicate fractions and mixed numbers with a ⏌ symbol. For example, the fraction $\frac{3}{4}$ is shown as 3⏌4,

3 $\boxed{a\frac{b}{c}}$ **4** $\boxed{=}$ → ⎡ *3⏌4* ⎤

and the mixed number $1\frac{7}{8}$ is shown as 1⏌7⏌8,

1 $\boxed{a\frac{b}{c}}$ **7** $\boxed{a\frac{b}{c}}$ **8** $\boxed{=}$ → ⎡ *1⏌7⏌8* ⎤

Improper fractions or fractions not in lowest terms can be simplified or written in decimal form. The improper fraction $\frac{12}{7}$ can be simplified as follows:

12 $\boxed{a\frac{b}{c}}$ **7** $\boxed{=}$ → ⎡ *1⏌5⏌7* ⎤ or $1\frac{5}{7}$

To write it as a decimal number press the fraction key again.

12 $\boxed{a\frac{b}{c}}$ **7** $\boxed{=}$ → ⎡ *1⏌5⏌7* ⎤ $\boxed{a\frac{b}{c}}$ → ⎡ *1.7142857* ⎤

To simplify the fraction $\frac{6}{16}$

6 $\boxed{a\frac{b}{c}}$ **16** $\boxed{=}$ → ⎡ *3⏌8* ⎤ or $\frac{3}{8}$ in lowest terms.

Perform arithmetic operations in the usual way. For example, $\frac{26}{8} - 1\frac{2}{3}$ is

26 $\boxed{a\frac{b}{c}}$ **8** $\boxed{-}$ **1** $\boxed{a\frac{b}{c}}$ **2** $\boxed{a\frac{b}{c}}$ **3** $\boxed{=}$ → ⎡ *1⏌7⏌12* ⎤ $\boxed{a\frac{b}{c}}$ → ⎡ *1.5833333* ⎤

$1\frac{7}{12}$

Notice that pressing the $\boxed{a\frac{b}{c}}$ key again will display the fraction as a decimal number.

For practice in calculating with fractions and decimals, work the following problems. Write your answer both as a fraction or mixed number and as a decimal rounded to two decimal places.

(a) $\frac{2}{3} + \frac{7}{8}$ (b) $1\frac{3}{4} - \frac{2}{5}$ (c) $\frac{25}{32} + 1.87$ (d) $8\frac{1}{5} \div 2\frac{1}{6}$

(e) $\frac{17}{20} \times \frac{1}{3}$

Check your solutions in **32**.

32 (a) **2** $\boxed{a^b_c}$ **3** $\boxed{+}$ **7** $\boxed{a^b_c}$ **8** $\boxed{=}$ → $1\rfloor13\rfloor24$ $\boxed{a^b_c}$ → 1.54166666 or 1.54 rounded

$1\frac{13}{24}$

(b) **1** $\boxed{a^b_c}$ **3** $\boxed{a^b_c}$ **4** $\boxed{-}$ **2** $\boxed{a^b_c}$ **5** $\boxed{=}$ → $1\rfloor7\rfloor20$ $\boxed{a^b_c}$ → 1.35

$1\frac{7}{20}$

(c) **25** $\boxed{a^b_c}$ **32** $\boxed{+}$ **1.87** $\boxed{=}$ → 2.6512500 or 2.65 rounded

(d) **8** $\boxed{a^b_c}$ **1** $\boxed{a^b_c}$ **5** $\boxed{\div}$ **2** $\boxed{a^b_c}$ **1** $\boxed{a^b_c}$ **6** $\boxed{=}$ → $3\rfloor51\rfloor65$ $\boxed{a^b_c}$ → 3.7846154

or 3.78 rounded

$3\frac{51}{65}$

(e) **17** $\boxed{a^b_c}$ **20** $\boxed{\times}$ **1** $\boxed{a^b_c}$ **3** $\boxed{=}$ → $17\rfloor60$ $\boxed{a^b_c}$ → 0.28333333

or 0.28 rounded

$\frac{17}{60}$

Ratio, Proportion, and Percent

Objective	Sample Problems	Where to Go for Help	
When you finish this unit you will be able to:		Page	Frame
1. Calculate ratios.	(a) Find the ratio of the pulley diameters.	137	**1**

	(b) A gasoline engine has a maximum cylinder volume of 520 cu cm and a compressed volume of 60 cu cm. Find the compression ratio.	138	**2**
2. Solve proportions.	(a) Solve for x. $\dfrac{8}{x} = \dfrac{12}{15}$	141	**6**
	(b) Solve for y. $\dfrac{4.4}{2.8} = \dfrac{y}{9.1}$		
3. Solve problems involving proportions.	(a) A mixture of weed killer must be prepared by combining three parts of a concentrate with every 16 parts water. How much water should be added to 12 ounces of concentrate?	144	**10**

Name _____

Date _____

Course/Section _____

4. Write fractions and decimal numbers as percents.

cent. _____ 150   **14**

(b) Write 0.46 as a percent. _____

(c) Write 5 as a percent. _____

(d) Write 0.075 as a percent. _____

5. Convert percents to decimal numbers.

(a) Write 35% as a decimal number _____ 152   **17**

(b) Write 0.25% as a decimal number. _____

(c) Write 112% as a decimal number. _____

6. Solve problems involving percent.
(a) Write $\frac{1}{4}$ as a per-

(a) Find $37\frac{1}{2}$% of 600. _____ 155   **21**

(b) Find 120% of 45. _____

(c) What percent of 80 is 5? _____ 159   **23**

(d) 12 is 16% of what number? _____ 160   **24**

(e) If a measurement is given as 2.778 ± 0.025 in., state the tolerance as a percent. _____ 173   **38**

(f) If a certain kind of solder is 52% tin, how many pounds of tin are needed to make 20 lb of solder? _____

(g) The paint needed for a redecorating job originally cost $84.75, but is on sale at 35% off. What is its sale price? _____ 164   **28**

(h) A shop motor rated at 2.0 hp is found to deliver only 1.6 hp when connected to a transmission system. What is the efficiency of the transmission? _____ 173   **37**

(i) If the voltage in a circuit is increased from 70 volts to 78 volts, what is the percent increase in voltage? _____ 176   **40**

(Answers to these preview problems are given on page 577.)

If you are certain that you can work *all* these problems correctly, turn to page 146 for a set of practice problems. If you cannot work one or more of the preview problems, turn to the page indicated to the right of the problem. Those who wish to master this material with the greatest success should turn to frame **1** and begin work there.

4 Ratio, Proportion, and Percent

WOULD YOU LIKE TO OPEN A SAVINGS ACCOUNT TODAY, MR. BUZZARD?

WE PAY A BIG, BIG, BIG 6% INTEREST!

LET'S SEE. INFLATION IS 10%...

I'LL PAY TAX ON THE INTEREST

SO I'LL ONLY LOSE A BIG, BIG, BIG 5% THAT WAY.

HOW DARE YOU FIGURE THAT OUT!

SLAM

Reprinted by permission: Tribune Media Services

4-1 RATIO AND PROPORTION

1 Machinists, mechanics, carpenters, and other trades workers use the ideas of ratio and proportion to solve very many technical problems. The compression ratio of an automobile engine, the gear ratio of a machine, scale drawings, the pitch of a roof, the mechanical advantage of a pulley system, and the voltage ratio in a transformer are all practical examples of the ratio concept.

Ratio A *ratio* is a comparison, using division, of two quantities of the same kind, both expressed in the same units. For example, the steepness of a hill can be written as the ratio of its height to its horizontal extent.

10 ft

80 ft

Steepness $= \dfrac{10 \text{ ft}}{80 \text{ ft}} = \dfrac{1}{8}$

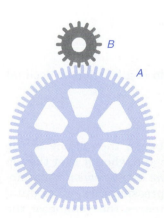

The ratio is usually written as a fraction in lowest terms, and you would read this ratio as either "one-eighth" or "one to eight."

The *gear ratio* of a gear system is defined as

$$\text{Gear ratio} = \frac{\text{number of teeth on the driving gear}}{\text{number of teeth on the driven gear}}$$

Find the gear ratio of the system shown if A is the driving gear and B is the driven gear, and where A has 64 teeth and B has 16 teeth.

Gear ratio = ?

Check your answer in **2**.

---

2 The ratio of the number of teeth on the driving gear to the number of teeth on the driven gear is

$$\text{Gear ratio} = \frac{64 \text{ teeth}}{16 \text{ teeth}} = \frac{4}{1}$$

Always reduce the fraction to lowest terms.

The gear ratio is 4 to 1. In technical work this is sometimes written as 4:1 and is read "four to one."

Typical gear ratios on a passenger car are

First gear: 3.545:1

Second gear: 1.904:1

Third gear: 1.310:1

Reverse: 3.250:1

A gear ratio of 3.545 means that the engine turns 3.545 revolutions for each revolution of the drive shaft. If the drive or rear-axle ratio is 3.722, the engine needs to make 3.545 × 3.722 or 13.194 revolutions in first gear to turn the wheels one full turn.

Here are a few important examples of the use of ratios in practical work.

Example 1:   Pulley Ratios

A pulley is a device that can be used to transfer power from one system to another. A pulley system can be used to lift heavy objects in a shop or to connect a power source to a piece of machinery. The ratio of the pulley diameters will determine the relative pulley speeds.

Find the ratio of the diameter of pulley A to the diameter of pulley B in the following drawing.

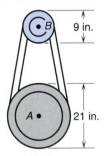

$$\frac{\text{Diameter of pulley } A}{\text{Diameter of pulley } B} = \frac{21 \text{ in.}}{9 \text{ in.}} \quad \boxed{\text{Same units}}$$

$$= \frac{7}{3}$$

The ratio is 7 to 3 or 7:3.

Notice that the units, inches, cancel from the ratio. A ratio is a fraction or decimal number and it has no units.

Example 2:   Compression Ratio

In an automobile engine there is a large difference between the volume of the cylinder space when a piston is at the bottom of its stroke and when it is at the top of its stroke. This difference in volumes is called the *engine displacement*. Automotive mechanics find it very useful to talk about the compression ratio of an automobile engine. The *compression ratio* of an engine compares the volume of the cylinder at maximum expansion to the volume of the cylinder at maximum compression.

$$\text{Compression ratio} = \frac{\text{expanded volume}}{\text{compressed volume}}$$

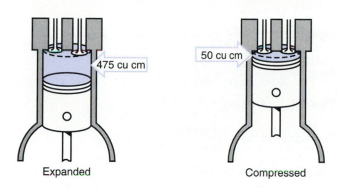

Expanded   Compressed

Find the compression ratio of a gasoline engine if each cylinder has a maximum volume of 475 cu cm and a minimum or compression volume of 50 cu cm.

$$\text{Compression ratio} = \frac{475 \text{ cu cm}}{50 \text{ cu cm}} = \frac{19}{2}$$

$$= 9.5$$

Compression ratios are always written so that the second number in the ratio is 1. This compression ratio would be written as $9\frac{1}{2}$ to 1.

Now, for some practice in calculating ratios, work the following problems.

(a)

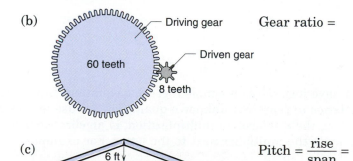

Ratio of pulley diameters $= \dfrac{A}{B} =$

(b)

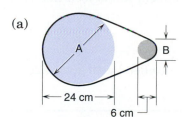

— Driving gear

— Driven gear

60 teeth

8 teeth

Gear ratio =

(c)

6 ft

30 ft

$$\text{Pitch} = \frac{\text{rise}}{\text{span}} =$$

(d) Find the rear-axle ratio of a car if the ring gear has 54 teeth and the pinion gear has 18 teeth.

$$\text{Rear-axle ratio} = \frac{\text{number of teeth on ring gear}}{\text{number of teeth on pinion gear}}$$

(e) A gasoline engine has a maximum cylinder volume of 34 cu in. and a compressed volume of 4 cu in. Find its compression ratio.

Check your answers in **3**.

---

3 (a) Ratio of pulley diameters $= \dfrac{24 \text{ cm}}{6 \text{ cm}}$ or   4 to 1.

(b) Gear ratio $= \dfrac{60 \text{ teeth}}{8 \text{ teeth}} = \dfrac{15}{2}$ or   15 to 2.

(c) Pitch $= \dfrac{6\text{-ft rise}}{30\text{-ft span}} = \dfrac{6}{30}$ or $\dfrac{1}{5}$.

The pitch is 1 to 5.

(d) Rear-axle ratio $= \dfrac{54 \text{ teeth}}{18 \text{ teeth}} = \dfrac{3}{1}$.

The rear-axle ratio is 3 to 1.

(e) Compression ratio $= \dfrac{34 \text{ cu in.}}{4 \text{ cu in.}} = \dfrac{17}{2}$ or   8.5 to 1.

Simple Equations   In order to use ratios to solve a variety of problems, you must be able to solve a simple kind of algebraic equation. Consider this puzzle: "I'm thinking of a number. When I multiply my number by 3, I get 15. What is my number?" Solve the puzzle and check your solution in **4**.

---

4 If you answered "five," you're correct. Think about how you worked it out. Most people take the answer 15 and do the "reverse" of multiplying by 3. That is, they divide by 3. In symbols the problem would look like this:

If $3 \times \square = 15$

then $\square = 15 \div 3 = 5$

Try another one: A number multiplied by 40 gives 2200. Find the number. Look for the answer in **5**.

---

5 Using symbols again,

$$40 \times \square = 2200$$

so $\square = 2200 \div 40 = \dfrac{2200}{40} = 55$

You have just solved a simple algebraic equation. But in algebra, instead of drawing boxes to represent unknown quantities, we use letters of the alphabet. Instead of using the $\times$ symbol for multiplication, in algebra we use a raised dot $\cdot$, or we simply write the multiplier next to the letter. For example,

$$40 \times \square = 2200$$

can be written $40 \cdot N = 2200$

or $40N = 2200$

To solve this type of equation, you divide 2200 by the multiplier of N, which is 40.

For practice, solve these equations.

(a) $25n = 275$ (b) $6M = 96$

(c) $100x = 550$ (d) $12y = 99$

(e) $88 = 8P$ (f) $15 = 30a$

Check your work in **6**.

---

6   (a) $n = \dfrac{275}{25} = 11$ (b) $M = \dfrac{96}{6} = 16$

(c) $x = \dfrac{550}{100} = 5.5$ (d) $y = \dfrac{99}{12} = 8.25$

(e) Notice that P, the unknown quantity, is on the *right* side of the equation. You must divide the number by the multiplier of the unknown, so

$$P = \frac{88}{8} = 11$$

(f) Again, a, the unknown quantity, is on the right side, so

$$a = \frac{15}{30} = 0.5$$

Careful ▶ In problems like (f) some students mistakenly think that they must always divide the larger number by the smaller. That is not always correct. Remember, for an equation of this simple form, always divide by the number that multiplies the unknown. ◀

Other types of equations will be solved in Chapter 7, but only this type is needed for our work with proportions and percent.

Proportions   A *proportion* is an equation stating that two ratios are equal. For example,

$\dfrac{1}{3} = \dfrac{4}{12}$ is a proportion

Notice that the equation is true because the fraction $\frac{4}{12}$ expressed in lowest terms is equal to $\frac{1}{3}$.

Note ▶ When you are working with proportions, it is helpful to have a way of stating them in words. For example, the proportion

$\dfrac{2}{5} = \dfrac{6}{15}$ can be stated in words as

"Two is to five as six is to fifteen." ◀

When one of the four numbers in a proportion is unknown, it is possible to find the value of that number. In the proportion

$$\frac{1}{3} = \frac{4}{12}$$

notice what happens when we multiply diagonally:

$\dfrac{1}{3} = \dfrac{4}{12} \rightarrow 1 \cdot 12 = 12$ $\dfrac{1}{3} = \dfrac{4}{12} \rightarrow 3 \cdot 4 = 12$

These diagonal products are called the *cross-products* of the proportion. If the proportion is a true statement, the cross-products will always be equal. Here are more examples:

$\dfrac{5}{8} = \dfrac{10}{16} \rightarrow 5 \cdot 16 = 80$ and $8 \cdot 10 = 80$

$$\frac{9}{12} = \frac{3}{4} \rightarrow 9 \cdot 4 = 36 \text{ and } 12 \cdot 3 = 36$$

$$\frac{10}{6} = \frac{5}{3} \rightarrow 10 \cdot 3 = 30 \text{ and } 6 \cdot 5 = 30$$

This very important fact is called the cross-product rule.

THE CROSS-PRODUCT RULE

If $\dfrac{a}{b} = \dfrac{c}{d}$ then $a \cdot d = b \cdot c$.

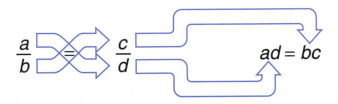

Find the cross-products for the proportion $\frac{3}{5} = \frac{12}{20}$. Check your answers in **7**.

---

7 The cross-products are $3 \cdot 20 = 60$ and $5 \cdot 12 = 60$.

The cross-product rule can be used to solve proportions, that is, to find the value of an unknown number in the proportion. For example, in the proportion

$$\frac{x}{4} = \frac{12}{16}$$

to solve the proportion means to find the value of the unknown quantity x that makes the equation true. To do this:

First, use the cross-product rule.

If $\quad \dfrac{x}{4} = \dfrac{12}{16} \quad$ then $\quad 16x = 48$

Then, solve this equation using the division technique.

$$x = \frac{48}{16} = 3$$

Finally, check your answer by replacing x with 3 in the original proportion equation.

$$\frac{3}{4} = \frac{12}{16}$$

Check 1: Find the cross-products

$3 \cdot 16 = 48 \quad$ and $\quad 4 \cdot 12 = 48 \quad$ The answer is correct.

Check 2: Write $\frac{12}{16}$ in lowest terms.

$$\frac{12}{16} = \frac{4 \cdot 3}{4 \cdot 4} = \frac{3}{4}$$

Now try this one:

$$\frac{6}{N} = \frac{15}{10}$$

Solve this proportion for N. Compare your answer to ours in **8**.

---

8 **Step 1** Apply the cross-product rule.

$$\text{If} \quad \frac{6}{N} = \frac{15}{10} \quad \text{then} \quad 6 \cdot 10 = 15N$$

$$\text{or} \quad 15N = 60$$

Step 2 Solve the equation.

$$N = \frac{60}{15} = 4$$

Step 3 **Check:** Substitute 4 for N in the original proportion.

$$\frac{6}{4} = \frac{15}{10}$$

The cross-products are equal: $6 \cdot 10 = 60$ and $4 \cdot 15 = 60$.

Notice that, in lowest terms,

$$\frac{6}{4} = \frac{3}{2} \quad \text{and} \quad \frac{15}{10} = \frac{3}{2}$$

Note ▶ The unknown quantity can appear in any one of the four positions in a proportion. No matter where the unknown appears, solve the proportion the same way: write the cross-products and solve the resulting equation. ◀

Here are more practice problems. Solve each proportion.

(a) $\dfrac{28}{40} = \dfrac{x}{100}$ (b) $\dfrac{12}{y} = \dfrac{8}{50}$ (c) $\dfrac{n}{6} = \dfrac{7}{21}$

(d) $\dfrac{12}{9} = \dfrac{32}{M}$ (e) $\dfrac{Y}{7} = \dfrac{3}{4}$ (f) $\dfrac{6}{5} = \dfrac{2}{T}$

(g) $\dfrac{2\frac{1}{2}}{3\frac{1}{2}} = \dfrac{w}{2}$ (h) $\dfrac{12}{E} = \dfrac{0.4}{1.5}$

Check your answers in **9**.

---

9 (a) $\dfrac{28}{40} = \dfrac{x}{100}$ $28 \cdot 100 = 40 \cdot x$ or $40x = 2800$

$$x = 2800 \div 40$$
$$x = 70$$

(b) $\dfrac{12}{y} = \dfrac{8}{50}$ $600 = 8y$ or $y = \dfrac{600}{8} = 75$

(c) $\dfrac{n}{6} = \dfrac{7}{21}$ $21n = 42$ $n = \dfrac{42}{21} = 2$

(d) $\dfrac{12}{9} = \dfrac{32}{M}$ $12M = 288$ $M = \dfrac{288}{12} = 24$

(e) $\dfrac{Y}{7} = \dfrac{3}{4}$ $4Y = 21$ $Y = \dfrac{21}{4} = 5\frac{1}{4}$ or 5.25

(f) $\dfrac{6}{5} = \dfrac{2}{T}$ $6T = 10$ $T = \dfrac{10}{6} = 1\dfrac{2}{3}$

(g) $\dfrac{2\frac{1}{2}}{3\frac{1}{2}} = \dfrac{w}{2}$ $3.5w = 5$ $w = \dfrac{5}{3.5} = 1\dfrac{3}{7}$

(h) $\dfrac{12}{E} = \dfrac{0.4}{1.5}$ $18 = 0.4E$ $E = \dfrac{18}{0.4} = 45$

Note ▶ You may have noticed that the two steps for solving a proportion can be simplified into one step. For example, in problem (a), to solve

$\dfrac{28}{40} = \dfrac{x}{100}$ we can write the answer directly as

$x = \dfrac{28 \cdot 100}{40}$

Always divide by the number that is diagonally opposite the unknown, and multiply the two remaining numbers that are diagonally opposite each other.

This shortcut comes in handy when using a calculator. For example, with a calculator, problem (h) becomes

12 ⊗ **1.5** ⊘ **.4** = →   45.

The numbers 12 and 1.5 are diagonally opposite each other in the proportion.

0.4 is diagonally opposite the unknown E. ◀

Practice this one-step process by solving the proportion

$\dfrac{120}{25} = \dfrac{6}{Q}$

Check your answer in **10**.

---

10 $Q = \dfrac{6 \cdot 25}{120} = 1.25$ Check it.

Problem Solving Using Proportions

We can use proportions to solve a variety of problems involving ratios.

If you are given the value of a ratio and one of its terms, it is possible to find the other term. For example, if the pitch of a roof is supposed to be 1 to 5, and the span is 20 ft, what must be the rise?

$\text{Pitch} = \dfrac{\text{rise}}{\text{span}}$

A ratio of 1 to 5 is equivalent to the fraction $\frac{1}{5}$; therefore, the equation becomes

$\dfrac{1}{5} = \dfrac{\text{rise}}{20}$

or

$\dfrac{1}{5} = \dfrac{R}{20}$

Find the cross-products.

$5R = 20$

Then solve for R.

$$R = \frac{20}{5}$$

$R = 4$ ft   The rise is 4 ft.

The algebra you learned earlier in this chapter will enable you to solve any ratio problem of this kind.

Try these problems.

(a) If the gear ratio on a mixing machine is 6:1 and the driven gear has 12 teeth, how many teeth are on the driving gear?

(b) The pulley system of an assembly belt has a pulley diameter ratio of 4. If the larger pulley has a diameter of 15 in., what is the diameter of the smaller pulley?

(c) The compression ratio of a Datsun 280Z is 8.3 to 1. If the compressed volume of the cylinder is 36 cu cm, what is the expanded volume of the cylinder?

(d) On a certain construction job, concrete is made using a volume ratio of 1 part cement to $2\frac{1}{2}$ parts sand and 4 parts gravel. How much sand should be mixed with 3 cu ft of cement?

Check your work in **11**.

---

11 (a) Gear ratio = $\dfrac{\text{number of teeth on driving gear}}{\text{number of teeth on driven gear}}$

A ratio of 6:1 is equivalent to the fraction $\dfrac{6}{1}$.

Therefore,

$$\frac{6}{1} = \frac{x}{12}$$

or $72 = x$ The driving gear has 72 teeth.

(b) Pulley ratio = $\dfrac{\text{diameter of larger pulley}}{\text{diameter of smaller pulley}}$

$$4 = \frac{15 \text{ in.}}{D} \qquad \text{or} \qquad \frac{4}{1} = \frac{15}{D}$$

$$4D = 15$$

$$D = \frac{15}{4} \qquad \text{or} \qquad D = 3\frac{3}{4} \text{ in.}$$

(c) Compression ratio = $\dfrac{\text{expanded volume}}{\text{compressed volume}}$

$$8.3 = \frac{V}{36 \text{ cu cm}} \qquad \text{or} \qquad \frac{83}{10} = \frac{V}{36}$$

$$V = \frac{83 \cdot 36}{10} = 298.8 \text{ cu cm}$$

 83 ⊗ **36** ⊘ **10** ⊜ → ☐ *298.8*

In any practical situation this answer would be rounded to 300 cu cm.

(d) Ratio of cement to sand = $\dfrac{\text{volume of cement}}{\text{volume of sand}}$

$$\dfrac{1}{2\frac{1}{2}} = \dfrac{3 \text{ cu ft}}{S}$$

Therefore,

$$S = 3 \cdot 2\frac{1}{2}$$

$$S = 7\frac{1}{2} \text{ cu ft}$$

Now turn to **12** for a set of practice problems on ratio and proportion.

12

Exercises 4-1 **Ratio and Proportion**

A. Complete the following tables.

1.

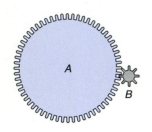

	Teeth on Driving Gear A	Teeth on Driven Gear B	Gear Ratio, $\dfrac{A}{B}$
(a)	35	5	
(b)	12	7	
(c)		3	2
(d)	21		$3\frac{1}{2}$
(e)	15		1 to 3
(f)		18	1 to 2
(g)		24	2:3
(h)	30		3:5
(i)	27	18	
(j)	12	30	

2.

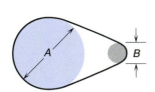

	Diameter of Pulley A	Diameter of Pulley B	Pulley Ratio, $\dfrac{A}{B}$
(a)	16 in.	6 in.	
(b)	15 in.	12 in.	
(c)		8 in.	2
(d)	27 cm		4.5
(e)		10 cm	4 to 1
(f)	$8\frac{1}{8}$ in.	$3\frac{1}{4}$ in.	
(g)	8.46 cm	11.28 cm	
(h)	20.41 cm		3.14 to 1
(i)		12.15 cm	1 to 2.25
(j)	4.45 cm		0.25

3.

	Rise	Span	Pitch
(a)	8 ft	12 ft	
(b)		24 ft	1 to 3
(c)	7 ft		1 to 4
(d)	14 ft 4 in.	25 ft 1 in.	
(e)	9 ft	15 ft	
(f)		20 ft	0.2
(g)	3 ft		0.15
(h)		30 ft 6 in.	1:6

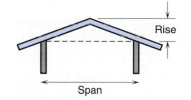

Rise

Span

B. Solve these proportion equations.

1. $\dfrac{3}{2} = \dfrac{x}{8}$

2. $\dfrac{6}{R} = \dfrac{5}{72}$

3. $\dfrac{y}{60} = \dfrac{5}{3}$

4. $\dfrac{2}{15} = \dfrac{8}{H}$

5. $\dfrac{5}{P} = \dfrac{30}{7}$

6. $\dfrac{1}{6} = \dfrac{17}{x}$

7. $\dfrac{A}{2.5} = \dfrac{13}{10}$

8. $\dfrac{27}{M} = \dfrac{3}{0.8}$

9. $\dfrac{2}{5} = \dfrac{T}{4.5}$

10. $\dfrac{0.12}{N} = \dfrac{2}{7}$

11. $\dfrac{138}{23} = \dfrac{18}{x}$

12. $\dfrac{3.25}{1.5} = \dfrac{A}{0.6}$

13. $\dfrac{x}{34.86} = \dfrac{1.2}{8.3}$

14. $\dfrac{2\frac{1}{2}}{R} = \dfrac{1\frac{1}{4}}{3\frac{1}{4}}$

15. $\dfrac{2 \text{ ft } 6 \text{ in.}}{4 \text{ ft } 3 \text{ in.}} = \dfrac{L}{8 \text{ ft } 6 \text{ in.}}$

16. $\dfrac{6.2 \text{ cm}}{x} = \dfrac{1.2 \text{ in.}}{11.4 \text{ in.}}$

17. $\dfrac{3 \text{ ft } 4 \text{ in.}}{4 \text{ ft } 2 \text{ in.}} = \dfrac{3.2 \text{ cm}}{x}$

18. $\dfrac{3\frac{1}{2} \text{ in.}}{W} = \dfrac{1.4}{0.05}$

C. Solve.

1. **Auto Mechanics**   The compression ratio in a certain engine is 6.6 to 1. If the expanded volume of a cylinder is 33 cu in., what is the compressed volume?

2. **Painting and Decorating**   If 1 gallon of paint covers 820 sq ft, how many gallons will be needed to cover 2650 sq ft with two coats? (Assume that you cannot buy a fraction of a gallon.)

3. **Machine Technology**   If 28 tapered pins can be machined from a steel rod 12 ft long, how many tapered pins can be made from a steel rod 9 ft long?

4. **Carpentry**   If 6 lb of nails are needed for each thousand lath, how many pounds are required for 4250 lath?

5. **Masonry**   For a certain kind of plaster work, 1.5 cu yd of sand are needed for every 100 sq yd of surface. How much sand will be needed for 350 sq yd of surface?

6. **Printing**   The paper needed for a printing job weighs 12 lb per 500 sheets. How many pounds of paper are needed to run a job requiring 12,500 sheets?

7. **Machine Technology**   A cylindrical oil tank 8 ft deep holds 420 gallons when filled to capacity. How many gallons remain in the tank when the depth of oil is $5\frac{1}{2}$ ft?

8. **Agricultural Technology**   A liquid fertilizer must be prepared by using one part of concentrate for every 32 parts of water. How much water should be added to 7 oz of concentrate?

9. **Welding**   A 10-ft bar of I-beam weighs 208 lb. What is the weight of a 6-ft length?

10. **Photography**   A photographer must mix a chemical in the ratio of 1 part chemical for every 7 parts of water. How many ounces of chemical should be used to make a 3-qt *total* mixture? (1 qt = 32 oz)

11. If you earn $409.60 for a 32-hr work week, how much would you earn for a 40-hr work week at the same hourly rate?

12. **Machine Technology**   A machinist can produce 12 parts in 40 min. How many parts can the machinist produce in 4 hr?

13. **Auto Mechanics**   The headlights on a car are set so the light beam drops 2 in. for each 25 ft measured horizontally. If the headlights are mounted 30 in. above the ground, how far ahead of the car will they hit the ground?

14. **Machine Technology**   A machinist wastes $2\frac{3}{4}$ lb of steel in fabricating 16 rods. How much steel will be wasted in producing 120 rods?

4-2 INTRODUCTION TO PERCENT

13   In many practical calculations, it is helpful to be able to compare two numbers, and it is especially useful to express the comparison in terms of a percent. Percent calculations are a very important part of any work in business, technical skills, or the trades.

The word *percent* comes from the Latin phrase *per centum* meaning "by the hundred" or "for every hundred." A number written as a percent is being compared with a second number called the standard or *base*. For example, what part of the base length is length *A?*

We could answer the question with a fraction or ratio, a decimal, or a percent. First, divide the base into 100 equal parts.

Then compare length A with it.

The length of A is 40 parts out of the 100 parts that make up the base.

A is $\frac{40}{100}$ or 0.40 or 40% of A.

$\frac{40}{100}$ = 40 %   40 % means 40 parts in 100 or $\frac{40}{100}$

For the following diagram, what part of the base length is length B?

Answer with a percent.

Turn to **14** to check your answer.

---

14 B is $\frac{60}{100}$ or 60% of the base.

The compared number may be larger than the base. For example,

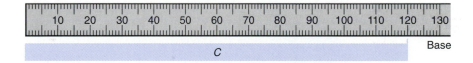

In this case, divide the base into 100 parts and extend it in length.

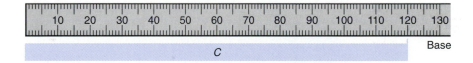

The length of C is 120 of the 100 parts that make up the base.

C is $\frac{120}{100}$ or 120 % of the base.

Ratios, decimals, and percents are all alternative ways to compare two numbers. For example, in the following drawing what fraction of the rectangle is shaded? What part of 12 is 3?

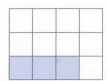

First, we can write the answer as a ratio:
$$\frac{3 \text{ shaded squares}}{12 \text{ squares}} = \frac{3}{12} = \frac{1}{4}$$

Second, by dividing 4 into 1 we can write this ratio or fraction as a decimal:

$$\frac{1}{4} = 0.25 \qquad 4\overline{)1.00}^{\,.25}$$

Finally, we can rewrite the fraction with a denominator of 100:

$$\frac{1}{4} = \frac{1 \times \boxed{25}}{4 \times \boxed{25}}$$

$$\frac{1}{4} = \frac{25}{100} \qquad \text{or} \qquad 25\%$$

Changing Decimal Numbers to Percents

Converting a number from fraction or decimal form to a percent is a very useful skill.

To write a decimal number as a percent, multiply the decimal number by 100%.

For example,

$$0.60 = 0.60 \times 100\% = 60\%$$

Multiply by 100%

$$0.375 = 0.375 \times 100\% = 37.5\%$$
$$3.4 = 3.4 \times 100\% = 340\%$$
$$0.02 = 0.02 \times 100\% = 2\%$$

Notice that multiplying by 100 is equivalent to shifting the decimal point two places to the right. For example,

$$0.60 = 0.60 = 60.\% \qquad \text{or} \qquad 60\%$$

Shift decimal point two places right

$$0.056 = 0.056 = 5.6\%$$

Rewrite the following decimal numbers as percents.

(a)   0.75   (b)   1.25   (c)   0.064

(d)   3   (e)   0.05   (f)   0.004

Check your answers in **15**.

---

15   (a) $0.75 = 0.75 \times 100\% = 75\%$:   0.75 becomes 75%

(b) $1.25 = 1.25 \times 100\% = 125\%$ 1.25 becomes 125%

(c) $0.064 = 0.064 \times 100\% = 6.4\%$ 0.064 becomes 6.4%

(d) $3 = 3 \times 100\% = 300\%$ 3 = 3.00   or   300%

(e) $0.05 = 0.05 \times 100\% = 5\%$ 0.05 becomes 5%

(f) $0.004 = 0.004 \times 100\% = 0.4\%$ 0.004 becomes 0.4%

Changing Fractions to Percents

> To rewrite a fraction as a percent, first change to decimal form by dividing, then multiply by 100%.

For example,

$\frac{1}{2}$ is 1 divided by 2, or $\quad 2\overline{)1.0}^{\,0.5}\quad$ so that

$\frac{1}{2} = 0.5 = 0.5 \times 100\% = 50\%$

$\quad$ └─Multiply by 100%

└─Change to a decimal

More examples:

$\frac{3}{4} = 0.75 = 0.75 \times 100\% = 75\%$

$\frac{3}{20} = 0.15 = 0.15 \times 100\% = 15\%\quad$ since $\frac{3}{20} \;=\; 20\overline{)3.00}^{\,0.15}$

$1\frac{7}{20} = \frac{27}{20} = 1.35 = 1.35 \times 100\% = 135\%\quad$ since $\frac{27}{20} \;=\; 20\overline{)27.00}^{\,1.35}$

Your turn. Rewrite $\frac{5}{16}$ as a percent.

Check your work in **16**.

Learning Help ▶ Here is an easy way to change certain fractions into percents. If the denominator of the fraction divides exactly into 100, change the given fraction into an equivalent fraction with a denominator of 100. The new numerator is the percent. For example,

$\frac{3}{20} = \frac{3 \times 5}{20 \times 5} = \frac{15}{100}\qquad$ so $\qquad \frac{3}{20} = 15\%$ ◀

---

16 $\quad \frac{5}{16}$ means $16\overline{)5.0000}^{\,0.3125}\qquad$ so that $\qquad \frac{5}{16} = 0.3125 = 31.25\%$

This is often written as $31\frac{1}{4}\%$.

Some fractions cannot be converted to an exact decimal. For example, $\frac{1}{3} = 0.3333\ldots$, where the 3s continue endlessly. We can round to get an approximate percent.

$\frac{1}{3} \approx 0.3333 \approx 33.33\%\qquad$ or $33\frac{1}{3}\%\qquad$ (The sign ≈ means "approximately equal to.")

The fraction $\frac{1}{3}$ is roughly equal to 33.33% and exactly equal to $33\frac{1}{3}\%$.

Rewrite the following fractions as percents.

(a) $\frac{4}{5}$ $\qquad$ (b) $\frac{2}{3}$ $\qquad$ (c) $3\frac{1}{8}$

(d) $\frac{5}{12}$ $\qquad$ (e) $\frac{5}{6}$ $\qquad$ (f) $1\frac{1}{3}$

Check your work in **17**.

17 (a) $\dfrac{4}{5} = 0.80 = 80\%$

Shift the decimal point
two places to the right

Change to a decimal
by dividing

(b) $\dfrac{2}{3} \approx 0.6667 \approx 66.67\%$ or exactly $66\dfrac{2}{3}\%$

(c) $\dfrac{1}{8}$ is $8)\overline{1.000}$ with quotient 0.125 so $3\dfrac{1}{8} = 3.125$

$3\dfrac{1}{8} = 3.125 = 312.5\%$

(d) $\dfrac{5}{12}$ is $12)\overline{5.000}$ with quotient $0.4166\ldots$ so

$\dfrac{5}{12} = 0.4166\ldots \approx 41.67\%$ or exactly $41\dfrac{2}{3}\%$

(e) $\dfrac{5}{6}$ is $6)\overline{5.000}$ with quotient $0.833\ldots$ so

$\dfrac{5}{6} \approx 0.833 \approx 83.3\%$ or exactly $83\dfrac{1}{3}\%$

(f) $1\dfrac{1}{3} = 1.333\ldots \approx 133.3\%$ or exactly $133\dfrac{1}{3}\%$

Changing Percents to Decimal Numbers

In order to use percent numbers when you solve practical problems, it is often necessary to change a percent to a decimal number.

> To change a percent to a decimal number, divide by 100%.

For example,

$50\% = \dfrac{50\%}{100\%} = \dfrac{50}{100} = 0.5$

Divide by 100%

$6\% = \dfrac{6\%}{100\%} = \dfrac{6}{100} = 0.06$

$0.2\% = \dfrac{0.2\%}{100\%} = \dfrac{0.2}{100} = 0.002$

Notice that in each of these examples division by 100% is the same as moving the decimal point two digits to the left. For example,

50% = 50.% = 0.50 or 0.5

Shift the decimal
point two
places left

5% = 05.% = 0.05

0.2% = 00.2% = 0.002

If a fraction is part of the percent number, write it as a decimal number before dividing by 100%. For example,

$$8\frac{1}{2}\% = 8.5\% = \frac{8.5\%}{100\%} = 0.085$$

Now try these. Write each percent as a decimal number.

(a) 4%   (b) 112%   (c) 0.5%

(d) $9\frac{1}{4}\%$ (e) 45%   (f) $12\frac{1}{3}\%$

Our answers are given in **18**.

---

18 (a) $4\% = \dfrac{4\%}{100\%} = \dfrac{4}{100} = 0.04$ since $100\overline{)4.00}^{\,0.04}$

This may also be done by shifting the decimal point: 4% = 04.% = 0.04.

(b) 112% = 112.% = 1.12

(c) 0.5% = 00.5% = 0.005

(d) $9\frac{1}{4}\% = 9.25\% = 09.25\% = 0.0925$

(e) 45% = 45.% = 0.45

(f) $12\frac{1}{3}\% \approx 12.33\% = 0.1233$

Did you notice that in problem (b) a percent greater than 100% gives a decimal number greater than 1?

100% = 1

200% = 2

300% = 3   and so on

Here is a table of the most often used fractions with their percent equivalents.

PERCENT EQUIVALENTS

Percent	Decimal	Fraction		Percent	Decimal	Fraction
5%	0.05	$\frac{1}{20}$		50%	0.50	$\frac{1}{2}$
$6\frac{1}{4}$%	0.0625	$\frac{1}{16}$		60%	0.60	$\frac{3}{5}$
$8\frac{1}{3}$%	$0.08\overline{3}$	$\frac{1}{12}$		$62\frac{1}{2}$%	0.625	$\frac{5}{8}$
10%	0.10	$\frac{1}{10}$		$66\frac{2}{3}$%	$0.\overline{6}$	$\frac{2}{3}$
$12\frac{1}{2}$%	0.125	$\frac{1}{8}$		70%	0.70	$\frac{7}{10}$
$16\frac{2}{3}$%	$0.1\overline{6}$	$\frac{1}{6}$		75%	0.75	$\frac{3}{4}$
20%	0.20	$\frac{1}{5}$		80%	0.80	$\frac{4}{5}$
25%	0.25	$\frac{1}{4}$		$83\frac{1}{3}$%	$0.8\overline{3}$	$\frac{5}{6}$
30%	0.30	$\frac{3}{10}$		$87\frac{1}{2}$%	0.875	$\frac{7}{8}$
$33\frac{1}{3}$%	$0.\overline{3}$	$\frac{1}{3}$		90%	0.90	$\frac{9}{10}$
$37\frac{1}{2}$%	0.375	$\frac{3}{8}$		100%	1.00	$\frac{10}{10}$
40%	0.40	$\frac{2}{5}$				

Work the following problems for practice in using percent.

A. Write each number as a percent.

1. 0.40	2. 0.10	3. 0.95	4. 0.03
5. 0.3	6. 0.015	7. 0.60	8. 7.75
9. 1.2	10. 4	11. 6.04	12. 9
13. $\frac{1}{4}$	14. $\frac{1}{5}$	15. $\frac{7}{20}$	16. $\frac{3}{8}$
17. $\frac{5}{6}$	18. $2\frac{3}{8}$	19. $3\frac{7}{10}$	20. $1\frac{4}{5}$

B. Write each percent as a decimal number.

1. 7%	2. 3%	3. 56%	4. 15%
5. 1%	6. $7\frac{1}{2}$%	7. 90%	8. 0.3%
9. 150%	10. $1\frac{1}{2}$%	11. $6\frac{3}{4}$%	12. $\frac{1}{2}$%
13. $12\frac{1}{4}$%	14. $125\frac{1}{5}$%	15. 1.2%	16. 240%

Check your answers in **19**.

---

19 A.
1. 40%	2. 10%	3. 95%	4. 3%
5. 30%	6. 1.5%	7. 60%	8. 775%

9. 120%	10. 400%	11. 604%	12. 900%
13. 25%	14. 20%	15. 35%	16. 37.5%
17. $83\frac{1}{3}$%	18. 237.5%	19. 370%	20. 180%

B.
1. 0.07	2. 0.03	3. 0.56	4. 0.15
5. 0.01	6. 0.075	7. 0.90	8. 0.003
9. 1.5	10. 0.015	11. 0.0675	12. 0.005
13. 0.1225	14. 1.252	15. 0.012	16. 2.4

Now turn to frame **20** for a set of problems involving percent calculations.

20

Exercises 4-2   **Introduction to Percent**

A. Convert to a percent.

1. 0.32	2. 1	3. 0.5	4. 2.1
5. $\frac{1}{4}$	6. 3.75	7. 40	8. 0.675
9. 2	10. 0.075	11. $\frac{1}{2}$	12. $\frac{1}{6}$
13. 0.335	14. 0.001	15. 0.005	16. $\frac{3}{10}$
17. $\frac{3}{2}$	18. $\frac{3}{40}$	19. $3\frac{3}{10}$	20. $\frac{1}{5}$

B. Convert to a decimal number.

1. 6%	2. 45%	3. 1%	4. 33%
5. 71%	6. 456%	7. $\frac{1}{2}$%	8. 0.05%
9. $6\frac{1}{4}$%	10. $8\frac{3}{4}$%	11. 30%	12. 2.1%
13. 800%	14. 8%	15. 0.25%	16. $16\frac{1}{3}$%

When you have had the practice you need, check your answers on page 589, then turn to frame **21**.

4-3 PERCENT PROBLEMS

21   In all your work with percent you will find that there are three basic types of problems. These three are related to all percent problems that arise in business, technology, or the trades. In this section we show you how to solve any percent problem, and we examine each of the three types of problems.

All percent problems involve three quantities:

B, the *base* or total amount, a standard used for comparison

P, the *percentage* or part being compared with the base

R, the *rate* or *percent*, a percent number

For any percent problem, these three quantities are related by the proportion

$$\frac{P}{B} = \frac{R}{100}$$

For example, the proportion

$$\frac{3}{4} = \frac{75}{100}$$

can be translated to the percent statement

   "Three is 75% of four."

Here, the percentage P is 3, the base B is 4, and the percent R is 75.

To solve any percent problem, we need to identify which of the quantities given in the problem is P, which is B, and which is R. Then we can write the percent proportion and solve for the missing or unknown quantity.

When P Is Unknown   Consider these three problems:

What is 30% of 50?
Find 30% of 50.
30% of 50 is what number?

These three questions are all forms of the same problem. They are all asking you to find P, the percentage.

We know that 30 is the percent R because it has the % symbol attached to it. The number 50 is the base B. To solve, write the percent proportion, substituting 50 for B and 30 for R.

$$\frac{P}{50} = \frac{30}{100}$$

Now solve using cross-products and division.

$100P = 30 \cdot 50$
$100P = 1500$

$$P = \frac{1500}{100} = 15$$

15 is 30% of 50.

Check: Substitute the answer back into the proportion.

$$\frac{15}{50} = \frac{30}{100} \qquad \text{or} \qquad 15 \cdot 100 = 30 \cdot 50 \qquad \text{which is correct}$$

The answer is reasonable: 30% is roughly one-third, and 15 is roughly one-third of 50.

The percent number R is easy to identify because it always has the % symbol attached to it. If you have trouble distinguishing the percentage P from the base B, notice that B is usually associated with the word "of" and P is usually associated with the word "is."

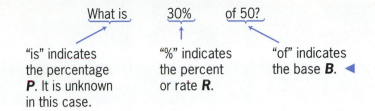

What is   30%   of 50?

"is" indicates the percentage P. It is unknown in this case.

"%" indicates the percent or rate R.

"of" indicates the base B. ◀

Now try this problem to test yourself.

Find $8\frac{1}{2}\%$ of 160.

Look for our solution in **22**.

---

22   **Step 1**   If your mental math skills are good, estimate the answer first. In this case, $8\frac{1}{2}\%$ of 160 should be a little less than 10% of 160, or 16.

Step 2   Now identify the three quantities, P, B, and R. R is obviously $8\frac{1}{2}$. The word "is" does not appear, but "of" appears with 160, so 160 is B. Therefore, P must be the unknown quantity.

Step 3   Set up the percent proportion.

$$\frac{P}{B} = \frac{R}{100} \rightarrow \frac{P}{160} = \frac{8\frac{1}{2}}{100}$$

Step 4   Solve.

$$100P = 8\frac{1}{2} \cdot 160$$
$$100P = 1360$$
$$P = 13.6$$

Step 5   Check your answer by substituting into the original proportion equation and comparing cross-products.

Notice that the calculated answer, 13.6, is a little less than 16, our rough estimate.

Solve the following percent problems.

(a) Find 2% of 140 lb.

(b) 35% of $20 is equal to what amount?

(c) What is $7\frac{1}{4}\%$ of $1000?

(d) Calculate $16\frac{2}{3}\%$ of 66.

(e) To account for waste, a carpenter needs to order 120% of the total wood used in making a cabinet. If 15 ft of white oak is actually used, how much should he order?

The step-by-step answers are given in **23**.

---

23 (a) $\dfrac{P}{140} = \dfrac{2}{100}$

$100P = 2 \cdot 140 = 280$

$P = 2.8 \text{ lb}$

Check: $\dfrac{2.8}{140} = \dfrac{2}{100}$

$280 = 280$

(b) $\dfrac{P}{20} = \dfrac{35}{100}$

$100P = 700$

$P = \$7$

Check: $\dfrac{7}{20} = \dfrac{35}{100}$

$700 = 700$

(c) $\dfrac{P}{1000} = \dfrac{7\frac{1}{4}}{100}$

$100P = 7\frac{1}{4} \cdot 1000 = 7250$

$P = \$72.50$

Check: $\dfrac{72.50}{1000} = \dfrac{7\frac{1}{4}}{100}$

$7250 = 7250$

(d) $\dfrac{P}{66} = \dfrac{16\frac{2}{3}}{100}$ $16\frac{2}{3} = \dfrac{50}{3}$

$100P = 66 \cdot \dfrac{50}{3} = \dfrac{\overset{22}{\cancel{66}}}{1} \cdot \dfrac{50}{\underset{1}{\cancel{3}}} = 22 \cdot 50$

$100P = 1100$

$P = 11$ Check it.

 16 $\boxed{a\frac{b}{c}}$ **2** $\boxed{a\frac{b}{c}}$ **3** $\boxed{\times}$ **66** $\boxed{\div}$ **100** $\boxed{=}$ → ▢ *11.*

(e) The key phrases in this problem are

"120% of the total . . ."   and   "15 ft . . . is used."

We are finding 120% of 15 ft. $B = 15 \text{ ft}$

$\dfrac{P}{15} = \dfrac{120}{100}$

$100P = 1800$

$P = 18 \text{ ft}$

Notice that in all problems where P is to be found, we end by dividing by 100 in the last step. Recall that the quick way to divide by 100 is to move the decimal point two places to the left. In problem (c), for example, $100P = 7250$.

The decimal point is after the zero in 7250. Move it two places to the left.

$7250 \div 100 = 72\widetilde{50}. = 72.50$ ◀

When R Is Unknown Consider the following problems:

> *5 is what percent of 8?*

or *Find what percent 5 is of 8.*

or *What percent of 8 is 5?*

Once again, these statements represent three different ways of asking the same question. In each statement the percent R is unknown. We know this because neither of the other two numbers has a % symbol attached.

To solve this kind of problem, first identify P and B. The word "of" is associated with the base B, and the word "is" is associated with the percentage P. In this case, $B = 8$ and $P = 5$.

Next, set up the percent proportion.

$$\frac{P}{B} = \frac{R}{100} \rightarrow \frac{5}{8} = \frac{R}{100}$$

Finally, solve for R.

$$8R = 5 \cdot 100 = 500$$

$$R = 62.5\%$$

Note ▶ When you solve for R, remember to include the % symbol with your answer. ◀

Now try these problems for practice.

(a) What percent of 40 lb is 16 lb?

(b) 65 is what percent of 25?

(c) Find what percent $9.90 is of $18.00.

(d) During reshaping, 6 lb of metal is removed from a casting originally weighing 80 lb. What percent of the metal is removed?

Check your work in **24**.

How can you have over 100% of something? All of it is all of it, right?

Percent relates size of something to a standard. If it is larger than the standard, it is more than 100% of the standard.

24 (a) $\dfrac{16}{40} = \dfrac{R}{100}$ 　　　　(b) $\dfrac{65}{25} = \dfrac{R}{100}$

$40R = 1600$ 　　　　　　　$25R = 6500$

$R = 40\%$ 　　　　　　　　$R = 260\%$

(c) $\dfrac{\$9.90}{\$18.00} = \dfrac{R}{100}$ 　　(d) $\dfrac{6}{80} = \dfrac{R}{100}$

$18R = 990$ 　　　　　　　$80R = 600$

$R = 55\%$ 　　　　　　　　$R = 7.5\%$

Notice that in problem (b) the percentage is larger than the base, resulting in a percent larger than 100%. Some students mistakenly believe that the percentage is always smaller than the base. This is not always true. When in doubt use the "is–of" method to determine P and B.

Learning Help ▶ In all problems where R is unknown, we end up multiplying by 100. Recall that the quick way to do this is to move the decimal point two places to the right. If there is no decimal point, attach two zeros. ◀

When _B_ Is Unknown This third type of percent problem requires that you find the total or base when the percent and the percentage are given. Problems of this kind can be stated as

　　　8.7 is 30% of what number?

or　　*30% of what number is 8.7?*

or　　*Find a number such that 30% of it is 8.7.*

Solve this problem in the same way as the previous ones.

First, identify P, B, and R. Because it carries the % symbol, we know that $R = 30$. Because the word "is" is associated with the number 8.7, we know that $P = 8.7$. The base B is unknown.

Next, set up the percent proportion.

$$\frac{P}{B} = \frac{R}{100} \longrightarrow \frac{8.7}{B} = \frac{30}{100}$$

Finally, solve the proportion.

$30B = 870$

　　$B = 29$ 　　　　**Check:** 　$\dfrac{8.7}{29} = \dfrac{30}{100}$

　　　　　　　　　　　　$8.7(100) = 30 \cdot 29$

　　　　　　　　　　　　which is correct

Ready for a few practice problems? Try these.

(a)　16% of what amount is equal to $5.76?

(b)　$2 is 8% of the cost. Find the cost.

(c)　Find a distance such that $12\frac{1}{2}\%$ of it is $26\frac{1}{4}$ ft.

(d)　The $2800 actually spent on a construction job was 125% of the original estimate. What was the original estimate?

(e)　A builder told his crew that the site preparation and foundation work on a house would represent about 5% of the total building time. If it took them 21 days to do this preliminary work, how long will the entire construction take?

The correct solutions are in **25**.

25 (a) $\dfrac{\$5.76}{B} = \dfrac{16}{100}$

$16B = 576$

$B = \$36$

(b) $\dfrac{2}{B} = \dfrac{8}{100}$

$8B = 200$

$B = \$25$

(c) $\dfrac{26\frac{1}{4}}{B} = \dfrac{12\frac{1}{2}}{100}$

$B = \dfrac{26.25 \times 100}{12.5} = 210$ ft

(d) $\dfrac{2800}{B} = \dfrac{125}{100}$

$B = \dfrac{2800 \times 100}{125} = \$2240$

 For problem (c): **26.25** ⊠ **100** ÷ **12.5** ⊟ → 210.

(e) We are asked to find the total construction time, and we are told that preliminary work is 5% of the total. Since preliminary work takes 21 days, we can restate the problem as

21 is 5% of the total

or $\dfrac{21}{B} = \dfrac{5}{100}$

$B = \dfrac{21 \times 100}{5} = 420$ days

So far we have solved sets of problems that were all of the same type. The key to solving percent problems is to be able to identify correctly the quantities P, B, and R.

Use the hints you have learned to identify these three quantities in the following problems, then solve each problem.

(a) What percent of 25 is 30?

(b) 12 is $66\frac{2}{3}\%$ of what amount?

(c) Find 6% of 2400.

(d) Television City asked you to make a 15% down payment on your purchase of a new TV set. If the down payment was $78, what was the cost of the set?

(e) A printer gives a 5% discount to preferred customers. If the normal cost of a certain job is $3400, what discount would a preferred customer receive?

(f) Twelve computer chips were rejected out of a run of 400. What percent were rejected?

Our solutions are in **26**.

---

26 (a) What percent   of 25   is 30?

            R B P

$\dfrac{30}{25} = \dfrac{R}{100}$

$25R = 3000$

$R = 120\%$

(b) 12   is · $66\frac{2}{3}\%$ of what?

   P R B

$\dfrac{12}{B} = \dfrac{66\frac{2}{3}}{100}$

$66\frac{2}{3} \cdot B = 1200$

$B = 18$

12 ⊠ **100** ÷ **66** $\boxed{a^b_c}$ **2** $\boxed{a^b_c}$ **3** ⊜ → *18.*

(c) Find   6%   of $2400.
     $\underbrace{}_{P}$ $\underbrace{}_{R}$ $\underbrace{}_{B}$

$$\frac{P}{2400} = \frac{6}{100}$$

$$100P = 6 \cdot 2400 = 14400$$

$$P = \$144$$

(d) In this problem, 15% is the rate, so $78 must be either the percentage or the base. Since $78 is the down payment, a part or percentage of the whole price, $78 must be P.

$$\frac{78}{B} = \frac{15}{100}$$

$$15B = 7800$$

$$B = \$520 \qquad \text{The total cost of the TV set was } \$520.$$

(e) 5% is the rate R, so $3400 must be either P or B. The discount is a percentage of the normal cost, so P is unknown and B is $3400.

$$\frac{P}{3400} = \frac{5}{100}$$

$$100P = 5 \cdot 3400 = 17000$$

$$P = \$170 \qquad \text{The preferred customer receives a discount of } \$170.$$

(f) The percent R is not given. The 12 chips are a part (P) of the whole (B) run of 400.

$$\frac{12}{400} = \frac{R}{100}$$

$$400R = 1200$$

$$R = 3\% \qquad 3\% \text{ of the chips were rejected.}$$

Now go to **27** for more practice on the three basic kinds of percent problems.

27

Exercises 4-3   **Percent Problems**

A. Solve.

1. 4 is _____ % of 5.

2. What percent of 25 is 16?

3. 20% of what number is 3?

4. 8 is what percent of 8?

5. 120% of 45 is _____.

6. 8 is _____ % of 64.

7. 3% of 5000 = _____.

8. 2.5% of what number is 2?

9. What percent of 54 is 36?

10. 60 is _____ % of 12.

11. 17 is 17% of _____.

12. 13 is what percent of 25?

13. $8\frac{1}{2}$% of $250 is _____.

14. 12 is _____ % of 2.

15. 6% of 25 is _____.

16. 60% of what number is 14?

17. What percent of 16 is 7?

18. 140 is _____ % of 105?

19. Find 24% of 10.

20. 45 is 12% of what number?

21. 30% of what number is equal to 12?

22. What is 65% of 5?

23. Find a number such that 15% of it is 750.

24. 16% of 110 is what number?

B. Solve.

1. 75 is $33\frac{1}{3}$% of _____.

2. What percent of 10 is 2.5?

3. 6% of $3.29 is _____.

4. 63 is _____ % of 35.

5. 12.5% of what number is 20?

6. $33\frac{1}{3}$% of $8.16 = _____.

7. 9.6 is what percent of 6.4?

8. $0.75 is _____ % of $37.50.

9. $6\frac{1}{4}$% of 280 is _____.

10. 1.28 is _____ % of 0.32.

11. 42.7 is 10% of _____.

12. 260% of 8.5 is _____.

13. $\frac{1}{2}$ is _____ % of 25.

14. $7\frac{1}{4}$% of 50 is _____.

15. 287.5% of 160 is _____.

16. 0.5% of _____ is 7.

17. 112% of _____ is 56.

18. $2\frac{1}{4}$% of 110 is _____.

C. Word Problems

1. If you answered 37 problems correctly on a 42-question test, what percent score would you have?

2. **Office Services** The profits from Ed's Plumbing Co. increased by $14,460, or 30%, this year. What profit did it earn last year?

3. **Machine Technology** A casting weighed 146 lb out of the mold. It weighed 138 lb after finishing. What percent of the weight was lost in finishing?

4. The Smiths bought a house for $184,500 and made a down payment of 20%. What was the actual amount of the down payment?

5. If your salary is $1960 per month and you get a 6% raise, what is the amount of your raise?

6. If state sales tax is 6%, what is the price before tax if the tax is $7.80?

7. A carpenter earns $24.50 per hour. If he receives a $7\frac{1}{2}$% pay raise, what is the amount of his hourly raise?

8. If 5.45% of your salary is withheld for Social Security, what amount is withheld from monthly earnings of $945.00?

9. **Printing** If an 8-in. by 10-in. photocopy is enlarged to 120% of its size, what are its new dimensions?

10. **Police Science** Year-end statistics for a community revealed that in 648 of the 965 residential burglaries, the burglar entered through an unlocked door or window. What percent of the entries were made in this way?

11. **Carpentry** Preparing an estimate for a decking job, a carpenter assumes about a 15% waste factor. If the finished deck contains 1230 lineal feet of boards, how many lineal feet should the carpenter order? (*Hint:* If 15% of the amount ordered is wasted, then 100% – 15% or 85% of that amount is used for the finished deck.)

When you have had the practice you need, check your answers on page 589, then continue in **28** with the study of some applications of percent to practical problems.

4-4 APPLICATIONS OF PERCENT CALCULATIONS

28 Now that you have seen and solved the three basic percent problems, you need some help in applying what you have learned to practical situations.

Discount An important kind of percent problem, especially in business, involves the idea of *discount*. In order to stimulate sales, a merchant may offer to sell some item at less than its normal price. A discount is the amount of money by which the normal price is reduced. It is a percentage or part of the normal price. The *list price* is the normal or original price before the discount is subtracted. The *sales price* is the new, reduced price. It is always less than the list price.

> Sales price = list price – discount

The discount rate (R) is a percent number that enables you to calculate the discount (the percentage P) as a part of the list price (the base B). Therefore, the percent proportion

$$\frac{\text{percentage}}{\text{base}} = \frac{\text{rate}}{100}$$

becomes

$$\frac{\text{discount } D}{\text{list price } L} = \frac{\text{discount rate } R}{100}$$

For example, the list price of a tool is $18.50. On a special sale it is offered at 20% off. What is the sales price?

Try working this problem, then turn to **29** to see our solution.

---

29 $$\frac{\text{discount}}{\text{list price}} = \frac{\text{discount rate}}{100}$$

$$\frac{D}{18.50} = \frac{20}{100}$$

$$100D = 18.50 \times 20 = 370$$

$$D = \$3.70$$

Sales price = list price – discount

$$= \$18.50 - \$3.70$$

$$= \$14.80$$

Think of it this way:

Ready for another problem?

After a 25% discount, the sales price of a power sander is $144. What was its list price?

Check your work in **30**.

---

30   We can use a pie chart diagram to help set up the correct percent proportion.

The sale price, $144, is 75% of the list price. Therefore, the correct proportion is

$$\frac{\text{sale price}}{\text{list price}} = \frac{\text{sale rate}}{100}$$

or $$\frac{144}{L} = \frac{75}{100}$$

Solving for L, we have

$$75L = 14400$$

$$L = \$192 \qquad \text{The list price if \$192.}$$

Check: $\dfrac{\text{discount}}{\text{list price}} = \dfrac{\text{discount rate}}{100}$

$$\frac{D}{192} = \frac{25}{100}$$

$$100D = 4800$$

$$D = \$48 \qquad \text{The discount is } \$48.$$

Sales price = list price − discount

$$= \$192 - \$48$$

$$= \$144 \qquad \text{which is the correct list price as given in the original problem.}$$

Here are a few problems to test your understanding of the idea of discount.

(a) A preinventory sale advertises all tools 20% off. What would be the sale price of a motor drill that cost $69.95 before the sale?

(b) A compressor is on sale for $476 and is advertised as "12% off regular price." What was its regular price?

(c) A set of four 740 by 15 automobile tires is on sale for 15% off list price. What is the sale price if their list price is $70.80 each?

(d) A kitchen sink that retails for $675 is offered to a plumbing contractor for $438.75. What discount rate is the contractor receiving?

Check your answers in **31**.

---

31 (a) $\dfrac{D}{L} = \dfrac{R}{100}$ becomes $\dfrac{D}{69.95} = \dfrac{20}{100}$

$$100D = 69.95 \times 20 = 1399$$

$$D = \$13.99$$

Sales price = list price − discount

$$= \$69.95 - \$13.99$$

$$= \$55.96$$

A pie chart provides a quicker way to do this problem.

If the discount is 20% of the list price, then the sale price is 80% of the list price.

$$\frac{\text{sale price}}{\text{list price}} = \frac{80\%}{100\%}$$

or $\dfrac{S}{69.95} = \dfrac{80}{100}$

$$100S = 5596$$

$$S = \$55.96$$

(b) Pie charts will help you set up the correct proportions.

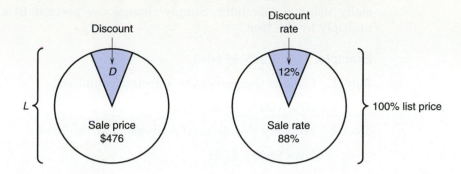

Use the sale price proportion.

$$\frac{\text{sale price}}{\text{list price}} = \frac{\text{sale rate}}{100}$$

$$\frac{476}{L} = \frac{88}{100}$$

$$88L = 47600$$

$$L = \$540.91 \qquad \text{rounded to the nearest cent}$$

(c) Use the discount proportion.

$$\frac{D}{L} = \frac{R}{100} \qquad \text{becomes} \qquad \frac{D}{4 \times 70.80} = \frac{15}{100}$$

Four tires at $70.80 each

$$\frac{D}{283.20} = \frac{15}{100}$$

$$100D = 4248$$

$$D = \$42.48$$

Sales price = list price – discount

$$= \$283.20 - \$42.48$$

$$= \$240.72$$

Using a calculator, first find the discount and store it in memory.

4 $\boxed{\times}$ **70.80** $\boxed{\times}$ **15** $\boxed{\div}$ **100** $\boxed{=}$ $\boxed{\text{STO}}$ $\rightarrow$ `42.48`

Then subtract it from the list price.

283.20 $\boxed{-}$ $\boxed{\text{RCL}}$ $\boxed{=}$ $\rightarrow$ `240.72`

(d) **First,** subtract to find the discount.

Discount = list price – sale price

$$= \$675 - \$438.75$$

$$= \$236.25$$

Then, set up the discount proportion.

$$\frac{D}{L} = \frac{R}{100} \qquad \text{becomes} \qquad \frac{236.25}{675} = \frac{R}{100}$$

$$675R = 23625$$

$$R = 35\% \qquad \text{The contractor receives a 35\% discount.}$$

Using the proportion method for solving percent problems makes it easier to distinguish among the different types of problems. However, many practical applications involve finding the percentage, and there is a quicker way to do this, especially using a calculator. Simply change the percent to a decimal number and multiply by the base.

Example: Find 20% of $640.

Step 1   Change the percent to a decimal number.

$$20\% = 20.\% = 0.20 = 0.2$$

Step 2   Multiply this decimal equivalent by the base.

$$0.2 \times \$640 = \$128$$

This is the same answer you would get using a proportion.

$$\frac{P}{640} = \frac{20}{100} \qquad \text{or} \qquad P = \frac{640 \times 20}{100} = 640 \times 0.2$$

Use this quick method to solve the following problems.

(a) Find 35% of 80.

(b) What is 4.5% of 650?

(c) What is 125% of 18?

(d) Find 0.3% of 5000.

Check your answers in **32**.

---

32   (a) $35\% = 35.\% = 0.35$
       $0.35 \times 80 = 28$

(b) $4.5\% = 04.5\% = 0.045$
       $0.045 \times 650 = 29.25$

(c) $125\% = 125.\% = 1.25$
       $1.25 \times 18 = 22.5$

(d) $0.3\% = 00.3\% = 0.003$
       $0.003 \times 5000 = 15$

Sales Tax   *Taxes* are almost always calculated as a percent of some total amount. *Property taxes* are written as some fraction of the value of the property involved. *Income taxes* are most often calculated from complex formulas that depend on many factors. We cannot consider either income or property taxes here.

A *sales tax* is an amount calculated from the actual price of a purchase and added to the buyer's cost. Retail sales tax rates are set by the individual states in the United States and vary from 0 to 8% of the sales price. A sales tax of 6% is often stated as "6 cents on the dollar," since 6% of $1.00 equals 6 cents.

If the retail sales tax rate is 7.25% in California, how much would you pay in Fresno for a pair of shoes costing $76.50?

Check your answer in **33**.

---

33   Using a proportion equation, we have

$$\frac{\text{sales tax}}{\text{cost}} = \frac{\text{tax rate}}{100}$$

or $\text{Sales tax} = \dfrac{\text{tax rate}}{100} \times \text{cost}$

$$= \frac{7.25}{100} \times 76.50$$

$$= \$5.55 \text{ rounded}$$

Using the equivalent quick method, we have

Sales tax = 7.25% of $76.50   7.25% = 07.25 = 0.0725
        = 0.0725 × $76.50
        = $5.54625 = $5.55   rounded

Actual cost = list price + sales tax
        = $76.50 + $5.55
        = $82.05

 .0725 ☒ 76.5 ⊜ → | 5.54625 | ⊞ 76.5 ⊜ → | 82.04625 |

Many stores provide their salesclerks with computerized cash registers that calculate sales tax automatically. However, all consumers and most small business or shop owners still need to be able to do the sales tax calculations.

Here are a few problems in calculating sales tax.

Find the tax and total cost for each of the following if the tax rate is as shown.

(a) A roll of plastic electrical tape at $2.79   (5%).

(b) A woodworker's vise priced at $57.50   (6%).

(c) A feeler gauge priced at $6.95   ($3\frac{1}{2}$%).

(d) A new car priced at $16,550   (4%).

(e) A portable paint compressor priced at $346.50   ($6\frac{1}{2}$%).

(f) A toggle switch priced at $1.89   ($7\frac{3}{4}$%).

Check your answers in **34**.

34

	Tax	Total Cost	Quick Calculation
(a)	14 cents	$2.93	0.05 × $2.79 = 13.95 cents ≈ 14 cents
(b)	$3.45	$60.95	0.06 × $57.50 = $3.45
(c)	$0.24	$7.19	0.035 × $6.95 = $0.24325 ≈ $0.24
(d)	$662.00	$17,212	0.04 × $16,550 = $662.00
(e)	$22.52	$369.02	0.065 × $346.50 = $22.5225 ≈ $22.52
(f)	$0.15	$2.04	0.0775 × $1.89 = $0.146475 ≈ $0.15

Interest   In modern society we have set up complex ways to enable you to use someone else's money. A *lender,* with money beyond his or her needs, supplies cash to a *borrower,* whose needs exceed his money. The money is called a *loan.*

Interest is the amount the lender is paid for the use of his money. Interest is the money you pay to use someone else's money. The more you use and the longer you use it, the more interest you must pay. *Principal* is the amount of money loaned or borrowed.

When you purchase a house with a bank loan, a car or a refrigerator on an installment loan, or gasoline on a credit card, you are using someone else's money, and you pay interest for that use. If you are on the other end of the money game, you may earn interest for money you invest in a savings account or in shares of a business.

Interest can be calculated from the following formulas.

$$\frac{\text{annual interest}}{\text{principal}} = \frac{\text{annual interest rate}}{100}$$

Total interest   = annual interest × time in years

For example, suppose that you have $500 in savings earning $5\frac{1}{2}\%$ interest annually in your local bank. How much interest do you receive in a year?

$$\frac{\text{annual interest } I}{\$500} = \frac{5\frac{1}{2}}{100}$$

$$100I = 2750$$

$$I = \$27.50$$

The total interest for one year is $27.50.

Using the quick calculation method to calculate $5\frac{1}{2}\%$ of $500,

$$\text{Annual interest } I = \$500 \times 5\frac{1}{2}\%$$

$$= 500 \times 0.055$$

$$= \$27.50$$

By placing your $500 in a bank savings account, you allow the bank to use it in various profitable ways, and you are paid $27.50 per year for its use.

Most of us play the money game from the other side of the counter. Suppose that you find yourself in need of cash and arrange to obtain a loan from a bank. You borrow $600 at 14% per year for 3 months. How much interest must you pay?

Try to set up and solve this problem exactly as we did in the previous problem.

Check your work in **35**.

35　$$\frac{\text{annual interest } I}{\$600} = \frac{14}{100}$$

$$I = \frac{600 \times 14}{100} = \$84$$

The time of the loan is 3 months, so the time in years is

$$\frac{3 \text{ months}}{12 \text{ months}} = \frac{3}{12} \text{ yr}$$

$$= 0.25 \text{ yr}$$

Total interest = annual interest × time in years

$$= \$84 \times 0.25$$

$$= \$21$$

By the quick calculation method,

Total interest = $600 \times 14\% \times \dfrac{3}{12}$ yr

$$= 600 \times 0.14 \times 0.25$$

$$= \$21$$

 600 ⊗ **.14** ⊗ **3** ÷ **12** = → ▭*21.*

Depending on how you and the bank decide to arrange it, you may be required to pay the total principal ($600) plus interest ($21) all at once, at the end of three months, or according to some sort of regular payment plan—for example, pay $207 each month for three months.

Commission　The simplest practical use of percent is in the calculation of a part or percentage of some total. For example, salespersons are often paid on the basis of their success at selling and receive a *commission* or share of the sales receipts. Commission is usually described as a percent of sales income.

$$\frac{\text{commission}}{\text{sales}} = \frac{\text{rate}}{100}$$

Suppose that your job as a lumber salesman pays 12% commission on all sales. How much do you earn from a sale of $400?

$$\frac{\text{commission } C}{400} = \frac{12}{100}$$

$$C = \frac{12 \times 400}{100}$$

$$C = \$48 \qquad \text{You earn \$48 commission.}$$

By the quick calculation method,

Commission = $400 \times 12\% = \$400 \times 0.12 = \$48$

Try this one yourself.

At the Happy Bandit Used Car Company each salesperson receives a 6% commission on sales. What would a salesperson earn if she sold a 1960 Airedale for $1299.95?

Check your work in **36**.

HOW DOES A CREDIT CARD WORK?

Many forms of small loans, such as credit card loans, charge interest by the month. These are known as "revolving credit" or "charge account" plans. (If you borrow very much money this way, *you* do the revolving and may run in circles for years trying to pay it back!) Essentially, you buy now and pay later. Generally, if you repay the full amount borrowed within 25 or 30 days, there is no charge for the loan.

After the first pay period of 25 or 30 days, you pay a percent of the unpaid balance each month, usually between $\frac{1}{2}$ and 2%. In addition, you must pay a yearly fee and some minimum amount each month, usually $10 or 10% of the unpaid balance, whichever is larger. Often, you are also charged a small monthly amount for insurance premiums. (The credit card company insures itself against your defaulting on the loan or disappearing, and you pay for their insurance.)

Let's see how it works. Suppose that you go on a short vacation and pay for gasoline, lodging, and meals with your Handy Dandy credit card. A few weeks later you receive a bill for $1000. You can pay it within 30 days and owe no interest or you can pay over several months as follows:

Month 1 $1\frac{1}{2}\%$ of $1000 = $1000 \times 0.015 = \$15.00$ owe $1015.00
pay $100 and carry $915.00 over to next month

Month 2 $1\frac{1}{2}\%$ of $915 = 13.73$ owe $928.73
pay $100 and carry $828.73 over to next month

. . . and so on.

A year later you will have repaid the $1000 loan and all interest.

The $1\frac{1}{2}\%$ per month interest rate seems small, but it is equivalent to between 15 and 18%. You pay at a high rate for the convenience of using the credit card and the no-questions-asked ease of getting the loan.

36

$$\frac{C}{\$1299.95} = \frac{6}{100}$$

$$C = \frac{6 \times 1299.95}{100}$$

$$C = \$77.9970 \quad \text{or} \quad \$78 \text{ rounded}$$

or

$$\text{Commission} = \$1299.95 \times 6\% = \$1299.95 \times 0.06$$
$$\approx \$78$$

Some situations require that you determine the total sales or the rate of commission. In the following two problems, concentrate on identifying each quantity in the commission proportion.

(a) A pharmaceutical sales rep generated $360,000 in sales last year and was paid a salary of $27,000. If she were to switch to straight commission compensation, what rate of commission would be needed for her commission to match her salary?

(b) A computer salesperson is paid a salary of $800 per month plus a 4% commission on sales. What amount of sales must he generate in order to earn $40,000 per year in total compensation?

Check your solutions in **37**.

---

37   (a)   Sales = $360,000
Commission = $27,000
Rate is unknown.

$$\frac{\$27,000}{\$360,000} = \frac{R}{100}$$

$$R = \frac{27,000 \times 100}{360000}$$

$$R = 7.5\% \quad \text{She needs to be paid a 7.5\% commission in order to match her salary.}$$

(b)   Salary = $800 per month × 12 months
= $9600 per year

Commission = total compensation − salary
= $40,000 − $9600 = $30,400

The commission needed is $30,400. Write the commission proportion. The sales, S, is not known.

$$\frac{\$30,400}{S} = \frac{4}{100}$$

$$S = \frac{30400 \times 100}{4}$$
$$S = \$760,000$$

He must sell $760,000 worth of computers to earn a total of $40,000 per year.

40000 $\boxed{-}$ **800** $\boxed{\times}$ **12** $\boxed{=}$ $\rightarrow$ `30400.` $\boxed{\times}$ **100** $\boxed{\div}$ **4** $\boxed{=}$ $\rightarrow$ `760000.`

Efficiency   When energy is converted from one form to another in any machine or conversion process, it is useful to talk about the *efficiency* of the machine or the process. The efficiency of a process is a fraction comparing the energy or power output to the energy or power input. It is usually expressed as a percent.

$$\frac{\text{output}}{\text{input}} = \frac{\text{efficiency}}{100}$$

For example, an auto engine is rated at 175 hp and is found to deliver only 140 hp to the transmission. What is the efficiency of the process?

$$\frac{\text{output}}{\text{input}} = \frac{\text{efficiency } E}{100}$$

$$\frac{140}{175} = \frac{E}{100}$$

$$E = \frac{140 \times 100}{175}$$

$$E = 80\%$$

Now you try one. A gasoline shop engine rated at 65 hp is found to deliver 56 hp through a belt drive to a pump. What is the efficiency of the drive system?

Check your work in **38**.

---

38   output = 56 hp,   input = 65 hp

$$\frac{56}{65} = \frac{E}{100}$$

$$E = \frac{56 \times 100}{65} = 86\% \text{ rounded}$$

56 $\boxed{\times}$ **100** $\boxed{\div}$ **65** $\boxed{=}$ $\rightarrow$ `86.153846`

The efficiency of any practical system will always be less than 100%—you can't produce an output greater than the input!

Tolerances   On technical drawings or other specifications, measurements are usually given with a *tolerance* showing the allowed error. For example, the fitting shown in this drawing has a length of

1.370 ± 0.015 in.

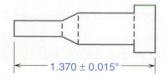

1.370 ± 0.015"

This means that the dimension shown must be between

1.370 + 0.015 in.   or   1.385 in.

and

1.370 − 0.015 in.   or   1.355 in.

The *tolerance limits* are 1.385 in. and 1.355 in.

Very often the tolerance is written as a percent.

$$\frac{\text{tolerance}}{\text{measurement}} = \frac{\text{percent tolerance}}{100}$$

In this case,

$$\frac{0.015}{1.370} = \frac{\text{percent tolerance } R}{100}$$

$$R = \frac{0.015 \times 100}{1.370}$$

$$R = 1.0948 \dots \quad \text{or} \quad 1.1\% \text{ rounded}$$

The dimension would be written 1.370 in. ± 1.1%.

Rewrite the following dimension with a percent tolerance:

1.426 ± 0.010 in. = _____

Check your work in **39**.

SMALL LOANS

Sooner or later everyone finds it necessary to borrow money. When you do, you will want to know beforehand how the loan process works. Suppose that you borrow $200 and the loan company specifies that you repay it at $25 per month plus interest at 3% per month on the unpaid balance. What interest do you actually pay?

Month 1	3% of $200 = $ 6.00	you pay	$25 + $6.00 = $ 31.00
Month 2	3% of $175 = $ 5.25	you pay	$25 + $5.25 = $ 30.25
Month 3	3% of $150 = $ 4.50	you pay	$25 + $4.50 = $ 29.50
Month 4	3% of $125 = $ 3.75	you pay	$25 + $3.75 = $ 28.75
Month 5	3% of $100 = $ 3.00	you pay	$25 + $3.00 = $ 28.00
Month 6	3% of $ 75 = $ 2.25	you pay	$25 + $2.25 = $ 27.25
Month 7	3% of $ 50 = $ 1.50	you pay	$25 + $1.50 = $ 26.50
Month 8	3% of $ 25 = $ 0.75	you pay	$25 + $0.75 = $ 25.75
	$27.00		$227.00

Total interest is $27.00

Total of eight loan payments

They might also set it up as eight equal payments of $227.00 ÷ 8 = $28.375 or $28.38 per month.

The 3% monthly interest rate seems small, but it amounts to about 20% per year.

A bank loan for $200 at 12% for 8 months would cost you

$$(12\% \text{ of } \$200) \times \frac{8}{12}$$

or $\qquad \$24 \times \dfrac{8}{12}$

or $\qquad \$16.00 \qquad$ Quite a difference.

The loan company demands that you pay more, and in return they are less worried about your ability to meet the payments. For a bigger risk, they want a higher rate of interest.

39

$$\frac{0.010}{1.426} = \frac{\text{percent tolerance } R}{100}$$

$$R = \frac{0.010 \times 100}{1.426}$$

$$R = 0.70126\ldots \qquad \text{or} \qquad 0.70\% \text{ rounded}$$

$1.426 \pm 0.010 \text{ in.} = 1.426 \text{ in.} \pm 0.70\%$

If the dimension is given in metric units, the procedure is exactly the same. For example, the dimension

315 ± 0.25 mm

can be converted to percent tolerance.

$$\frac{0.25}{315} = \frac{R}{100}$$

$$R = \frac{0.25 \times 100}{315}$$

$$R = 0.07936\ldots \qquad \text{or} \qquad 0.08\% \text{ rounded}$$

The dimension is 315 mm $\pm 0.08\%$

 .25 ⊗ 100 ÷ 315 ⊜ → `0.0793651`

If the tolerance is given as a percent, it is easy to use the proportion to calculate the tolerance as a dimension value.

$$\frac{\text{tolerance}}{\text{measurement}} = \frac{\text{percent tolerance}}{100}$$

For example, 2.450 in. $\pm 0.2\%$ can be converted to a dimension value as follows:

$$\frac{\text{tolerance } T}{2.450} = \frac{0.2}{100}$$

$$T = \frac{0.2 \times 2.45}{100}$$

$$T = 0.0049 \text{ in.} \qquad \text{or} \qquad 0.005 \text{ in. rounded}$$

Therefore,

2.450 in. $\pm 0.2\%$ = 2.450 in. ± 0.005 in.

 .2 ⊗ 2.45 ÷ 100 ⊜ → `0.0049`

Now, for some practice in working with tolerances, convert the following measurement tolerances to percents, and vice versa. Round the tolerances to three decimal places and round the percents to two decimal places.

Measurement	Tolerance	Percent Tolerance
2.345 in.	± 0.001 in. ± 0.002 in. ± 0.005 in. ± 0.010 in. ± 0.125 in.	
274 mm	± 0.10 mm ± 0.20 mm ± 0.50 mm	
3.475 in.		± 0.10% ± 0.20%
123 mm		± 0.05% ± 0.15%

Check your answers in **40**.

40

Measurement	Tolerance	Percent Tolerance
2.345 in.	± 0.001 in. ± 0.002 in. ± 0.005 in. ± 0.010 in. ± 0.125 in.	± 0.04% ± 0.09% ± 0.21% ± 0.43% ± 5.33%
274 mm	± 0.10 mm ± 0.20 mm ± 0.50 mm	± 0.04% ± 0.07% ± 0.18%
3.475 in.	± 0.003 in. ± 0.007 in.	± 0.10% ± 0.20%
123 mm	± 0.062 mm ± 0.185 mm	± 0.05% ± 0.15%

Percent Change

In many situations in technical work and the trades, you may need to find a *percent increase* or *percent decrease* in a given quantity. For example, suppose that the output of a certain electrical circuit is 20 amperes (A), and the output increases by 10%. What is the new value of the output?

$$\frac{\text{amount of increase}}{\text{original amount}} = \frac{\text{percent increase}}{100}$$

$$\frac{\text{amount of increase } i}{20} = \frac{10}{100}$$

$$100i = 200$$

$$i = 2 \text{ A}$$

New value = original amount + increase

$$= 20 \text{ A} + 2\text{A} = 22 \text{ A}$$

A quicker way to solve the problem is to add 100% to the percent increase and use this total for the rate R. The revised proportion becomes

$$\frac{\text{new value } V}{\text{original amount}} = \frac{100\% + \text{percent increase}}{100}$$

$$\frac{V}{20} = \frac{100\% + 10\%}{100}$$

$$\frac{V}{20} = \frac{110}{100}$$

$$V = \frac{20 \times 110}{100}$$

$$V = 22 \text{ A}$$

Percent decrease can be calculated in a very similar way. For example, suppose that the output of a certain machine at the Acme Gidget Co. is 600 gidgets per day. If the output decreases by 8%, what is the new output?

Percent *decrease* involves subtracting, and, as with percent increase, we can subtract either at the beginning or at the end of the calculation. Subtracting at the end, we have

$$\frac{\text{amount of decrease}}{\text{original amount}} = \frac{\text{percent decrease}}{100}$$

$$\frac{\text{amount of decrease } d}{600} = \frac{8}{100}$$

$$d = \frac{8 \times 600}{100}$$

$$d = 48 \text{ gidgets per day}$$

Now subtract to find the new value.

New value = original amount – amount of decrease

$$= 600 - 48 = 552 \text{ gidgets per day}$$

We can also solve the problem by subtracting at the beginning of the problem.

$$\frac{\text{new value}}{\text{original amount}} = \frac{100\% - \text{percent decrease}}{100}$$

$$\frac{\text{new value } V}{600} = \frac{100\% - 8\%}{100}$$

$$\frac{V}{600} = \frac{92}{100}$$

$$100V = 55200$$

$$V = 552 \text{ gidgets per day}$$

Try these problems involving percent change. In each, add or subtract the percents at the beginning of the problem.

(a)   What is your new pay rate if your old rate of $6.72 per hour is increased 15%?

(b)   Normal line voltage, 115 volts, drops 3.5% during a system malfunction. What is the reduced voltage?

(c)   Because of changes in job specifications, the cost of a small construction job is increased 25% from the original cost of $1275. What is the cost of the job now?

(d) By using automatic welding equipment, the time for a given job can be reduced by 40% from its original 30 hours. How long will the job take using the automatic equipment?

Check your answers in **41**.

---

41 (a) $\dfrac{\text{new pay rate}}{\$672} = \dfrac{115}{100}$ ← $\boxed{100\% + 15\%}$

New pay rate $= \dfrac{6.72 \times 115}{100} = \$7.728$ or $\$7.73$ rounded

(b) $\dfrac{\text{reduced voltage}}{115 \text{ volts}} = \dfrac{96.5}{100}$ ← $\boxed{100\% - 3.5\%}$

Reduced voltage $= \dfrac{115 \times 96.5}{100} = 110.975$ or 111 volts rounded

 115 ⊠ **96.5** ⊝ **100** ⊜ → $\boxed{110.975}$

(c) $\dfrac{\text{new cost}}{\$1275} = \dfrac{125}{100}$

New cost $= \dfrac{1275 \times 125}{100} = \$1593.75$

(d) $\dfrac{\text{new time}}{30} = \dfrac{60}{100}$

New time $= \dfrac{30 \times 60}{100} = 18$ hours

In trade and technical work, changes in a measured quantity are often specified as a percent increase or decrease. For example, if the reading on a pressure valve increases from 30 psi to 36 psi, the percent increase can be found by using the following proportion:

$$\begin{array}{l}\text{The change} \rightarrow \\ \text{Original value} \longrightarrow\end{array} \dfrac{36 \text{ psi} - 30 \text{ psi}}{30 \text{ psi}} = \dfrac{\text{percent increase}}{100}$$

$$\dfrac{6}{30} = \dfrac{R}{100}$$

$$R = \dfrac{100 \times 6}{30} = 20\%$$

The reading on the pressure valve increased by 20%.

 100 ⊠ ⦇ **36** ⊖ **30** ⦈ ⊝ **30** ⊜ → $\boxed{20.}$

Example: The length of a heating duct is reduced from 110 in. to 96 in. because of design changes. The percent decrease is calculated as follows:

$$\begin{array}{l}\text{The change} \rightarrow \\ \text{Original value} \longrightarrow\end{array} \dfrac{110 \text{ in.} - 96 \text{ in.}}{110 \text{ in.}} = \dfrac{\text{percent decrease}}{100}$$

$$\dfrac{14}{110} = \dfrac{R}{100}$$

$$R = \dfrac{100 \times 14}{110} = 12.7272\ldots \quad \text{or} \approx 13\%$$

The length of the duct was reduced by about 13%.

Careful ▶ Students often are confused about which number to use for the base in percent change problems. The difficulty usually arises when they think in terms of which is the bigger or smaller value. Always use the *original* value as the base. In percent increase problems, the original or base number will be the smaller number, and in percent decrease problems, the original or base number will be the larger number. ◀

Now try these problems to sharpen your ability to work with percent changes.

(a) If 8 in. is cut from a 12-ft board, what is the percent decrease in length?

(b) What is the percent increase in voltage when the voltage increases from 65 volts to 70 volts?

(c) The measured value of the power output of a motor is 2.7 hp. If the motor is rated at 3 hp, what is the percent difference between the measured value and the expected value?

Check your answers in **42**.

---

42 (a) Use 12 ft = 144 in.

$$\text{The change} \longrightarrow \frac{8 \text{ in.}}{144 \text{ in.}} = \frac{\text{percent decrease}}{100}$$
$$\text{Original value} \longrightarrow$$

$$R = \frac{100 \times 8}{144} = 5.555\ldots \qquad \text{or} \qquad 5.6\% \text{ rounded}$$

The length of the board decreased by about 5.6%.

(b) To find the amount of increase, subtract 65 volts from 70 volts.

$$\frac{70 \text{ volts} - 65 \text{ volts}}{65 \text{ volts}} = \frac{\text{percent increase}}{100}$$
$$\text{Original value} \longrightarrow$$

$$\frac{5}{65} = \frac{R}{100}$$

$$R = \frac{100 \times 5}{65} = 7.6923\ldots \qquad \text{or} \qquad 7.7\% \text{ rounded}$$

The voltage increased by about 7.7%.

(c) To find the amount of decrease, subtract 2.7 hp from 3 hp.

$$\frac{3 \text{ hp} - 2.7 \text{ hp}}{3 \text{ hp}} = \frac{\text{percent difference}}{100}$$

The expected value is used as the base.

$$\frac{0.3}{3} = \frac{R}{100}$$

$$R = \frac{100 \times 0.3}{3} = 10\%$$

There is a 10% difference between the measured value and the expected value.

Now, turn to **43** for a set of problems involving percent calculations.

43
Exercises 4-4   **Applications of Percent Calculations**

1. **Electronics**   A CB radio is rated at 7.5 watts, and actual measurements show that it delivers 4.8 watts to its antenna. What is its efficiency?

2. **Electrical Technology** An electric motor uses 6 kilowatts at an efficiency of 63%. How much power does it deliver?

3. **Electrical Technology** An engine supplies 110 hp to an electric generator, and the generator delivers 70 hp of electrical power. What is the efficiency of the generator?

4. **Electronics** Electrical resistors are rated in ohms and color coded to show both their resistance and percent tolerance. For each of the following resistors, find its tolerance limits and actual tolerance.

Resistance (ohms)	Limits (ohms)	Tolerance (ohms)
5300 ± 5%	_____ to _____	_____
2750 ± 2%	_____ to _____	_____
6800 ± 10%	_____ to _____	_____
5670 ± 20%	_____ to _____	_____

5. **Auto Mechanics** A 120-hp automobile engine delivers only 81 hp to the driving wheels of the car. What is the efficiency of the transmission and drive mechanism?

6. **Electronics** An electrical resistor is rated at 500 ohms ± 10%. What is the highest value its resistance could have within this tolerance range?

7. **Roofing** On a roofing job 21 of 416 shingles had to be rejected for minor defects. What percent is this?

8. If you earn 12% commission on sales of $4200, what actual amount do you earn?

9. **Metalworking** On a cutting operation 2 sq ft of sheet steel is wasted for every 16 sq ft used. What is the percent waste?

10. **Metalworking** Four pounds of a certain bronze alloy is one-sixth tin, 0.02 zinc, and the rest copper. Express the portion of each metal in (a) percents and (b) pounds.

11. **Metalworking** An iron casting is made in a mold with a hot length of 16.40 in. After cooling, the casting is 16.25 in. long. What is the shrinkage in percent?

12. **Manufacturing** Because of friction, a pulley block system is found to be only 83% efficient. What actual load can be raised if the theoretical load is 2000 lb?

13. **Manufacturing** A small gasoline shop engine develops 65 hp at 2000 rpm. At 2400 rpm its power output is increased by 25%. What actual horsepower does it produce at 2400 rpm?

14. **Building Construction** On the basis of past experience, a contractor expects to find 4.5% broken bricks in every truck load. If he orders 2000 bricks, will he have enough to complete a job requiring 1900 bricks?

15. **Machine Technology** Specifications call for a hole in a machined part to be 2.315 in. in diameter. If the hole is measured to be 2.318 in., what is the machinist's percent error?

16. **Welding** A welding shop charges the customer 115% of the cost of labor and materials. If a bill totals $36.75, what is the cost of labor and materials?

17. Complete the following table.

Measurement	Tolerance	Percent Tolerance
3.425 in.	± 0.001 in.	(a)
3.425 in.	± 0.015 in.	(b)
3.425 in.	(c)	± 0.20%

18. If you receive a pay increase of 12%, what is your new pay rate, assuming that your old rate was $8.65?

19. **Manufacturing**   The pressure in a hydraulic line increases from 40 psi to 55 psi. What is the percent increase in pressure?

20. **Painting and Decorating**   The cost of the paint used in a redecorating job is $65.70. This is a reduction of 20% from its initial cost. What was the original cost?

21. What sales tax, at a rate of $6\frac{1}{2}$%, must you pay on the purchase of a computer hard disk drive costing $256.75?

22. **Interior Design**   An interior designer working for a department store is paid a weekly salary of $350 plus a 20% commission on total sales. What would be his weekly pay if the total sales income was $1875.00?

23. **Electronics**   Murph's electronic supply shop offered a digital oscilloscope on sale at 25% off. If the original price was $1750, what was the sale price?

24. **Upholstery**   Chris paid for the initial expenses in setting up her upholstery shop with a bank loan for $17,000 for 5 years at $12\frac{1}{2}$% annual interest.
    (a)   What total interest will she pay?
    (b)   What will be her monthly payments, assuming equal monthly installments?

25. **Woodworking**   A small bench grinder is on sale for $245.45 and is marked as being 30% off. What was the original list price?

26. Driving into a 20-mph breeze cuts gas mileage by about 12%. What will be your gas mileage in a 20-mph wind if it is 24 miles/gallon with no wind?

27. **Auto Mechanics**   Supplier A offers a mechanic a part at 35% off the retail price of $68.40. Supplier B offers him the same part at 20% over the wholesale cost of $35.60. Which is the better deal, and by how much?

28. **Auto Mechanics**   An auto mechanic purchases the parts necessary for a repair for $126.40 and sells them to the customer at a 35% markup. How much does the mechanic charge the customer?

29. **Auto Mechanics**   A new tire had a tread depth of $\frac{3}{8}$ in. After one year of driving the tire had $\frac{15}{64}$ in. of tread remaining. What percent of the tread was worn?

30. **Printing**   A printer agrees to give a nonprofit organization a 10% discount on a printing job. The normal price to the customer is $1020, and the printer's cost is $850. What is the printer's percent profit over cost after the discount is subtracted?

31. **Wastewater Technology**   A treatment plant with a capacity of 20 MGD has a normal daily flow equal to 60% of capacity. When it rains the flow increases by 30% of the normal flow. What is the flow on a rainy day?

32. **Police Science**   Last year a community had 23 homicides. This year they had 32 homicides. What was the percent of increase?

33. **Fire Science** The pressure reading coming out of a pump is 180 lb. Every 50-ft section reduces the pressure by about 3%. What will the nozzle pressure be at the end of six 50-ft sections? (*Hint:* One reduction of 18% is *not* equivalent to six reductions of 3% each. You must do this the long way.)

34. **Machine Technology** Complete the following table of tolerances for some machine parts.

Measurement	Tolerance	Maximum	Minimum	Percent Tolerance
1.58 in.	± 0.002 in.			
	± 0.005 in.			
				± 0.15%
0.647 in.	± 0.004 in.			
	± 0.001 in.			
				± 0.20%
165.00 mm	± 0.15 mm			
	± 0.50 mm			
				± 0.05%
35.40 mm	± 0.01 mm			
	± 0.07 mm			
				± 0.05%
				± 0.08%
				± 0.10%

35. **Electrical Technology** An electrician purchases some outdoor lighting for a customer at a wholesale price of $928.80. The customer would have paid the retail price of $1290. What was the wholesale discount?

36. In problem 35, what will the total retail price be after 7.75% tax is added?

37. **Building Construction** A contractor takes a draw of $26,500 from a homeowner to start a remodeling job. He uses $12,450 to purchase materials and deposits the remaining money in a three-month CD paying 6.5% annual interest. How much interest does he earn on the remaining funds during the three-month term?

38. **Food Services** A wholesale food sales rep is paid a base salary of $1200 per month plus a $3\frac{1}{2}$% commission on sales. How much sales must she generate in order to earn $35,000 per year in total compensation?

39. **Roofing** A sales rep for roofing materials figures he can generate about $1,800,000 in business annually. What rate of commission does he need in order to earn $50,000?

When you have completed these exercises, check your answers on page 590, then turn to **44** for a set of practice problems on percent.

Ratio, Proportion, and Percent

44   Answers are given on page 590.

A.   Complete the following tables.

1.

	Diameter of Pulley A	Diameter of Pulley B	Pulley Ratio, $\dfrac{A}{B}$
(a)	12 in.		2 to 5
(b)	95 cm	38 cm	
(c)		15 in.	5 to 3

2.

	Teeth on Driving Gear A	Teeth on Driven Gear B	Gear Ratio, $\dfrac{A}{B}$
(a)	20	60	
(b)		10	6.5
(c)	56		4 to 3

B.   Solve the following proportions.

1. $\dfrac{5}{6} = \dfrac{x}{42}$ 2. $\dfrac{8}{15} = \dfrac{12}{x}$ 3. $\dfrac{x}{12} = \dfrac{15}{9}$

4. $\dfrac{3\frac{1}{2}}{x} = \dfrac{5\frac{1}{4}}{18}$ 5. $\dfrac{1.6}{5.2} = \dfrac{4.4}{x}$ 6. $\dfrac{x}{12.4} = \dfrac{4 \text{ ft } 6 \text{ in.}}{6 \text{ ft } 3 \text{ in.}}$

C.   Write each number as a percent.

1. 0.72   2. 0.06   3. 0.6   4. 0.358   5. 1.3

6. 3.03   7. 4   8. $\dfrac{7}{10}$ 9. $\dfrac{1}{6}$ 10. $2\dfrac{3}{5}$

D.   Write each percent as a decimal number.

1. 4%   2. 37%   3. 11%   4. 94%

5. $1\frac{1}{4}\%$ 6. 0.09%   7. $\frac{1}{5}\%$ 8. 1.7%

9. $3\frac{7}{8}\%$ 10. 8.02%   11. 115%   12. 210%

Name _____

Date _____

Course/Section _____

E. Solve.

1. 3 is _____ % of 5.

2. 5% of $120 is _____.

3. 25% of what number is 1.4?

4. 16 is what percent of 8?

5. 105% of 40 is _____.

6. 1.38 is _____ % of 1.15?

7. $7\frac{1}{4}$% of _____ is $2.10.

8. 250% of 50 is _____.

9. 0.05% of _____ is 4.

10. $8\frac{1}{4}$% of 1.2 is _____.

F. Solve:

1. **Metalworking** Extruded steel rods shrink 12% in cooling from furnace temperature to room temperature. If a standard tie rod is 34 in. exactly when it is formed, how long will it be after cooling?

2. **Metalworking** Cast iron contains up to 4.5% carbon, and wrought iron contains up to 0.08% carbon. How much carbon is in a 20-lb bar of each metal?

3. (a) A real estate saleswoman sells a house for $284,500. Her usual commission is 1.5%. How much does she earn on the sale?
   (b) All salespeople in the Ace Junk Store receive $260 per week plus a 2% commission. If you sold $1975 worth of junk in a week what would be your income?
   (c) A salesman at the Wasteland TV Company sold five color TV sets last week and earned $128.70 in commissions. If his commission is 6%, what does a color TV set cost?

4. **Carpentry** What is the selling price of a radial arm saw with a list price of $265.50 if it is on sale at a 35% discount?

5. If the retail sales tax in your state is 6%, what would be the total cost of each of the following items?
   (a) An $8.98 pair of pliers.
   (b) An $11.10 adjustable wrench.
   (c) 69 cents' worth of washers.
   (d) A $33.60 textbook.
   (e) A $445.95 shop table.

6. A computer printer sells for $376 after a 12% discount. What was its original or list price?

7. **Building Construction** How many running feet of matched 1-in. by 6-in. boards will be required to lay a subfloor in a house that is 28 ft by 26 ft? Add 20% to the area to allow for waste and matching.

8. **Sheet Metal Technology** In the Easy Does It Metal Shop, one sheet of metal is wasted for every 25 purchased. What percent of the sheets are wasted?

9. **Machine Technology** Complete the following table.

Measurement	Tolerance	Percent Tolerance
1.775 in.	± 0.001 in.	(a)
1.775 in.	± 0.05 in.	(b)
1.775 in.	(c)	± 0.50%
310 mm	± 0.1 mm	(d)
310 mm	(e)	± 0.20%

10. **Electronics**   If 16 in. is cut from a 6-ft cable, what is the percent decrease in length?

11. **Manufacturing**   A production job is bid at $6275, but cost overruns amount to 15%. What is the actual job cost?

12. **Metalworking**   After heating, a metal rod expanded 3.5%, to 15.23 cm. What was its original length?

13. **Electrical Technology**   An electrical resistor is rated at 4500 ohms ± 3%. Express this tolerance in ohms and state the actual range of resistance.

14. **Auto Mechanics**   A 140-hp automobile engine delivers only 96 hp to the driving wheels of the car. What is the efficiency of the transmission and drive mechanism?

15. **Metalworking**   A steel casting has a hot length of 26.500 in. After cooling, the length is 26.255 in. What is the shrinkage expressed as a percent? Round to one decimal place.

16. **Auto Mechanics**   The parts manager for an automobile dealership can buy a part at a 25% discount off the retail price. If the retail price is $6.75, how much does he pay?

17. **Manufacturing**   An electric shop motor rated at 2.5 hp is found to deliver 1.8 hp to a vacuum pump when it is connected through a belt drive system. What is the efficiency of the drive system?

18. **Welding**   In order to purchase a truck for his mobile welding service, Jerry arranges a 36-month loan of $19,000 at 12.75% annual interest.
    (a)   What total interest will he pay?
    (b)   What monthly payment will he make? (Assume equal monthly payments on interest and principal.)

19. **Refrigeration**   The motor to run a refrigeration system is rated at 12 hp, and the system has an efficiency of 76%. What is the effective power output of the system?

20. **Interior Design**   An interior designer is able to purchase a sofa from a design center for $1457. Her client would have paid $1880 for the same sofa at a furniture store. What rate of discount was the designer receiving off the retail price?

21. **Manufacturing**   The commission rate paid to a manufacturer's rep varies according to the type of equipment sold. If his monthly statement showed $124,600 in sales, and his total commission was $4438, what was his average rate of commission to the nearest tenth of a percent?

22. **Masonry**   Six square feet of a certain kind of brick wall contains 78 bricks. How many bricks are needed for 150 square feet?

23. If 60 miles per hour is equivalent to 88 feet per second, express
    (a)   45 miles per hour in feet per second.
    (b)   22 feet per second in miles per hour.

24. **Auto Mechanics**   The headlights on a car are mounted at the height of 28 in. The light beam must illuminate the road for 400 ft. That is, the beam must hit the ground 400 ft ahead of the car.
    (a)   What should be the drop ratio of the light beam in inches per foot?
    (b)   What is this ratio in inches per 25 ft?

25. **Printing**   A photograph measuring 8 in. wide by 12 in. long must be reduced to a width of $3\frac{1}{2}$ in. in order to fit on a printed page. What will be the corresponding length of the reduced photograph?

26. **Machine Technology**   If 250 ft of wire weighs 22 lb, what will be the weight of 100 ft of the same wire?

27. **Agricultural Technology**   A sprayer discharges 600 cc of herbicide in 45 sec. For what period of time should the sprayer be discharged in order to apply 2000 cc of herbicide?

28. **Auto Mechanics**   An auto mechanic currently rents a 560-sq-ft shop for $850 per month. How much will his monthly rent be if he moves to a comparably priced 780-sq-ft shop? (Round to the nearest dollar.)

29. **Machine Technology**   Six steel parts weigh 1.8 lb. How many of these parts are in a box weighing 142 lb if the box itself weighs 7 lb?

30. **Sports Technology**   Use the following five steps to determine the NFL passing rating for quarterback Bubba Grassback.
    (a) Bubba completed 176 passes in 308 attempts. Calculate his pass completion rate as a percent. Subtract 30 from this number and multiply the result by 0.05.
    (b) He has 11 touchdown passes in 308 attempts. Calculate the percent of his attempts that resulted in touchdowns. Multiply that number by 0.2.
    (c) He has thrown 13 interceptions. Calculate the percent of his 308 attempts that have resulted in interceptions. Multiply this percent by 0.25 and then subtract the result from 2.375.
    (d) Bubba passed for 2186 yards. Calculate the average number of yards gained per passing attempt. Subtract 3 from this number and then multiply the result by 0.25.
    (e) Add the four numbers obtained in parts (a), (b), (c), and (d).
    (f) Divide the total in part (e) by six and then multiply by 100. This is Bubba's NFL quarterback passing rating.
    (*Note:* Ratings for starting quarterbacks usually range from a low of about 50 to a high of about 110.)

Objective	Sample Problems		Where to Go for Help	
			Page	Frame
When you finish this unit you will be able to:				
1. Do arithmetic with measurement numbers.	(a) $4.0 \text{ hr} \times 35 \text{ mph}$	_____	191	**3**
	(b) $1\frac{1}{4} \text{ hr} + 40 \text{ min} + 2 \text{ hr} - 5 \text{ min}$	_____		
	(c) $60 \text{ mi} \div 20 \text{ mph}$	_____		
2. Convert from one measurement unit to another. (Round to two significant digits if necessary.)	(a) $2\frac{1}{2}$ weeks	$=$ _____ min	203	**13**
	(b) 46 gal	$=$ _____ bbl		
	(c) 120 cu ft	$=$ _____ cu yd		
	(d) 60 lb	$=$ _____ oz		
	(e) 23 ft $7\frac{1}{2}$ in.	$=$ _____ in.		
	(f) 12 lb 14 oz	$=$ _____ oz		
3. Work with metric units. (Round to two significant digits.)	(a) Estimate the weight in pounds of a casting weighing 45 kg.	_____ lb	213	**23**
	(b) 20 m	$\approx$ _____ ft		
	(c) 3 in.	$\approx$ _____ cm		
	(d) 40°F	$\approx$ _____ °C	222	**31**
	(e) 1500 mm = _____ cm	$=$ _____ cm	213	**23**

Name _____

Date _____

Course/Section _____

4. Use the common technical measuring instruments.

Read a length rule, micrometer, and vernier caliper.

228   **35**

(Answers to these preview problems are given on page 577.)

If you are certain that you can work *all* these problems correctly, turn to page 249 for a set of practice problems. If you cannot work one or more of the preview problems, turn to the page indicated to the right of the problem. For those who wish to master this material with the greatest success, turn to frame **1** and begin work there.

5 Measurement

PEANUTS reprinted by permission of UFS, Inc.

5-1 WORKING WITH MEASUREMENT NUMBERS

1 Most of the numbers used in practical or technical work either come from measurements or lead to measurements. A carpenter must measure a wall to be framed; a machinist may weigh materials; an electronic technician may measure circuit resistance or voltage output. At one time or another each of these technicians must make measurements, do calculations based on measurements, perhaps round measurement numbers, and worry about the units of measurement. Of course, they must also work with measurement numbers from drawings, blueprints, or other sources.

Measurement numbers are never exact. There is a built-in limit of precision to every measurement, either because of the person doing the measuring or because of the limitations of the measuring device. For example, look at this line.

If we measured the length of this line using a rough ruler, we would write that its length is

Length = 2 in.

To the nearest inch

But with a better measuring device, we might be able to say with confidence that its length is

Length = 2.1 in.

To the nearest tenth of an inch

Using a vernier caliper, we might find that the length is

Length = 2.13 in.

To the nearest hundredth of an inch

We could go on using more and more precise measuring devices until we found ourselves looking through a microscope trying to decide where the ink line begins. If someone told us that the length of that line was

Length = 2.1304756 in.

To the nearest 0.0000001 inch

we would know immediately that the length had been measured with incredible precision—they couldn't get a number like that by using a carpenter's rule!

Which of the following measurements has the greatest precision?

(a) 4.15 sec   (b) 4.1 sec   (c) 4.1452 sec

Choose the correct answer, then turn to **2** and continue.

---

2 Measurement (c), 4.1452 sec, is the most precise. Measurement (b) is given to the nearest tenth of a second and could have been obtained with a hand timer. Measurement (a) is given to the nearest hundredth of a second and could have been obtained with an electronic stopwatch. But (c) is precise to the nearest ten-thousandth of a second and this would require some fancy electronic timing.

Tolerance Very often the technical measurements and numbers shown in drawings and specifications have a *tolerance* attached so that you will know the precision needed. For example, if a dimension on a machine tool is given as 2.835 ± 0.004 in., this means that the dimension should be 2.835 in. to within 4 thousandths of an inch. In other words, the dimension may be no more than 2.839 in. and no less than 2.831 in. These are the limits of size that are allowed or tolerated.

Suppose that the length of a piece of pipe is measured to be 8 ft. Two bits of information are given in this measurement: the size of the measurement (8) and the units of the measurement (ft).

Units A number usually has no practical physical meaning for a technician or scientist unless there are units attached to it.

Note ▶ The unit name for any measurement number *must* be included when you use that number, write it, or talk about it. ◀

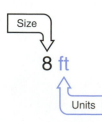

For example, we know immediately that 20 is a number, 20 ft is a length, 20 lb is a weight or force, and 20 sec is a time interval.

The measurement unit compares the size of the quantity being measured to some standard. For example, if the piece of pipe is measured to be 8 ft in length, it is 8 times the standard 1-ft length.

Length = 8 ft = 8 × 1 ft

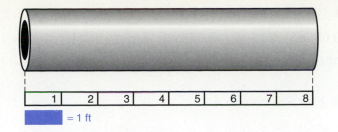

= 1 ft

Complete the following.

(a) 6 in. = 6 × _____

(b) 10 lb = _____ × _____

(c) $5 = 5 × _____

(d) 14.7 psi = _____ × _____

Check your answers in **3**.

---

3 (a) 6 in. = 6 × 1 in.

(b) 10 lb = 10 × 1 lb

(c) $5 = 5 × $1

(d) 14.7 psi = 14.7 × 1 psi

Addition and Subtraction of Measurement Numbers

When measurement numbers are added, subtracted, multiplied, or divided, we must be especially careful to keep track of the units as well as the numbers. For example, to do the addition

4.1 sec + 3.5 sec + 1.27 sec = ____?____

First, check to be certain that all numbers have the same units. In the given problem all numbers have units of seconds. If one or more of the numbers are given in other units, you will need to convert so that all have the same units.

Second, add the numerical parts.

$$\begin{array}{r} 4.1 \\ 3.5 \\ + 1.27 \\ \hline 8.87 \end{array}$$

Notice that we line up the decimal points and add as we would with any decimal numbers.

4.1 ⊞ **3.5** ⊞ **1.27** ⊟ → | *8.87* |

Third, attach the common units to the sum.

8.87 sec

If some of the numbers to be added are in fraction form, you should convert them to decimal form before adding. For example, the sum 3.2 lb + $1\frac{3}{4}$ lb + 2.4 lb becomes

$$\begin{array}{r} 3.2 \ \text{lb} \\ 1.75 \ \text{lb} \\ + 2.4 \ \ \text{lb} \\ \hline 7.35 \ \text{lb} \end{array}$$

You may need to round the answer if any numbers were converted from fraction to decimal form. We will look at rounding later in this chapter, but first you need to work some practice problems in adding measurement numbers.

Add.

(a) 4 in. + 5 in. + 17 in. = ____

(b) 12 in. + 3 in. + 7 in. = ____

(c) 1.8 sec + 3.5 sec = ____

(d) $3\frac{1}{2}$ in. + 1.7 in. + 2.75 in. = ____

(e) 1.375 in. + 3.50 in. + 2.833 in. = _____   (f) 4 lb + $2\frac{1}{2}$ lb + $4\frac{1}{4}$ lb = _____

(g) 0.05 in. + 0.15 in. + 0.0075 in. = _____   (h) 5 in. + 17 in. + 16 in. = _____

(i) $4\frac{1}{8}$ in. + 12 in. + $11\frac{1}{4}$ in. = _____   (j) $2\frac{1}{5}$ lb + $1\frac{1}{4}$ lb + $\frac{1}{4}$ lb = _____

Check your answers in **4**.

---

4 (a) 26 in.   (b) 22 in.

(c) 5.3 sec   (d) 7.95 in.

(e) 7.708 in.   (f) 10.75 lb

(g) 0.2075 in.   (h) 38 in.

(i) $27\frac{3}{8}$ in.   (j) $3\frac{7}{10}$ lb

The subtraction of measurement numbers is very similar to their addition. For example, in the subtraction

7.425 in. – 3.5 in. = _____?_____

First, check to be certain that both numbers have the same units. In this problem both numbers have inch units. If one of the numbers was given in other units, we would convert so that they have the same units. (You'll learn more about converting measurement units later in this chapter.)

Second, subtract the numerical parts.

$$\begin{array}{r} 7.425 \\ -\ 3.500 \\ \hline 3.925 \end{array}$$ ⟵ Attach zeros if necessary

 7.425 ⊟ **3.5** ⊟ → **3.925**

Third, attach the common units to the difference.   3.925 in.

If one of the numbers in the subtraction is in fraction form, you may need to convert it to a decimal before subtracting.

Now for a few practice subtraction problems, try these.

(a) 5.7 sec – 1.9 sec = _____   (b) 11.75 in. – 3.825 in. = _____

(c) 4.2 lb – 1.5 lb = _____   (d) 2.15 in. – 1.005 in. = _____

(e) 14 in. – 8 in. = _____   (f) $4\frac{1}{2}$ lb – 1.7 lb = _____

(g) 12 in. – 3.7 in. = _____   (h) 19.2 oz – 11.5 oz = _____

(i) 30.4 mph – 8.5 mph = _____   (j) 30 psi – 14.7 psi = _____

Check your work in **5**.

---

5 (a) 3.8 sec   (b) 7.925 in.   (c) 2.7 lb   (d) 1.145 in.

(e) 6 in.   (f) 2.8 lb   (g) 8.3 in.   (h) 7.7 oz

(i) 21.9 mph   (j) 15.3 psi

Multiplication and Division of Measurement Numbers

With both addition and subtraction of measurement numbers, the answer to the arithmetic calculation has the same units as the numbers being added or subtracted. This is not true when we multiply or divide measurement numbers. For example, to multiply

4.30 ft × 3.60 ft = _____   (Round to one decimal place.)

First, multiply the numerical parts:

$$
\begin{array}{r}
4.3 \\
\times\ 3.6 \\
\hline
15.48
\end{array}
$$

Second, multiply the units: 1 ft × 1 ft = 1 ft^2 or   1 square foot

It may help if you remember that this multiplication is actually the following:

$$4.3 \text{ ft} \times 3.6 \text{ ft} = (4.3 \times \textbf{1 ft}) \times (3.6 \times \textbf{1 ft})$$

$$= (4.3 \times 3.6) \times (\textbf{1 ft} \times \textbf{1 ft})$$

$$= 15.48 \qquad \times \textbf{1 sq ft}$$

$$= 15.48 \text{ sq ft}$$

The rules of arithmetic say that we can do the multiplications in any order we wish, so we multiply the units part separately from the number part.

Notice that the product 1 ft × 1 ft can be written either as 1 ft^2 or as 1 sq ft.

If both numbers are lengths but are given in different units, it is best to convert to the same units. For example, to multiply 2 ft × 5 in., convert it to 24 in. × 5 in.

Third, round the answer. The problem statement tells us to round to the nearest tenth, so the answer should be rounded to 15.5 sq ft to agree with the numbers being multiplied.

In most calculations in the trades, the rounding is obvious from the numbers given. To simplify the process here, we will include rounding instructions with most problems.

Example:

8.1 in. × 1.8 in. × 2.5 in. ___?___   (Round to the nearest whole number.)

First, multiply the numbers:

8.1 × 1.8 × 2.5 = 36.45

Second, multiply the units: 1 in. × 1 in. × 1 in. = 1 in.^{3} or 1 cubic inch

36.45 cu in.

Third, round to the nearest whole number: 36 cu in.

Try these practice problems. (Round to the nearest tenth if necessary.)

(a)   5 ft × 4 ft = _____   (b)   1.4 in. × 2.5 in. = _____

(c)   12 in. × 5 in. = _____   (d)   2 in. × 5 in. × 7 in. = _____

(e) $9\frac{1}{3}$ ft × 12.0 ft = _____   (f) $6\frac{3}{4}$ in. × 8.4 in. = _____

(g)   30 mph × 2 hr = _____   (h)   17.0 ft/sec × 2.1 sec = _____

Now check your work in **6**.

VISUALIZING UNITS

It is helpful to have a visual understanding of measurement units and the "dimension" of a measurement.

 The length unit **inch,** in., specifies a **one**-dimensional or linear measurement. It gives the length of a straight line.

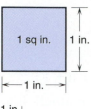 The area unit **square inch,** sq in., specifies a **two**-dimensional measurement. Think of a square inch as giving the area of a square whose sides are one inch in length.

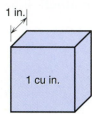 The volume unit **cubic inch,** cu in., specifies a **three**-dimensional measurement. Think of a cubic inch as the volume of a cube whose edges are each one inch in length.

6 (a) 20 sq ft (b) 3.5 sq in.

(c) 60 sq in. (d) 70 cu in.

(e) 112 sq ft (f) 56.7 sq in.

(g) 60 miles (h) 35.7 ft

Did problems (g) and (h) give you any special difficulty?
Do them this way:

mph means $\frac{\text{miles}}{\text{hr}}$

(g) $30 \text{ mph} \times 2 \text{ hr} = 30 \, \dfrac{\text{miles}}{\text{hr}} \times 2 \text{ hr}$

$$= \left(\frac{30 \times 1 \text{ mile}}{1 \text{ hr}} \right) \times (2 \times 1 \text{ hr})$$

$$= 60 \times 1 \text{ mile}$$

$$= 60 \text{ mi}$$

(h) The unit ft/sec means "feet per second" or $\dfrac{\text{ft}}{\text{sec}}$.

$$17 \text{ ft/sec} \times 2.1 \text{ sec} = \left(\frac{17 \times 1 \text{ ft}}{1 \text{ sec}} \right) \times (2.1 \times 1 \text{ sec})$$

$$= 17 \times 1 \text{ ft} \times 2.1$$

$$= 35.7 \times 1 \text{ ft}$$

$$= 35.7 \text{ ft}$$

Division with measurement numbers requires that you take the same care with the units that is needed in multiplication. For example, to divide

$24 \text{ mi} \div 1.5 \text{ hr} = \underline{\quad ? \quad}$

First, divide the numerical parts as usual.

$24 \div 1.5 = 16$

Second, divide the units.

$$1 \text{ mi} \div 1 \text{ hr} = \frac{1 \text{ mi}}{1 \text{ hr}} \qquad \text{or} \qquad 1 \text{ mph}$$

In this case the unit $\frac{1 \text{ mi}}{1 \text{ hr}}$ is usually written mi/hr or mph—miles per hour—and the answer to the problem is 16 mph. Remember that

$$24 \text{ mi} \div 1.5 \text{ hr} = \frac{24 \times 1 \text{ mi}}{1.5 \times 1 \text{ hr}} = \left(\frac{24}{1.5}\right) \times \left(\frac{1 \text{ mi}}{1 \text{ hr}}\right)$$

$$= 16 \times 1 \text{ mph} = 16 \text{ mph}$$

Units such as sq ft or mph, which are made up of a combination of simpler units, are called *compound* units.

Example:

$48 \text{ lb} \div 1.8 \text{ sq in.} = \underline{\quad ? \quad}$ (Round to the nearest whole number of units.)

First, divide the numbers $48 \div 1.8 = 26.666 \ldots \approx 27$ rounded. (Remember, the sign $\approx$ means "approximately equal to.")

Second, divide the units $1 \text{ lb} \div 1 \text{ sq in.} = \dfrac{1 \text{ lb}}{1 \text{ sq in.}} = 1 \text{ lb/sq in.} = 1 \text{ psi}$

$$\left(\text{The unit } psi \text{ means "pounds per square inch" or } \frac{\text{lb}}{\text{sq in.}}.\right)$$

The answer is 27 psi.

Now try a few practice problems.

(a) $4 \text{ ft} \div 2 \text{ sec} = \underline{\qquad}$

(b) $20 \text{ mi} \div 1.25 \text{ hr} = \underline{\qquad}$

(c) $24 \text{ sq ft} \div 3 \text{ ft} = \underline{\qquad}$

(d) $98 \text{ cents} \div 1.50 \text{ gal} = \underline{\qquad}$ (Round to the nearest cent.)

(e) $40 \text{ lb} \div 2 \text{ cu ft} = \underline{\qquad}$

(f) $2\frac{1}{4} \text{ sq ft} \div \frac{1}{4} \text{ ft} = \underline{\qquad}$

(g) $70 \text{ mi} \div 35 \text{ mph} = \underline{\qquad}$

(h) $3\frac{1}{2} \text{ volts} \div 0.5 \text{ ohm} = \underline{\qquad}$

Check your answers in **7**.

---

7 (a) 2 ft/sec

(b) 16 mph

(c) 8 ft

(d) 65 cents/gal (rounded)

(e) 20 lb/cu ft

(f) 9 ft

(g) 2 hr

(h) 7 volts/ohm = 7 amperes

Did problems (f), (g), or (h) seem difficult? Do them this way.

(f) $2\frac{1}{4} \text{ sq ft} \div \frac{1}{4} \text{ ft} = 2.25 \text{ sq ft} \div 0.25 \text{ ft}$

$$= \left(\frac{2.25}{0.25}\right) \times \left(\frac{\text{sq ft}}{\text{ft}}\right) = 9 \text{ ft}$$

$$\frac{\text{sq ft}}{\text{ft}} = \frac{\text{ft}^2}{\text{ft}} = \frac{\text{ft} \times \text{ft}}{\text{ft}} = \text{ft}$$

Problem (c) is done in a similar way.

(g) $70 \text{ mi} \div 35 \text{ mph} = 70 \text{ mi} \div 35\dfrac{\text{mi}}{\text{hr}}$

$$= \left(\dfrac{70}{35}\right) \times \left(\dfrac{\text{mi}}{\text{mi/hr}}\right) = 2 \text{ hr}$$

To divide by the fraction mi/hr, invert and multiply:

$$\dfrac{\text{mi}}{\text{mi/hr}} = \cancel{\text{mi}} \times \dfrac{\text{hr}}{\cancel{\text{mi}}} = \text{hr}$$

(h) People who work with electrical units use the unit *ampere,* abbreviated A, for volts/ohm.

$$\dfrac{1 \text{ volt}}{1 \text{ ohm}} = 1 \text{ ampere}$$

Rounding Measurement Numbers

You have already seen many problems in which a division does not "come out even" or where a multiplication of measurement numbers produces an answer that has more digits than either of its factors. In such cases we must approximate or *round off* the answer. For example, suppose that you need to find the decimal equivalent of the fraction $\frac{12}{7}$. Punch this division into your trusty calculator and you'll get

$$\dfrac{12}{7} = 1.7142857$$

 12 $\boxed{\div}$ **7** $\boxed{=}$ $\rightarrow$ `1.7142857`

which is correct to seven decimal digits. But in any practical situation you will not need all those digits. You need to round this number to one or two decimal places in order to use it in any practical work.

To *one* decimal digit $\dfrac{12}{7} \approx 1.7$ 　　　　To *two* decimal digits $\dfrac{12}{7} \approx 1.71$

To *three* decimal digits $\dfrac{12}{7} \approx 1.714$ 　　To *four* decimal digits $\dfrac{12}{7} \approx 1.7143$

To *five* decimal digits $\dfrac{12}{7} \approx 1.71429$

If you need to review the rules for rounding, return to Chapter 3, starting on page 111.

Accuracy and Precision

The **accuracy** of a measuring instrument is an indication of how closely it can be expected to agree with some calibration standard. When you hear or use the word *accuracy,* you should be thinking of measurement standards and instrument calibration. For example, a thermometer may be calibrated as accurate to the nearest 0.1°C, and a voltmeter to ± 2% of full-scale deflection.

The **precision** of a measuring instrument is an indication of its ability to discriminate, and is usually stated as the smallest scale division of the instrument. A common ruler, divided into sixteenths of an inch, is said to have a precision of $\frac{1}{16}$ in. It can be used to discriminate between two lengths that differ by at least $\frac{1}{16}$ in. An ac voltmeter might have a precision of 0.02 volt and could be used to discriminate between two voltages that differ by that amount. The following table gives the accuracy and precision of some common length measuring instruments.

LENGTH MEASUREMENT DEVICES

	Precision	Accuracy
Ruler	$\frac{1}{16}$ in.	$\pm\frac{1}{32}$ in.
Steel rule	0.02 in.	±0.01 in.
Steel tape	0.1 in.	±0.005 in.
Vernier caliper	0.001 in.	±0.002 in.
Micrometer	0.001 in.	±0.0005 in.
Vernier micrometer	0.0001 in.	±0.0003 in.

A thermometer may be calibrated to an accuracy of ±1°C, but its smallest scale division, or precision, may be 0.2°C. It will provide a measure of the actual temperature only to within ±1°C, but it can register small temperature changes of as little as 0.2°C.

When we do calculations with measurement numbers, our answers cannot be *more* accurate or more precise than the numbers we used to calculate them. For example, calculate the speed of a flywheel if it turns 45 rotations in 1.3 minutes. Run these numbers through your calculator and you will find the following:

$$45 \text{ rotations} \div 1.3 \text{ min} = \frac{45 \text{ rotations}}{1.3 \text{ min}}$$

$$= 34.615385 \ldots \text{ rpm}$$

 45 ÷ **1.3** = → 34.615385

Of course, the calculator can give the answer to eight or ten digits, but a ten-digit answer is actually *not* correct. We cannot have an answer accurate to ten digits when the time, 1.3 minutes, is accurate to only two digits. The answer must be rounded to agree in accuracy with the measurement number. The rotation speed is 35 rpm, rounded to two digits.

In this problem the time, 1.3 minutes, was probably measured, perhaps with a stopwatch. The number of rotations was probably counted directly, perhaps with a mechanical counter, and it is an *exact* number. Numbers that result from simple counting or that are defined are considered to be exact. The calculation should be rounded to agree with the approximate or measurement number.

**Rounding
Sums and
Differences**

When adding or subtracting measurement numbers, round the answer to the same precision as the least precise of the measurement numbers in the calculation.

Examples:

(a) 4.6 + 2.145 = 6.745 ≈ 6.7
Round to the nearest tenth to agree with the least precise number, 4.6, which is stated to the nearest tenth.

(b) 0.24 − 0.028 = 0.212 ≈ 0.21
Round the answer to the nearest hundredth to agree with the least precise number, 0.24.

(c) 2473 + 321.2 = 2794.2 ≈ 2794
Round to the nearest whole number to agree with the least precise measurement, 2473.

Try these problems for practice in adding and subtracting measurement numbers.

(a) 10.5 in. + 8.72 in.

(b) 2.46 cu ft + 3.517 cu ft + 4.8 cu ft

(c) 47.5 sec − 21.844 sec

(d) 56 lb − 4.8 lb

(e) 325 sq ft + 11.8 sq ft + 86.06 sq ft

(f) 0.027 volts + 0.18 volts + 0.009 volts

Check your work in **8**.

---

8 (a) 10.5 in. + 8.72 in. = 19.22 in. ≈ 19.2 in.

(b) 2.46 cu ft + 3.517 cu ft + 4.8 cu ft = 10.777 cu ft ≈ 10.8 cu ft

(c) 47.5 sec − 21.844 sec = 25.656 sec ≈ 25.7 sec

(d) 56 lb − 4.8 lb = 51.2 lb ≈ 51 lb

(e) 325 sq ft + 11.8 sq ft + 86.06 sq ft = 422.86 sq ft ≈ 423 sq ft

(f) 0.027 volts + 0.18 volts + 0.009 volts = 0.216 volts ≈ 0.22 volts

Significant Digits   To multiply or divide measurement numbers we need to consider the concept of significant digits. The *significant digits* in a number are those that represent an actual measurement result. Digits other than zero are always significant. A zero is significant (a) when it appears between two nonzero digits, (b) when it is at the right end of a decimal number, or (c) when it is marked as significant.

For example, a tachometer reading of 84.2 rpm is accurate to three significant digits. A pressure reading of 3206 torr is accurate to four significant digits, and a reading of 4002 torr is accurate to four significant digits, but a reading of $5\overline{0}00$ is accurate to only two significant digits. A voltage of 240 volts is accurate to two significant digits, but a measurement of $24\overline{0}$ volts is accurate to three significant digits. The overbar tells us that the digit marked is to be considered significant. A current of 0.2 ampere is accurate to one significant digit, but a current given as 0.200 is accurate to three significant digits.

More Examples:

5-Digit Accuracy	4-Digit Accuracy	3-Digit Accuracy	2-Digit Accuracy	1-Digit Accuracy
5.4276	5.428	5.43	5.4	5
325.99	326.0	326	330	300
26,299	$26,3\overline{0}0$	26,300	26,000	30,000
0.023901	0.02390	0.0239	0.024	0.02
0.0013042	0.001304	0.00130	0.0013	0.001

To say that a digit is "significant" means that your measuring instrument actually measured that digit. A measurement given as 240 volts means you measured it to the nearest 10 volts. A measurement given as $24\overline{0}$ means that you measured to the nearest volt. A measurement given as 240.0 means that you measured to the nearest tenth of a volt.

Rounding Products and Quotients   When multiplying or dividing measurement numbers, round the answer to the same number of significant digits as the least accurate number used in the calculation.

Examples:

(a) $4.2 \times 8.7 = 36.54 \approx 37$
Round the answer to two significant digits to agree with the factors in the multiplication.

(b) $0.23 \times 1.456 = 0.33488 \approx 0.33$
Round the answer to two significant digits to agree with the two significant digits in 0.23.

(c) $0.04 \div 1.56 \approx 0.025641 \approx 0.03$
Round to one significant digit to agree with 0.04.

(d) $8.40 \times 2.351 = 19.7484 \approx 19.7$
Round to three significant digits to agree with the three significant digits in 8.40.

(e) $0.024 \div 325.0 \approx 0.00007385 \approx 0.000074$
Round to two significant digits to agree with the two significant digits in 0.024.

Now try these problems for practice in multiplying and dividing measurement numbers.

(a) 7.2 in. × 0.48 in.   (b) 0.08 ft × 3.15 ft

(c) 82 ft ÷ 3.492 sec   (d) 2.7 ft × 2.7 ft

(e) 0.074 in. ÷ 8.16 sec   (f) 2450 ton/sq ft × 3178.4 sq ft

Check your work in **9**.

---

9 (a) 7.2 in. × 0.48 in. = 3.456 sq in. ≈ 3.5 sq in.

(b) 0.08 ft × 3.15 ft = 0.252 sq ft ≈ 0.3 sq ft

(c) 82 ft ÷ 3.492 sec ≈ 23.482245 ft/sec ≈ 23 ft/sec

(d) 2.7 ft × 2.7 ft = 7.29 sq ft ≈ 7.3 sq ft

(e) 0.074 in. ÷ 8.16 sec ≈ 0.00906863 in./sec ≈ 0.0091 in./sec

(f) 2450 ton/sq ft × 3178.4 sq ft = 7787080 ton ≈ 7,790,000 ton

Careful ▶ When dividing by hand it is important that you carry the arithmetic to one digit more than will be retained after rounding. ◀

Follow these rounding rules for all calculations shown in this book unless specific instructions for rounding are given.

Decimal Equivalents

Most of the measurements made by technical workers in a shop are made in fractions. On many shop drawings and specifications, the dimensions may be given in decimal form.

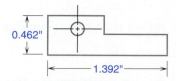

The steel rule usually used in the shop or on the job may be marked in 8ths, 16ths, 32nds, or even 64ths of an inch. A common problem is to rewrite the decimal number to the nearest 32nd or 64th of an inch and to determine how much error is

involved in using the fraction number rather than the decimal number. For example, in the drawing shown, what is the fraction, to the nearest 32nd of an inch, equivalent to 0.462 in.? The rule is to multiply the decimal number by the fraction $\frac{32}{32}$.

$$0.462 \text{ in.} = 0.462 \text{ in.} \times \boxed{\frac{32}{32}}$$

$$= \frac{0.462 \text{ in.} \times 32}{32}$$

$$= \frac{14.784}{32} \text{ in.} \quad \text{Round the top number to the nearest unit.}$$

$$\approx \frac{15}{32} \text{ in.} \quad \text{rounded to the nearest 32nd of an inch.}$$

The error involved is the difference between

$$\frac{15}{32} \text{ in.} \quad \text{and} \quad \frac{14.784}{32} \text{ in.,} \quad \text{or} \quad \frac{15 - 14.784}{32} \text{ in.} = \frac{0.216}{32} \text{ in.}$$

$$= 0.00675 \text{ in.}$$

$$\approx 0.0068 \text{ in.} \quad \text{rounded}$$

The error involved in using $\frac{15}{32}$ in. instead of 0.462 in. is about 68 ten-thousandths of an inch. In problems of this kind we would usually round the error to the nearest ten-thousandth.

Fraction: **.462** ⊗ **32** ⊜ (STO) → 14.784   Round to $\frac{15}{32}$.

Error: **15** ⊖ (RCL) ⊜ ⊘ **32** ⊜ → 0.00675   About 0.0068 in.

Now you try it. Change 1.392 in. to a fraction expressed in 64ths of an inch, and find the error.

Check your work in **10**.

---

10 $1.392 \text{ in.} = 1.392 \text{ in.} \times \boxed{\frac{64}{64}}$

$$= \frac{1.392 \times 64}{64} \text{ in.}$$

$$= \frac{89.088}{64} \text{ in.}$$

$$\approx \frac{89}{64} \text{ in.} \quad \text{rounded to the nearest unit}$$

$$\approx 1\frac{25}{64} \text{ in.}$$

The error in using the fraction rather than the decimal is

$$\frac{89.088}{64} \text{ in.} - \frac{89}{64} \text{ in.} = \frac{0.088}{64} \text{ in.}$$

$$\approx 0.0014 \text{ in.} \quad \text{or about fourteen ten-thousandths of an inch}$$

Fraction: **1.392** ⊗ **64** ⊜ → 89.088   Round to $\frac{89}{64}$.

Error: ⊖ **89** ⊜ ⊘ **64** ⊜ → 0.001375   About 0.0014 in.

Try these problems for practice.

(a) Express to the nearest 16th of an inch.

    (1) 0.438 in.   (2) 0.30 in.

    (3) 0.18 in.   (4) 2.70 in.

(b) Express to the nearest 32nd of an inch.

    (1) 1.650 in.   (2) 0.400 in.

    (3) 0.720 in.   (4) 2.285 in.

(c) Express to the nearest 64th of an inch and find the error, to the nearest ten-thousandth, involved in using the fraction in place of the decimal number.

    (1) 0.047 in.   (2) 2.106 in.

    (3) 1.640 in.   (4) 0.517 in.

Check your answers in **11**.

A DIFFERENT WAY TO ROUND NUMBERS

Some engineers and technicians use the following rounding rule rather than the ones we presented in Chapter 3, page 111.

Example

1. Determine the place to which the number is to be rounded. Mark it with a $_\wedge$.

    Round 4.786 to three digits.

              4.78$_\wedge$6

2. When the digit to the right of the mark is greater than 5, increase the digit to the left by 1.

    4.78$_\wedge$6   becomes 4.79
    1.701$_\wedge$72 becomes 1.702

3. When the digit to the right of the mark is less than 5, drop the digits to the right or replace them with zeros.

    3.81$_\wedge$2   becomes 3.81
    14$_\wedge$39   becomes 1400

4. When the digit to the right of the mark is equal to 5, round the left digit to the nearest *even* value.

    1.41$_\wedge$5   becomes 1.42
    37.00$_\wedge$5 becomes 37.00
    53$_\wedge$50   becomes 5400

11 (a) (1) $\frac{7}{16}$ in.   (2) $\frac{5}{16}$ in.   (3) $\frac{3}{16}$ in.   (4) $2\frac{11}{16}$ in.

(b) (1) $1\frac{21}{32}$ in.   (2) $\frac{13}{32}$ in.   (3) $\frac{23}{32}$ in.   (4) $2\frac{9}{32}$ in.

(c) (1) $\frac{3}{64}$ in. The error is about 0.0001 in.

    (2) $2\frac{7}{64}$ in. The error is about 0.0034 in.

    (3) $1\frac{41}{64}$ in. The error is about 0.0006 in.

    (4) $\frac{33}{64}$ in. The error is about 0.0014 in.

For a set of practice problems over this section of Chapter 5, turn to **12**.

A. Add or subtract as shown.

1.   4 in. + 7 in.

2.   14 lb + 36 lb

3. $3\frac{1}{4}$ in. + 1.7 in. + 4.05 in.

4.   16.0 psi + 15.6 psi + 7.0 psi

5.   0.008 in. + 0.016 in. + 0.310 in.

6.   7.68 ft + 1.25 ft

7.   64.5 mph − 17.0 mph

8.   4.7 gal − 1.9 gal

9. $\frac{1}{4}$ in. − 0.17 in.

10.   37.0 psi − 11.6 psi

B. Multiply or divide as shown. Round to the appropriate number of significant digits.

1.   6 ft × 7 ft

2.   3.1 in. × 1.7 in.

3.   3.0 ft × 2.407 ft

4.   21 mph × 1.2 hr

5. $1\frac{1}{4}$ ft × 1.5 ft

6.   48 in. × 8.0 in.

7.   2.1 ft × 1.7 ft × 1.3 ft

8.   18 sq ft ÷ 2.1 ft

9.   45 mi ÷ 7.3 hr

10.   \$1.37 ÷ 3.2 lb

11. $40\frac{1}{2}$ mi ÷ 25 mph

12.   2.0 sq ft ÷ 1.073 ft

C. Convert as shown.

1.   Write each of the following fractions as a decimal number rounded to two decimal places.

(a) $1\frac{7}{8}$ in.
(b) $4\frac{3}{64}$ in.
(c) $3\frac{1}{8}$ sec
(d) $\frac{1}{16}$ in.

(e) $\frac{3}{32}$ in.
(f) $1\frac{3}{16}$ lb
(g) $2\frac{17}{64}$ in.
(h) $\frac{19}{32}$ in.

(i) $\frac{3}{8}$ lb

2.   Write each of the following decimal numbers as a fraction rounded to the nearest 16th of an inch.

(a)   0.921 in.
(b)   2.55 in.
(c)   1.80 in.
(d)   3.69 in.

(e)   0.802 in.
(f)   0.306 in.
(g)   1.95 in.
(h)   1.571 in.

(i)   0.825 in.

3.   Write each of the following decimal numbers as a fraction rounded to the nearest 32nd of an inch.

(a)   1.90 in.
(b)   0.85 in.
(c)   2.350 in.
(d)   0.666 in.

(e)   2.091 in.
(f)   0.285 in.
(g)   0.600 in.
(h)   0.685 in.

(i)   1.525 in.

4.   Write each of the following decimal numbers as a fraction rounded to the nearest 64th of an inch, and find the error, to the nearest ten-thousandth, involved in using the fraction in place of the decimal number.

(a)   0.235 in.
(b)   0.515 in.
(c)   1.80 in.

(d)   2.420 in.
(e)   3.175 in.
(f)   2.860 in.

(g)   1.935 in.
(h)   0.645 in.
(i)   0.480 in.

D. Practical Problems

1. **Machine Technology** Specifications call for drilling a hole 0.637 ± 0.005 in. in diameter. Will a $\frac{41}{64}$-in. hole be within the required tolerance?

2. **Metalworking** According to standard American wire size tables, 3/0 wire is 0.4096 in. in diameter. Write this size to the nearest 32nd of an inch. What error is involved in using the fraction rather than the decimal?

3. **Auto Mechanics** An auto mechanic converts a metric part to 0.473 in. The part comes only in fractional sizes given to the nearest 64th of an inch. What would be the closest size?

4. **Manufacturing** What size wrench, to the nearest 32nd of an inch, will fit a bolt 0.748 in. in diameter?

5. **Sheet Metal Technology** Find the missing distance x in the drawing.

6. **Auto Mechanics** An auto mechanic uses shims to adjust the clearance in various parts of a car. If the intake valve is supposed to have an 0.008-in. clearance, and the mechanic measures it to be exactly $\frac{1}{32}$ in., what size shim does she need?

7. **Auto Mechanics** The wear on a cylinder has increased its bore to 3.334 in. The most likely original bore was the nearest 32nd-in. size less than this. What was the most likely original bore?

8. **Interior Design** If the rectangular floor of a room has an area of 455 sq ft and a length of 26 ft, what is its width?

9. A truck, moving continuously, averages 52.5 mph for 6.75 hr. How far does it travel during this time?

10. **Industrial Technology** The volume of a cylindrical tank can be calculated as its base area multiplied by its height. If the tank has a capacity of 2760 cu ft of liquid and has a base area of 120 sq ft, what is its height?

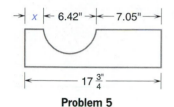

Problem 5

When you have completed these problems, check your answers on pages 591–592, then turn to **13** for some information on units and unit conversion.

5-2 UNITS AND UNIT CONVERSION

13 When you talk about a 4-ft by 8-ft wall panel, a $1\frac{1}{4}$-in. bolt, 3 lb of solder, or a gallon of paint, you are comparing an object with a *standard*. The units or standards used in science, technology, or the trades have two important characteristics: they are convenient in size and they are standardized. Common units such as the foot, pound, gallon, or hour were originally chosen because they were related either to natural quantities or to body measurements.

A pinch of salt, a drop of water, a handful of sand, or a tank of gas are all given in *natural* units. The width of a finger, the length of a foot, or a stride length are all distances given in *body-related* units. In either case, there is a need to define and standardize these units before they are useful for business, science, or technology.

In biblical times, the *digit* was a length unit defined as the width of a person's finger. Four digits was called a *hand,* a unit still used to measure horses. Four hands was a *cubit,* the distance from fingertip to elbow, about 18 in. One *foot* was defined as the length of a man's foot. A *yard* was the distance from nose to tip of outstretched arm. All these lengths depend on whose finger, hand, arm, or foot is used, of course. Standardization began in the fifteenth century, and one yard was defined as the length of a standard iron bar kept in a London vault. One foot was defined as exactly one-third the length of the bar, and one inch was one-twelfth of a foot. (The word *inch* comes from Latin and Anglo-Saxon words meaning "one-twelfth.") Today the inch is legally defined worldwide in terms of metric units.

Because there are a great many units available for writing the same quantity, it is necessary that you be able to convert a given measurement from one unit to another. For example, a length of exactly one mile can be written as

1 mile = 1760 yards, for the traffic engineer
        = 5280 ft, for a landscape architect
        = 63,360 in., in a science problem
        = 320 rods, for a surveyor
        = 8 furlongs, for a horse racing fan
        = 1609.344 meters, in metric units

In this section we will show you a quick and mistake-proof way to convert measurements from one unit to another. Whatever the units given, if a new unit is defined, you should be able to convert the measurement to the new units.

Try the following simple unit conversion. Convert 4 yards to feet.

4 yd = _____ ft

Check your answer in **14**.

---

14 You should know that 1 yard is defined as exactly 3 feet; therefore,

4 yd = 4 × 3 ft = 12 ft

Easy.

Try another, more difficult, conversion. Surveyors use a unit of length called a *rod,* defined as exactly $16\frac{1}{2}$ ft.

Convert: 11 yd = _____ rods.

Check your answer in **15**.

---

15 The correct answer is 2 rods.

You might have thought to first convert 11 yd to 33 ft and then noticed that 33 ft divided by $16\frac{1}{2}$ equals 2. There are 2 rods in a 33-ft length.

Unity Fractions   Most people find it difficult to reason through a problem in this way. To help you, we have devised a method of solving *any* unit conversion problem quickly with no chance of error. This is the *unity fraction* method.

The first step in converting a number from one unit to another is to set up a unity fraction from the definition linking the two units. For example, to convert the distance 4 yd to foot units, take the equation relating yards to feet:

1 yd = 3 ft

and form the fractions $\frac{1 \text{ yd}}{3 \text{ ft}}$ and $\frac{3 \text{ ft}}{1 \text{ yd}}$.

These fractions are called unity fractions because they are both equal to 1. Any fraction whose top and bottom are equal has the value 1.

Practice this step by forming pairs of unity fractions from each of the following definitions.

(a)   12 in. = 1 ft

(b)   1 lb = 16 oz

(c)   3.8 liters = 1 gallon

Check your answers in **16**.

---

16 (a) 12 in. = 1 ft. The unity fractions are $\dfrac{12 \text{ in.}}{1 \text{ ft}}$ and $\dfrac{1 \text{ ft}}{12 \text{ in.}}$.

(b) 1 lb = 16 oz. The unity fractions are $\dfrac{1 \text{ lb}}{16 \text{ oz}}$ and $\dfrac{16 \text{ oz}}{1 \text{ lb}}$.

(c) 3.8 liters = 1 gallon. The unity fractions are $\dfrac{3.8 \text{ liters}}{1 \text{ gallon}}$ and $\dfrac{1 \text{ gallon}}{3.8 \text{ liters}}$.

The second step in converting units is to multiply the original number by one of the unity fractions. For example, to convert the distance 4 yd to foot units, multiply this way:

$$4 \text{ yd} = (4 \times 1 \text{ yd}) \times \left(\frac{3 \text{ ft}}{1 \text{ yd}}\right)$$

$$= 4 \times 3 \text{ ft}$$

$$= 12 \text{ ft}$$

You must choose one of the two fractions. Multiply by the fraction that allows you to cancel out the *yard* units you do not want and to keep the *feet* units you do want.

Did multiplying by a fraction give you any trouble? Remember that any number A can be written as $\dfrac{A}{1}$, so that we have

$$A \times \frac{B}{C} = \frac{A}{1} \times \frac{B}{C} \text{ or } \frac{A \times B}{C}$$

This   means   this

Example:

$$6 \times \frac{3}{4} = \overset{3}{\underset{2}{\cancel{6}}} \times \frac{3}{4}$$

$$= \frac{9}{2} \text{ or } 4\frac{1}{2}$$

(If you need to review the multiplication of fractions, turn back to page 65.)

Now try this problem. Use unity fractions to convert 44 oz to pounds. Remember that 1 lb = 16 oz.

Check your work in **17**.

---

17 Use the equation 1 lb = 16 oz to set up the unity fractions $\dfrac{1 \text{ lb}}{16 \text{ oz}}$ and $\dfrac{16 \text{ oz}}{1 \text{ lb}}$.

Multiply by the first fraction to convert ounces to pounds.

$$44 \text{ oz} = (44 \text{ oz}) \times \left(\frac{1 \text{ lb}}{16 \text{ oz}}\right)$$

$$= \frac{44 \times 1 \text{ lb}}{16}$$

$$= \frac{44}{16} \text{ lb or } 2\frac{3}{4} \text{ lb}$$

Note ▶ When measurement units are converted by division, the remainder often is expressed as the number of original units. For example, instead of giving this last answer as $2\frac{3}{4}$ lb, we can express it as 2 lb 12 oz. ◀

It may happen that several unity fractions are needed in the same problem. For example, if you know that

$$1 \text{ rod} = 16\tfrac{1}{2} \text{ ft} \quad \text{and} \quad 1 \text{ yd} = 3 \text{ ft}$$

then to convert 11 yd to rods multiply as follows:

$$11 \text{ yd} = (11 \text{ yd}) \times \left(\frac{3 \text{ ft}}{}\right) \times \left(\frac{1 \text{ rod}}{16\tfrac{1}{2} \text{ ft}}\right)$$

$$= \frac{11 \times 3 \times 1 \text{ rod}}{16\tfrac{1}{2}}$$

$$= \frac{33}{16\tfrac{1}{2}} \text{ rod}$$

$$11 \text{ yd} = 2 \text{ rods}$$

 11 ✕ **3** ÷ **16.5** = → ⬚ *2.*

Remember: To divide by a fraction, invert and multiply.

$$\frac{33}{16\tfrac{1}{2}} = 33 \div 16\tfrac{1}{2} = 33 \div \tfrac{33}{2} = 33 \times \tfrac{2}{33} = 2$$

Invert and multiply

Be careful to set up the unity fraction so that the unwanted units cancel. At first it helps to write both fractions and then choose the one that fits. As you become an expert at unit conversion, you will be able to write down the correct fraction at first try.

Try these practice problems. Use unity fractions to convert.

(a) 6.25 ft = _____ in.

(b) $5\tfrac{1}{4}$ yd = _____ ft

(c) 2.1 mi = _____ yd   (1760 yd = 1 mi) (Round to the nearest hundred.)

(d) 13 gallons = _____ barrel or bbl (31.5 gal = 1 bbl)   (Round to two significant digits.)

(e) 12.5 bbl = _____ cu ft   (1 gal = 231 cu in.
1 cu ft = 1728 cu in.) (Round to the nearest tenth.)

(f) 32 psi = _____ atm (1 atm = 14.7 psi) (Round to two significant digits.)

(g) 87 in. = _____ ft _____ in.

(h) 9 lb 7 oz = _____ oz

The solutions are in **18**.

---

18   (a) $6.25 \text{ ft} = (6.25 \text{ ft}) \times \left(\dfrac{12 \text{ in.}}{1 \text{ ft}}\right)$ (b) $5\tfrac{1}{4} \text{ yd} = (5\tfrac{1}{4} \text{ yd}) \times \left(\dfrac{3 \text{ ft}}{1 \text{ yd}}\right)$

$\qquad\qquad = 6.25 \times 12 \text{ in.}$ $\qquad\qquad\qquad = 5\tfrac{1}{4} \times 3 \text{ ft}$

$\qquad\qquad = 75 \text{ in.}$ $\qquad\qquad\qquad = 15\tfrac{3}{4} \text{ ft}$

(c) $2.1 \text{ mi} = (2.1 \text{ mi}) \times \left(\dfrac{1760 \text{ yd}}{1 \text{ mi}}\right)$ (d) $13 \text{ gal} = (13 \text{ gal}) \times \left(\dfrac{1 \text{ bbl}}{31.5 \text{ gal}}\right)$

$\qquad\qquad = 2.1 \times 1760 \text{ yd}$ $\qquad\qquad\qquad = \dfrac{13}{31.5} \text{ bbl}$

$\qquad\qquad = 3696 \text{ yd}$ $\qquad\qquad\qquad = 0.41269 \ldots \text{ bbl}$

$\qquad\qquad \approx 3700 \text{ yd} \quad \text{rounded}$ $\qquad\qquad\qquad \approx 0.41 \text{ bbl rounded}$

(e) $12.5 \text{ bbl} = (12.5 \text{ bbl}) \times \left(\dfrac{31.5 \text{ gal}}{}\right) \times \left(\dfrac{231 \text{ cu in.}}{}\right) \times \left(\dfrac{1 \text{ cu ft}}{1728 \text{ cu in.}}\right)$

$$= \dfrac{12.5 \times 31.5 \times 231}{1728} \text{ cu ft}$$

$$= 52.63671 \ldots \text{ cu ft}$$

$$\approx 52.6 \text{ cu ft} \quad \text{rounded}$$

 12.5 ⊗ **31.5** ⊗ **231** ⊘ **1728** ⊜ → $\boxed{52.636719}$

(f) $32 \text{ psi} = (32 \text{ psi}) \times \left(\dfrac{1 \text{ atm}}{14.7 \text{ psi}}\right)$

$$= \dfrac{32}{14.7} \text{ atm}$$

$$\approx 2.1768707 \text{ atm}$$

$$\approx 2.2 \text{ atm} \quad \text{rounded}$$

(g) $87 \text{ in.} = (87 \text{ in.}) \times \left(\dfrac{1 \text{ ft}}{12 \text{ in.}}\right)$

$$= \dfrac{87}{12} \text{ ft}$$

$$= 7\text{R}3 \text{ ft} = 7 \text{ ft } 3 \text{ in.}$$

(h) **First,** convert the pounds to ounces:

$9 \text{ lb} = (9 \text{ lb}) \times \left(\dfrac{16 \text{ oz}}{1 \text{ lb}}\right)$

$$= 9 \times 16 \text{ oz}$$

$$= 144 \text{ oz}$$

Then, add the ounces portion of the original measurement:

$9 \text{ lb } 7 \text{ oz} = 144 \text{ oz} + 7 \text{ oz}$
$$= 151 \text{ oz}$$

Compound Units   In the English system of units commonly used in the United States in technical work, the basic units for length, weight, and time are named with a single word or abbreviation: ft, lb, and sec. We often define units for other similar quantities in terms of these. For example, 1 mile = 5280 ft, 1 ton = 2000 lb, 1 hr = 3600 sec, and so on. We may also define units for more complex quantities using these basic units. For example, the common unit of speed is miles per hour or mph, and it involves both distance (miles) and time (hour) units. Units named using the product or quotient of two or more simpler units are called *compound units*.

If a car travels 75 miles at a constant rate in 1.5 hours, it is moving at a speed of

$$\dfrac{75 \text{ miles}}{1.5 \text{ hours}} = \dfrac{75 \times 1 \text{ mi}}{1.5 \times 1 \text{ hr}}$$

$$= \dfrac{75}{1.5} \times \dfrac{1 \text{ mi}}{1 \text{ hr}}$$

$$= 50 \text{ mi/hr}$$

We write this speed as 50 mi/hr or 50 mph, and read it as "50 miles per hour." Ratios and rates are often written with compound units.

Which of the following are expressed in compound units?

(a) Diameter, 4.65 in.

(b) Density, 12 lb/cu ft

(c)   Gas usage, mi/gal or mpg

(d)   Current, 4.6 amperes

(e)   Rotation rate, revolution/min or rpm

(f)   Pressure, lb/sq in. or psi

(g)   Cost ratio, cents/lb

(h)   Length, 5 yd 2 ft

(i)   Area, sq ft

(j)   Volume, cu in.

Check your answers in **19**.

---

19   Compound units are used in quantities (b), (c), (e), (f), (g), (i), and (j). Two different units are used in (h), but they have not been combined by multiplying or dividing.

Unity fractions can also be used to convert compound units. For example, to convert 50 mph to units of ft/sec:

$$50 \text{ mph} = \left(\frac{50 \times 1 \text{ mi}}{}\right) \times \left(\frac{5280 \text{ ft}}{1 \text{ mi}}\right) \times \left(\frac{1 \text{ hr}}{}\right)$$

$$= \frac{50 \times 5280}{3600} \text{ ft/sec}$$

$$\approx 73.333333 \text{ ft/sec}$$

$$\approx 73 \text{ ft/sec} \quad \text{rounded}$$

Use this procedure to convert a rotation rate of 160 revolutions per minute (rpm) to revolutions per second (rps). (Round to the nearest tenth.)

160 rpm = _____ rps.

Check your work in **20**.

---

20 $$160 \text{ rpm} = \left(\frac{160 \times 1 \text{ rev}}{1 \text{ min}}\right) \times \left(\frac{1 \text{ min}}{}\right)$$

$$= \frac{160 \text{ rev}}{60 \text{ sec}}$$

$$\approx 2.6666667 \text{ rps}$$

$$\approx 2.7 \text{ rps} \quad \text{rounded}$$

In Chapters 8 and 9 we study the geometry needed to calculate the area and volume of the various plane and solid figures used in technical work. Here we look at the many different units used to measure area and volume.

Area   When the area of a surface is found, the measurement or calculation is given in "square units." If the lengths are measured in inches, the area is given in square inches or sq in.; if the lengths are measured in feet, the area is given in square feet or sq ft.

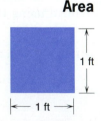

Area = 1 ft × 1 ft
     = 1 sq ft

To see the relation between sq ft area units and sq in. area units, rewrite the above area in inch units.

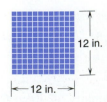

Area = 12 in. × 12 in.
     = 144 × 1 in. × 1 in.
     = 144 sq in.

Therefore,

1 sq ft = 144 sq in.

Use this information to do the following conversion:

15 sq ft = _____ sq in.   (Round to the nearest hundred.)

Check your work in **21**.

LUMBER MEASURE

Carpenters and workers in the construction trades use a special unit to measure the amount of lumber. The volume of lumber is measured in *board feet*. One board foot (bf) of lumber is a piece having an area of 1 sq ft and a thickness of 1 in. or less.

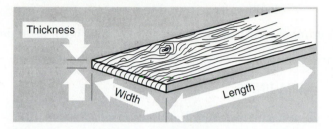

> Number of board feet = thickness in inches × width in feet × length in feet

To find the number of board feet in a piece of lumber, multiply the length in feet by the width in feet by the thickness in inches. A thickness of less than 1 inch should be counted as 1 inch. If the lumber is dressed or finished, use the full size or rough stock dimension to calculate board feet.

For example a 2 by 4 used in framing a house would actually measure $1\frac{1}{2}$ in. by $3\frac{1}{2}$ in., but we would use the dimensions 2 in. by 4 in. in calculating board feet. A 12-ft length of 2 in. by 4 in. would have a volume of

$$12 \text{ ft} \times \frac{4}{12} \text{ ft} \times 2 \text{ in.} = 8 \text{ board ft or 8 bf}$$

In lumber measure the phrase "per foot" means "per board foot." The phrase "per running foot" means "per foot of length."

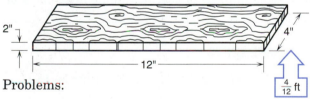

Problems:

1. Find the number of board feet in each of the following pieces of lumber. Be sure to convert the middle dimension, width, to feet.

   (a) $\frac{3}{4}$ in. × 6 in. × 4 ft   (b) 2 in. × 6 in. × 16 ft

   (c) $\frac{1}{2}$ in. × 8 in. × 12 ft   (d) 4 in. × 4 in. × 8 ft

   (e) 1 in. × 6 in. × 14 ft   (f) $\frac{7}{8}$ in. × 12 in. × 3 ft

2. How many board feet are there in a shipment containing 80 boards, 2 in. by 6 in., each 16 ft long?

3. To floor a small building requires 245 boards, each $1\frac{1}{2}$ in. by 12 in. by 12 ft long. How many board feet should be ordered?

The answers are on page 592.

---

21 $15 \text{ sq ft} = (15 \times 1 \text{ sq ft}) \times \left(\dfrac{144 \text{ sq in.}}{1 \text{ sq ft}} \right)$

$= 15 \times 144 \text{ sq in.}$

$= 2160 \text{ sq in.}$ or 2200 sq in. rounded

 15 $\times$ **144** $=$ →   *2160.*

If you do not remember the conversion number, 144, try it this way:

$15 \text{ sq ft} = (15 \times 1 \text{ ft} \times 1 \text{ ft}) \times \left(\dfrac{12 \text{ in.}}{1 \text{ ft}} \right) \times \left(\dfrac{12 \text{ in.}}{1 \text{ ft}} \right)$

$= 15 \times 12 \times 12 \times 1 \text{ in.} \times 1 \text{ in.}$

$= 2160 \text{ sq in.}$

$= 2200 \text{ sq in.}$ rounded

A number of other convenient area units have been defined.

AREA UNITS

1 square yard (sq yd) = 9 sq ft = 144 sq in.

1 square rod (sq rod) = 30.25 sq yd

1 acre = 160 sq rod   = 4840 sq yd

1 sq mile (sq mi)   = 640 acres

Volume   Volume units are given in "cubic units." If the lengths are measured in inches, the volume is given in cubic inches or cu in.; if the lengths are measured in feet, the volume is given in cubic feet or cu ft.

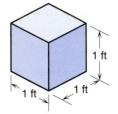

Volume = 1 ft × 1 ft × 1 ft
    = 1 cu ft

In inch units this same volume is 12 in. × 12 in. × 12 in. = 1728 cu in., so that 1 cu ft = 1728 cu in.

A very large number of special volume units have been developed for various uses.

VOLUME UNITS

1 cubic yard (cu yd) = 27 cu ft = 1728 cu in.

1 gallon (gal)   = 231 cu in.   or   1 cu ft ≈ 7.48 gal

1 barrel (bbl)   = 31.5 gal

1 bushel (bu)   = 2150.42 cu in.

1 fluid ounce (fl oz) = 1.805 cu in.

1 pint (pt)   = 28.875 cu in. (liquid measure)

Dozens of other volume units are in common use: cup, quart, peck, teaspoonful, and others. The cubic inch, cubic foot, and gallon are most used in technical work.

In the next section we study metric units, and you will find only a few basic units connected by simple conversion numbers rather than the confusion of having many different length, area, and volume units.

Now, turn to **22** for a set of problems involving unit conversion.

A.   Convert the units as shown.

1. $4\frac{1}{4}$ ft   = _____ in.   2.   33 yd   = _____ ft

3.   3.4 mi   = _____ ft   4.   17 lb   = _____ oz

5.   126 gal   = _____ bbl   6. $8\frac{1}{2}$ mi   = _____ yd

7.   46 psi   = _____ atm   8.   7.5 lb   = _____ oz

9.   18.5 bbl   = _____ gal   10.   32 in.   = _____ ft

11.   2640 ft   = _____ mi   12.   19 yd   = _____ rods

13.   26 ft   = _____ yd   14.   230 oz   = _____ lb

15.   5.1 gal   = _____ cu in.   16.   114 gal   = _____ cu in.

17.   1400 yd   = _____ mile   18.   0.85 cu ft   = _____ gal

19.   6 ft 11 in.   = _____ in.   20.   12 lb 14 oz   = _____ oz

21.   326 oz = _____ lb _____ oz   22.   29 ft = _____ yd _____ ft

B.   Convert the units as shown.

1.   6 sq ft   = _____ sq in.   2. $2\frac{1}{2}$ acre   = _____ sq ft

3.   17.25 bbl   = _____ cu ft   4.   385 sq ft   = _____ sq yd

5.   972 cu in.   = _____ cu ft   6.   325 rpm   = _____ rps

7. $60\ \dfrac{\text{lb}}{\text{cu ft}}$ = _____ $\dfrac{\text{lb}}{\text{cu in.}}$ 8.   30 cu ft   = _____ bbl

9.   14,520 sq ft   = _____ acre   10. $2500\ \dfrac{\text{ft}}{\text{min}}$ = _____ $\dfrac{\text{in.}}{\text{sec}}$

11.   20 sq mi   = _____ acre   12.   65 mph   = _____ ft/sec

13.   2.1 sq yd   = _____ sq in.   14.   42 sq yd   = _____ sq ft

15.   14,000 cu in.   = _____ cu yd   16.   8.3 cu yd   = _____ cu ft

C.   Solve:

1.   **Wastewater Technology** Hydrological engineers often measure very large amounts of water in acre-feet. 1 acre-ft of water fills a rectangular volume of base 1 acre and height 1 ft. Convert 1 acre-ft to

   (a)   _____ cu ft   (b)   _____ gal   (c)   _____ bbl

   (d)   _____ cu yd

2. **Carpentry** The board foot defined on page 209 is a volume unit. Use the definition given to complete the following.

   (a) 1 bd ft = _____ cu ft   (b) 1 bd ft = _____ cu in.

3. Martian greeb workers use the gronk, smersh, and pflug as length units. There are 3 gronks in a smersh, and 2 pflugs equal one smersh. Translate a length of 6 gronks to pflugs.

4. **Painting and Decorating** The area covered by a given volume of paint is determined by the consistency of the paint and the nature of the surface. The measure of covering power is called the *spreading rate* of the paint and is expressed in units of square feet per gallon. The following table gives some typical spreading rates.

Paint	Surface	Spreading Rate	
		One Coat	Two Coats
Oil-based paint	Wood, smooth	580	320
	Wood, rough	360	200
	Plaster, smooth	300	200
	Plaster, rough	250	160
Latex paint	Wood	500	350
Varnish	Wood	400	260

Use this table to answer the following questions.

   (a) How many gallons of oil-based paint are needed to cover 2500 sq ft of rough wood with two coats?

   (b) At $11.98 per gallon, how much would it cost to paint a room of 850 sq ft area with latex paint? The surfaces are finished wood.

   (c) How many square feet of rough plaster will 8 gallons of oil-based paint cover with one coat?

   (d) How many gallons of varnish are needed to cover 600 sq ft of cabinets with two coats?

   (d) At $14.98 per gallon, how much will it cost to paint 750 sq ft of smooth plaster with two coats of oil-based paint?

5. The earliest known unit of length to be used in a major construction project is the "megalithic yard" used by the builders of Stonehenge in southwestern Britain about 2600 B.C. If 1 megalithic yard = 2.72 ± 0.05 ft, convert this length to inches. (Round to the nearest tenth of an inch.)

6. In the Bible (Genesis, Chapter 7) Noah built an ark 300 cubits long, 50 cubits wide, and 30 cubits high. In I Samuel, Chapter 17, it is reported that the giant Goliath was "six cubits and a span" in height. If 1 cubit = 18 in. and 1 span = 9 in.:

   (a) What were the dimensions of the ark in English units?

   (b) How tall was Goliath?

7. **Interior Decorating** A homeowner needing carpeting has calculated that 1265 sq ft of carpet must be ordered. However, carpet is sold in square yards. Make the necessary conversion for her. (Round up to the nearest whole number.)

8. **Metalworking** Cast aluminum has a density of 160.0 lb/cu ft. Convert this density to units of lb/cu in. and round to one decimal place.

9. **Machine Technology** A $\frac{5}{16}$-in. twist drill with a periphery speed of 50 ft/min has a cutting speed of 611 rpm. Convert this speed to rps and round to one decimal place.

10. **Welding** How many pieces $8\frac{3}{8}$ in. long can be cut from a piece of 20-ft stock? Allow for $\frac{3}{16}$ in. kerf (waste due to the width of a cut).

11. **Wastewater Technology** A sewer line has a flow rate of 1.25 MGD (million gallons per day). Convert this to cfs (cubic feet per second).

12. **Wastewater Technology** A reservoir has a capacity of 9000 cu ft. How long will it take to fill the reservoir at the rate of 250 gallons per minute?

13. **Sheet Metal Technology** A machinist must cut 12 strips of metal that are each 3 ft $2\frac{7}{8}$ in. long. What is the total length needed in inches?

14. **Marine Technology** A scientist wishes to treat a 430,000-barrel (bbl) oil spill with a bacterial culture. If the culture must be applied at the rate of 1 oz of culture per 100 cu ft of oil, how many ounces of culture will the scientist need? (Round to two significant digits.)

Check your answers on page 592, then continue in **23** with the study of metric units.

5-3 METRIC UNITS

23 For many technical or trades workers, the ability to use metric units is important. In many ways, working with meters, kilograms, and liters is easier and more logical than using our traditional units of feet, pounds, or quarts. In this section we define and explain the most useful metric units, show how they are related to the corresponding English units, explain how to convert from one unit to another, and provide practice problems designed to help you use the metric system and to "think metric."

The most important common units in the metric system are those for length or distance, speed, weight, volume, area, and temperature. Time units—year, day, hour, minute, second—are the same in the metric as in the English system.

Length The basic unit of length in the metric system is the *meter*, pronounced *meet-ur* and abbreviated m. (The word is sometimes spelled *metre* but pronounced exactly the same.) One meter is roughly equal to one yard.

The meter is the appropriate unit to use in measuring your height, the width of a room, length of lumber, or the height of a building.

Estimate the following lengths in meters:

1 foot

1 yard

1 meter

(a) The length of the room in which you are now sitting = _____ m.

(b) Your height = _____ m.

(c) The length of a Ping-Pong table = _____ m.

(d) The length of a football field = _____ m.

Guess as closely as you can, then turn to **24** and continue.

---

24 (a) A small room might be 3 or 4 m long, and a large one might be 6 or 8 m long.

(b) Your height is probably between $1\frac{1}{2}$ and 2 m. Note that 2 m is roughly 6 ft 7 in.

(c) A Ping-Pong table is a little less than 3 m in length.

(d) A football field is about 100 m long.

All other length units used in the metric system are defined in terms of the meter, and these units differ from one another by multiples of ten. For example, the *centimeter*, pronounced *cent-a-meter* and abbreviated *cm*, is defined as exactly one-hundredth of a meter. The *kilometer*, pronounced *kill-o-meter* and abbreviated *km*, is defined as exactly 1000 meters. The *millimeter*, abbreviated *mm*, is defined as exactly $\frac{1}{1000}$ meter.

Because metric units increase or decrease in multiples of ten, they may be named using prefixes attached to a basic unit. For length units we have:

Metric Length Unit	Prefix	Multiplier	Money Analogy	Memory Aid
kilometer	kilo-	1000×1 meter	$1000	*kilo*watt
hectometer	hecto-	100×1 meter	$ 100	
decameter	deca-	10×1 meter	$ 10	*deca*de = 10 years
meter	—	1 meter	$ 1	
decimeter	deci-	0.1×1 meter	10¢	*deci*mal = $\frac{1}{10}$
centimeter	centi-	0.01×1 meter	1¢	*cent*ury = 100 years
millimeter	milli-	0.001×1 meter	$\frac{1}{10}$¢	*mill*ennium = 1000 years

The following conversion facts are especially useful for workers in the trades.

> 1 centimeter (cm) = $\frac{1}{100}$ of a meter = 0.01 meter or 1 m = 100 cm
>
> 1 km = 1000 m   so   1 m = 0.001 km
>
> 1 mm = 0.1 cm = 0.001 m   so   1 m   = 1000 mm
>
> 1 cm = 10 mm

The millimeter (mm) unit of length is very often used on technical drawings, and the centimeter (cm) is handy for shop measurements.

1 cm is roughly the width of a paper clip.

1 mm is roughly the thickness of the wire in a paper clip.

1 cm

1 mm

1 meter is roughly 10% more than a yard.

One very great advantage of the metric system is that we can easily convert from one metric unit to another. There are no hard-to-remember conversion factors: 36 in. in a yard, 5280 ft in a mile, 220 yards in a furlong, or whatever.

For example, with English units, if we convert a length of 137 in. to feet or yards, we find

$$137 \text{ in.} = 11\frac{5}{12} \text{ ft} = 3\frac{29}{36} \text{ yard}$$

Here we must divide by 12 and then by 3 to convert the units. But in the metric system we simply shift the decimal point. The same length in the metric system is

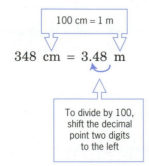

100 cm = 1 m

348 cm = 3.48 m

To divide by 100, shift the decimal point two digits to the left

Of course we may also use unity fractions:

$$348 \text{ cm} = 348 \text{ cm} \times \left(\frac{1 \text{ m}}{100 \text{ cm}} \right)$$

$$= \frac{348}{100} \text{ m}$$

$$= 3.48 \text{ m}$$

Try it. Complete the following unit conversions.

(a)   147 cm = _____ m   (b)   3.1 m = _____ cm

(c)   2.1 km = _____ m   (d)   307 m = _____ km

(e)   7 cm = _____ mm   (f)   20.5 mm = _____ cm

Check your answers in **25**.

---

25   (a) $147 \text{ cm} = 147 \text{ cm} \times \left(\frac{1 \text{ m}}{100 \text{ cm}} \right)$

$$= \frac{147}{100} \text{ m}$$

$$= 1.47 \text{ m} \qquad \text{The decimal point shifts two places to the left.}$$

(b) $3.1 \text{ m} = 3.1 \text{ m} \times \left(\frac{100 \text{ cm}}{1 \text{ m}} \right)$

$$= 3.1 \times 100 \text{ cm}$$

$$= 310 \text{ cm} \quad \text{The decimal point shifts two places right.}$$

(c) $2.1 \text{ km} = 2.1 \text{ km} \times \left(\frac{1000 \text{ m}}{1 \text{ km}} \right)$

$$= 2100 \text{ m}$$

(d) $307 \text{ m} = 307 \text{ m} \times \left(\frac{1 \text{ km}}{1000 \text{ m}} \right)$

$$= 0.307 \text{ km}$$

(e) $7 \text{ cm} = 7 \text{ cm} \times \left(\frac{10 \text{ mm}}{1 \text{ cm}} \right)$

$$= 70 \text{ mm}$$

(f) $20.5 \text{ mm} = 20.5 \text{ mm} \times \left(\frac{1 \text{ cm}}{10 \text{ mm}} \right)$

$$= 2.05 \text{ cm}$$

**Metric–English
Conversion**   Because metric units are the only international units, all English units are defined in terms of the metric system. For length measurements,

1 inch

1 centimeter

1 in. = 2.54 cm
1 ft = 30.48 cm
1 yd = 91.44 cm
1 cm = 0.3937 in.

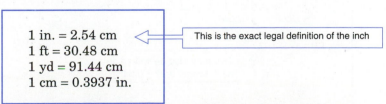

This is the exact legal definition of the inch

To shift from English to metric units or from metric to English units, use either unity fractions or the conversion factors given in the table.

APPROXIMATE CONVERSION FACTORS: LENGTH

When you know	you can find	if you multiply by
inches, in.	millimeters, mm	25.40
inches, in.	centimeters, cm	2.54
feet, ft	centimeters, cm	30.48
feet, ft	meters, m	0.3048
yards, yd	meters, m	0.9144
miles, mi	kilometers, km	1.6093
millimeters, mm	inches, in.	0.03937
centimeters, cm	inches, in.	0.3937
centimeters, cm	feet, ft	0.0328
meters, m	feet, ft	3.2808
meters, m	yards, yd	1.0936
kilometers, km	miles, mi	0.6214

Convert the following measurements as shown. (Round to one decimal place.)

(a) 4.3 yd = _____ m

(b) 16.0 in. = _____ cm

(c) $2\frac{1}{2}$ in. = _____ mm

(d) 6.0 ft = _____ m

(e) 1.2 m = _____ ft

(f) 37.0 cm = _____ in.

(g) 4.2 m = _____ yd

(h) 140 cm = _____ ft

(i) 18 ft 6 in. = _____ m

(j) $2\frac{1}{4}$ in. = _____ mm

Check your answers in **26**.

DUAL DIMENSIONING

Some companies involved in international trade use "dual dimensioning" on their technical drawings and specifications. With dual dimensioning both inch and metric dimensions are given. For example, a part might be labeled like this:

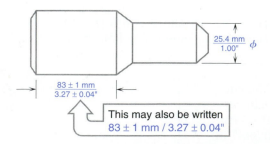

Notice that the metric measurement is written first or on top of the fraction bar. Diameter dimensions are marked with the symbol φ.

26 (a) 3.9 m (b) 40.6 cm

(c) 63.5 mm (d) 1.8 m

(e) 3.9 ft (f) 14.6 in.

(g) 4.6 yd (h) 4.6 ft

(i) 5.6 m (j) 57.2 mm

In problem (a) $4.3 \text{ yd} \times \dfrac{91.44 \text{ cm}}{1 \text{ yd}} \times \dfrac{1 \text{ m}}{100 \text{ cm}} = \dfrac{4.3 \times 91.44}{100} \text{ m} = 3.93192 \text{ m} \approx 3.9 \text{ m}.$

Or, using the conversion factor, $4.3 \text{ yd} \times 0.9144 \dfrac{\text{m}}{\text{yd}} \approx 3.9 \text{ m}.$

In problem (e) $1.2 \text{ m} \times 3.2808 \dfrac{\text{ft}}{\text{m}} = 3.93696 \text{ ft} \approx 3.9 \text{ ft}.$

Area When the area of a surface is calculated, the units will be given in "square length units." If the lengths are measured in feet, the area will be given in square feet; if the lengths are measured in meters, the area will be given in square meters.

For example, the area of a carpet 3 m wide and 4 m long would be

$3 \text{ m} \times 4 \text{ m} = 12$ square meters
$= 12$ sq m

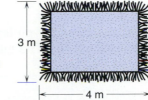

Areas roughly the size of carpets and room flooring are usually measured in square meters. Larger areas are usually measured in *hectares,* a metric surveyor's unit. One hectare is the area of a square 100 m on each side. The hectare is roughly $2\frac{1}{2}$ acres, or about double the area of a football field.

1 hectare = 100 m × 100 m
$\qquad\qquad$ = 10,000 square meters
$\qquad\qquad$ = 2.471 acres

To convert, use the following approximate conversion factors.

APPROXIMATE CONVERSION FACTORS: AREA

When you know	you can find	if you multiply by
square inches, in.^{2}	square centimeters, cm^2	6.452
square feet, ft^2	square meters, m^2	0.093
square yards, yd^2	square meters, m^2	0.836
acres	hectares, ha	0.4047
acres	square meters, m^2	4047
square centimeters, cm^2	square inches, in.^{2}	0.155
square meters, m^2	square feet, ft^2	10.764
square meters, m^2	square yards, yd^2	1.196
square meters, m^2	hectares, ha	0.0001
hectares, ha	acres	2.471

Use this information to solve the following problems. (Round to the nearest tenth.)

(a) 15 sq ft = _____ m^2

(b) 0.21 m^2 = _____ sq ft

(c) $8\frac{1}{2}$ acres = _____ ha

(d) 10 cm^2 = _____ sq in.

(e) 78 sq yd = _____ sq m

(f) 960,000 m^2 = _____ ha

Check your answers in **27**.

---

27 (a) 1.4 m^2 (b) 2.3 sq ft   (c) 3.4 ha

(d) 1.6 sq in.   (e) 65.2 sq m   (f) 96 ha

Speed In the metric system, ordinary highway speeds are measured in kilometers per hour, abbreviated km/hr or kmh. To "think metric" while you drive, remember that

> 100 kmh is approximately equal to 62 mph.

Machinists measure pulley, drill, and lathe speeds in centimeters per second, abbreviated cm/sec. The following table gives useful factors.

APPROXIMATE CONVERSION FACTORS: SPEED

When you know	you can find	if you multiply by
inches per second, ips	centimeters per second, cm/sec	2.54
feet per second, fps	centimeters per second, cm/sec	30.48
feet per second, fps	meters per second, m/sec	0.3048
feet per minute, fpm	centimeters per second, cm/sec	0.5080
miles per hour, mph	kilometers per hour, kmh	1.6093
centimeters per second, cm/sec	inches per second, ips	0.3937
centimeters per second, cm/sec	feet per second, fps	0.0328
meters per second, m/sec	feet per second, fps	3.2808
kilometers per hour, kmh	miles per hour, mph	0.6214

Use these conversion factors to find the following. (Round your answers to the nearest whole number.)

(a) 35 mph = _____ kmh

(b) 30 m/sec = _____ ft/sec

(c) 120 cm/sec = _____ in./sec

(d) 18 ips = _____ cm/sec

(e) 280 kmh = _____ mph

(f) The cutting speed for a soft steel part in a lathe is 165 ft/min. Express this in cm/sec.

Check your answers in **28**.

28   (a)   56 kmh   (b)   98 ft/sec   (c)   47 in./sec

      (d)   46 cm/sec   (e)   174 mph   (f)   84 cm/sec

Volume

When the volume of an object is calculated or measured, the units will normally be given in "cubic length units." If the lengths are measured in feet, the volume will be given in cubic feet; if the lengths are measured in meters, the volume will be given in cubic meters.

For example, the volume of a box 2 m high by 3 m wide by 4 m long would be

$2 \text{ m} \times 3 \text{ m} \times 4 \text{ m} = 24$ cubic meters

and this is often written as 24 m^3.

The cubic meter is an appropriate unit for measuring the volume of large amounts of water, sand, or gravel, or the volume of a room. However, a more useful metric volume unit for most practical purposes is the *liter*, pronounced *leet-ur*. You may find this volume unit spelled either *liter* in the United States or *litre* elsewhere in the world.

One liter is defined as 1000 cm^3, the volume of a cube 10 cm on each edge.

A volume of 1 liter is slightly larger than 1 quart. In fact, in the United States, 1 quart is legally defined as exactly 0.94635295 liter.

> 1 liter = $1000 \text{ cm}^3 \approx 1.0567$ qt
>
> 1 gal = 4 qt
>
> 1 qt $\approx$ 0.946 liter

One liter is roughly 6% more than one quart.

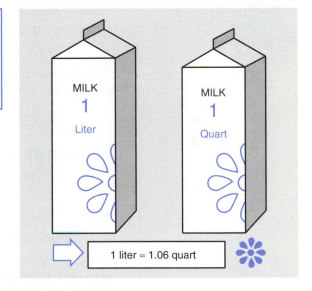

Calculate each of the following volumes in liters.

(a)   A rough iron casting 20 cm by 15 cm by 32 cm

(b)   A rectangular tank 15 in. by 18 in. by 16 in.   (Round to nearest liter.)

(c)   A 25-gallon container of fuel oil   (Round to nearest liter.)

Check your work in **29**.

29 (a)   Volume = 20 cm × 15 cm × 32 cm

$$= 9600 \text{ cm}^3$$

$$= 9600 \text{ cm}^3 \times \frac{1 \text{ liter}}{1000 \text{ cm}^3}$$

$$= 9.6 \text{ liters}$$

 20 ⊗ **15** ⊗ **32** ⊘ **1000** ⊜ → | 9.6 |

(b)   Convert all dimensions to cm using the conversion factor 1 in. = 2.54 cm or the table on page 216.

Volume = 15 in. × 18 in. × 16 in.

$$= 38.1 \text{ cm} \times 45.72 \text{ cm} \times 40.64 \text{ cm}$$

$$\approx 70792 \text{ cm}^3$$

$$\approx 71 \text{ liters, rounded.}$$

 **15** ⊗ **2.54** ⊗ **18** ⊗ **2.54** ⊗ **16** ⊗ **2.54** ⊘ **1000** ⊜ → | 70.792116 |

(c)   25 gal = 100 quarts

$$\approx 100 \text{ qt} \times \frac{1 \text{ liter}}{1.0567 \text{ qt}}$$

$$\approx 95 \text{ liters}$$

Volumes smaller than the liter are usually measured in units of cubic centimeters, abbreviated cm³ or cu cm.

1 cubic centimeter = 1 cm × 1 cm × 1 cm

$$= 1 \text{ cm}^3 \quad \text{or} \quad 1 \text{ cu cm}$$

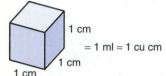

= 1 ml = 1 cu cm

Because 1000 cubic centimeters are needed to make up 1 liter, a cubic centimeter is $\frac{1}{1000}$ of a liter or a *milli-liter*.

To convert volumes from English to metric units or from metric to English units, use either unity fractions or the following conversion factors.

APPROXIMATE CONVERSION FACTORS: VOLUME

When you know	you can find	if you multiply by
cubic inches, in.³	cubic centimeters, cm³	16.387
cubic feet, ft³	liters	28.317
quarts (liquid), qt	liters	0.946
gallons, gal	liters	3.785
cubic yards, yd³	cubic meters, m³	0.765
fluid ounce, fl oz	cubic centimeters, cm³	29.574
cubic centimeters, cm³	cubic inches, in.³	0.061
liters	cubic feet, ft³	0.035
liters	quarts, qt	1.057
liters	gallons, gal	0.264
cubic meters, m³	cubic yards, yd³	1.307

Convert the following measurements as shown. Round the answer to the same number of significant digits as in the original number.

(a)   13.4 gal = _____ liter   (b)   18.2 cu ft = _____ liter

(c)   9.1 fl oz = _____ cu cm   (d)   16.0 cu yd = _____ m^3

(e)   5.1 liters = _____ qt   (f)   140 cm^3 = _____ cu in.

(g)   24.0 m^3 = _____ cu yd   (h)   64.0 liters = _____ gal

(i)   12.4 qt = _____ liter   (j)   4.2 in.^{3} = _____ cm^3

Check your answers in **30**.

---

30   (a)   50.7 liters   (b)   515 liters   (c)   270 cm^3 (d)   12.2 m^3

(e)   5.4 qt   (f)   8.5 cu in.   (g)   31.4 cu yd   (h)   16.9 gal

(i)   11.7 liters   (j)   69 cm^3

Weight   The weight of an object is the gravitational pull of the earth exerted on that object. The mass of an object is related to how it behaves when pushed or pulled. For scientists the difference between mass and weight can be important. For all practical purposes they are the same.

The basic unit of mass or metric weight in the International Metric System is the *kilogram,* pronounced *kill-o-gram* and abbreviated *kg.* The kilogram is defined as the mass of a standard platinum–iridium cylinder kept at the International Bureau of Weights and Measures near Paris. By law in the United States, the pound is defined as exactly equal to the weight of 0.45359237 kg, so that

> 1 kg weighs about 2.2046 lb

One kilogram weighs about 10% more than 2 lb.

The kilogram was originally designed so that it was almost exactly equal to the weight of one liter of water.

Two smaller metric units of mass are also defined. The *gram,* abbreviated *g,* and the *milligram,* abbreviated *mg,* are defined as follows:

> 1 gram = $\frac{1}{1000}$ of a kilogram = 0.001 kilogram
> 1 milligram = $\frac{1}{1000}$ of a gram = 0.001 gram

The gram is very small unit of weight, equal to roughly $\frac{1}{500}$ of a pound, or the weight of a paper clip. The milligram is often used by druggists and nurses to measure very small amounts of medication or other chemicals, and it is also used in scientific measurements.

To convert weights from English to metric units or from metric to English units, use either unity fractions or the following conversion factors.

APPROXIMATE CONVERSION FACTORS: WEIGHT

When you have	you can find	if you multiply by
pounds, lb	kilograms, kg	0.454
ounces, oz	grams, g	28.350
tons, T	kilograms, kg	907.19
tons, T	metric tons, t	0.907
kilograms, kg	pounds, lb	2.2046
grams, g	ounces, oz	0.0353
kilograms, kg	tons, T	0.0011
metric tons, t	tons, T	1.102

Convert the following measurements as shown. Round to the same number of significant digits.

(a) 150 lb = _____ kg

(b) 6.0 oz = _____ g

(c) 2.50 T = _____ kg

(d) $10\overline{0}$ kg = _____ lb

(e) 1.50 kg = _____ oz

(f) 3.5 t = _____ T

(g) 150 g = _____ lb

(h) 4 lb 5 oz = _____ kg

Check your work in **31.**

---

31 (a) 68 kg

(b) 170 g

(c) 2270 kg

(d) 220 lb

(e) 52.9 oz

(f) 3.9 tons

(g) 0.33 lb

(h) 2.0 kg

In problem (e) $1.50 \text{ kg} \times \dfrac{2.2046 \text{ lb}}{1 \text{ kg}} \times \dfrac{16 \text{ oz}}{1 \text{ lb}}$.

In problem (h) the answer 1.96 is 2.0 when rounded to two significant digits.

Temperature   The Fahrenheit temperature scale is commonly used in the United States for weather reports, cooking, and other practical work. On this scale water boils at 212°F and freezes at 32°F. The metric or *Celsius* temperature scale is a simpler scale originally designed for scientific work but now used worldwide for most temperature measurements. On the Celsius scale water boils at 100°C and freezes at 0°C.

Notice that the range from freezing to boiling is covered by 180 degrees on the Fahrenheit scale and 100 degrees on the Celsius scale. The size and meaning of a temperature degree is different for the two scales. A temperature of zero does not mean "no temperature"—it is simply another point on the scale. Temperatures less than zero are possible, and they are labeled with negative numbers.

Because zero does not correspond to the same temperature on both scales, converting Celsius to Fahrenheit or Fahrenheit to Celsius is not as easy as converting length or weight units. The simplest way to convert from one scale to the other is to use the chart on page 223.

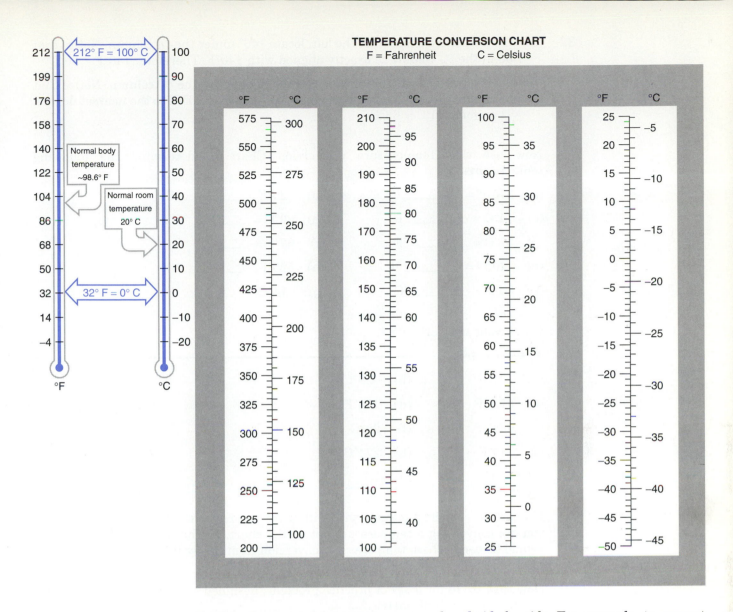

TEMPERATURE CONVERSION CHART

F = Fahrenheit   C = Celsius

On this chart equal temperatures are placed side by side. For example, to convert 50°F to Celsius:

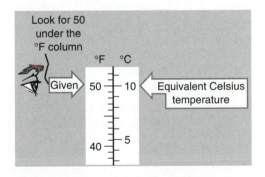

Use this temperature conversion chart to convert

(a)   140°F to Celsius

(b)   20°C to Fahrenheit

Check your work in **32**.

32 (a) On the second chart from the left, locate 140° under the °F column. Notice that 60 in the °C column is exactly aligned with 140°F. Therefore, 140°F = 60°C.

(b) On the third chart from the left, locate 20°C under the °C column. Notice that 68 in the °F column is approximately aligned with 20°C, to the nearest degree. Therefore, 20°C ≈ 68°F.

Now use this Temperature Conversion Chart to determine the following temperatures.

(a) 375°F = _____ °C   (b) −20°F = _____ °C

(c) 14°C = _____ °F   (d) 80°C = _____ °F

(e) 37°C = _____ °F   (f) 68°F = _____ °C

(g) −40°C = _____ °F   (h) 525°F = _____ °C

(i) 80°F = _____ °C   (j) −14°C = _____ °F

Check your answers in **33**.

---

33 (a) 191°C   (b) −29°C

(c) 57.5°F   (d) 176°F

(e) 98.6°F   (f) 20°C

(g) −40°F   (h) 274°C

(i) 27°C   (j) 7°F

If you must convert a temperature with more accuracy than this chart allows, or if you are converting a temperature not on the chart, algebraic conversion formulas should be used. These temperature conversion formulas will be explained in Chapter 7 when we study basic algebra.

Now turn to **34** for a set of practice problems designed to help you use the metric system and to "think metric."

34
Exercises 5-3 **Metric Units**

A. **Think Metric.** For each problem, circle the measurement closest to the first one given. No calculations are needed.

Remember: (1) A meter is a little more (about 10%) than a yard.
(2) A kilogram is a little more (about 10%) than 2 pounds.
(3) A liter is a little more (about 6%) than a quart.

1. 30 cm   (a) 30 in.   (b) 75 in.   (c) 1 ft

2. 5 ft   (a) 1500 cm   (b) 1.5 m   (c) 2 m

3. 1 yd   (a) 90 cm   (b) 110 cm   (c) 100 cm

4. 2 m   (a) 6 ft 6 in.   (b) 6 ft   (c) 2 yd

5. 3 km   (a) 3 mi   (b) 2 mi   (c) 1 mi

6. 200 km   (a) 20 mi   (b) 100 mi   (c) 120 mi

7.	50 kmh	(a)	30 mph	(b)	50 mph	(c)	60 mph
8.	55 mph	(a)	30 kmh	(b)	60 kmh	(c)	90 kmh
9.	100 m	(a)	100 yd	(b)	100 ft	(c)	1000 in.
10.	400 lb	(a)	800 kg	(b)	180 kg	(c)	250 kg
11.	6 oz	(a)	1.7 g	(b)	17 g	(c)	170 g
12.	5 kg	(a)	2 lb	(b)	5 lb	(c)	10 lb
13.	50 liters	(a)	12 gal	(b)	120 gal	(c)	50 gal
14.	6 qt	(a)	2 liters	(b)	3 liters	(c)	6 liters
15.	14 sq ft	(a)	1.3 m^2	(b)	13 m^2	(c)	130 m^2
16.	8 sq in.	(a)	50 cm^2	(b)	5 cm^2	(c)	500 cm^2
17.	100°F	(a)	38°C	(b)	212°C	(c)	32°C
18.	60°C	(a)	20°F	(b)	140°F	(c)	100°F
19.	212°F	(a)	100°C	(b)	400°C	(c)	50°C
20.	0°C	(a)	100°F	(b)	32°F	(c)	−30°F

B. **Think Metric.** Choose the closest estimate.

1.	Diameter of a penny	(a)	3 cm	(b)	1.5 cm	(c)	5 cm
2.	Length of a man's foot	(a)	3 m	(b)	3 cm	(c)	30 cm
3.	Tank of gasoline	(a)	50 liters	(b)	5 liters	(c)	500 liters
4.	Volume of a wastebasket	(a)	10 liters	(b)	100 liters	(c)	10 cu cm
5.	One-half gallon of milk	(a)	1 liter	(b)	2 liters	(c)	$\frac{1}{2}$ liter
6.	Hot day in Phoenix, Arizona	(a)	100°C	(b)	30°C	(c)	45°C
7.	Cold day in Minnesota	(a)	−80°C	(b)	−10°C	(c)	20°C
8.	200-lb barbell	(a)	400 kg	(b)	100 kg	(c)	40 kg
9.	Length of paper clip	(a)	10 cm	(b)	35 mm	(c)	3 mm
10.	Your height	(a)	17 m	(b)	17 cm	(c)	1.7 m
11.	Your weight	(a)	80 kg	(b)	800 kg	(c)	8 kg
12.	Boiling water	(a)	100°C	(b)	212°C	(c)	32°C

C. Convert to the units shown. (Round to the appropriate number of significant digits.)

1. 3.0 in.   = _____ cm   2. 4.2 ft   = _____ cm

3. 2.0 mi   = _____ km   4. 20 ft 6 in. = _____ m
                                         (three sig. digits)

5. $9\frac{1}{4}$ in.   = _____ cm   6. 31 kg   = _____ lb
    (two sig. digits)

7. 152 lb   = _____ kg   8. 3 lb 4 oz = _____ kg
                                         (two sig. digits)

9. 8.5 kg   = _____ lb   10. 10.4 oz   = _____ g

11. 3.15 gal = _____ liters   12. 2.5 liters = _____ qt

13. 6.5 qt   = _____ liters   14. 61 mph   = _____ kmh

15. 8.0 kmh = _____ mph   16. 230°F   = _____ °C

17. 10°C   = _____ °F   18. 35 in./sec = _____ cm/sec

19. 63 ft/sec = _____ m/sec   20. 2.0 m   = _____ ft

21. 285 sq ft = _____ m^2 22. 512 cu in. = _____ cm^3

23. 2.75 m^3 = _____ cu yd   24. 46.2 cm^2 = _____ sq in.

D. Practical Problems. (Round as indicated.)

1. **Welding** Welding electrode sizes are presently given in inches. Convert the following set of electrode sizes to millimeters. (1 in. = 25.4 mm)

in.	mm
0.030	
0.035	
0.040	
0.045	

in.	mm
$\frac{1}{16}$	
$\frac{5}{64}$	
$\frac{3}{32}$	
$\frac{1}{8}$	

in.	mm
$\frac{5}{32}$	
$\frac{3}{16}$	
$\frac{3}{8}$	
$\frac{11}{64}$	

(Round to the nearest one-thousandth of a millimeter.)

2. One cubic foot of water weighs 62.4 lb. (a) What is the weight in pounds of 1 liter of water? (b) What is the weight in kilograms of 1 cubic foot of water? (c) What is the weight in kilograms of 1 liter of water? (Round to one decimal place.)

3. At the Munich Olympic Games in 1972 Frank Shorter of the United States won the 26.2-mile marathon race in 2 hr 12 min 19.8 sec. What was his average speed in mph and kmh? (Round to the nearest tenth.)

4. The metric unit of pressure is the *pascal,* where

$1\ \dfrac{\text{lb}}{\text{in.}^2}$ (psi) ≈ 6894 pascal (Pa), and $1\ \dfrac{\text{lb}}{\text{ft}^2}$ (psf) ≈ 47.88 Pa

Convert the following pressures to metric units.

(a) 15 psi = _____ Pa   (b) 100 psi = _____ Pa

(c) 40 psf = _____ Pa   (d) 2500 psf = _____ Pa

(Round to the nearest thousand pascal.)

5. Find the difference between

(a) 1 mile and 1 kilometer   _____ km

(b) 3 miles and 5000 meters   _____ ft

(c) 120 yards and 110 meters   _____ ft

(d) 1 quart and 1 liter   _____ qt

(e) 10 pounds and 5 kilograms   _____ lb

6. The Mamiya RB 67 camera uses 120 film and produces $2\frac{1}{4}$-in. by $2\frac{3}{4}$-in. negatives. How would this negative size be given in millimeters in Japan?

7. Which is performing more efficiently, a car getting 12 km per liter of gas or one getting 25 miles per gallon of gas?

8. **Painting and Decorating** Industrial paint sells for $5.25 per gallon. What would 1 liter cost?

9. **Roofing** In flooring and roofing, the unit "one square" is sometimes used to mean 100 sq ft. Convert this unit to the metric system. (Round to one decimal place.)

1 square = _____ m^2

10. A *cord* of wood is a volume of wood 8 ft long, 4 ft wide, and 4 ft high. (a) What are the equivalent metric dimensions for a "metric cord" of wood? (b) What is the volume of a cord in cubic meters? (Round to the nearest tenth.) (*Hint:* volume = length × width × height.)

11. Translate these well-known phrases into metric units.

(a) A miss is as good as _____ km. (1 mile = _____ km)

(b) A _____ cm TV screen. (30 in. = _____ cm)

(c) An _____ cm by _____ cm sheet of paper. ($8\frac{1}{2}$ in. = _____ cm; 11 in. = _____ cm)

(d) He was beaten within _____ cm of his life. (1 in. = _____ cm)

(e) Race in the Indy _____ km. (500 miles = _____ km)

(f) Take it with _____ grams of salt. (*Hint:* 437.5 grains = 1 oz.)

(g) _____ grams of prevention is worth _____ grams of cure. (1 oz = _____ g; 1 lb = _____ g)

(h) See the world in _____ grams of sand. (1 grain = _____ g)

(i) Peter Piper picked _____ liters of pickled peppers. (*Hint:* 1 peck = 8 quarts.)

(j) A cowboy in a _____ liter hat. (10 gal = _____ liter)

(k) He's in _____ ml of trouble. (1 peck = _____ ml)

12. **Auto Mechanics** In an automobile, cylinder displacement is measured in either cubic inches or liters. What is the cylinder displacement in liters of a 400-cu in. engine? (Round to the nearest liter.)

When you have completed these exercises, check your answers on pages 592–593, then turn to **35** to learn about direct measurement.

5-4 DIRECT MEASUREMENTS

35

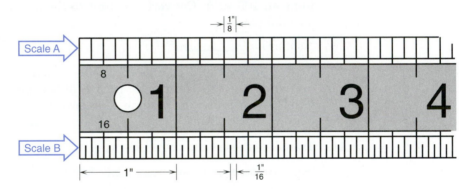

The simplest kind of measuring device to use is one in which the output is a digital display. For example, in a digital clock or digital volt-meter, the measured quantity is translated into electrical signals, and each digit of the measurement number is displayed.

A digital display is easy to read, but most technical work involves direct measurement where the numerical reading is displayed on some sort of scale. Reading the scale correctly requires skill and the ability to estimate or *interpolate* accurately between scale divisions.

Length Measurements

Length measurement involves many different instruments that produce either fraction or decimal number readouts. Carpenters, electricians, plumbers, roofers, and machinists all use rulers, yardsticks, or other devices to make length measurements. Each instrument is made with a different scale that determines the precision of the measurement that can be made. The more precise the measurement has to be, the smaller the scale divisions needed.

For example, on the carpenter's rule shown, scale A is graduated in eighths of an inch. Scale B is graduated in sixteenths of an inch.

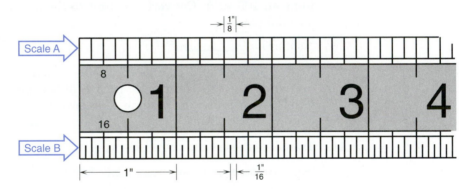

A carpenter or roofer may need to measure no closer than $\frac{1}{8}$ or $\frac{1}{16}$ in., but a machinist will often need a scale graduated in 32nds or 64ths of an inch. The finely engraved steel rule shown is a machinist's rule.

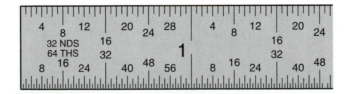

Some machinists' rules are marked in tenths or hundredths of an inch rather than in the usual divisions of $\frac{1}{2}$, $\frac{1}{4}$, $\frac{1}{8}$, $\frac{1}{16}$, and so on.

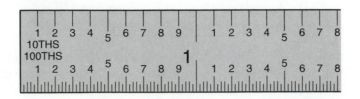

Notice that on all these rules only a few scale markings are actually labeled with numbers. It would be easy to read a scale marked like this:

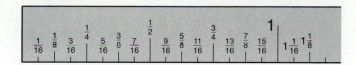

Any useful scale has divisions so small that this kind of marking is impossible. An expert in any technical field must be able to identify the graduations on the scale correctly even though the scale is *not* marked. To use any scale, first determine the size of the smallest division by counting the number of spaces in a 1-in. interval. Here is an example:

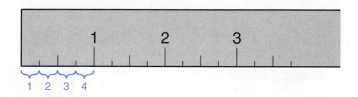

There are four spaces in the first 1-in. interval. The smallest division is therefore $\frac{1}{4}$ in., and we can label the scale like this:

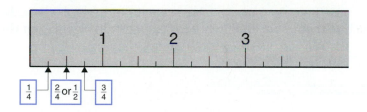

Try it. Label all marks on the following scales. The units are inches.

(a)

Smallest division = _____

(b)

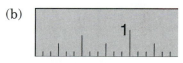

Smallest division = _____

(c)

Smallest division = _____

(d)

Smallest division = _____

(e)

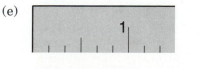

Smallest division = _____

(f)

Smallest division = _____

Check your work in **36**.

36

(a) Smallest division = $\frac{1}{8}$ in.

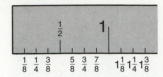

(b) Smallest division = $\frac{1}{12}$ in.

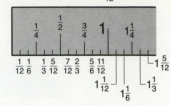

(c) Smallest division = $\frac{1}{5}$ in.

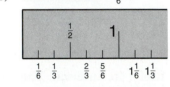

(d) Smallest division = $\frac{1}{10}$ in.

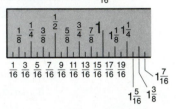

(e) Smallest division = $\frac{1}{6}$ in.

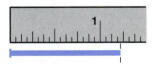

(f) Smallest division = $\frac{1}{16}$ in.

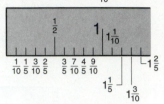

To use the rule once you have mentally labeled all scale markings, first count inches, then count the number of smallest divisions from the last whole-inch mark.

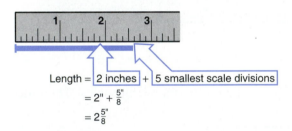

Length = 2 inches + 5 smallest scale divisions

$$= 2" + \frac{5"}{8}$$

$$= 2\frac{5"}{8}$$

If the length of the object being measured falls between scale divisions, record the nearest scale division. For example, in this measurement:

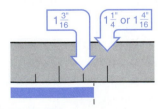

the endpoint of the object falls between $1\frac{3}{16}$ and $1\frac{4}{16}$ in.

To the nearest scale division, the length is $1\frac{3}{16}$ in. When a scale with $\frac{1}{16}$-in. divisions is used, we usually write the measurement to the nearest $\frac{1}{16}$ in., but we may estimate it to the nearest half of a scale division or $\frac{1}{32}$ in.

For practice in using length rulers, find the dimensions given on the following inch rules. Express all fractions in lowest terms.

(a)

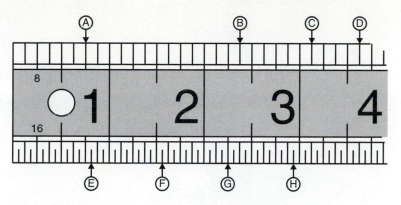

(b)

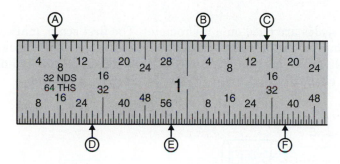

(c)   Express answers in decimal form.

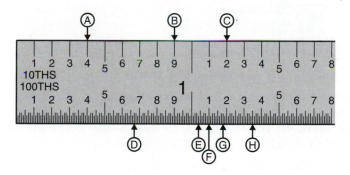

Check your answers in **37**.

---

37 (a) A. $\frac{3}{4}$ in. B. $2\frac{3}{8}$ in. C. $3\frac{1}{8}$ in. D. $3\frac{5}{8}$ in.

E. $\frac{13}{16}$ in. F. $1\frac{9}{16}$ in. G. $2\frac{1}{4}$ in. H. $2\frac{15}{16}$ in.

(b) A. $\frac{7}{32}$ in. B. $1\frac{3}{32}$ in. C. $1\frac{15}{32}$ in.

D. $\frac{28}{64}$ in. $= \frac{7}{16}$ in. E. $\frac{29}{32}$ in. F. $1\frac{37}{64}$ in.

(c) A. $\frac{4}{10}$ in. $= 0.4$ in. B. $\frac{9}{10}$ in. $= 0.9$ in. C. $1\frac{2}{10}$ in. $= 1.2$ in.

D. $\frac{67}{100}$ in. $= 0.67$ in. E. $1\frac{4}{100}$ in. $= 1.04$ in. F. $1\frac{10}{100}$ in. $= 1.10$ in.

G. $1\frac{18}{100}$ in. $= 1.18$ in. H. $1\frac{35}{100}$ in. $= 1.35$ in.

Micrometers   A steel rule can be used to measure lengths of $\pm\frac{1}{32}$ in. or $\pm\frac{1}{64}$ in., or as fine as $\pm\frac{1}{100}$ in. To measure lengths with greater accuracy than this, more specialized instruments are needed. A *micrometer* can be used to measure lengths to ±0.001 in., one one-thousandth of an inch.

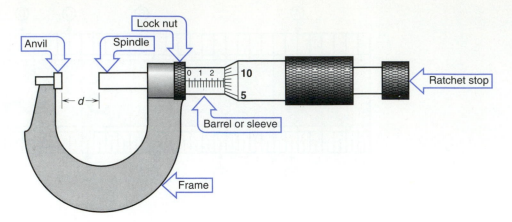

Two scales are used on a micrometer. The first scale, marked on the sleeve, records the movement of a screw machined accurately to 40 threads per inch. The smallest divisions on this scale record movements of the spindle of one-fortieth of an inch or 0.025 in.

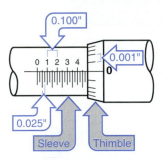

The second scale is marked on the rotating spindle. One complete turn of the thimble advances the screw one turn or $\frac{1}{40}$ in. The thimble scale has 25 divisions, so that each mark on the thimble scale represents a spindle movement of $\frac{1}{25} \times \frac{1}{40}$ in. = $\frac{1}{1000}$ in. or 0.001 in.

To read a micrometer, follow these steps.

Step 1   Read the largest numeral visible on the sleeve and multiply this number by 0.100 in.

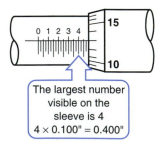

The largest number visible on the sleeve is 4
$4 \times 0.100" = 0.400"$

Step 2   Read the number of additional scale spaces visible on the sleeve. (This will be 0, 1, 2, or 3.) Multiply this number by 0.025 in.

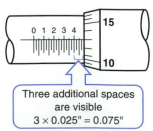

Three additional spaces are visible
$3 \times 0.025" = 0.075"$

Step 3 Read the number on the thimble scale opposite the horizontal line on the sleeve and multiply this number by 0.001 in.

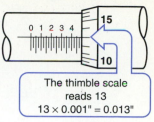

The thimble scale
reads 13
13 × 0.001" = 0.013"

Step 4 Add these three products to find the measurement.

$$\begin{array}{r} 0.400 \text{ in.} \\ 0.075 \text{ in.} \\ +\ 0.013 \text{ in.} \\ \hline 0.488 \text{ in.} \end{array}$$

Careful ▶ To count as a "space" in Step 2, the right end marker must be visible. ◀

Here are more examples.

Example 1:

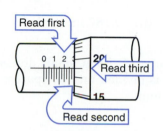

Read first
Read third
Read second

Step 1 The largest number completely visible on the sleeve is 2.

2 × 0.100 in. = 0.200 in.

Step 2 Three additional spaces are visible on the sleeve.

3 × 0.025 in. = 0.075 in.

Step 3 The thimble scale reads 19.

19 × 0.001 in. = 0.019 in.

Step 4 Add: 0.200 in.
        0.075 in.
        0.019 in.
Length = 0.294 in.

Example 2:

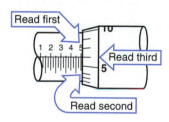

Read first
Read third
Read second

Step 1 5 × 0.100 in. = 0.500 in.

Step 2 No additional spaces are visible on the sleeve.

0 × 0.025 in. = 0.000 in.

Step 3 6 × 0.001 in. = 0.006 in.

Step 4 Add: 0.500 in.
        0.000 in.
        0.006 in.
Length = 0.506 in.

For practice, read the following micrometers.

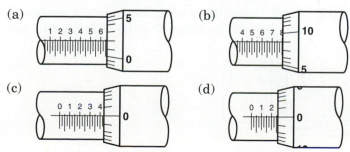

(a)   1 2 3 4 5 6   5   0

(b)   4 5 6 7 8   10   5

(c)   0 1 2 3 4   0

(d)   0 1 2   0

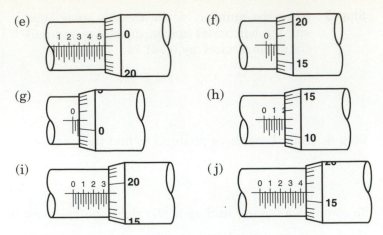

(e)   (f)

(g)   (h)

(i)   (j)

Check your micrometer readings in **38**.

---

38   (a)   0.652 in.   (b)   0.809 in.   (c)   0.425 in.   (d)   0.250 in.   (e)   0.549 in.

(f)   0.092 in.   (g)   0.051 in.   (h)   0.187 in.   (i)   0.344 in.   (j)   0.441 in.

Metric micrometers are easier to read. Each small division on the sleeve represents 0.5 mm. The thimble scale is divided into 50 spaces; therefore, each division is $\frac{1}{50} \times 0.5$ mm $= \frac{1}{100}$ mm or 0.01 mm.

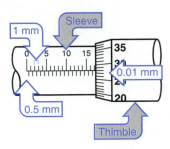

Turning the thimble scale of a metric micrometer by one scale division advances the spindle one one-hundredth of a millimeter, or 0.001 cm.

To read a metric micrometer, follow these steps.

Step 1   Read the largest mark visible on the sleeve and multiply this number by 1 mm.

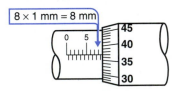

Step 2   Read the number of additional half spaces visible on the sleeve. (This number will be either 0 or 1.) Multiply this number by 0.5 mm.

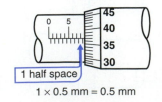

1 × 0.5 mm = 0.5 mm

Step 3 Read the number on the thimble scale opposite the horizontal line on the sleeve and multiply this number by 0.01 mm.

The thimble scale reads 37.
37×0.01 mm $= 0.37$ mm

Step 4 Add these three products to find the measurement value.

8.00 mm
0.50 mm
0.37 mm
—————
8.87 mm

Another Example:

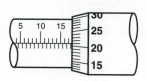

17×1 mm $\quad = 17$ mm
1×0.5 mm $\quad = 0.5$ mm
21×0.01 mm $= 0.21$ mm
Length $= 17$ mm $+ 0.5$ mm $+ 0.21$ mm
$\qquad = 17.71$ mm

Read these metric micrometers.

(a)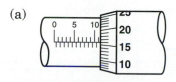

(b)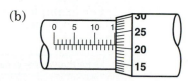

(c)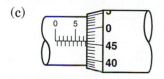

(d)

Check your micrometer readings in **39**.

---

39 (a) 11.16 mm   (b) 15.21 mm

(c) 7.96 mm   (d) 21.93 mm

Vernier Micrometers When more accurate length measurements are needed, the *vernier* micrometer can be used to provide an accuracy of $\pm$ 0.0001 in.

On the vernier micrometer, a third or vernier scale is added to the sleeve. This new scale appears as a series of ten lines parallel to the axis of the sleeve above the usual scale.

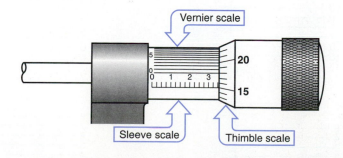

The ten vernier marks cover a distance equal to nine thimble scale divisions—as shown on this spread-out diagram of the three scales.

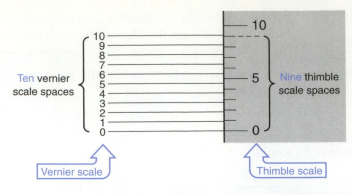

The difference between a vernier scale division and a thimble scale division is one-tenth of a thimble scale division, or $\frac{1}{10}$ of 0.001 in. = 0.0001 in.

To read the vernier micrometer, follow the first three steps given for the ordinary micrometer, then:

Step 4 Find the number of the line on the vernier scale that exactly lines up with any line on the thimble scale. Multiply this number by 0.0001 in. and add this amount to the other three distances.

For example, for the vernier micrometer shown:

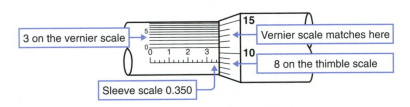

Sleeve scale: 3×0.100 in. = 0.300 in.   When adding these four lengths,
                  2×0.025 in. = 0.050 in.   be careful to align the decimal
Thimble scale: 8×0.001 in. = 0.008 in.   digits correctly.
Vernier scale: 3×0.0001 in. = $\underline{0.0003 \text{ in.}}$
                  Length   = 0.3583 in.

The vernier scale provides a way of measuring accurately the distance to an additional tenth of the finest scale division.

Now for some practice in using the vernier scale, read each of the following.

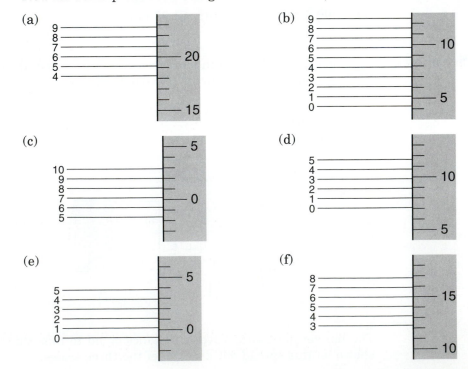

Check your vernier readings in **40**.

---

40 (a) 7   (b) 3   (c) 9

     (d) 1   (e) 3   (f) 6

Now read the following vernier micrometers.

(a) 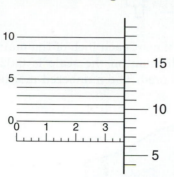   (b)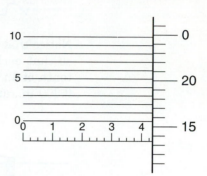

(c) 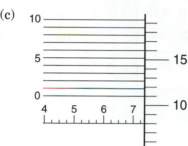   (d)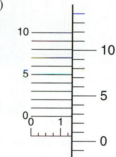

(e) 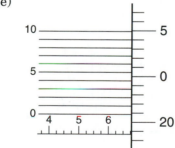   (f)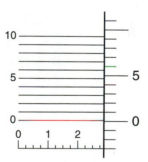

Check your micrometer readings in **41**.

---

41 (a) 0.3568 in.   (b) 0.4397 in.   (c) 0.7330 in.

     (d) 0.1259 in.   (e) 0.6939 in.   (f) 0.2971 in.

Vernier Calipers   The vernier caliper is another length measuring instrument that uses the vernier principle and can measure lengths to ±0.001 in.

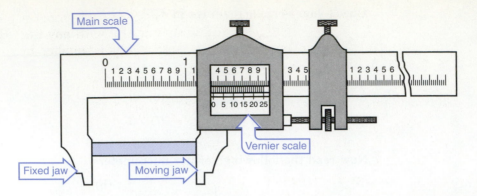

Notice that on this instrument each inch of the main scale is divided into 40 parts, so that each smallest division on the scale is 0.025 in. Every fourth division or tenth of an inch is numbered. The numbered marks represent 0.100 in., 0.200 in., 0.300 in., and so on.

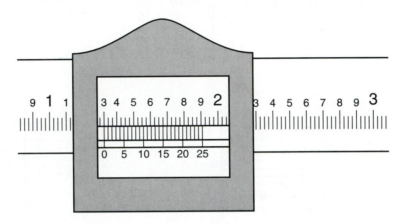

The sliding vernier scale is divided into 25 equal parts numbered by 5s. These 25 divisions on the vernier scale cover the same length as 24 divisions on the main scale.

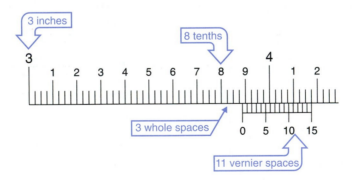

To read the vernier caliper above, follow these steps.

Step 1 Read the number of whole-inch divisions on the main scale to the left of the vernier zero. This gives the number of whole inches.

3 in. = 3.000 in.

Step 2 Read the number of tenths on the main scale to the left of the vernier zero. Multiply this number by 0.100 in.

8 tenths
8×0.100 in. $= 0.800$ in.

Step 3 Read the number of additional whole spaces on the main scale to the left of the vernier zero. Multiply this number by 0.025 in.

3 spaces
3×0.025 in. $= 0.075$ in.

Step 4 On the sliding vernier scale, find the number of the line that exactly lines up with any line on the main scale. Multiply this number by 0.001 in.

11 lines up exactly.
11×0.001 in. $= 0.011$ in.

Step 5 Add the four numbers.

3.000 in.
0.800 in.
0.075 in.
+0.011 in.
3.886 in.

Another Example:

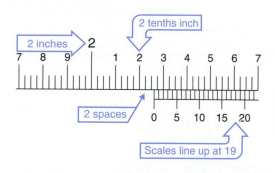

2 in. $= 2.000$ in.
2×0.100 in. $= 0.200$ in.
2×0.025 in. $= 0.050$ in.
19×0.001 in. $= \underline{0.019}$ in.
2.269 in.

Read the following vernier calipers.

(a)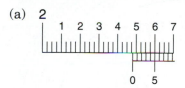

(b)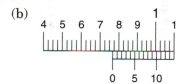

(c)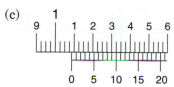

(d)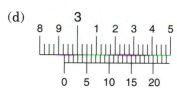

(e)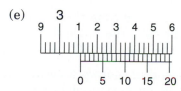

(f)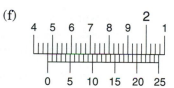

Check your caliper readings in **42**.

What does "vernier" mean—as in vernier scale or vernier caliper?

The vernier scale was invented about 350 years ago by Pierre Vernier, a French mathematician.

42 (a) 2.478 in. (b) 0.763 in. (c) 1.084 in.

(d) 2.931 in. (e) 3.110 in. (f) 1.470 in.

Protractors A *protractor* is used to measure and draw angles. The simplest kind of protractor, shown here, is graduated in degrees with every tenth degree labeled with a number.

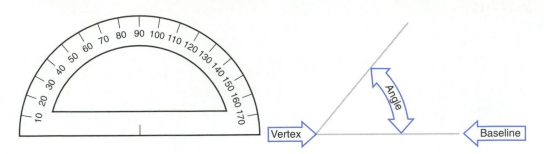

In measuring or drawing angles with a protractor, be certain that the vertex of the angle is exactly at the center of the protractor base. One side of the angle should be placed along the 0°–180° baseline.

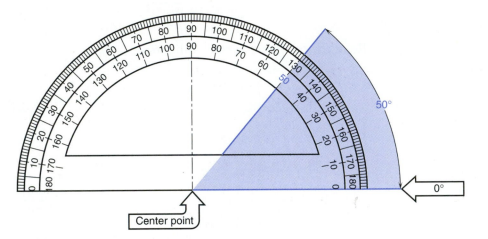

Notice that the protractor can be read from the 0° point on either the right or left side. Measuring from the right side, the angle is read as 50° on the inner scale. Measuring from the left side, the angle is 130° on the outer scale.

The *bevel protractor* is very useful in shopwork because it can be adjusted with an accuracy of ±0.5°. The angle being measured is read directly from the movable scale.

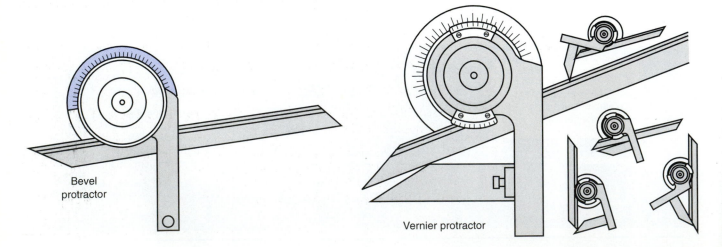

Bevel protractor

Vernier protractor

For very accurate angle measurements, the *vernier protractor* can be used with an accuracy of $\pm\frac{1}{60}$ of a degree or 1 minute, since 60 minutes = 1 degree.

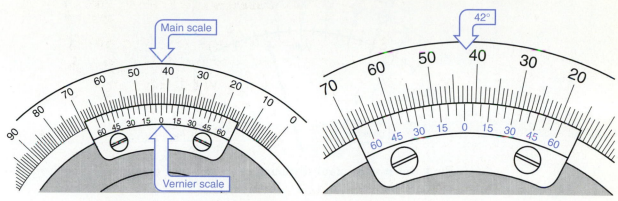

To read the vernier protractor shown, follow these steps.

Step 1 Read the number of whole degrees on the main scale between the zero on the main scale and the zero on the vernier scale.

The mark on the vernier scale just to the right of 0° is 42°.

Step 2 Find the vernier scale line on the left of the vernier zero that lines up exactly with a main scale marking. The vernier protractor in the example reads 42°30′.

The vernier scale lines up with a mark on the main scale at 30. This gives the fraction of a degree in minutes.

Read the following vernier protractors.

(a)

(b)

(c)

(d)

Check your angle readings in **43**.

---

43   (a)   73°46′   (b)   11°15′   (c)   61°5′   (d)   27°25′

Meters   Many measuring instruments convert the measurement into an electrical signal and then display that signal by means of a pointer and a decimal scale. Instruments of this kind, called *meters,* are used to measure electrical quantities such as voltage, current, or power, and other quantities such as air pressure, flow rates, or speed. As with length scales, not every division on the meter scale is labeled, and practice is required in order to read meters quickly and correctly.

The *range* of a meter is its full-scale or maximum reading. The *main* divisions of the scale are the numbered divisions. The *small* divisions are the smallest marked portions of a main division. For example, on the following meter scale,

The *range* is 10 volts;

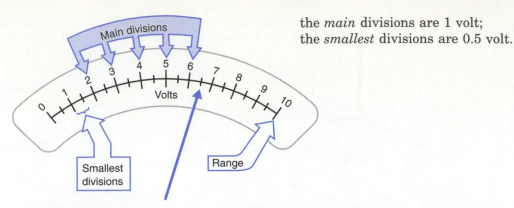

the *main* divisions are 1 volt;
the *smallest* divisions are 0.5 volt.

Find the range, main divisions, and smallest divisions for each of the following meter scales.

(a) Range = _____
main divisions = _____
smallest divisions = _____

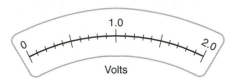

(b) Range = _____
main divisions = _____
smallest divisions = _____

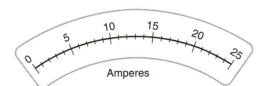

(c) Range = _____
main divisions = _____
smallest divisions = _____

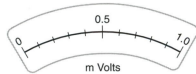

(d) Range = _____
main divisions = _____
smallest divisions = _____

Check your answers in **44**.

---

44 (a) Range = 2.0 V   (b) Range = 25 A
main divisions = 1.0 V   main divisions = 5 A
smallest divisions = 0.1 V   smallest divisions = 1 A

(c) Range = 1.0 mV   (d) Range = 500 psi
main divisions = 0.5 mV   main divisions = 50 psi
smallest divisions = 0.1 mV   smallest divisions = 10 psi

To read a meter scale, either we can choose the small scale division nearest to the pointer, or we can estimate between scale markers. For example, this meter

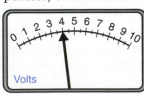

reads exactly 4.0 volts.

But this meter

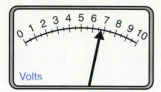

reads 6.5 volts to the nearest scale marker, or 6.6 volts if we estimate between scale markers. The smallest division is 0.5 volt.

On the following meter scale

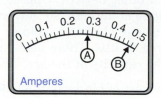

the smallest scale division is $\frac{1}{5}$ of 0.1 ampere or 0.02 ampere. The reading at A is therefore 0.2 plus 4 small divisions or 0.28 ampere. The reading at B is 0.4 plus $3\frac{1}{2}$ small divisions or 0.47 ampere.

Read the following meters. Estimate between scale divisions where necessary.

(a)

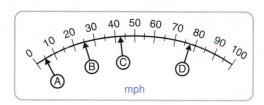

(b)

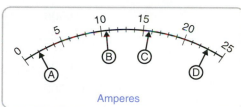

(c)

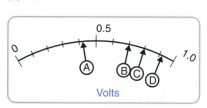

(d)

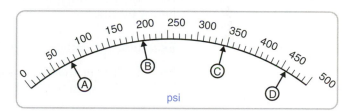

(e)

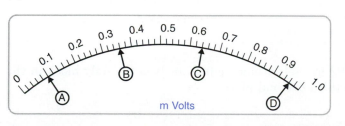

(f)

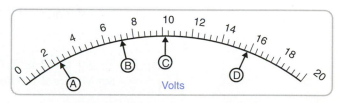

Check your meter readings in **45**. (Your estimated answers may differ from ours.)

---

45 (a) A. 5   B. 25   C. 43   D. 78 mph

(b) A. 2   B. 10.6   C. 15.8   D. 23 amperes

(c)	A.	0.42	B.	0.7	C.	0.8	D.	0.93 volt
(d)	A.	82	B.	207	C.	346	D.	458 psi
(e)	A.	0.1	B.	0.36	C.	0.64	D.	0.96 millivolt
(f)	A.	2.8	B.	7.2	C.	10.2	D.	15.7 volts

Notice in meter (f) that the smallest division is $\frac{1}{5}$ of 2 volts or 0.4 volt. If your answers differ slightly from ours, the difference is probably due to the difficulty of estimating the position of the arrows on the diagrams.

Gauge Blocks *Gauge blocks* are extremely precise rectangular steel blocks used for making, checking, or inspecting tools and other devices. Opposite faces of each block are ground and lapped to a high degree of flatness, so that when two blocks are fitted together they will cling tightly. Gauge blocks are sold in standard sets of from 36 to 81 pieces in precisely measured widths. These pieces may be combined to give any decimal dimension within the capacity of the set.

A standard 36-piece set would consist of blocks of the following sizes (in inches):

0.0501	0.051	0.050	0.110	0.100	1.000
0.0502	0.052	0.060	0.120	0.200	2.000
0.0503	0.053	0.070	0.130	0.300	
0.0504	0.054	0.080	0.140	0.400	
0.0505	0.055	0.090	0.150	0.500	
0.0506	0.056				
0.0507	0.057				
0.0508	0.058				
0.0509	0.059				

To combine gauge blocks to produce any given dimension, work from digit to digit, right to left. This procedure will produce the desired length with a minimum number of blocks.

Example: Combine gauge blocks from the set given to equal 1.8324 in.

Step 1 To get the 4 in 1.8324 in., select block 0.0504 in. It is the only block in the set with a 4 in the fourth decimal place.

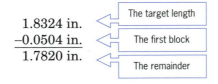

$$
\begin{array}{r}
1.8324 \text{ in.} \\
-0.0504 \text{ in.} \\
\hline
1.7820 \text{ in.}
\end{array}
$$

The target length
The first block
The remainder

Step 2 To get the 2 in 1.782 in., select block 0.052 in. It is the only block in the set with a 2 in the third decimal place.

$$
\begin{array}{r}
1.782 \text{ in.} \\
-0.052 \text{ in.} \\
\hline
1.730 \text{ in.}
\end{array}
$$

The new target length
the second block
the remainder

Step 3 To get the 3 in 1.73 in., select block 0.130 in.

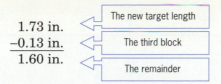

1.73 in. ← The new target length
−0.13 in. ← The third block
1.60 in. ← The remainder

Step 4 To get 1.6, select blocks 1.000, 0.500, and 0.100 in. The six blocks can be combined or "stacked" to total the required 1.8324-in. length.

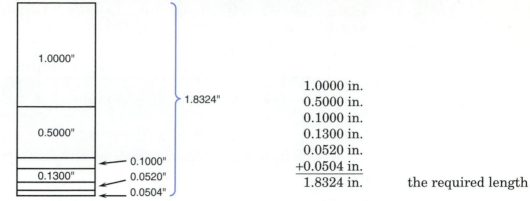

1.0000 in.
0.5000 in.
0.1000 in.
0.1300 in.
0.0520 in.
+0.0504 in.
1.8324 in.   the required length

Stack sets of gauge blocks that total each of the following.

(a) 2.0976 in.   (b) 1.8092 in.   (c) 3.6462 in.

(d) 0.7553 in.   (e) 1.8625 in.   (f) 0.7474 in.

Check your answers in **46**.

---

46 (a) 1.000 in. + 0.500 in. + 0.400 in. + 0.090 in. + 0.057 in. + 0.0506 in.

(b) 1.000 in. + 0.500 in. + 0.200 in. + 0.059 in. + 0.0502 in.

(c) 1.000 in. + 2.000 in. + 0.4000 in. + 0.140 in. + 0.056 in. + 0.0502 in.

(d) 0.500 in. + 0.100 in. + 0.050 in. + 0.055 in. + 0.0503 in.

(e) 1.000 in. + 0.500 in. + 0.200 in. + 0.060 in. + 0.052 in. + 0.0505 in.

(f) 0.500 in. + 0.140 in. + 0.057 in. + 0.0504 in.

Now turn to **47** for a set of practice problems designed to help you become more expert in making direct measurements.

47
Exercises 5-4   **Direct Measurements**

A.   Find the lengths marked on the following rules.

1.

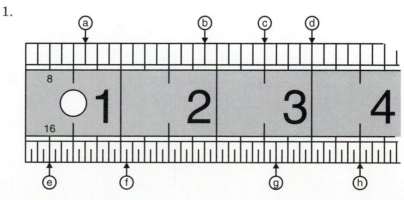

2.

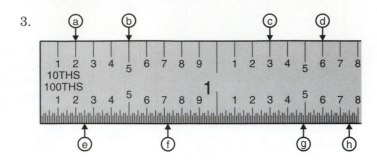

3.

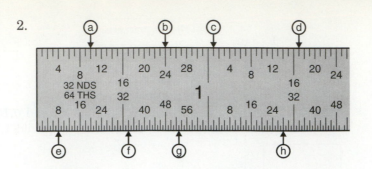

B. Read the following micrometers.

1.

2.
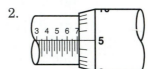

3.
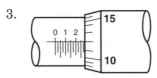

4.
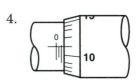

5.
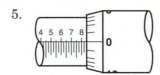

6.
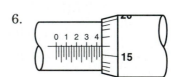

7.
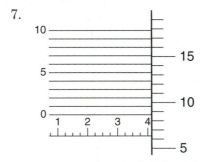

8.
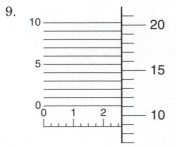

9.

10.
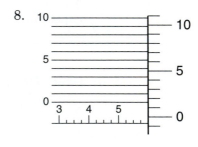

11.

12.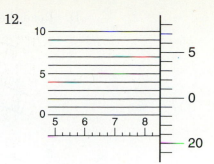

13.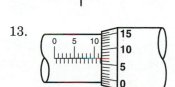

14.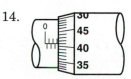

15.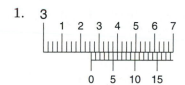

16.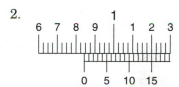

17.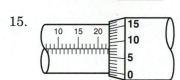

18.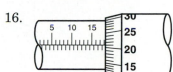

15.

16.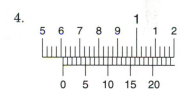

17.

18.

15.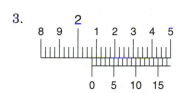

16.

17.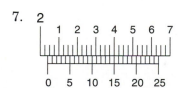

18.

15.

16.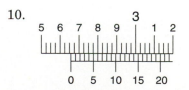

C.   Read the following vernier calipers.

1.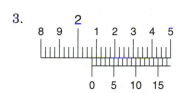

2.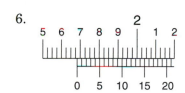

3.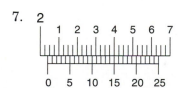

4.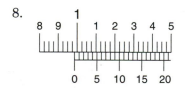

5.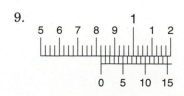

6.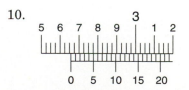

7.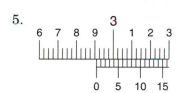

8.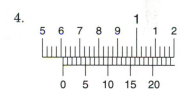

9.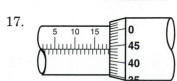

10.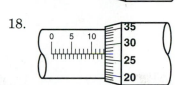

D. Read the following vernier protractors.

1.

2.

3.

4.

5.

6.

7.

8.

E. Find combinations of gauge blocks that produce the following lengths. (Use the 36-block set given on page 244.)

1. 2.3573 in.   2. 1.9467 in.

3. 0.4232 in.   4. 3.8587 in.

5. 1.6789 in.   6. 2.7588 in.

7. 0.1539 in.   8. 3.0276 in.

9. 2.1905 in.   10. 1.5066 in.

Check your answers on page 593, then turn to **48** for a set of practice problems on measurement.

Measurement

48 Answers are given on page 594.

A. Perform the following calculations with measurement numbers.
Round to the appropriate number of significant digits.

1. 5.75 sec − 2.38 sec

2. 3.1 ft × 2.2 ft

3. 4.5 sq in. ÷ 6.1 in.

4. 3.2 cm × 1.2 cm

5. $7\frac{1}{2}$ in. + $6\frac{1}{4}$ in. + $3\frac{1}{8}$ in.

6. 9 lb 6 oz − 5 lb 8 oz

7. 765 mi ÷ 36.2 gal

8. 1.5 cm + 2.3 cm + 5.8 cm

9. 34.38 psi − 16.73 psi

10. $7\frac{1}{4}$ ft − 4.1 ft

11. 7.064 in. − 3.195 in.

12. 22.01 psi × 18.35 sq in.

13. 8 ft 7 in. + 11 ft 11 in. + 9 in.

14. $2\frac{15}{64}$ in. + $3\frac{1}{8}$ in. + $7\frac{1}{32}$ in.

B. Convert the following measurement numbers to new units as shown. Round to
two decimal places whenever necessary.

1. 6 in. = _____ cm

2. 50 mph = _____ ft/sec

3. 22°C = _____ °F

4. 2.6 liters = _____ qt

5. $26\frac{1}{2}$ cu ft = _____ cu in.

6. 41°F = _____ °C

7. 14.3 kg _____ lb

8. 123.7 psi = _____ atm

9. $14\frac{1}{2}$ mi = _____ km

10. 350 cu in. = _____ cm^3

11. 14.3 bbl = _____ cu ft

12. 160 sq ft = _____ m^2

13. 24,680 cu in. = _____ bu

14. 6.9 liters = _____ gal

15. 6.8 sq ft = _____ sq in.

16. $12\frac{1}{4}$ qt = _____ liters

Name _____

Date _____

Course/Section _____

17. 3.4 ft = _____ cm   **18.** 45 in./sec = _____ cm/sec

19. 85 kmh = _____ mph   **20.** 28 lb = _____ kg

C. Think metric. Choose the closest estimate.

1. Weight of a pencil   **(a)** 4 kg   **(b)** 4 g   **(c)** 40 g

2. Size of a house   **(a)** 6 m^2 **(b)** 200 m^2 **(c)** 1200 m^2

3. Winter in New York   **(a)** –8°C   **(b)** –40°C   **(c)** 20°C

4. Volume of a 30-gal trash can   **(a)** 0.1 m^3 **(b)** 2.0 m^3 **(c)** 5.0 m^3

5. Height of an oak tree   **(a)** 500 mm   **(b)** 500 cm   **(c)** 50 km

6. Capacity of a 50-gal water heater **(a)** 25 liters   **(b)** 100 liters **(c)** 180 liters

D. Read the following measuring devices.

1. Rulers

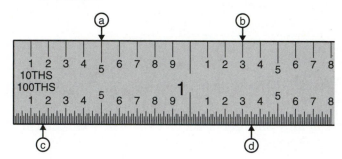

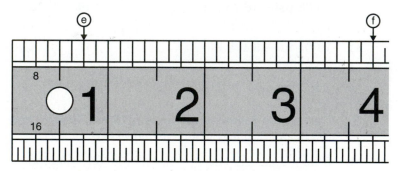

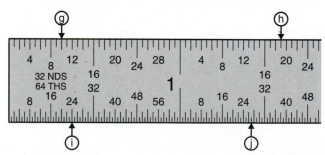

2. Micrometers. Problems (a) and (b) are English, (c) and (d) are metric, and (e) and (f) are English with a vernier scale.

(a) 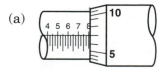   (b)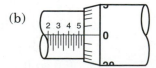

(c)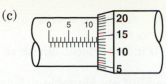

(d)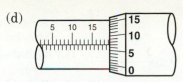

(e)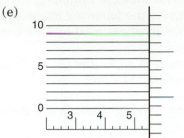

(f)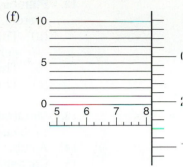

3. Vernier calipers.

(a)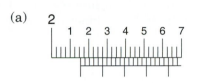

(b)

(c)

(d)

(e)

(f)

E. Practical Problems

1. **Machine Technology** A piece $6\frac{1}{32}$ in. long is cut from a steel bar $28\frac{5}{64}$ in. long. How much is left?

2. An American traveling in Italy notices a road sign saying 85 kmh. What is this speed in miles per hour? (Round to the nearest mph.)

3. **Auto Mechanics** An automotive technician testing cars finds that a certain bumper will prevent damage up to 9 ft/sec. Convert this to miles per hour. (Round to one decimal place.)

4. **Machine Technology** The gap in the piece of steel shown in the figure should be 2.4375 in. wide. Which of the gauge blocks in the set on page 244 should be used for testing this gap?

5. What size wrench, to the nearest 32nd of an inch, will fit a bolt 0.455 in. in diameter?

6. **Metalworking** The melting point of a casting alloy is 260°C. What temperature is this on the Fahrenheit scale?

7. **Sheet Metal Technology** What area of sheet metal in square inches is needed to construct a vent with a surface area of 6.5 sq ft?

8. How many gallons of gas are needed to drive a car 350 miles if the car normally averages 18.5 mi/gal? (Round to one decimal place.)

9. **Metalworking** A certain alloy appears bright red at a temperature of 560°F. What Celsius temperature does this correspond to? (Round to the nearest degree.)

Problem 4

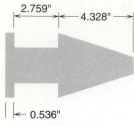

2.759"

4.328"

0.536"

Problem 11

10. **Auto Mechanics**   A Nissan sports car has a 1600-cc engine. What would this be in cubic inches?

11. **Metalworking**   Find the total length of the piece of steel shown in the figure.

12. **Plumbing**   The Uniform Plumbing Code is being converted to the metric system. Convert the following measurements.
    (a)   The amount of gas used by a furnace (136 cu ft/hour) to cubic meters per hour.
    (b)   The length of a gas line (25 ft) to meters.
    (c)   The diameter of water pipe ($2\frac{1}{2}$ in.) to centimeters.
    (Round to the nearest hundredth.)

13. **Metalworking**   Wrought iron has a density of 480.0 lb/cu ft. Convert this to lb/cu in.

14. A cubic foot of water contains 7.48 gallons. What volume is this in liters?

15. **Carpentry**   What are the dimensions of a metric "two by four" (a piece of wood 2 in. by 4 in.) in centimeters?

16. **Architectural Technology**   A common scale used by architects and drafters in making blueprints is $\frac{1}{4}$ in. = 1 ft. What length in meters would correspond to a drawing dimension of $1\frac{1}{4}$ in.?

Objective	Sample Problems	Where to Go for Help	
When you finish this unit you will be able to:		Page	Frame
1. Understand the meaning of signed numbers.	(a) Which is smaller, −12 or −14? (b) Mark the following points on a number line: $-5\frac{1}{4}$, +1.0, −0.75, $+3\frac{1}{2}$. (c) Represent a loss of $250 as a signed number.	255	**1**
2. Add signed numbers.	(a) $(-3) + (-5)$ (b) $18\frac{1}{2} + (-6\frac{3}{4})$ (c) $-21.7 + 14.2$ (d) $(-3) + 15 + 8 + (-20) + 6 + (-9)$	257	**4**
3. Subtract signed numbers.	(a) $4 - 7$ (b) $-11 - (-3)$ (c) $6.2 - (-4.7)$ (d) $-\frac{5}{8} - 1\frac{1}{4}$	263	**8**
4. Multiply and divide signed numbers.	(a) $(-12) \times 6$ (b) $(-54) \div (-9)$ (c) $(-4.73)(-6.5)$ (two significant digits) (d) $\frac{3}{8} \div (-2\frac{1}{2})$ (e) $60 \div (-0.4) \times (-10)$	268	**11**

Name _____

Date _____

Course/Section _____

5. Work with exponents.

   (a) 4^3 _____   272   **15**

   (b) $(2.05)^2$ _____

6. Use the order of operations.

   (a) $2 + 3^2$ _____   275   **20**

   (b) $(4 \times 5)^2 + 4 \times 5^2$ _____

   (c) $6 + 8 \times 2^3 \div 4$ _____

7. Find square roots.

   (a) $\sqrt{169}$ _____   276   **21**

   (b) $\sqrt{14.5}$ (to two decimal places)   _____

(Answers to these preview problems are given on page 578.)

If you are certain that you can work *all* these problems correctly, turn to page 283 for a set of practice problems. If you cannot work one or more of the preview problems, turn to the page indicated to the right of the problem. Those who wish to master this material should turn to frame **1** and begin work there.

6 Pre-Algebra

© DRABBLE reprinted by permission of UFS, Inc.

6-1 ADDITION OF SIGNED NUMBERS

1

The Meaning of Signed Numbers

What kind of number would you use to name each of the following: a golf score two strokes below par, a $2 loss in a poker game, a debt of $2, a loss of 2 yards on a football play, a temperature 2 degrees below zero, the year 2 B.C.?

The answer is that they are all *negative numbers,* all equal to –2.

We can better understand the meaning of negative numbers if we can picture them on a number line. First we will display the familiar whole numbers as points along a number line:

Notice that the numbers become larger as you move to the right along the number line, and they become smaller as you move to the left. The numbers to the *right* of zero, *larger* than zero, are considered to be *positive numbers*. The negative numbers are all the numbers *less than* zero, and we must extend the number line to the left in order to picture them:

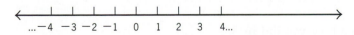

Think of the negative part of the number line as a reflection of the positive part through an imaginary mirror placed through zero. This means that ordering the negative numbers is the opposite of ordering the positive numbers. For example, 4 is larger than 1, but –4 is smaller than –1.

Circle the larger of each pair of numbers.

(a) −5 or −3   (b) 9 or 6   (c) −15 or −22

(d) 12 or 31   (e) −8 or 2   (f) 43 or −6

Check your answers in **2**.

---

2 (a) −3   (b) 9   (c) −15   (d) 31

    (e) 2   (f) 43

To help in solving the preceding problems, you may extend the number line in both directions. This is perfectly legitimate, since the number line is infinitely long. Notice we have marked only the positions of the positive and negative whole numbers. These numbers are called the *integers*. The positions of fractions and decimals, both positive and negative, can be located between the integers.

Try marking the following points on a number line.

(a) $-3\frac{1}{2}$ (b) $\frac{1}{4}$ (c) −2.4   (d) 1.75   (e) −0.6

Compare your answer with ours in **3**.

3

When we work with both positive and negative numbers and need to use names to distinguish them from each other, we often refer to them as *signed numbers*. Signed numbers are a bookkeeping concept. They were used in bookkeeping by the ancient Greeks, Chinese, and Hindus more than 2000 years ago. Chinese merchants wrote positive numbers in black and negative numbers in red in their account books. (Being "in the red" still means losing money or having a negative income.) The Hindus used a dot or circle to show that a number was negative. We use a minus sign (−) to show that a number is negative.

For emphasis, the "+" symbol is sometimes used to denote a positive number, although the absence of any symbol automatically implies that a number is positive. Because the + and − symbols are used also for addition and subtraction, we will use parentheses when writing problems involving addition and subtraction of signed numbers.

Before we attempt these operations, test your understanding of signed numbers by representing each of the following situations with the appropriate signed number.

(a) Two seconds before liftoff.

(b) A profit of eight dollars

(c) An elevator going up six floors

(d) Diving 26 feet below the surface of the ocean

(e) The year 50 B.C.

(f) Twelve degrees above zero

See how you did in **4**.

---

4 (a) −2 sec   (b) +$8   (c) +6   (d) −26 ft

    (e) −50   (f) +12°

Addition of Signed Numbers

Addition of signed numbers may be pictured using the number line. A positive number may be represented on the line by an arrow to the right or positive direction. Think of a positive arrow as a "gain" of some quantity such as money. The sum of two numbers is the net "gain."

For example, the sum of 4 + 3 or (+4) + (+3) is

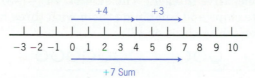

The numbers +4 and +3 are represented by arrows directed to the right. The +3 arrow begins where the +4 arrow ends. The sum 4 + 3 is the arrow that begins at the start of the +4 arrow and ends at the tip of the +3 arrow. In terms of money, a "gain" of $4 followed by a "gain" of $3 produces a net gain of $7.

A negative number may be represented on the number line by an arrow to the left or negative direction. Think of a negative arrow as a loss of some quantity.

For example, the sum (−4) + (−3) is −7.

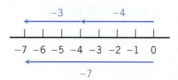

The numbers −4 and −3 are represented by arrows directed to the left. The −3 arrow begins at the tip of the −4 arrow. Their sum, −7, is the arrow that begins at the start of the −4 arrow and ends at the tip of the −3 arrow. A loss of $4 followed by another loss of $3 makes a total loss of $7.

Try setting up number line diagrams for the following two sums.

(a) 4 + (−3)   (b) (−4) + 3

Check your work in **5**.

---

5 (a) 4 + (−3) = 1

The number 4 is represented by an arrow directed to the right. The number −3 is represented by an arrow starting at the tip of +4 and directed to the left. A gain of $4 followed by a loss of $3 produces a net gain of $1.

(b) (−4) + 3 = −1

The number −4 is represented by an arrow directed to the left. The number 3 is represented by an arrow starting at the tip of −4 and directed to the right. A loss of $4 followed by a gain of $3 produces a net loss of $1.

ANOTHER WAY TO PICTURE ADDITION OF SIGNED NUMBERS

If you are having trouble with the number line, you might find it easier to picture addition of signed numbers using ordinary poker chips. Suppose that we designate white chips to represent positive integers and blue chips to represent negative integers. The sum of (+4) + (+3) can be found by taking four white chips and combining them with three more white chips. The result is seven white chips or +7:

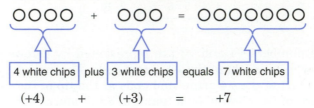

Similarly, (−4) + (−3) can be illustrated by taking four *blue* chips and combining them with three additional *blue* chips. We end up with seven blue chips or −7:

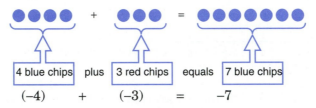

To illustrate 4 + (−3), take four white chips and three blue chips. Realizing that a white and a blue will "cancel" each other out ((+1) + (−1) = 0), we rearrange them as shown below, and the result is one white chip or +1:

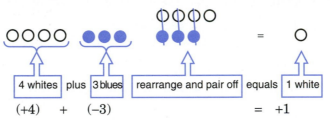

Finally, to represent (−4) + 3, picture the following:

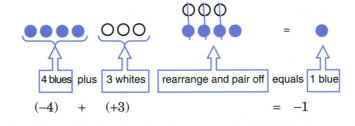

Number lines and poker chips provide a good way to see what is happening when we add signed numbers, but we need a nice easy rule that will allow us to do the addition more quickly.

Absolute Value

To add or subtract signed numbers you should work with their absolute value. The **absolute value** of a number is the distance along the number line between that number and the zero point. We indicate the absolute value by enclosing the number with vertical lines. The absolute value of any number is always a nonnegative number. For example,

$|4| = 4$ 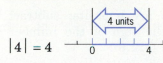 The distance between 0 and 4 is 4 units, so the absolute value of 4 is 4.

$|-5| = 5$ 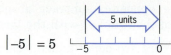 The distance between 0 and −5 is 5 units, so the absolute value of −5 is 5.

To add two signed numbers, follow these steps.

Step 1 Find the absolute value of each number.

Step 2 If the two numbers have the **same sign,** add their absolute values and give the sum the sign of the original numbers.

Example: $9 + 3 = +12$

Add absolute values: $9 + 3 = 12$

Absolute value = 9
Absolute value = 3

Sign is positive since both numbers are positive

Example: $(-6) + (-9) = -15$

Add absolute values: $6 + 9 = 15$

Absolute value = 6
Absolute value = 9

Sign is negative since both numbers are negative

If the two numbers have **opposite signs,** subtract the absolute values and give the difference the sign of the number with the larger absolute value.

Example: $7 + (-12) = -5$

Subtract absolute values: $12 - 7 = 5$

Absolute value = 7
Absolute value = 12

Same sign as the number with the larger absolute value (−12)

Example: $22 + (-14) = +8$

Subtract absolute values: $22 - 14 = 8$

Absolute value = 22
Absolute value = 14

Same sign as the number with the larger absolute value (+22)

Try it. Use this method to do the following calculations.

(a) $7 + (-2)$ (b) $-6 + (-14)$ (c) $(-15) + (-8)$

(d) $23 + 13$ (e) $-21.6 + 9.8$ (f) $-6\frac{3}{8} + 14\frac{3}{4}$

(g) $-5 + 8 + 11 + (-21) + (-9) + 12$

Check your answers in **6**.

---

6 (a) $7 + (-2) = +5$ Signs are opposite; subtract absolute values. Answer has the sign of the number with the larger absolute value (7).

(b) $-6 + (-14) = -20$ Signs are the same; add absolute values. Answer has the sign of the original numbers.

(c) $(-15) + (-8) = -23$ Signs are the same; add absolute values. Answer has the sign of the original numbers.

(d) $23 + 13 = +36$ Signs are the same; add absolute values. Answer has the sign of the original numbers.

(e) $-21.6 + 9.8 = -11.8$ Signs are opposite; subtract absolute values. Answer has the sign of the number with the larger absolute value (-21.6).

(f) $-6\frac{3}{8} + 14\frac{3}{4} = +8\frac{3}{8}$ Signs are opposite; subtract absolute values. Answer has the sign of the number with the larger absolute value $(+14\frac{3}{4})$.

(g) When more than two numbers are being added, rearrange them so that numbers with like signs are grouped together:

$$-5 + 8 + 11 + (-21) + (-9) + 12 = \underbrace{(8 + 11 + 12)} + \underbrace{[(-5) + (-21) + (-9)]}$$

Rearrange: Positives   Negatives

Find the positive total: $8 + 11 + 12 = +31$

Find the negative total: $(-5) + (-21) + (-9) = -35$

Add the positive total to the negative total: $31 + (-35) = -4$

USING A CALCULATOR

To work with signed numbers on a calculator, we need only one new key, the ⊞ key. This key allows us to enter a negative number on our display. For example, to enter "−4," simply press 4, then ⊞. You must press ⊞ *after* entering the number. Try the following addition problems using a calculator.

(a) $(-2675) + 1437$ (b) $(-6.975) + (-5.2452)$

(c) $0.026 + (-0.0045)$ (d) $(-683) + (-594) + 438 + (-862)$

Check your answers below.

(a) **2675** ⊞ ⊞ **1437** ⊟ → $\quad -1238$

(b) **6.975** ⊞ ⊞ **5.2452** ⊞ ⊟ → $\quad -12.2202$

(c) **.026** ⊞ **.0045** ⊞ ⊟ → $\quad 0.0215$

(d) **683** ⊞ ⊞ **594** ⊞ ⊞ **438** ⊞ **862** ⊞ ⊟ → $\quad -1701$

For more practice in addition of signed numbers, turn to **7** for a set of exercises.

7
Exercises 6-1   **Addition of Signed Numbers**

A. Circle the smaller number.

1. -4 or 7 2. -17 or -14 3. 23 or 18 4. 13 or -15

5. -9 or -10 6. -31 or 41 7. -65 or -56 8. 47 or 23

9. -47 or -23 10. -3 or 0 11. 6 or 0 12. -1.4 or -1.39

13. 2.7 or -2.8 14. $-3\frac{1}{4}$ or $-3\frac{1}{8}$ 15. -11.4 or $-11\frac{1}{4}$ 16. -0.1 or -0.001

B. For each problem, make a separate number line and mark the points indicated.

1. $-8.7, -9.4, -8\frac{1}{2}, -9\frac{1}{2}$

2. $-1\frac{3}{4}, +1\frac{3}{4}, -\frac{1}{4}, +\frac{1}{4}$

3. $-0.2, -1\frac{1}{2}, -2.8, -3\frac{1}{8}$

4. $+4.1, -1.4, -3.9, +1\frac{1}{4}$

C. Represent each of the following situations with a signed number.

1. A temperature of six degrees below zero

2. A debt of three hundred dollars

3. An airplane rising to an altitude of 12,000 feet

4. The year A.D. 240

5. A golf score of nine strokes below par

6. A quarterback getting sacked for a seven-yard loss

7. Diving eighty feet below the surface of the ocean

8. Winning fifteen dollars at poker

9. Taking five steps backward

10. Ten seconds before takeoff

11. A profit of six thousand dollars

12. An elevation of thirty feet below sea level

D. Add.

1. $-1 + 6$
2. $-11 + 7$
3. $-13 + (-2)$
4. $-6 + (-19)$
5. $31 + (-14)$
6. $23 + (-28)$
7. $21 + 25$
8. $12 + 19$
9. $-8 + (-8)$
10. $-7 + (-9)$
11. $-34 + 43$
12. $-56 + 29$
13. $45 + (-16)$
14. $22 + (-87)$
15. $-18 + (-12)$
16. $-240 + 160$
17. $-3.6 + 2.7$
18. $5.7 + (-7.9)$

19. $-12.43 + (-15.66)$

20. $-90.15 + (-43.69)$

21. $2\frac{5}{8} + (-3\frac{7}{8})$

22. $-4\frac{1}{4} + 6\frac{3}{4}$

23. $-9\frac{1}{2} + (-4\frac{1}{2})$

24. $-6\frac{2}{3} + (-8\frac{2}{3})$

25. $-2\frac{1}{2} + 1\frac{3}{4}$

26. $-5\frac{5}{8} + 8\frac{3}{4}$

27. $9\frac{7}{16} + (-7\frac{7}{8})$

28. $-13\frac{3}{10} + (-8\frac{4}{5})$

29. $-6 + 4 + (-8)$

30. $22 + (-38) + 19$

31. $-54 + 16 + 22$

32. $11 + (-6) + (-5)$

33. $-7 + 12 + (-16) + 13$

34. $23 + (-19) + 14 + (-21)$

35. $-36 + 45 + 27 + (-18)$

36. $18 + (-51) + (-8) + 26$

37. $-250 + 340 + (-450) + (-170) + 320 + (-760) + 880 + 190 + (-330)$

38. $-1 + 3 + (-5) + 7 + (-9) + 11 + (-13) + (15) + (-17) + 19$

39. $23 + (-32) + (-14) + 55 + (-66) + 44 + 28 + (-31)$

40. $-12.64 + 22.56 + 16.44 + (-31.59) + (-27.13) + 44.06 + 11.99$

41. $(-2374) + (5973)$

42. $(6.875) + (-11.765)$

43. $(-46,720) + (-67,850) + 87,950$

44. $(0.0475) + (-0.0875) + (-0.0255)$

E. Word Problems

1. **Printing** The Jiffy Print Shop has a daily work quota that it tries to meet. During the first two weeks of last month its daily comparisons to this quota showed the following: +$30, –$27, –$15, +$8, –$6, +$12, –$22, +$56, –$61, +$47. (In each case "+" means they were above the quota and "–" means they were below.) How far below or above their quota was the two-week total?

2. During a series of plays in a football game, a running back carried the ball for the following yardages: 7, –3, 2, –1, 8, –5, and –2. Find his total yards and his average yards per carry.

3. On a cold day in Fairbanks the temperature stood at 12 degrees below zero at 2 P.M. Over the next 5 hours the temperature rose 4 degrees, dropped 2 degrees, rose 5 degrees, dropped 1 degree, and finally, dropped 3 degrees. What was the temperature at 7 P.M.?

4. Pedro's checking account contained $489 on June 1. He then made the following transactions: deposited $150, withdrew $225, withdrew $34, deposited $119, withdrew $365, and deposited $750. What was his new balance after these transactions?

5. A struggling young company reported the following results for the four quarters of the year: A loss of $257,000, a loss of $132,000, a profit of $87,000, a profit of $166,000. What was the net result for the year?

6. **Plumbing**   In determining the fitting allowance for a pipe configuration, a plumber has negative allowances of $-2\frac{1}{2}$ in. and $-1\frac{7}{16}$ in. and two positive allowances of $\frac{1}{2}$ in. each. What is the net fitting allowance for the job?

When you have completed these exercises, check your answers on page 594 then turn to **8** to learn how to subtract signed numbers.

6-2 SUBTRACTION OF SIGNED NUMBERS

8

Opposites   When the number line is extended to the left to include the negative numbers, every positive number can be paired with a negative number whose absolute value, or distance from zero, is the same. These pairs of numbers are called *opposites*. For example, –3 is the opposite of +3, +5 is the opposite of –5, and so on.

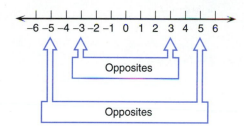

Note ▶   Zero has no opposite because it is neither positive nor negative. ◀

The concept of opposites will help us express a rule for subtraction of signed numbers.

Subtraction of Signed Numbers   We know from basic arithmetic that $5 - 2 = 3$. From the rules for addition of signed numbers, we know that $5 + (-2) = 3$ also. Therefore,

$(+5) - (+2) = (+5) + (-2) = 3$

This means that subtracting (+2) is the same as adding its opposite, (–2). Our knowledge of arithmetic and signed numbers verifies that subtracting and adding the opposite are equivalent for all the following cases:

$5 - 5 = 5 + (-5) = 0$
$5 - 4 = 5 + (-4) = 1$
$5 - 3 = 5 + (-3) = 2$
$5 - 2 = 5 + (-2) = 3$
$5 - 1 = 5 + (-1) = 4$
$5 - 0 \qquad\quad = 5$ (zero has no opposite)

But what happens with a problem like $5 - 6 = ?$ If we rewrite this subtraction as addition of the opposite like the others, we have

$5 - 6 = 5 + (-6) = -1$

Notice that this result fits right into the pattern established by the previous results. The answers *decrease* as the numbers being subtracted *increase*: $5 - 3 = 2$, $5 - 4 = 1$, $5 - 5 = 0$, so it seems logical that $5 - 6 = -1$.

What happens if we subtract a negative? For example, how can we solve the problem $5 - (-1) = ?$ If we rewrite this subtraction as addition of the opposite, we have

$5 - (-1) = 5 + (+1) = 6$

| subtracting −1 | becomes | adding the opposite of −1 or +1 |

Notice that this too fits in with the pattern established by the previous results. The answers *increase* as the numbers being subtracted *decrease:* $5 - 2 = 3$, $5 - 1 = 4$, $5 - 0 = 5$, so again it seems logical that $5 - (-1) = 6$.

We can now state a rule for subtraction of signed numbers. To subtract two signed numbers, follow these steps.

Step 1 Rewrite the subtraction as an addition of the opposite.

Step 2 Use the rules for addition of signed numbers to add.

Example: $3 - 7 = (+3) - (+7) = 3 + (-7) = -4$

| subtracting +7 | | adding −7 |

| is the same as |

Example: $-2 - (-6) = -2 + (+6) = +4$

| subtracting −6 | | adding +6 |

| is the same as |

Example: $-8 - (-3) = -8 + (+3) = -5$

| subtracting −3 | | adding +3 |

| is the same as |

Example: $-4 - 9 = -4 - (+9) = (-4) + (-9) = -13$

| subtracting +9 | | adding −9 |

| is the same as |

Careful ▶ Never change the sign of the first number. You change the sign of the *second* number only *after* changing subtraction to addition. ◀

Try these:

(a) $-7 - 5$ (b) $8 - 19$ (c) $4 - (-4)$ (d) $-12 - (-8)$

(e) $-3 - 9 - (-7) + (-8) + (+2)$ (f) $(-26.75) - (-44.38) - 16.96$

Check your answers in **9**.

9 (a) $-7 - 5 = -7 + (-5) = -12$ (b) $8 - 19 = (+8) + (-19) = -11$

(c) $4 - (-4) = (+4) + (+4) = +8$ (d) $-12 - (-8) = -12 + (+8) = -4$

(e) $-3 - 9 - (-7) + (-8) + (+2) = -3 + (-9) + (+7) + (-8) + (+2) = -11$

> These last two operations are additions, so they remain unchanged when the problem is rewritten.

 (f) **26.75** $\boxed{+/-}$ $\boxed{-}$ **44.38** $\boxed{+/-}$ $\boxed{-}$ **16.96** $\boxed{=}$ $\rightarrow$ *0.67*

SUBTRACTION WITH CHIPS

If you are having trouble visualizing subtraction of signed numbers, try the poker chip method that we used to show addition. Subtraction means "taking away" chips. First, a simple example: $7 - 3$ or $(+7) - (+3)$.

Begin with 7 white (positive) chips.	Take away 3 white chips.	Four white chips remain.

What about the problem: $(+3) - (+7) = ?$ How can you remove 7 white chips if you have only 3 white chips? Here we must be very clever. Remember that a positive chip and a negative chip "cancel" each other $[(+1) + (-1) = 0]$. So if we add both a white (positive) and a blue (negative) chip to a collection, we do not really change its value. Do $(+3) - (+7)$ this way:

Start with 3 white chips.	Add in 4 more whites and 4 more blues	Now we can remove 7 white chips.	Four blue chips remain.

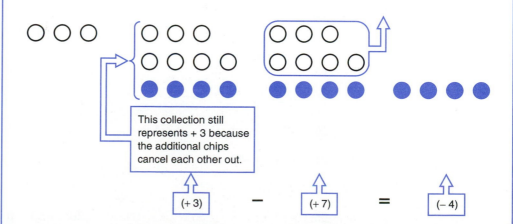

> This collection still represents + 3 because the additional chips cancel each other out.

Our rules for subtraction verify that this is correct.

Finally, let's try subtracting a negative. Consider the problem –2 – (–5) = ?

| Begin with –2: 2 blue chips. | To subtract –5 we must remove 5 blue chips. We have only 2 blues, so add 3 more blues and 3 more whites to keep the value the same. | Now remove 5 blues. | Three whites remain. |

 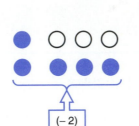 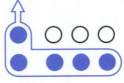

(– 2) – (– 5) = (+ 3)

Our rules for subtraction verify that this is correct.

Now turn to **10** for a set of problems on subtraction of signed numbers.

10

Exercises 6-2   **Subtraction of Signed Numbers**

A. Find the value of the following.

1. 3 – 5
2. 7 – 20
3. –4 – 12
4. –9 – 15
5. 6 – (–3)
6. 14 – (–7)
7. –11 – (–5)
8. –22 – (–10)
9. 12 – 4
10. 23 – 9
11. –5 – (–16)
12. –14 – (–17)
13. 12 – 30
14. 30 – 65
15. –23 – 8
16. –18 – 11
17. 2 – (–6)
18. 15 – (–12)
19. 44 – 18
20. 70 – 25
21. –34 – (–43)
22. –27 – (–16)
23. 0 – 19
24. 0 – 26
25. –17 – 0
26. –55 – 0
27. 0 – (–8)
28. 0 – (–14)

29. $-56 - 31$

30. $-45 - 67$

31. $22 - 89$

32. $17 - 44$

33. $-12 - 16 + 18$

34. $23 - (-11) - 30$

35. $19 + (-54) - 7$

36. $-21 + 6 - (-40)$

37. $-2 - 5 - (-7) + 6$

38. $5 - 12 + (-13) - (-7)$

39. $-3 - 4 - (-5) + (-6) + 7$

40. $2 - (-6) - 10 + 4 + (-11)$

41. $(-\frac{3}{8}) - \frac{1}{4}$

42. $2\frac{1}{2} - (-3\frac{1}{4})$

43. $6\frac{9}{16} - 13\frac{1}{16}$

44. $(-\frac{5}{6}) - (-\frac{2}{3})$

45. $6.5 - (-2.7)$

46. $(-0.275) - 1.375$

47. $(-86.4) - (-22.9)$

48. $0.0025 - 0.075$

49. $22.75 - 36.45 - (-54.72)$

50. $(-7.35) - (-4.87) - 3.66$

51. $3\frac{1}{2} - 6.6 - (-2\frac{1}{4})$

52. $(-5.8) - (-8\frac{1}{8}) - 3.9$

B. Word Problems

1. The height of Mt. Whitney is +14,410 ft. The lowest point in Death Valley is −288 ft. What is the difference in altitude between these two points?

2. The highest temperature ever recorded on Earth was +136°F at Libya in North Africa. The coldest was −127°F at Vostok in Antarctica. What is the difference in temperature between these two extremes?

3. At 8 P.M. the temperature was −2°F in Anchorage, Alaska. At 4 A.M. it was −37°F. By how much did the temperature drop in that time?

4. A vacation trip involved a drive from a spot in Death Valley 173 ft below sea level (−173 ft) to the San Bernardino Mountains 4250 ft above sea level (+4250 ft). What was the vertical ascent for this trip?

5. **Auto Mechanics** Dave's Auto Shop was $2465 in the "red" (−$2465) at the end of April—they owed this much to their creditors. At the end of December they were $648 in the red. How much money did they make in profit between April and December to achieve this turnaround?

6. A golfer had a total score of six strokes over par going into the last round of a tournament. He ended with a total score of three strokes under par. What was his score relative to par during the final round?

7. **Roofing** Pacific Roofing was $3765 in the "red" (see problem 5) at the end of June. At the end of September they were $8976 in the "red." Did they make or lose money between June and September? How much?

8. **Sheet Metal Technology** To establish the location of a hole relative to a fixed zero point, a machinist must make the following calculation:

$y = 5 - (3.750 - 0.500) - 2.375$

Find y.

Check your answers on page 595, then turn to 11 to learn about multiplication and division of signed numbers.

**Multiplication of
Signed Numbers**

As we learned earlier, multiplication is the shortcut for repeated addition. For example,

$$5 \times 3 = 3 + 3 + 3 + 3 + 3 = 15$$

| Five times three | means | three added five times |

Notice that the *product of two positive numbers* is always a *positive* number.

We can use repeated addition to understand multiplication involving negative numbers. Consider the problem $(+5) \times (-3)$.

$$(+5) \times (-3) = (-3) + (-3) + (-3) + (-3) + (-3) = -15$$

| Five times negative three | | negative three added five times |

Notice that the product of a *positive number* and a *negative number* is a *negative* number. The order in which you multiply two numbers does not matter, so a negative times a positive is also a negative. For example,

$$(-5) \times (+3) = (+3) \times (-5) = (-5) + (-5) + (-5) = -15$$

To understand the product of a negative times a negative, look at the pattern in the following sequence of problems:

$$(-5) \times (+3) = -15$$
$$(-5) \times (+2) = -10$$
$$(-5) \times (+1) = -5$$
$$(-5) \times \quad 0 = 0$$
$$(-5) \times (-1) = ?$$

Each time we *decrease* the multiplicand (the factor on the right) by 1, we *increase* the product by 5. So to preserve this pattern, the product of $(-5) \times (-1)$ must be +5. In other words, *a negative times a negative is a positive.*

**Rules for
Multiplication**

In summary, to multiply two signed numbers, follow these steps.

Step 1 Find the product of their absolute values.

Step 2 Attach a sign to the answer:

- If both numbers have the **same** sign, the answer is **positive.**

    $$(+) \times (+) = + \qquad (-) \times (-) = +$$

- If the two numbers have **opposite** signs, the answer is **negative.**

    $$(+) \times (-) = - \qquad (-) \times (+) = -$$

Use these rules on the following problems.

(a) $(-4) \times (+7)$ (b) $(-8) \times (-6)$ (c) $(+5) \times (+9)$

(d) $(+10) \times (-3)$ (e) $(-3) \times (-3)$ (f) $(-2) \times (-2) \times (-2)$

(g) $(-2) \times (-3) \times (+4) \times (-5) \times (-6)$ (h) $\left(-2\frac{1}{3}\right) \times \left(\frac{9}{14}\right)$

(i) $(-3.25) \times (-1.40)$

Check your answers in **12**.

---

12 (a) $(-4) \times (+7) = -28$

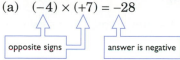

(b) $(-8) \times (-6) = +48$

(c) $+45$ (same sign, answer is positive)

(d) -30 (opposite signs, answer is negative)

(e) $+9$

(f) -8

(g) $+720$

(h) $-1\frac{1}{2}$

(i) **3.25** [+/-] [×] **1.4** [+/-] [=] → ☐ **4.55**

Did you discover an easy way to do problems (f) and (g) where more than two signed numbers are being multiplied? You could go left to right, multiply two numbers, then multiply the answer by the next number. You would then need to apply the sign rules for each separate product. However, since every *two* negative factors results in a positive, an *even* number of negative factors gives a positive answer, and an *odd* number of negative factors gives a negative answer. In a problem with more than two factors, count the total number of negative signs to determine the sign of the answer.

In problem (f) there are three negative factors—an odd number—so the answer is negative. Multiply the absolute values as usual to get the numerical part of the answer, 8. In problem (g) there are four negative factors—an even number—so the answer is positive. The product of the absolute values of all five numbers gives us the numerical part of the answer, 720.

Division of Signed Numbers

Recall that division is the reverse of the process of multiplication. The division

$24 \div 4 = \square$ can be written as the multiplication $24 = 4 \times \square$

and in this case the number $\square$ is 6. Notice that *a positive divided by a positive equals a positive*. We can use this "reversing" process to verify that the rest of the sign rules for division are exactly the same as those for multiplication. To find $(-24) \div (+4)$, rewrite the problem as follows:

$(-24) \div (+4) = \square$ $-24 = (+4) \times \square$

From our knowledge of signed number multiplication, $\square = -6$. Therefore, *a negative divided by a positive equals a negative*. Similarly,

$(+24) \div (-4) = \square$ means $(+24) = (-4) \times \square$ so that $\square = -6$

A *positive divided by a negative is also a negative.* Finally, let's try a negative divided by a negative:

$(-24) \div (-4) = \square$ means $(-24) = (-4) \times \square$ so that $\square = +6$

A *negative divided by a negative equals a positive.*

When dividing two signed numbers, determine the sign of the answer the same way you do for multiplication:

- If both numbers have the **same** signs, the answer is **positive.**

  $(+) \div (+) = +$ $(-) \div (-) = +$

- If the two numbers have **opposite** signs, the answer is **negative.**

  $(+) \div (-) = -$ $(-) \div (+) = -$

Divide the absolute values of the two numbers to get the numerical value of the quotient.

Try these division problems:

(a) $(-36) \div (+3)$ (b) $(-72) \div (-9)$ (c) $(+42) \div (-6)$

(d) $(-5.4) \div (-0.9)$ (e) $\left(\frac{3}{8}\right) \div \left(-\frac{1}{4}\right)$ (f) $(-3.75) \div (6.4)$

(Round to two decimal places.)

Check your answers in **13**.

---

13   (a) -12 (b) $+8$ (c) -7 (d) $+6$ (e) $-1\frac{1}{2}$ (f) -0.59

Enter (f) this way on a calculator:

3.75 $\boxed{+/-}$ $\boxed{\div}$ **6.4** $\boxed{=}$ $\rightarrow$ *−0.5859375*

For more practice in multiplication and division of signed numbers, turn to **14** for a set of problems.

14

Exercises 6-3   **Multiplication and Division of Signed Numbers**

A. Multiply or divide as indicated.

1. $(-7) \times (9)$	2. $8 \times (-5)$	3. $(-7) \times (-11)$
4. $(-4) \times (-9)$	5. $28 \div (-7)$	6. $(-64) \div 16$
7. $(-72) \div (-6)$	8. $(-44) \div (-11)$	9. $(-8.4) \div 2.1$
10. $(-10.8) \div (-1.2)$	11. $\left(-4\frac{1}{2}\right) \times \left(-2\frac{3}{4}\right)$	12. $\frac{2}{3} \times \left(-\frac{3}{4}\right)$
13. $(-3.4) \times (-1.5)$	14. $(-4) \div (16)$	15. $(-5) \div (-25)$
16. $(-13) \times (-4)$	17. $15 \times (-6)$	18. $(-18) \times 3$
19. $(-12) \times (-7)$	20. $(-120) \div 6$	21. $140 \div (-5)$
22. $(-200) \div (-10)$	23. $(-350) \div (-70)$	24. $2.3 \times (-1.5)$
25. $(-1.6) \times (5.0)$	26. $\left(-6\frac{2}{5}\right) \times \left(-1\frac{1}{4}\right)$	27. $\left(-\frac{5}{8}\right) \times \left(-\frac{4}{15}\right)$
28. $(-2.25) \div 0.15$	29. $8\frac{1}{2} \div \left(-3\frac{1}{3}\right)$	30. $\left(-\frac{5}{6}\right) \div \left(-\frac{3}{4}\right)$

31. $(-4.8) \div (-12)$

32. $(-2) \times \left(-5\frac{7}{8}\right)$

33. $(-3) \times \left(6\frac{1}{2}\right)$

34. $(-0.07) \times (-1.1)$

35. $3.2 \times (-1.4)$

36. $(-0.096) \div 1.6$

37. $(-5) \times (-6) \times (-3)$

38. $(-8) \times 4 \times (-7)$

39. $(-24) \div (-6) \times (-5)$

40. $35 \div (-0.5) \times (-10)$

41. $(-12) \times (-0.3) \div (-9)$

42. $0.2 \times (-5) \div (-0.02)$

B. Multiply or divide using a calculator and round as indicated.

1. $(-3.87) \times (4.98)$ (three significant digits)

2. $(-5.8) \times (-9.75)$ (two significant digits)

3. $(-0.075) \times (-0.025)$ (two significant digits)

4. $(648) \times (-250)$ (two significant digits)

5. $(-4650) \div (1470)$ (three significant digits)

6. $(-14.5) \div (-3.75)$ (three significant digits)

7. $(-0.58) \div (-2.5)$ (two significant digits)

8. $0.0025 \div (-0.084)$ (two significant digits)

9. $(-4.75) \times 65.2 \div (-6.8)$ (two significant digits)

10. $78.4 \div (-8.25) \times (-22.6)$ (three significant digits)

11. $(-2400) \times (-450) \times (-65)$ (two significant digits)

12. $(-65,000) \div (2.75) \times 360$ (two significant digits)

C. Practical Problems

1. To convert 17°F to degrees Celsius, it is necessary to perform the calculation

$$\frac{5}{9} \times (17 - 32)$$

What is the Celsius equivalent of 17°F?

2. In the first four months of the year, a certain country had an average monthly trade deficit (a *negative* quantity) of $16.5 million. During the final eight months of the year, the country had an average monthly trade surplus (a *positive* quantity) of $5.3 million. What was the overall trade balance for the year? (Use the appropriate sign to indicate an overall deficit or surplus.)

3. Due to the relative sizes of two connected pulleys A and B, pulley A will turn $1\frac{1}{2}$ times more than pulley B and in the opposite direction. If positive numbers are used to indicate counterclockwise revolutions and negative numbers are used for clockwise revolutions, express as a signed number the result for:
   (a) Pulley A when pulley B turns 80 revolutions counterclockwise.
   (b) Pulley B when pulley A turns 60 revolutions clockwise.

4. **Machinist** A certain metal rod changes in length by 1.4 mm for every 1°C change in temperature. If it expands when heated and contracts when cooled, express its change in length as a signed number as the temperature changes from 12°C to 4°C.

5. An airplane descends from 42,000 ft to 18,000 ft in 4 minutes. Express its rate of change in altitude in feet per minute as a signed number.

6. To convert −8°C to degrees Fahrenheit, you must do the following calculation:

$$\frac{9 \times (-8)}{5} + 32$$

What is the Fahrenheit equivalent of −8°C?

Check your answers on page 595, then turn to **15** to learn about exponents and square roots.

6-4 EXPONENTS AND SQUARE ROOTS

15 When the same number appears many times in a multiplication, writing the product may become monotonous, tiring, and even inaccurate. It is easy, for example, to miscount the twos in

$$32{,}768 = 2 \times 2 \times 2 \times 2 \times 2 \times 2 \times 2 \times 2 \times 2 \times 2 \times 2 \times 2 \times 2 \times 2 \times 2$$

or the tens in

$$100{,}000{,}000{,}000 = 10 \times 10 \times 10 \times 10 \times 10 \times 10 \times 10 \times 10 \times 10 \times 10 \times 10$$

Products of this sort are usually written in a shorthand form as 2^{15} and 10^{11}. In this *exponential* form the raised integer 15 shows the number of times that 2 is to be used as a factor in the multiplication. For example,

$$2 \times 2 = 2^2 \qquad \text{Product of \underline{two} factors of 2}$$
$$2 \times 2 \times 2 = 2^3 \qquad \text{Product of \underline{three} factors of 2}$$
$$2 \times 2 \times 2 \times 2 = 2^4 \qquad \text{Product of \underline{four} factors of 2}$$

Four 2s

Write $3 \times 3 \times 3 \times 3 \times 3$ in exponential form.

$$3 \times 3 \times 3 \times 3 \times 3 = \underline{\hspace{2cm}}$$

Check your answer in **16**.

---

16 $\underbrace{3 \times 3 \times 3 \times 3 \times 3}_{\text{Five factors of 3}} = 3^5$

Exponents In this expression, 3 is called the *base,* and the integer 5 is called the *exponent.* The exponent 5 tells you how many times the base 3 must be used as a factor in the multiplication.

Find the value of $4^3 = \underline{\hspace{2cm}}$

Pick an answer and turn to the frame shown after the answer:

(a) 12   Go to **17**.

(b) 64   Go to **18**.

(c) 81   Go to **19**.

17 Your answer is incorrect; 4^3 is *not* equal to 12.

The raised 3 in 4^3 tells you to multiply 4 by itself. Use the number 4 *three* times as a factor in a multiplication.

$$4^3 = \underbrace{4 \times 4 \times 4}$$

means use three factors of 4

Once you have set up the multiplication in this way it is easy to do it.

$4 \times 4 \times 4 = (4 \times 4) \times 4 = 16 \times 4 = ?$

Now, hop back to **16** and choose the correct answer.

---

18 Excellent!

$4^3 = 4 \times 4 \times 4 = (4 \times 4) \times 4 = 16 \times 4 = 64$

It is important that you be able to read exponential forms correctly.

2^2 is read "two to the second power" or "two squared"

2^3 is read "two to the third power" or "two cubed"

2^4 is read "two to the fourth power"

2^5 is read "two to the fifth power"   and so on

Students studying basic electronics, or those going on to another mathematics or science class, will find that it is important to understand exponential notation.

Do the following problems for practice in using exponents.

(a)   Write in exponential form.

$5 \times 5 \times 5 \times 5$ = _____   base = _____   exponent = _____

7×7 = _____   base = _____   exponent = _____

$10 \times 10 \times 10 \times 10 \times 10$ = _____   base = _____   exponent = _____

$3 \times 3 \times 3 \times 3 \times 3 \times 3 \times 3$ = _____   base = _____   exponent = _____

$9 \times 9 \times 9$ = _____   base = _____   exponent = _____

$1 \times 1 \times 1 \times 1$ = _____   base = _____   exponent = _____

(b)   Write as a product of factors and multiply out.

2^6 = _____ = _____   base = _____   exponent = _____

10^7 = _____ = _____   base = _____   exponent = _____

$(-3)^4$ = _____ = _____   base = _____   exponent = _____

5^2 = _____ = _____   base = _____   exponent = _____

$(-6)^3$ = _____ = _____   base = _____   exponent = _____

4^5 = _____ = _____   base = _____   exponent = _____

12^3 = _____ = _____   base = _____   exponent = _____

1^5 = _____ = _____   base = _____   exponent = _____

The correct answers are given in **20**. Go there when you finish these problems.

---

19 This answer is not correct. Apparently, you found the product $3 \times 3 \times 3 \times 3$.

Do it this way:

$4^3 = \underline{4 \times 4 \times 4}$

3 factors of 4

The raised 3 tells you how many factors of 4 are to be multiplied together.

$4 \times 4 \times 4 = (4 \times 4) \times 4 = 16 \times 4 = ?$

Complete this problem and return to **16** to continue.

---

20 (a) $5 \times 5 \times 5 \times 5$ $\qquad = \underline{5^4}\quad$ base $= \underline{\quad 5\quad}$ exponent $= \underline{\quad 4\quad}$

$\quad 7 \times 7$ $\qquad\qquad\qquad\qquad = \underline{7^2}\quad$ base $= \underline{\quad 7\quad}$ exponent $= \underline{\quad 2\quad}$

$\quad 10 \times 10 \times 10 \times 10 \times 10$ $\qquad = \underline{10^5}\quad$ base $= \underline{\quad 10\quad}$ exponent $= \underline{\quad 5\quad}$

$\quad 3 \times 3 \times 3 \times 3 \times 3 \times 3 \times 3$ $\qquad = \underline{3^7}\quad$ base $= \underline{\quad 3\quad}$ exponent $= \underline{\quad 7\quad}$

$\quad 9 \times 9 \times 9$ $\qquad\qquad\qquad = \underline{9^3}\quad$ base $= \underline{\quad 9\quad}$ exponent $= \underline{\quad 3\quad}$

$\quad 1 \times 1 \times 1 \times 1$ $\qquad\qquad = \underline{1^4}\quad$ base $= \underline{\quad 1\quad}$ exponent $= \underline{\quad 4\quad}$

(b) $2^6 = 2 \times 2 \times 2 \times 2 \times 2 \times 2$ $\qquad = \underline{64}\quad$ base $= \underline{\quad 2\quad}$ exponent $= \underline{\quad 6\quad}$

$\quad 10^7 = 10 \times 10 \times 10 \times 10 \times 10 \times 10 \times 10 = 10,000,000$

base $= \underline{\quad 10\quad}$ exponent $= \underline{\quad 7\quad}$

$\quad (-3)^4 = (-3) \times (-3) \times (-3) \times (-3)$ $\qquad = \underline{+81}\quad$ base $= \underline{\quad -3\quad}$ exponent $= \underline{\quad 4\quad}$

$\quad 5^2 = 5 \times 5$ $\qquad\qquad\qquad\qquad = \underline{25}\quad$ base $= \underline{\quad 5\quad}$ exponent $= \underline{\quad 2\quad}$

$\quad (-6)^3 = (-6) \times (-6) \times (-6)$ $\qquad = \underline{-216}\quad$ base $= \underline{\quad -6\quad}$ exponent $= \underline{\quad 3\quad}$

Note ▶ Notice that $(-3)^4 = +81$, but $(-6)^3 = -216$. The rules for multiplying signed numbers tell us that a negative number raised to an **even** power is positive, but a negative number raised to an **odd** power is negative. ◀

$4^5 = 4 \times 4 \times 4 \times 4 \times 4$ $\quad = \underline{1024}\quad$ base $= \underline{\quad 4\quad}$ exponent $= \underline{\quad 5\quad}$

$12^3 = 12 \times 12 \times 12$ $\qquad = \underline{1728}\quad$ base $= \underline{\quad 12\quad}$ exponent $= \underline{\quad 3\quad}$

$1^5 = 1 \times 1 \times 1 \times 1 \times 1$ $\quad = \underline{1}\quad$ base $= \underline{\quad 1\quad}$ exponent $= \underline{\quad 5\quad}$

Careful ▶ Be careful when there is a negative sign in front of the exponential expression. For example,

$3^2 = 3 \times 3 = 9 \quad$ and $\quad (-3)^2 = (-3) \times (-3) = 9$

but

$-3^2 = -(3^2) = -(3 \times 3) = -9$

To calculate $-(-2)^3$ first calculate

$(-2)^3 = (-2) \times (-2) \times (-2) = -8$

Then

$$-(-2)^3 = -(-8) = 8 \blacktriangleleft$$

Any power of 1 is equal to 1, of course.

$$1^2 = 1 \times 1 = 1$$
$$1^3 = 1 \times 1 \times 1 = 1$$
$$1^4 = 1 \times 1 \times 1 \times 1 = 1 \text{ and so on}$$

Notice that when the base is ten, the product is easy to find.

$$10^2 = 10 \times 10 = 100$$
$$10^3 = 10 \times 10 \times 10 = 1000$$
$$10^4 = 10000$$
$$10^5 = \underline{100000}$$

$$\quad\quad\quad\quad \text{5 zeros}$$

The exponent number is always exactly equal to the number of zeros in the final product when the base is 10.

Continue the pattern to find the value of 10^0 and 10^1.

$10^5 = 100000$	$10^2 = 100$
$10^4 = 10000$	$10^1 = 10$
$10^3 = 1000$	$10^0 = 1$

For any base, powers of 1 or 0 are easy to find.

$2^1 = 2$	$2^0 = 1$
$3^1 = 3$	$3^0 = 1$
$4^1 = 4$	$4^0 = 1 \quad$ and so on

Order of Operations with Exponents

In Chapter 1 we discussed the order of operations when you add, subtract, multiply, or divide. To evaluate an arithmetic expression containing exponents, first calculate the value of the exponential factor, then perform the other operations using the order of operations already given. If the calculations are to be performed in any other order, parentheses will be used to show the order. The revised order of operations is:

1. Simplify all operations within **parentheses.**

2. Simplify all **exponents,** left to right.

3. Perform all **multiplications** and **divisions,** left to right.

4. Perform all **additions** and **subtractions,** left to right.

Learning Help ▶

If you think you will have trouble remembering this order of operations, try memorizing this phrase: **P**lease **E**xcuse **M**y **D**ear **A**unt **S**ally. The first letter of each word should remind you of **P**arentheses, **E**xponents, **M**ultiply/**D**ivide, **A**dd/**S**ubtract. ◀

Here are some examples showing how to apply this expanded order of operations.

Example: To calculate $12 + 54 \div 3^2$

First, simplify the exponent factor	$= 12 + 54 \div 9$
Second, divide	$= 12 + 6$
Finally, add	$= 18$

Example: To calculate $237 - (2 \times 3)^3$

First, work within parentheses $\quad = 237 - (6)^3$
Second, simplify the exponent factor $= 237 - 216$
Finally, subtract $\quad\quad\quad\quad\quad\quad = 21$

Calculate.

(a) $2^4 \times 5^3$

(b) $6 + 3^2$

(c) $(6 + 3)^2$

(d) $(24 \div 6)^3$

(e) $2^2 + 4^2 \div 2^3$

(f) $4 + 3 \times 2^4 \div 6$

Check your answers in **21**.

---

21 (a) $2^4 \times 5^3 = 16 \times 125 = 2000$

(b) $6 + 3^2 = 6 + 9 = 15$

(c) $(6 + 3)^2 = (9)^2 = 81$

(d) $(24 \div 6)^3 = (4)^3 = 64$

(e) $2^2 + 4^2 \div 2^3 = 4 + 16 \div 8 = 4 + 2 = 6$ (Do the division before adding.)

(f) $4 + 3 \times 2^4 \div 6 = 4 + 3 \times 16 \div 6 = 4 + 48 \div 6 = 4 + 8 = 12$

Square Roots   What is interesting about the numbers

1, 4, 9, 16, 25, 36, 49, 64, 81, 100, . . .?

Do you recognize them?

These numbers are the squares or second powers of the counting numbers,

$1^2 = 1$
$2^2 = 4$
$3^2 = 9$
$4^2 = 16$, and so on. 1, 4, 9, 16, 25, . . . are called *perfect squares*.

If you have memorized the multiplication table for one-digit numbers, you will recognize them immediately. The number 3^2 is read "three squared." What is "square" about $3^2 = 9$? The name comes from an old Greek idea about the nature of numbers. Ancient Greek mathematicians called certain numbers "square numbers" or "perfect squares" because they could be represented by a square array of dots.

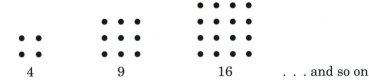

The number of dots along the side of the square was called the "root" or origin of the square number. We call it the *square root*. For example, the square root of 16 is 4, since $4 \times 4 = 16$.

What is the square root of 64?

(a)   32   Go to **22**.

(b)   8   Go to **23**.

PERFECT SQUARES

$1^2 = 1$	$6^2 = 36$	$11^2 = 121$	$16^2 = 256$
$2^2 = 4$	$7^2 = 49$	$12^2 = 144$	$17^2 = 289$
$3^2 = 9$	$8^2 = 64$	$13^2 = 169$	$18^2 = 324$
$4^2 = 16$	$9^2 = 81$	$14^2 = 196$	$19^2 = 361$
$5^2 = 25$	$10^2 = 100$	$15^2 = 225$	$20^2 = 400$

22 Sorry, you are not correct.

You simply cannot divide 64 by 2 to find its square root!

The square root of 64 is some number $\square$ such that $\square \times \square = 64$. For example, the square root of 25 is equal to 5 because $5 \times 5 = 25$. To "square" means to multiply by itself, and to find a square root means to find a number that, when multiplied by itself, gives the original number.

Go back to **21** and try again.

---

23 Right! The square root of 64 is equal to 8 because $8 \times 8 = 64$.

The sign $\sqrt{}$ is used to indicate the square root.

$\sqrt{16} = 4$ read it "square root of 16"

$\sqrt{9} = 3$ read it "square root of 9"

$\sqrt{169} = 13$ read it "square root of 169"

Find $\sqrt{81} = $ _____ $\sqrt{361} = $ _____ $\sqrt{289} = $ _____

Try using the table of perfect squares if you do not recognize these.

Check your answers in **24**.

---

24 $\sqrt{81} = 9$ **Check:** $9 \times 9 = 81$

$\sqrt{361} = 19$ **Check:** $19 \times 19 = 361$

$\sqrt{289} = 17$ **Check:** $17 \times 17 = 289$

Always check your answer as shown.

How do you find the square root of any number? You can consult the table of square roots given on page 611, which lists the square roots of all whole numbers from 1 to 200, rounded to four decimal places. But using a calculator is the easiest way to find the square or other power of a number, or to find a square root.

To square a number using a calculator, enter the number, then press the $\boxed{x^2}$ key. For example, 38^2 can be found by entering

 38 $\boxed{x^2}$ → *1444.*

Notice that it is not necessary to use the $\boxed{=}$ key.

To raise a number to any other power, enter the base number, press the $\boxed{y^x}$ or $\boxed{x^y}$ key, enter the exponent number, and finally press the $\boxed{=}$ key. For example, to calculate 3.8^5 enter **3.8** $\boxed{y^x}$ **5** $\boxed{=}$ → *792.35168*

To calculate a square root enter the number and then press the $\boxed{\sqrt{}}$ key. There is no need to press the $\boxed{=}$ key. For example, to calculate $\sqrt{237}$ enter

237 $\boxed{\sqrt{}}$ → *15.394804.*

Perform the following calculations.

(a) 68^2 (b) $\sqrt{760}$ (Round to two significant digits.)   (c) 65^3

(d) 2.75^4 (Round to three significant digits.)

(e) 0.475^2 (Round to three significant digits.)

(f) $\sqrt{5.74}$ (Round to three significant digits.)

Compare your answers with ours.

(a) **68** $\boxed{x^2}$ → *4624.*

(b) **760** $\boxed{\sqrt{}}$ → *27.568098* or 28, rounded

(c) **65** $\boxed{y^x}$ **3** $\boxed{=}$ → *274625.*

(d) **2.75** $\boxed{y^x}$ **4** $\boxed{=}$ → *57.191406* or 57.2, rounded

(e) **.475** $\boxed{x^2}$ → *0.225625* or 0.226, rounded

(f) **5.74** $\boxed{\sqrt{}}$ → *2.3958297* or 2.40, rounded

Careful ▶ Notice that you cannot take the square root of a negative number. There is no number $\square$ such that $\square \times \square = -81$.

If $\square$ is positive, then $\square \times \square$ is positive, and if $\square$ is negative, than $\square \times \square$ is also positive. You cannot multiply a number by itself and get a negative result. (In order to perform higher-level algebra, mathematicians have invented imaginary numbers, which are used to represent the square roots of negative numbers. Students who go on to higher-level courses will learn about these numbers later.) ◀

Square roots have the same place in the order of operations as exponents. Simplify square roots before performing multiplication, division, addition, or subtraction.

Examples:

$3\sqrt{49} = 3 \times \sqrt{49} = 3 \times 7 = 21$ Note that $3\sqrt{49}$ is a shorthand

$2 + \sqrt{64} = 2 + 8 = 10$ way of writing $3 \times \sqrt{49}$.

$\dfrac{\sqrt{81}}{3} = \dfrac{9}{3} = 3$

If an arithmetic operation appears under the square root symbol, perform this operation before taking the square root. For example,

$\sqrt{9 + 16} = \sqrt{25} = 5$

Careful ▶ Notice $\sqrt{9 + 16}$ is not the same as $\sqrt{9} + \sqrt{16} = 3 + 4 = 7$. ◀

Calculate:

(a) $100 - \sqrt{100}$ (b) $\dfrac{4^2 \times \sqrt{36}}{3}$ (c) $\sqrt{169 - 144}$ (d) $\sqrt{64} + \sqrt{121}$

(e) $6 + 4\sqrt{25}$ (f) $49 - 3^2\sqrt{4}$ (g) $\dfrac{\sqrt{225} - \sqrt{81}}{\sqrt{36}}$ (h) $\sqrt{\dfrac{225 - 81}{36}}$

Check your answers in **25**.

25 (a) $100 - \sqrt{100} = 100 - 10 = 90$

(b) $\dfrac{4^2 \times \sqrt{36}}{3} = \dfrac{16 \times (6)}{3} = \dfrac{96}{3} = 32$

(c) $\sqrt{169 - 144} = \sqrt{25} = 5$

(d) $\sqrt{64} + \sqrt{121} = 8 + 11 = 19$

(e) $6 + 4\sqrt{25} = 6 + 4 \cdot 5 = 6 + 20 = 26$

(f) $49 - 3^2\sqrt{4} = 49 - 9\sqrt{4} = 49 - 9 \cdot 2 = 49 - 18 = 31$

(g) $\dfrac{\sqrt{225} - \sqrt{81}}{\sqrt{36}} = \dfrac{15 - 9}{6} = \dfrac{6}{6} = 1$

(h) $\sqrt{\dfrac{225 - 81}{36}} = \sqrt{\dfrac{144}{36}} = \sqrt{4} = 2$

SIMPLIFYING SQUARE ROOTS

When the square root is not an exact whole number, a calculator or table can be used to find a decimal approximation of its value. For example, a plumber calculating the length of a connecting pipe might approximate $\sqrt{12}$ as about 3.5. But in more mathematical or theoretical work, $\sqrt{12}$ might be expressed in what is known as Simplified Radical Form. Follow these steps to write the simplified radical form of a square root.

Step 1 Find the largest perfect square that exactly divides the number.

The largest perfect square that exactly divides 12 is 4.

Step 2 Use this perfect square to factor the number.

$\sqrt{12} = \sqrt{4 \cdot 3}$

Step 3 Since $\sqrt{a \cdot b} = \sqrt{a} \cdot \sqrt{b}$ for any two positive numbers a and b, we can rewrite **Step 2** as a product.

$\sqrt{4 \cdot 3} = \sqrt{4} \cdot \sqrt{3}$
$= 2 \cdot \sqrt{3}$

Step 4 Take the square root of the perfect square factor.

Therefore $2\sqrt{3}$ is the simplest radical form for $\sqrt{12}$.

Write the following square roots in simplest radical form.

(a) $\sqrt{18}$ (b) $\sqrt{75}$ (c) $\sqrt{72}$ (d) $\sqrt{48}$ (e) $\sqrt{500}$ (f) $\sqrt{162}$

Check your work below.

(a) $\sqrt{18} = \sqrt{9 \cdot 2} = \sqrt{9} \cdot \sqrt{2} = 3\sqrt{2}$

(b) $\sqrt{75} = \sqrt{25 \cdot 3} = \sqrt{25} \cdot \sqrt{3} = 5\sqrt{3}$

(c) $\sqrt{72} = \sqrt{36 \cdot 2} = \sqrt{36} \cdot \sqrt{2} = 6\sqrt{2}$

(d) $\sqrt{48} = \sqrt{16 \cdot 3} = \sqrt{16} \cdot \sqrt{3} = 4\sqrt{3}$

(e) $\sqrt{500} = \sqrt{100 \cdot 5} = \sqrt{100} \cdot \sqrt{5} = 10\sqrt{5}$

(f) $\sqrt{162} = \sqrt{81 \cdot 2} = \sqrt{81} \cdot \sqrt{2} = 9\sqrt{2}$

In problems (c), (d), and (f) it is important to find the *largest* perfect square that divides each number.

Now complete **26** for a set of problems on exponents and square roots.

Exercises 6-4   **Exponents and Square Roots**

A. Find the value of these.

1. 2^4	2. 3^2	3. 4^3
4. 5^3	5. 10^3	6. 7^2
7. 2^8	8. 6^2	9. 8^3
10. 3^4	11. 5^4	12. 10^5
13. $(-2)^3$	14. $(-3)^5$	15. 9^3
16. 6^0	17. 5^1	18. 1^4
19. 2^5	20. 8^2	21. $(-14)^2$
22. $(-21)^2$	23. 15^3	24. 16^2
25. $2^2 \times 3^3$	26. $2^6 \times 3^2$	27. $3^2 \times 5^3$
28. $2^3 \times 7^3$	29. $6^2 \times 5^2 \times 3^1$	30. $2^{10} \times 3^2$
31. $2 + 4^2$	32. $14 - 2^3$	33. $(8 - 2)^3$
34. $(14 \div 2)^2$	35. $3^3 - 8^2 \div 2^3$	36. $12 - 5 \times 2^5 \div 2^4$

B. Calculate. (Round to two decimal places if necessary.)

1. $\sqrt{81}$	2. $\sqrt{144}$	3. $\sqrt{36}$	4. $\sqrt{16}$
5. $\sqrt{25}$	6. $\sqrt{9}$	7. $\sqrt{256}$	8. $\sqrt{400}$
9. $\sqrt{225}$	10. $\sqrt{49}$	11. $\sqrt{324}$	12. $\sqrt{121}$
13. $\sqrt{4.5}$	14. $\sqrt{500}$	15. $\sqrt{12.4}$	16. $\sqrt{700}$
17. $\sqrt{210}$	18. $\sqrt{321}$	19. $\sqrt{810}$	20. $\sqrt{92.5}$
21. $\sqrt{1000}$	22. $\sqrt{2000}$	23. $\sqrt{25000}$	24. $\sqrt{2500}$
25. $\sqrt{150}$	26. $\sqrt{300}$	27. $\sqrt{3000}$	28. $\sqrt{30000}$
29. $\sqrt{1.25}$	30. $\sqrt{1.008}$	31. $3\sqrt{20}$	32. $4 + 2\sqrt{5}$
33. $55 - \sqrt{14}$	34. $\dfrac{3^2\sqrt{24}}{3 \times 2}$	35. $\sqrt{2^5 - 3^2}$	

36. $5.2^2 - 3\sqrt{52 - 5^2}$

C. Practical Problems

1. Find the length of the side of a square whose area is 184.96 sq ft. (*Hint:* The area of a square is equal to the square of one of its sides. The side of a square equals the square root of its area.)

2. A square building covers an area of 1269 sq ft. What is the length of each side of the building? Round to the nearest tenth.

3. **Mechanical Engineering** The following calculation must be performed to find the stress (in psi) on a plunger. Find this stress to the nearest thousand psi.

$$\frac{3000\sqrt{34}}{0.196}$$

4. **Electronics** The resistance (in ohms) in a certain silicon diode is found using the expression

$$\sqrt{15 \times 10^6 \times 24}$$

Calculate this resistance to the nearest thousand ohms.

5. **Fire Protection** The velocity (in fps) of water discharged from a hose with a nozzle pressure of 62 psi is given by

$$12.14 \sqrt{62}$$

Calculate and round to the nearest whole number.

6. **Police Science** Skid marks can be used to determine the maximum speed of a car involved in an accident. If a car leaves a skid mark of 175 ft on a road with a 35% coefficient of friction, its speed (in mph) can be estimated using

$$\sqrt{30 \times 0.35 \times 175}$$

Find this speed to the nearest whole number.

Check your answers on p. 596, then turn to **27** for a set of *practice problems* on Chapter 6.

Pre-Algebra

27 Answers are given on pages 596–597.

A. Rewrite each group of numbers in order from smallest to largest.

1. −13, 7, 4, −8, −2

2. −0.6, 0, 0.4, −0.55, −0.138, −0.96

3. −150, −140, −160, −180, −120, 0

4. $-\dfrac{5}{8}, \dfrac{3}{4}, \dfrac{1}{2}, -\dfrac{1}{2}, -\dfrac{3}{4}, \dfrac{3}{8}$

B. Add or subtract as indicated.

1. −16 + 6

2. −5 + (−21)

3. 7 − 13

4. −4 − (−4)

5. 23 + (−38)

6. −13 − 37

7. 20 − (−5)

8. 11 + (−3)

9. −2.6 + 4.9

10. $2\frac{1}{4} - 4\frac{3}{4}$

11. −$5.26 − $3.89

12. $-2\frac{3}{8} + \left(-7\frac{1}{2}\right)$

13. −9 − 6 + 4 + (−13) − (−8)

14. $-12.4 + 6\frac{1}{2} - \left(-1\frac{3}{4}\right) + (-18) - 5.2$

C. Multiply or divide as indicated.

1. 6 × (−10)

2. (−7) × (−5)

3. −48 ÷ (−4)

4. (−320) ÷ 8

5. $\left(-1\frac{1}{8}\right) \times \left(-6\frac{1}{4}\right)$

6. $\left(-5\frac{1}{4}\right) \div 7$

Name _____

Date _____

Course/Section _____

7. $6.4 \div (-16)$ **8.** $(-2.4) \times 15$

9. $(-4) \times (-5) \times (-6)$ **10.** $72 \div (-4) \times (-3)$

11. $(2.4735) \times (-6.4)$ (two significant digits)

12. $(-0.675) \div (-2.125)$ (three significant digits)

D. **Find the numerical value of each expression.**

1. $12.5 \times (-6.3 + 2.8)$ **2.** $(-3) \times 7 - 4 \times (-9)$

3. $\dfrac{14 - 46}{(-4) \times (-2)}$ **4.** $\left(-4\frac{1}{2} - 3\frac{3}{4}\right) \div 8$

5. 2^6 **6.** 3^5 **7.** 17^2 **8.** 23^3

9. 0.5^2 **10.** 1.2^3 **11.** 0.02^2 **12.** 0.03^3

13. 0.001^3 **14.** 2.01^2 **15.** 4.02^2 **16.** $(-3)^3$

17. $(-2)^4$ **18.** $\sqrt{1369}$ **19.** $\sqrt{784}$ **20.** $\sqrt{4.41}$

21. $\sqrt{0.16}$ **22.** $\sqrt{5.29}$ **23.** $5.4 + 3.3^2$

24. $2.1^2 + 3.1^3$ **25.** 4.5×5.2^2 **26.** $4.312 \div 1.4^2$

27. $(3 \times 6)^2 + 3 \times 6^2$ **28.** $28 - 4 \times 3^2 \div 6$ **29.** $2 + 2\sqrt{16}$

30. $5\sqrt{25} - 4.48$ **31.** $\sqrt{0.49} - \sqrt{0.04}$ **32.** $10 - 3\sqrt{1.21}$

Round to two decimal digits.

33. $\sqrt{80}$ **34.** $\sqrt{106}$ **35.** $\sqrt{310}$ **36.** $\sqrt{1.8}$

37. $\sqrt{4.20}$ **38.** $\sqrt{3.02}$ **39.** $\sqrt{1.09}$ **40.** $\sqrt{0.08}$

41. $7 - 3\sqrt{2}$ **42.** $\sqrt{6} - \sqrt{5}$ **43.** $3\sqrt{2} - 2\sqrt{3}$ **44.** $3\sqrt{2} + 2\sqrt{3}$

E. **Practical Problems**

1. **Electrician** The current in a circuit changes from -2 amperes to 5 amperes. What is the change in current?

2. Find the side of a square whose area is 225 sq ft. (The side of a square is the square root of its area.)

3. In Hibbing, Minnesota, the temperature dropped 8°F overnight. If the temperature read -14°F before the plunge, what did it read afterward?

4. **Plumbing** Morgan Plumbing had earnings of $3478 in the last week of January and expenses of $4326. What was the net profit or loss? (Indicate the answer with the appropriate sign.)

5. **Electrician**   The current capacity (in amperes) of a service using a three-phase system for 124 kilowatts of power and a line voltage of 420 volts is given by the calculation

$$\frac{124}{420\sqrt{3}}$$

Determine this current to two decimal places.

6. **Fire Protection**   The flow rate (in gallons per minute) of water from a $1\frac{1}{2}$-in. nozzle at 65 psi is calculated from the expression

$$29.7 \times (1.5)^2 \sqrt{65}$$

Find the flow rate to the nearest gallon per minute.

7. **Civil Engineer**   The crushing load (in tons) for a square pillar 8 in. thick and 12 ft high is given by the formula

$$\frac{25 \times (8)^4}{12^2}$$

Find this load to the nearest hundred tons.

8. **Automotive Services**   An auto shop was \$14,560 in the "red" at the beginning of the year and \$47,220 in the "black" at the end of the year. How much profit did the shop make during the year?

9. **Electronics**   In determining the bandwidth of a high-fidelity amplifier, a technician would first perform the following calculation to find the rise time of the signal:

$$\sqrt{3200^2 - 22.5^2}$$

Calculate this rise time to the nearest hundred. The time is in nanoseconds.

10. **Electronics**   The following calculation is used to determine one of the voltages for a sweep generator:

$$10 - 9.5 \div 0.03$$

Calculate this voltage. Round to one significant digit.

11. Use the formula

$$C = \frac{5}{9} \times (F - 32)$$

to convert the following Fahrenheit temperatures to Celsius rounded to the nearest degree.
(a) $-4°F$ (b) $24°F$

12. Use the formula

$$F = \frac{9 \times C}{5} + 32$$

to convert the following Celsius temperatures to Fahrenheit.
(a) $-21°C$ (b) $-12°C$

Objective	Sample Problems		Where to Go for Help	
			Page	Frame
When you finish this unit you will be able to:				
1. Evaluate formulas and literal expressions.	If $x = 2$, $y = 3$, $a = 5$, $b = 6$, find the value of		291	1
	(a) $2x + y$	_____		
	(b) $\dfrac{1 + x^2 + 2a}{y}$	_____		
	(c) $A = x^2 y$	$A = $ _____		
	(d) $T = \dfrac{2(a + b + 1)}{3x}$	$T = $ _____		
	(e) $P = abx^2$	$P = $ _____		
2. Perform the basic algebraic operations.	(a) $3ax^2 + 4ax^2 - ax^2$	$= $ _____	301	10
	(b) $5x - 3y - 8x + 2y$	$= $ _____		
	(c) $5x - (x + 2)$	$= $ _____		
	(d) $3(x^2 + 5x) - 4(2x - 3)$	$= $ _____		
3. Solve linear equations in one unknown and solve formulas.	(a) $3x - 4 = 11$	$x = $ _____	318	28
	(b) $2x = 18$	$x = $ _____	307	17
	(c) $2x + 7 = 43 - x$	$x = $ _____	329	36
	(d) Solve the following formula for N: $S = \dfrac{N}{2} + 26$	$N = $ _____		
	(e) Solve the following formula for A: $M = \dfrac{(A - B)L}{8}$.	$A = $ _____		

Name _____

Date _____

Course/Section _____

4. Translate simple English phrases and sentences into algebraic expressions and equations.

Write each phrase as an algebraic expression or equation.

(a) Four times the area.

_____ 340 **43**

(b) Current squared times resistance.

_____

(c) Efficiency is equal to 100 times the output divided by the input.

_____

(d) Resistance is equal to 12 times the length divided by the diameter squared.

_____

(e) An electrician collected $1265.23 for a job that included $840 labor, which was not taxed, and the rest for parts, which were taxed, at 6%. Determine how much of the total was tax.

_____

(f) A plumber's helper earns $18 per hour plus $27 per hour overtime. If he works 40 regular hours during a week, how much overtime does he need in order to earn $1000?

_____

5. Multiply and divide simple algebraic expressions.

(a) $2y \cdot 3y$

_____ 364 **64**

(b) $(6x^4y^2)(-2xy^2)$

_____

(c) $3x(y - 2x)$

_____

(d) $\dfrac{10x^7}{-2x^2}$

_____

(e) $\dfrac{6a^2b^4}{18a^5b^2}$

_____

(f) $\dfrac{12m^4 - 9m^3 + 15m^2}{3m^2}$

_____

6. Solve problems involving ratio and proportion.

(a) An architectural drawing of the living room of a house is $5\frac{1}{2}$ in. long and $4\frac{1}{2}$ in. wide. If the actual length of the living room is 22 ft, what is the actual width?

_____ 350 **52**

(b) A large production job is completed by 4 machines in 6 hours. The same size job must be finished in 2 hours the next time it is ordered. How many machines should be used to accomplish this task? _____

7. Use scientific notation.

Write in scientific notation.

(a) 0.000184 _____ 369 **69**

(b) 213,000 _____

Calculate.

(c) $(3.2 \times 10^{-6}) \times (4.5 \times 10^2)$ _____

(d) $(1.56 \times 10^{-4}) \div (2.4 \times 10^3)$ _____

(Answers to these preview problems are given on page 578.)

If you are certain that you can work *all* these problems correctly, turn to page 377 for the set of practice problems. If you cannot work one or more of the preview problems, turn to the page indicated to the right of the problem. Those who wish to master this material should turn to frame **1** and begin work there.

7 Basic Algebra

PEANUTS reprinted by permission of UFS, Inc.

1 In this chapter you will study algebra, but not the very formal algebra that deals with theorems, proofs, sets, and abstract problems. Instead, we shall study practical or applied algebra as actually used by technical and trades workers. Let's begin with a look at the language of algebra and algebraic formulas.

7-1 ALGEBRAIC LANGUAGE AND FORMULAS

The most obvious difference between algebra and arithmetic is that in algebra letters are used to replace or to represent numbers. A mathematical statement in which letters are used to represent numbers is called a *literal* expression. Algebra is the arithmetic of literal expressions—a kind of symbolic arithmetic.

Any letters will do, but in practical algebra we use the normal lower- and upper-case letters of the English alphabet. It is helpful in practical problems to choose the letters to be used on the basis of their memory value: t for time, D for diameter, C for cost, A for area, and so on. The letter used reminds you of its meaning.

Multiplication

Most of the usual arithmetic symbols have the same meaning in algebra that they have in arithmetic. For example, the addition (+) and subtraction (−) signs are used in exactly the same way. However, the multiplication sign (×) of arithmetic looks like the letter x and to avoid confusion we have other ways to show multiplication in algebra. The product of two algebraic quantities a and b, "a times b," may be written using

A raised dot	$a \cdot b$
Parentheses	$a(b)$ or $(a)b$ or $(a)(b)$
Nothing at all	ab

Obviously, this last way of showing multiplication won't do in arithmetic; we cannot write "two times four" as "24"—it looks like twenty-four. But it is a quick and easy way to write a multiplication in algebra.

Placing two quantities side by side to show multiplication is not new and it is not only an algebra gimmick; we use it every time we write 20 cents or 4 feet.

20 cents = 20 × 1 cent
4 feet   = 4 × 1 foot

Write the following multiplications using no multiplication symbols.

(a) 8 times a = _____   (b) m times p = _____

(c) 2 times s times t = _____   (d) 3 times x times x = _____

Check your answers in **2**.

---

2   (a) $8a$ (b) mp (c) $2st$ (d) $3x^2$

Did you notice in problem (d) that powers are written just as in arithmetic:

$x \cdot x = x^2$

$x \cdot x \cdot x = x^3$

$x \cdot x \cdot x \cdot x = x^4$ and so on.

Parentheses   Parentheses ( ) are used in arithmetic to show that some complicated quantity is to be treated as a unit. For example,

$2 \cdot (13 + 14 - 6)$

means that the number 2 multiplies *all* of the quantity in the parentheses.

In exactly the same way in algebra, parentheses ( ), brackets [ ], or braces { } are used to show that whatever is enclosed in them should be treated as a single quantity. An expression such as

$(3x^2 - 4ax + 2by^2)^2$

should be thought of as (something)^2. The expression

$(2x + 3a - 4) - (x^2 - 2a)$

should be thought of as (first quantity) – (second quantity). Parentheses are the punctuation marks of algebra. Like the period, comma, or semicolon in regular sentences, they tell you how to read an expression and get its correct meaning.

Division   In arithmetic we would write "48 divided by 2" as

$2\overline{)48}$ or $48 \div 2$ or $\frac{48}{2}$

But the first two ways of writing division are used very seldom in algebra. Division is usually written as a fraction.

"x is divided by y" is written $\frac{x}{y}$

"$(2n + 1)$ divided by $(n - 1)$" is written $\frac{(2n + 1)}{(n - 1)}$ or $\frac{2n + 1}{n - 1}$

Write the following using algebraic notation.

(a) 8 times $(2a + b)$ = _____   (b) $(a + b)$ times $(a - b)$ = _____

(c) x divided by y^2 = _____   (d) $(x + 2)$ divided by $(2x - 1)$ = _____

Turn to **3** to check your answers.

3 (a) $8(2a + b)$ 　　　　(b) $(a + b)(a - b)$

(c) $\dfrac{x}{y^2}$ 　　　　(d) $\dfrac{x + 2}{2x - 1}$

Algebraic Expressions

The word "expression" is used very often in algebra. An *expression* is a general name for any collection of numbers and letters connected by arithmetic signs. For example,

$$x + y \qquad 2x^2 + 4 \qquad 3(x^2 - 2ab)$$

$$\dfrac{D}{T} \qquad \sqrt{x^2 + y^2} \qquad \text{and} \qquad (b - 1)^2$$

are all algebraic expressions.

If the algebraic expression has been formed by multiplying quantities, each multiplier is called a *factor* of the expression.

Expression	Factors
ab	a and b
$2x(x + 1)$	$2x$ and $(x + 1)$
$(R - 1)(2R + 1)$	$(R - 1)$ and $(2R + 1)$

The algebraic expression can also be a sum or difference of simpler quantities or *terms*.

Expression	Terms	
$x + 4y$	x and $4y$	← The first term is x The second term is $4y$
$2x^2 + xy$	$2x^2$ and xy	
$A - R$	A and R	

Now let's check to see if you understand the difference between terms and factors. In each algebraic expression below, tell whether the portion quoted is a term or a factor.

(a) $2x^2 - 3xy$ 　　　　　　y is a _____

(b) $7x - 4$ 　　　　　　$7x$ is a _____

(c) $4x(a + 2b)$ 　　　　　　$4x$ is a _____

(d) $-2y(3y - 5)$ 　　　　　　$(3y - 5)$ is a _____

(e) $3x^2y + 8y^2 - 9$ 　　　　$3x^2y$ is a _____

Check your answers in **4**.

---

4 (a) factor 　　　(b) term 　　　(c) factor 　　　(d) factor 　　　(e) term

Evaluating Formulas

One of the most useful algebraic skills for any technical or practical work involves finding the value of an algebraic expression when the letters are given numerical values. A *formula* is a rule for calculating the numerical value of one quantity from the values of the other quantities. The formula or rule is usually written in mathematical form because algebra gives a brief, convenient to use, and easy to remember form for the rule. Here are a few examples of rules and formulas used in the trades.

1. **Rule:** The voltage across a simple resistor is equal to the product of the current through the resistor and the value of its resistance.

   Formula: $V = iR$

2. **Rule:** The cost of setting type is equal to the total ems set in the job multiplied by the hourly wage divided by the rate at which the type is set, in ems per hour.

   Formula: $C = \dfrac{TH}{E}$

3. **Rule:** The number of standard bricks needed to build a wall is about 21 times the volume of the wall.

   Formula: $N = 21\ LWH$

Evaluating a formula or algebra expression means to find its value by substituting numbers for the letters in the expression. For example, in retail stores the following formula is used:

$M = R - C$ where M is the markup on an item, R is the retail selling price, and C is the original cost

Find M if $R = \$25$ and $C = \$21$.

$M = $ _____

Check your work in **5**.

---

5 $M = \$25 - \$21 = \$4$

Easy? Of course. Simply substitute the numbers for the correct letters and then do the arithmetic. A formula is a recipe for a calculation.

Try another. Automotive engineers use the following formula to calculate the horsepower rating of an engine.

$H = \dfrac{D^2N}{2.5}$ where D is the diameter of a cylinder in inches, and N is the number of cylinders

Find H when $D = 3\frac{1}{2}$ in. and $N = 6$.

Check your work in **6**.

---

6 $H = \dfrac{(3.5)^2(6)}{2.5}$

3.5 $\boxed{x^2}$ $\boxed{\times}$ **6** $\boxed{\div}$ **2.5** $\boxed{=}$ → $\quad$ *29.4*

$H = 29.4$ hp or roughly 29 horsepower.

To be certain you do it correctly, follow this two-step process:

Step 1   Place the numbers being substituted in parentheses, and then substitute them in the formula.

Step 2   Do the arithmetic carefully *after* the numbers are substituted.

Example: Find $A = x + 2y$ for $x = 5$ ft, $y = 3$ ft.

Step 1	$A = (5 \text{ ft}) + 2(3 \text{ ft})$	Put 5 ft and 3 ft in parentheses.
Step 2	$A = 5 \text{ ft} + 6 \text{ ft}$	Do the arithmetic
	$A = 11 \text{ ft}$	

Example: Find $B = x^2 - y$ for $x = 4$, $y = -3$.

Step 1	$B = (4)^2 - (-3)$	Put the numbers in parentheses.
Step 2	$B = 16 + 3$	Do the arithmetic.
	$B = 19$	

Using parentheses in this way may seem like extra work for you, but it is the key to avoiding mistakes when evaluating formulas.

Evaluate the following formulas.

(a) $D = 2R$ for $R = 3.45$ in.

(b) $W = T - C$ for $T = 1420$ lb, $C = 385$ lb

(c) $P = 0.433H$ for $H = 11.4$ in. Round to three significant digits.

(d) $A = bh - 2$ for $b = 1.75$ in., $h = 4.20$ in.

(e) $P = 2(8 + x)$ for $x = -3$

Use the two-step process and check your work in **7**.

---

7 (a) $D = 2(3.45) = 6.90$ in.   (b) $W = (1420) - (385)$
                                                                   $= 1420 - 385$
                                                                  $= 1035$ lb

(c) $P = 0.433(11.4)$ (d) $A = (1.75)(4.2) - 2$
         $= 4.9362$ or 4.94 in., rounded $= 7.35 - 2$
                                                                 $= 5.35$ sq in.

 For problem (d),

1.75 ⊠ **4.2** ⊟ **2** ⊜ →   **5.35**

(e) $P = 2(8 + (-3))$
     $P = 2(8 - 3)$
     $P = 2 \cdot 5$
     $P = 10$

The order of operations for arithmetic calculations should be used when evaluating formulas. Remember:

1. Do any operations inside parentheses.

2. Find all powers and roots.

3. Do all multiplications or divisions left to right.

4. Do all additions or subtractions left to right.

1. Evaluate the following formulas using the standard order of operations.

     $a = 3$, $b = 4$, $c = 6$

(a) $3ab$ $=$ _____   (b) $2a^2c$ $=$ _____

(c) $2a^2 - b^2 =$ _____   (d) $3a^2bc - ab$ $=$ _____

(e) $b + 2a$ $=$ _____   (f) $a + 3(2b - c)$ $=$ _____

(g) $2(a + b) =$ _____   (h) $(a^2 - 1) - (2 + b) =$ _____

2. To convert Fahrenheit F to Celsius C temperature, use the following formula:

$$C = \frac{5(F - 32)}{9}$$

To convert Celsius C to Fahrenheit F temperature, use this formula:

$$F = \frac{9C}{5} + 32$$

Use these formulas to find the following temperatures. Round to the nearest degree.

(a) 650°F = _____ °C

(b) −160°C = _____ °F

(c) 398°C = _____ °F

(d) 2200°F = _____ °C

(e) 80°F = _____ °C

(f) 52.5°C = _____ °F

Check your work in **8**.

_____

8 1. (a) $3(3)(4) = 36$

(b) $2(3)^2(6) = 2 \cdot 9 \cdot 6 = 108$

(c) $2(3)^2 - (4)^2 = 2 \cdot 9 - 16$
$$= 18 - 16$$
$$= 2$$

(d) $3(3)^2(4)(6) - (3)(4) = (3 \cdot 9 \cdot 4 \cdot 6) - (3 \cdot 4)$
$$= 648 - 12$$
$$= 636$$

 3 ⊗ **3** $\boxed{x^2}$ ⊗ **4** ⊗ **6** ⊟ **3** ⊗ **4** ⊜ → *636.*

(e) $(4) + 2(3) = 4 + 2 \cdot 3$
$$= 4 + 6$$
$$= 10$$

(f) $(3) + 3(2(4) - (6)) = 3 + 3(2 \cdot 4 - 6)$
$$= 3 + 3(8 - 6)$$
$$= 3 + 3 \cdot 2 = 3 + 6$$
$$= 9$$

(g) $2((3) + (4)) = 2 \cdot (3 + 4)$
$$= 2 \cdot 7$$
$$= 14$$

(h) $((3)^2 - 1) - (2 + (4)) = (9 - 1) - (2 + 4)$
$$= 8 - 6$$
$$= 2$$

2. (a) $C = \dfrac{5 \times (650 - 32)}{9}$

(b) $F = \dfrac{9 \times (-160)}{5} + 32$

$$= \frac{5 \times (618)}{9}$$

$$= \frac{-1440}{5} + 32$$

$$= \frac{3090}{9}$$

$$= -288 + 32$$

$$\approx 343°C$$

$$= -256°F$$

(c) $\quad F = \dfrac{9 \times 398}{5} + 32$

$\quad = \dfrac{3582}{5} + 32$

$\quad = 716.4 + 32$

$\quad \approx 748°F$

(d) $\quad C = \dfrac{5 \times (2200 - 32)}{9}$

$\quad = \dfrac{5 \times 2168}{9}$

$\quad = \dfrac{10840}{9}$

$\quad \approx 1204°C$

(e) $\quad C = \dfrac{5 \times (80 - 32)}{9}$

$\quad = \dfrac{5 \times 48}{9}$

$\quad = \dfrac{240}{9}$

$\quad \approx 27°C$

(f) $\quad F = \dfrac{9 \times 52.5}{5} + 32$

$\quad = \dfrac{472.5}{5} + 32$

$\quad = 94.5 + 32$

$\quad \approx 127°F$

For problem (e) using a calculator.

 5 ⊗ ⟮ 80 ⊖ 32 ⟯ ÷ 9 ⊜ → **26.666666**

Careful ▶ Avoid the temptation to combine steps when you evaluate formulas. Take it slowly and carefully, follow the standard order of operations, and you will arrive at the correct answer. Rush through problems like these and you usually make mistakes. ◀

Now turn to **9** for a set of practice problems in evaluating formulas.

9

Exercises 7-1

Evaluating Formulas

A. Find the value of each of these formulas for $x = 2$, $y = 3$, $z = 4$, $R = 5$.

1. $A = 3x$
2. $D = 2R - y$
3. $T = x^2 + y^2$
4. $H = 2x + 3y - z$
5. $K = 3z - x^2$
6. $Q = 2xyz - 10$
7. $F = 2(x + y^2) - 3$
8. $W = 3(y - 1)$
9. $L = 3R - 2(y^2 - x)$
10. $A = R^2 - y^2 - xz$
11. $B = 3R - y + 1$
12. $F = 3(R - y + 1)$

B. Find the value of each of the following formulas. (Round to the nearest whole number if necessary.)

1. $x = 2(x + y) - 1$ for $x = 2, y = 4$
2. $V = (L + W)(2L + W)$ for $L = 7.5$ ft, $W = 5.0$ ft
3. $I = PRT$ for $P = 150, R = 0.05, T = 2$
4. $H = 2(a^2 + b^2)$ for $a = 2$ cm, $b = 1$ cm

5. $T = \dfrac{(A+B)H}{2}$ for $A = 3.26$ m, $B = 7.15$ m, $H = 4.4$ m

6. $V = \dfrac{\pi D^2 H}{4}$ for $\pi = 3.14$, $D = 6.25$ in., $H = 7.2$ in.

7. $P = \dfrac{NR(T+273)}{V}$ for $N = 5$, $R = 0.08$, $T = 27$, $V = 3$

8. $W = D(AB - \pi R^2)H$ for $D = 9$ lb/in.^{3}, $A = 6.3$ in., $B = 2.7$ in., $\pi = 3.14$, $R = 2$ in., $H = 1.0$ in.

9. $V = LWH$ for $L = 16.25$ m, $W = 3.1$ m, $H = 2.4$ m

10. $V = \pi R^2 A$ for $\pi = 3.14$, $R = 3.2$ ft, $A = 0.425$ ft

C. Practical Problems

1. The perimeter of a rectangle is given by the formula $P = 2L + 2W$, where L is the length of the rectangle and W is its width. Find P when L is $8\frac{1}{2}$ in. and W is 11 in.

2. **Electronics**   The current in a simple electrical circuit is given by the formula $i = V/R$, where V is the voltage and R is the resistance of the circuit. Find the current in a circuit whose resistance is 10 ohms and which is connected across a 120-volt power source.

3. **Electrical Technology**   Find the power used in an electric light bulb, $P = i^2 R$, if the current $i = 0.80$ ampere and the resistance $R = 150$ ohms. P will be in watts.

4. Find the surface area of a sphere, $A = 4\pi R^2$, when π equals roughly 3.14 and $R = 10.0$ cm. (Round to the nearest 10 cm^2.)

5. The Fahrenheit temperature F is related to the Celsius temperature C by the formula $F = \frac{9}{5}C + 32$. Find the Fahrenheit temperature when $C = 40°$.

6. **Machine Technology**   The volume of a round steel bar depends on its length L and diameter D according to the formula $V = \pi D^2 L/4$. Find the volume of a bar 20.0 in. long and 3.0 in. in diameter. Use $\pi \approx 3.14$. (Round to the nearest 10 cu in.)

7. If D dollars is invested at p percent interest for t years, the amount A of the investment is

$$A = D\left(1 + \frac{pt}{100}\right)$$

Find A if $D = \$1000$, $p = 9\%$, and $t = 5$ years.

8. **Electronics**   The total resistance R of two resistances a and b connected in parallel is $R = ab/(a + b)$. What is the total resistance if $a = 200$ ohms and $b = 300$ ohms?

9. The area of a trapezoid is $T = \dfrac{(A+B)H}{2}$, where A and B are the lengths of its parallel sides and H is the height. Find the area of the trapezoid shown.

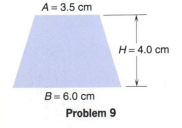

$A = 3.5$ cm

$H = 4.0$ cm

$B = 6.0$ cm

Problem 9

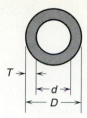

Problem 10

10. **Plumbing** Use the formula $T = \frac{1}{2}(D - d)$ to find the wall thickness T of tubing having the following dimensions.

D, outside diameter	d, inside diameter
(a) 2.125 in	1.500 in.
(b) 0.785 in.	0.548 in.
(c) 1.400 cm	0.875 cm
(d) $\frac{15}{16}$ in.	$\frac{5}{8}$ in.

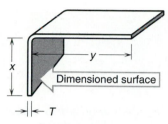

Dimensioned surface

Problem 11

11. **Sheet Metal Technology** To make a right-angle inside bend in sheet metal, the length of sheet used is given by the formula $L = x + y + \frac{1}{2}T$. Find L when $x = 6\frac{1}{4}$ in., $y = 11\frac{7}{8}$ in., $T = \frac{1}{4}$ in.

Problem 12

12. **Sheet Metal Technology** The length of a chord of a circle is given by the formula $L = 2RH - H^2$, where R is the radius of the circle and H is the height of the arc above the chord. If you have a portion of a circular disk for which $R = 4\frac{1}{4}$ in. and $H = 5\frac{1}{2}$ in., what is the length of the chord? (See the figure.)

13. **Building Technology** The rope capacity of a drum is given by the formula $L = ABC(A + D)$. How many feet of $\frac{1}{2}$-in. rope can be wound on a drum where $A = 6$ in., $C = 30$ in., $D = 24$ in., and $B = 1.05$ for $\frac{1}{2}$-in rope? (Round to the nearest hundred inches.)

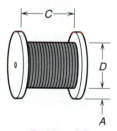

Problem 13

14. **Manufacturing** A millwright uses the following formula to find the required length of a pulley belt (See the figure.)

$$L = 2C + 1.57(D + d) + (D + d)/4C.$$

Find the length of belt needed if

$C = 36$ in. between pulley centers
$D = 24$-in. follower
$d = 4$-in. driver

Round to the nearest 10 in.

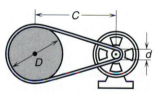

Problem 14

15. **Electrical Technology** An electrician uses a bridge circuit to locate a ground in an underground cable several miles long. The formula

$$\frac{R_1}{L - x} = \frac{R_2}{x} \qquad \text{or} \qquad x = \frac{R_2 L}{R_1 + R_2}$$

is used to find x, the distance to the ground. Find x if $R_1 = 750$ ohms, $R_2 = 250$ ohms, and $L = 4000$ ft.

16. **Machine Technology** The cutting speed of a lathe is the rate, in feet per minute, that the revolving workpiece travels past the cutting edge of the tool. Machinists use the following formula to calculate cutting speed:

$$\text{Cutting speed, } C = \frac{3.1416DN}{12}$$

where D is the diameter of the work and N is the turning rate in rpm. Find the cutting speed if a steel shaft 3.25 in. in diameter is turned at 210 rpm. (Round to the nearest whole number.)

17. **Electronics** Find the power load P in kilowatts of an electrical circuit that takes a current I of 12 amperes at a voltage V of 220 volts if

$$P = \frac{VI}{100}$$

18. **Aeronautical Mechanics** A jet engine developing T pounds of thrust and driving an airplane at V mph has a thrust horsepower, H, given approximately by the formula

$$H = \frac{TV}{375} \quad \text{or} \quad V = \frac{375H}{T}$$

Find the airspeed V if $H = 16{,}000$ hp and $T = 10{,}000$ lb.

19. **Electronics** The resistance R of a conductor is given by the formula

$$R = \frac{PL}{A} \quad \text{or} \quad L = \frac{AR}{P}$$

where $P =$ coefficient of resistivity
$L =$ length of conductor
$A =$ cross-sectional area of the conductor

Find the length in cm of No. 16 Nichrome wire needed to obtain a resistance of 8 ohms. For this wire $P = 0.000113$ ohm-cm and $A = 0.013$ cm^2. (Round to the nearest centimeter.)

20. **Auto Mechanics** The pressure P and total force F exerted on a piston of diameter D are approximately related by the equation

$$P = \frac{1.27F}{D^2} \quad \text{or} \quad F = \frac{PD^2}{1.27}$$

Find the total force on a piston of diameter 3.25 in. if the pressure exerted on it is 150 lb/sq in. (Round to the nearest 50 lb.)

21. **Printing** The formula for the number of type lines L set solid in a form is

$$L = \frac{12D}{T}$$

where D is the depth in picas and T is the point size of the type. How many lines of 10-point type can be set in a space 30 picas deep?

22. **Plumbing** To determine the number N of smaller pipes that provide the same total flow as a larger pipe, use the formula

$$N = \frac{D^2}{d^2}$$

where D is the diameter of the larger pipe and d is the diameter of each smaller pipe. How many $1\frac{1}{2}$-in. pipes will it take to produce the same flow as a $2\frac{1}{2}$-in. pipe?

23. Find the volume of a sphere of radius $R = 1.0076$ in.:

$$\text{Volume } V = \frac{4\pi R^3}{3} \quad \text{Use } 3.1416 \text{ for } \pi.$$

(Round to 0.0001 cu in.)

24. **Machine Technology** How many minutes will it take a lathe to make 17 cuts each 24.5 in. in length on a steel shaft if the tool feed F is 0.065 in. per revolution and the shaft turns at 163 rpm? Use the formula

$$T = \frac{LN}{FR}$$

where T is the cutting time (min), N is the number of cuts, L the length of cut (in.), F the tool feed rate (in./rev), and R the rpm rate of the workpiece. Round to 0.1 minute.

25. **Machine Technology**   Find the area of each of the following circular holes using the formula

$$A = \frac{\pi D^2}{4}$$

   (a) $D = 1.0004$ in.   (b) $D = \frac{7}{8}$ in.   (c) $D = 4.1275$ in.   (d) $D = 2.0605$ cm

   Use 3.1416 for π and round to 0.0001 unit.

26. Suppose that a steel band was placed tightly around the earth at the equator. If the temperature of the steel is raised 1°F, the metal will expand 0.000006 in. each inch. How much space would there be between the earth and the steel band if the temperature was raised 1°F? Use the formula

$$D \text{ (in ft)} = \frac{(0.000006)(\text{diameter of the earth})(5280)}{\pi}$$

   where diameter of the earth = 7917 miles. Use $\pi \approx 3.1415927$.

   (Round to two decimal places.)

27. **Plumbing**   When a cylindrical container is lying on its side, the following formula can be used for calculating the volume of liquid in the container:

$$V = \frac{4}{3}h^2 L \sqrt{\frac{d}{h} - 0.608}$$

   Use this formula to calculate the volume of water in such a tank 6.0 ft long (L), 2.0 ft in diameter (d), and filled to a height (h) of 0.75 ft. Round to two significant digits.

28. **Meteorology**   The following formula is used to calculate the wind chill factor W in degrees Celsius:

$$W = 33 - \frac{(10.45 + 10\sqrt{V} - V)(33 - T)}{22.04}$$

   where T is the air temperature in degrees Celsius, and V is the wind speed in meters per second. Determine the wind chill for the following situations:

   (a)   Air temperature 7°C and wind speed 20 meters per second.

   (b)   Air temperature 28°F and wind speed 22 mph. (*Hint:* Use the conversion factors from Section 5-3 and the temperature formulas given earlier in this chapter to convert to the required units. Convert your final answer to degrees Fahrenheit.)

When you have completed these exercises, check your answers on page 597, then turn to **10** to learn how to add and subtract algebraic expressions.

7-2   ADDING AND SUBTRACTING ALGEBRAIC EXPRESSIONS

10   In the preceding section we learned how to find the value of an algebraic expression after substituting numbers for the letters. There are other useful ways of using algebra where we must manipulate algebraic expressions *without* first substituting numbers for letters. In this section we learn how to simplify algebraic expressions by adding and subtracting terms.

Combining Like Terms

Two algebraic terms are said to be *like terms* if they contain exactly the same literal part. For example, the terms

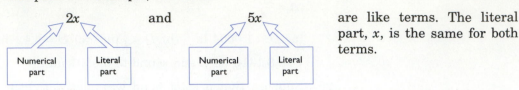

and are like terms. The literal part, x, is the same for both terms.

The number multiplying the letters is called the *numerical coefficient* of the term.

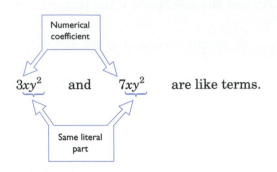

$3xy^2$ and $7xy^2$ are like terms.

But

$3x^2$ and $2x$ are unlike terms. Their literal parts, x and x^2, are different.

You can add and subtract like terms but not unlike terms. To add or subtract like terms, add or subtract their numerical coefficients and keep the same literal part.

Example:

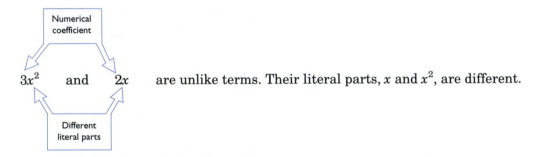

We are adding like quantities

$$\boxed{x} + \boxed{x} \quad + \boxed{x} + \boxed{x} + \boxed{x} = \boxed{x} + \boxed{x} + \boxed{x} + \boxed{x} + \boxed{x}$$

Another Example:

$$8 - 3 = 5$$

$$8a^2x - 3a^2x = 5a^2x$$

Same literal part

Try these problems for practice.

(a) $12d^2 + 7d^2$ = _____

(b) $2ax - ax - 5ax$ = _____

(c) $3(y + 1) + 9(y + 1)$ = _____

(d) $8x^2 + 2xy - 2x^2 =$ _____

(e) $x - 6x + 2x$ = _____

(f) $4xy - xy + 3xy$ = _____

Turn to **11** to check your answers.

---

11 (a) $12d^2 + 7d^2 = 19d^2$

(b) $2ax - ax - 5ax = -4ax$ The term ax is equal to $1 \cdot ax$.

(c) $3(y + 1) + 9(y + 1) = 12(y + 1)$

(d) $8x^2 + 2xy - 2x^2 = 6x^2 + 2xy$

We cannot combine the x^2-term and the xy-term because the literal parts are not the same. They are unlike terms.

(e) $x - 6x + 2x = -3x$

(f) $4xy - xy + 3xy = 6xy$

In general, to simplify a series of terms being added or subtracted, first group together like terms, then add or subtract. For example,

$3x + 4y - x + 2y + 2x - 8y$

Be careful not to lose the negative sign on 8y.

becomes

$(3x - x + 2x) + (4y + 2y - 8y)$ after grouping like terms,

Be careful not to lose the negative sign on x.

$(3x - x + 2x) + (4y + 2y - 8y) = 4x + (-2y)$ or $4x - 2y$

Adding a negative   is the same as   subtracting a positive.

It is simpler to write the final answer as a subtraction rather than as an addition.

Simplify the following expressions by adding and subtracting like terms.

(a) $5x + 4xy - 2x - 3xy$

(b) $3ab^2 + a^2b - ab^2 + 3a^2b - a^2b$

(c) $x + 2y - 3z - 2x - y + 5z - x + 2y - z$

(d) $17pq - 9ps - 6pq + ps - 6ps - pq$

(e) $4x^2 - x^2 + 2x + 2x^2 + x$

Check your answers in **12**.

---

12 (a) $(5x - 2x) + (4xy - 3xy) = 3x + xy$

(b) $(3ab^2 - ab^2) + (a^2b + 3a^2b - a^2b) = 2ab^2 + 3a^2b$

(c) $(x - 2x - x) + (2y - y + 2y) + (-3z + 5z - z) = -2x + 3y + z$

(d) $(17pq - 6pq - pq) + (-9ps + ps - 6ps) = 10pq - 14ps$

(e) $(4x^2 - x^2 + 2x^2) + (2x + x) = 5x^2 + 3x$

Expressions with Parentheses

Parentheses are used in algebra to group together terms that are to be treated as a unit. Adding and subtracting expressions usually involves working with parentheses. For example, to add

$(a + b) + (a + d)$

First, remove all parentheses: $(a + b) + (a + d)$

$$= a + b + a + d$$

Second, add like terms:
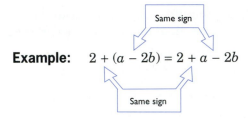

$$= a + a + b + d$$

$$= 2a + b + d$$

Removing parentheses can be a tricky business. Remember these two rules:

Rule 1 If the parenthesis has a plus sign in front, simply remove the parentheses.

Example: $1 + (3x + y) = 1 + 3x + y$

Example: $2 + (a - 2b) = 2 + a - 2b$

with "Same sign" arrows

Rule 2 If the parenthesis has a negative sign in front, change the sign of each term inside, then remove the parentheses.

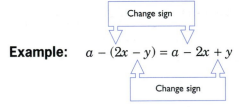

Example: $2 - (x + 2y) = 2 - x - 2y$

Example: $a - (2x - y) = a - 2x + y$

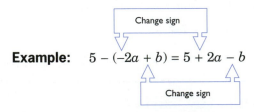

Example: $5 - (-2a + b) = 5 + 2a - b$

Note ▶ In using Rule 2 you are simply rewriting subtraction as addition of the opposite. However, you must add the opposite of *all* terms inside parentheses. ◀

Simplify the following expressions by using these two rules to remove parentheses.

(a) $x + (2y - a^2)$

(b) $4 - (x^2 - y^2)$

(c) $-(x + 1) + (y + a)$

(d) $ab - (a - b)$

(e) $(x + y) - (p - q)$

(f) $-(-x - 2y) - (a + 2b)$

(g) $3 + (-2p - q^2)$

(h) $-(x - y) + (-3x^2 + y^2)$

Check your answers in **13**.

---

13 (a) $x + 2y - a^2$

(b) $4 - x^2 + y^2$

(c) $-x - 1 + y + a$

(d) $ab - a + b$

(e) $x + y - p + q$

(f) $x + 2y - a - 2b$

(g) $3 - 2p - q^2$

(h) $-x + y - 3x^2 + y^2$

A third rule is needed when a multiplier is in front of the parentheses.

Rule 3 If the parenthesis has a multiplier in front, multiply each term inside the parentheses by the multiplier.

Example: $+2(a + b) = +2a + 2b$

Think of this as $(+2)(a + b) = (+2)a + (+2)b$
$$= +2a + 2b$$

Each term inside the parentheses is multiplied by +2.

Example: $-2(x + y) = -2x - 2y$

Think of this as $(-2)(x + y) = (-2)x + (-2)y$

Each term inside the parentheses is multiplied by –2.

Example: $-(x - y) = (-1)(x - y)$
$$= (-1)(x) + (-1)(-y) = -x + y$$

Example: $2 - 5(3a - 2b) = 2 + (-5)(3a - 2b) = 2 + (-5)(3a) + (-5)(-2b)$
$$= 2 + (-15a) + (+10b) \quad \text{or} \quad 2 - 15a + 10b$$

Careful ▶ In the last example it would be incorrect first to subtract the 5 from the 2. The order of operations requires that we multiply before we subtract. ◀

When you multiply negative numbers, you may need to review the arithmetic of negative numbers starting in frame **11** on page 268.

Notice that we must multiply *every* term inside the parentheses by the number outside the parentheses. Once the parentheses have been removed, you can add and subtract like terms, as explained in operation 1.

Simplify the following expressions by removing parentheses. Use the three rules.

(a) $2(2x - 3y)$

(b) $1 - 4(x + 2y)$

(c) $a - 2(b - 2x)$

(d) $x^2 - 3(x - y)$

(e) $p - 2(-y - 2x)$

(f) $3(x - y) - 2(2x^2 + 3y^2)$

Check your answers in **14**.

---

14 (a) $4x - 6y$

(b) $1 - 4x - 8y$

(c) $a - 2b + 4x$

(d) $x^2 - 3x + 3y$

(e) $p + 2y + 4x$

(f) $3x - 3y - 4x^2 - 6y^2$

Once you can simplify expressions by removing parentheses, it is easy to add and subtract like terms. For example,

$$(3x - y) - 2(x - 2y) = 3x - y - 2x + 4y \qquad \text{Simplify by removing parentheses.}$$
$$= \underbrace{3x - 2x}\; \underbrace{- y + 4y} \qquad \text{Group like terms.}$$
$$= x + 3y \qquad \text{Combine like terms.}$$

Try these problems for practice.

(a) $(3y + 2) + 2(y + 1)$ (b) $(2x + 1) + 3(4 - x)$

(c) $(a + b) - (a - b)$ (d) $2(a + b) - 2(a - b)$

(e) $2(x - y) - 3(y - x)$ (f) $2(x^3 + 1) - 3(x^3 - 2)$

(g) $(x^2 - 2x) - 2(x - 2x^2)$ (h) $-2(3x - 5) - 4(x - 1)$

Check your answers in **15**.

---

15   (a) $5y + 4$ (b) $-x + 13$ (c) $2b$ (d) $4b$

(e) $5x - 5y$ (f) $-x^3 + 8$ (g) $5x^2 - 4x$ (h) $-10x + 14$

Some step-by-step solutions:

(a) $(3y + 2) + 2(y + 1) = 3y + 2 + 2y + 2$
$$= 3y + 2y + 2 + 2$$
$$= 5y + 4$$

(b) $(2x + 1) + 3(4 - x) = 2x + 1 + 12 - 3x$
$$= 2x - 3x + 1 + 12$$
$$= -x + 13$$

(c) $(a + b) - (a - b) = a + b - a - (-b)$
$$= a + b - a + b$$
$$= a - a + b + b = 0 + 2b = 2b$$

(h) $-2(3x - 5) - 4(x - 1) = -6x - 2(-5) - 4x - 4(-1)$
$$= -6x + 10 - 4x + 4$$
$$= -6x - 4x + 10 + 4 = -10x + 14$$

Now turn to **16** for additional practice on addition and subtraction of algebraic expressions.

16

Exercises 7-2   **Adding and Subtracting Algebraic Expressions**

A.   Simplify by adding or subtracting like terms.

  1. $3y + y + 5y$ 2. $4x^2y + 5x^2y$

  3. $E + 2E + 3E$ 4. $ax - 5ax$

  5. $9B - 2B$ 6. $3m - 3m$

  7. $3x^2 - 5x^2$ 8. $4x + 7y + 6x + 9y$

  9. $6R + 2R^2 - R$ 10. $1.4A + 0.05A - 0.8A^2$

11. $x - \frac{1}{2}x - \frac{1}{4}x - \frac{1}{8}x$ 12. $x + 2\frac{1}{2}x - 5\frac{1}{2}x$

13. $2 + W - 4.1W - \frac{1}{2}$ 14. $q - p - 1\frac{1}{2}p$

15. $2xy + 3x + 4xy$
16. $ab + 5ab - 2ab$
17. $x^2 + x^2y + 4x^2 + 3x$
18. $1.5p + 0.3pq + 3.1p$

B. Simplify by removing parentheses and, if possible, combining like terms.

1. $3x^2 + (2x - 5)$
2. $6 + (-3a + 8b)$
3. $8m + (4m^2 + 2m)$
4. $9x + (2x - 5x^2)$
5. $2 - (x + 5y)$
6. $7x - (4 + 2y)$
7. $3a - (8 - 6b)$
8. $5 - (w - 6z)$
9. $4x - (10x^2 + 7x)$
10. $12m - (6n + 4m)$
11. $15 - (3x - 8)$
12. $-12 - (5y - 9)$
13. $-(x - 2y) + (2x + 6y)$
14. $-(3 + 5m) + (11 - 4m)$
15. $-(14 + 5w) - (2w - 3z)$
16. $-(16x - 8) - (-2x + 4y)$
17. $3(3x - 4y)$
18. $4(5a + 6b)$
19. $-8(7m + 6)$
20. $-2(6x - 3)$
21. $x - 5(3 + 2x)$
22. $4y - 2(8 + 3y)$
23. $9m - 7(-2m + 6)$
24. $w - 5(4w - 3)$
25. $3 - 4(2x + 3y)$
26. $6 - 2(3a - 5b)$
27. $12 - 2(3w - 8)$
28. $2 - 11(7 + 5a)$
29. $2(3x + 4y) - 4(6x^2 - 5y^2)$
30. $-4(5a - 6b) + 7(2ab - 4b^2)$
31. $(22x - 14y) + 3(8y - 6x)$
32. $8(3w - 5z) + (9z - 6w)$
33. $6(x + y) - 3(x - y)$
34. $-5(2x - 3x^2) + 2(9x^2 + 8x)$
35. $4(x^2 - 6x + 8) - 6(x^2 + 3x - 5)$
36. $2(3a - 5b + 6ab) - 8(2a + b - 4ab)$

When you have completed these exercises, check your answers on page 597, then turn to **17** to learn how to solve algebraic equations.

_____

7-3 SOLVING SIMPLE EQUATIONS

17 An arithmetic equation such as $3 + 2 = 5$ means that the number named on the left $(3 + 2)$ is the same as the number named on the right (5).

An algebraic equation such as $x + 3 = 7$ is a statement that the sum of some number x and 3 is equal to 7. If we choose the correct value for x, the number $x + 3$ will be equal to 7.

x is a *variable,* a symbol that stands for a number in an equation, a blank space to be filled. Many numbers might be put in the space, but only one makes the equation a true statement.

Find the missing numbers in the following arithmetic equations.

(a) $37 + \underline{\hspace{1cm}} = 58$
(b) $\underline{\hspace{1cm}} - 15 = 29$
(c) $4 \times \underline{\hspace{1cm}} = 52$
(d) $28 \div \underline{\hspace{1cm}} = 4$

Puzzle them out, then turn to **18**.

_____

18 (a) $37 + \boxed{21} = 58$ (b) $\boxed{44} - 15 = 29$

(c) $4 \times \boxed{13} = 52$ (d) $28 \div \boxed{7} = 4$

We could have written these equations as follows, with variables instead of blanks.

$$37 + A = 58 \qquad B - 15 = 29 \qquad 4C = 52 \qquad \frac{28}{D} = 4$$

Of course, any letters would do in place of *A, B, C,* and *D* in these algebraic equations.

How did you solve these equations? You probably "eyeballed" them—mentally juggled the other information in the equation until you found a number that made the equation true. Solving algebraic equations is very similar except that we can't "eyeball" it entirely. We need certain and systematic ways of solving the equation that will produce the correct answer quickly every time.

In this section you will learn first what a solution to an algebraic equation is—how to recognize it if you stumble over it in the dark—then how to solve linear equations.

Solution   Each value of the variable that makes an equation true is called a *solution* of the equation. For example, the solution of $x + 3 = 7$ is $x = 4$.

Check: $(4) + 3 = 7$

Another example: The solution of the equation $2x - 9 = 18 - 7x$ is $x = 3$.

Check: $\begin{aligned} 2(3) - 9 &= 18 - 7(3) \\ 6 - 9 &= 18 - 21 \\ -3 &= -3 \end{aligned}$

For certain equations more than one value of the variable may make the equation true. For example, the equation $x^2 + 6 = 5x$ is true for $x = 2$,

Check: $\begin{aligned} (2)^2 + 6 &= 5(2) \\ 4 + 6 &= 5 \cdot 2 \\ 10 &= 10 \end{aligned}$

and it is also true for $x = 3$.

Check: $\begin{aligned} (3)^2 + 6 &= 5(3) \\ 9 + 6 &= 5 \cdot 3 \\ 15 &= 15 \end{aligned}$

Determine whether each given value of the variable is a solution to the equation.

(a) For $4 - x = -3$ is $x = -7$ a solution?

(b) For $3x - 8 = 22$ is $x = 10$ a solution?

(c) For $-2x + 15 = 5 + 3x$ is $x = -2$ a solution?

(d) For $3(2x - 8) = 5 - (6 - 4x)$ is $x = 11.5$ a solution?

(e) For $3x^2 - 5x = -2$ is $x = \frac{2}{3}$ a solution?

Check your answers in **19**.

19 (a) No—the left side equals 11.

(b) Yes. (c) No—the left side equals 19, the right side equals –1.

(d) Yes. (e) Yes.

Here is the check for (e): $3\left(\dfrac{2}{3}\right)^2 - 5\left(\dfrac{2}{3}\right) = -2$

$$3\left(\dfrac{4}{9}\right) - 5\left(\dfrac{2}{3}\right) = -2$$

$$\dfrac{4}{3} - \dfrac{10}{3} = -2$$

$$-\dfrac{6}{3} = -2$$

$$-2 = -2$$

Equivalent Equations

Equations with the exact same solution are called *equivalent* equations. The equations $2x + 7 = 13$ and $3x = 9$ are equivalent because substituting the value 3 for x makes them both true.

We say that an equation with the variable x is *solved* if it can be put in the form

$x = \square$ where $\square$ is some number

For example, the solution to the equation

$2x - 1 = 7$ is $x = 4$

because $2(4) - 1 = 7$

or $8 - 1 = 7$

is a true statement.

Solving Equations

Equations as simple as the previous one are easy to solve by guessing, but guessing is not a very dependable way to do mathematics. We need some sort of rule that will enable us to rewrite the equation to be solved ($2x - 1 = 7$, for example) as an equivalent solution equation ($x = 4$).

The general rule is to treat every equation as a balance of the two sides.

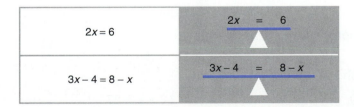

Any changes made in the equation must not disturb this balance.

Any operation performed on one side of the equation must also be performed on the other side.

Two kinds of balancing operations may be used.

Example:

1. Adding or subtracting a number on both sides of the equation does not change the balance

and

2. Multiplying or dividing both sides of the equation by a number (but not zero) does not change the balance.

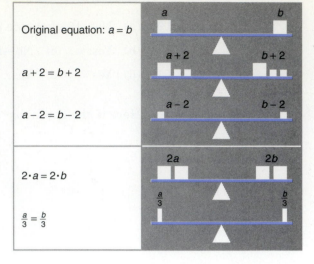

Original equation: $a = b$

$a + 2 = b + 2$

$a - 2 = b - 2$

$2 \cdot a = 2 \cdot b$

$\frac{a}{3} = \frac{b}{3}$

Let's work through an example.

Solve: $x - 4 = 2$.

Step 1 We want to change this equation to an equivalent equation with only x on the left, so we add 4 to each side of the equation.

$$x - 4 + 4 = 2 + 4$$

Step 2 Combine terms.

$$x \underbrace{- 4 + 4}_{0} = 2 + 4$$

$$x = 6 \quad \text{Solution}$$

Check: $(6) - 4 = 2$
$$2 = 2$$

Use these balancing operations to solve the equation

$8 + x = 14$

Check your work in **20**.

---

20 **Solve:** $8 + x = 14$

Step 1 We want to change this equation to an equivalent equation with only x on the left, so we subtract **8** from each side of the equation.

$$8 + x - 8 = 14 - 8$$

Step 2 Combine terms.

$$x \underbrace{+ 8 - 8}_{0} = 14 - 8 \qquad \text{where } 8 + x = x + 8$$

$$x = 6 \quad \text{Solution}$$

Check: $8 + (6) = 14$
$$14 = 14$$

Solve these in the same way.

(a) $x - 7 = 10$ (b) $12 + x = 27$ (c) $x + 6 = 2$

(d) $8.4 = 3.1 + x$ (e) $6.7 + x = 0$ (f) $\frac{1}{4} = x - \frac{1}{2}$

(g) $-11 = x + 5$ (h) $x - 5.2 = -3.7$

The complete step-by-step solutions are given in **21**.

---

21 (a) **Solve:** $\qquad x - 7 = 10$

$\qquad\qquad x - 7 \boxed{+ 7} = 10 \boxed{+ 7}$ $\qquad\qquad$ Add 7 to each side

$\qquad\qquad x \underbrace{- 7 \boxed{+ 7}}_{0} = 10 \boxed{+ 7}$ $\qquad\qquad$ Combine terms.

$\qquad\qquad\qquad x = 17$ $\quad$ Solution

Check: $(17) - 7 = 10$

$\qquad\qquad\qquad 10 = 10$

(b) **Solve:** $\qquad 12 + x = 27$

$\qquad\qquad 12 + x \boxed{- 12} = 27 \boxed{- 12}$ $\qquad\qquad$ Subtract 12 from each side.

$\qquad\qquad x \underbrace{+ 12 \boxed{- 12}}_{0} = 27 \boxed{- 12}$ $\qquad\qquad$ Combine terms.
$\qquad\qquad\qquad\qquad\qquad\qquad\qquad$ (Note that $12 + x = x + 12$.)

$\qquad\qquad\qquad x = 15$ $\quad$ Solution

Check: $12 + (15) = 27$

$\qquad\qquad\qquad 27 = 27$

(c) **Solve:** $\qquad x + 6 = 2$

$\qquad\qquad x + 6 \boxed{- 6} = 2 \boxed{- 6}$ $\qquad\qquad$ Subtract 6 from each side.

$\qquad\qquad x + 6 \boxed{- 6} = 2 \boxed{- 6}$ $\qquad\qquad$ Combine terms.

$\qquad\qquad\qquad 0$

$\qquad\qquad\qquad x = -4$ $\quad$ Solution $\qquad$ The solution is a negative number. Remember, any number, positive or negative, may be the solution of an equation.

Check: $(-4) + 6 = 2$

$\qquad\qquad\qquad 2 = 2$

(d) **Solve:** $\qquad 8.4 = 3.1 + x$ $\qquad\qquad$ The variable x is on the right side of this equation.

$\qquad\qquad 8.4 \boxed{- 3.1} = 3.1 + x \boxed{- 3.1}$ $\qquad\qquad$ Subtract 3.1 from each side.

$\qquad\qquad 8.4 - 3.1 = x$ $\qquad\qquad$ Combine terms.

$\qquad\qquad\qquad 5.3 = x$ $\qquad\qquad$ Decimal numbers often appear in practical problems.

$\qquad\qquad$ or $\quad x = 5.3$ $\quad$ Solution $\qquad$ $5.3 = x$ is the same as $x = 5.3$.

Check: $8.4 = 3.1 + (5.3)$

$\qquad\qquad 8.4 = 8.4$

(e) **Solve;** $\qquad 6.7 + x = 0$

$\qquad\qquad 6.7 + x \boxed{- 6.7} = 0 \boxed{- 6.7}$ $\qquad\qquad$ Subtract 6.7 from each side.

$\qquad\qquad\qquad x = -6.7$ $\quad$ Solution $\qquad$ A negative-number answer is reasonable.

Check: $6.7 + (-6.7) = 0$

$\qquad\qquad 6.7 - 6.7 = 0$

(f) **Solve:** $\dfrac{1}{4} = x - \dfrac{1}{2}$

$\dfrac{1}{4} + \dfrac{1}{2} = x - \dfrac{1}{2} + \dfrac{1}{2}$ Add $\dfrac{1}{2}$ to each side.

$\underbrace{}_{0}$ $\dfrac{1}{4} + \dfrac{1}{2} = \dfrac{3}{4}$

$\dfrac{3}{4} = x$ or $x = \dfrac{3}{4}$ Solution

Check: $\dfrac{1}{4} = \left(\dfrac{3}{4}\right) - \dfrac{1}{2}$

$\dfrac{1}{4} = \dfrac{1}{4}$

(g) **Solve:** $-11 = x + 5$

$-11 - 5 = x + 5 - 5$ Subtract 5 from each side.

$-16 = x$ or $x = -16$ Solution

Check: $-11 = (-16) + 5$
$-11 = -11$

(h) **Solve:** $x - 5.2 = -3.7$

$x - 5.2 + 5.2 = -3.7 + 5.2$ Add 5.2 to each side.

$x = 1.5$

Check: $(1.5) - 5.2 = -3.7$
$-3.7 = -3.7$

Learning Hint ▶ Notice that to solve these equations, we always do the opposite of whatever operation is being performed on the variable. If a number is being added to x in the equation, we must subtract that number in order to solve. If a number is being subtracted from x in the equation, we must add that number in order to solve. Finally, whatever we do to one side of the equation, we must also do to the other side of the equation. ◀

In the equations solved so far, the variable has appeared as simply x. In some equations the variable may be multiplied or divided by a number, so that it appears as $2x$ or $x/3$ for example. To solve such an equation, do the operation opposite of the one being performed on the variable. If the variable appears as $2x$, divided by 2; if the variable appears as $x/3$, multiply by 3, and so on.

Try this problem.

$27 = -3x$

Go to **22** for the solution.

---

22 **Solve:** $27 = -3x$ Notice that the x-term is on the right side of the equation. It is x multiplied by -3. To change $-3x$ into $1x$ or x, we must perform the opposite operation, that is, divide by -3.

Step 1 $\dfrac{27}{-3} = \dfrac{-3x}{-3}$ Divide each side by -3.

Step 2 $-9 = \left(\dfrac{-3}{-3}\right)x$
$-9 = x$ or $x = -9$

Check: $27 = -3(-9)$
$27 = 27$

Here is a slightly different problem.

Solve: $\frac{x}{3} = 13$

Look in **23** for the solution.

---

23 **Solve:** $\frac{x}{3} = 13$

In this equation the variable x is divided by 3.

Step 1 $\frac{x}{3} \cdot 3 = 13 \cdot 3$

We want to change this equation to an equivalent equation with x alone on the left, so performing the opposite operation, we multiply both sides by 3.

$$\frac{3x}{3} = \left(\frac{3}{3}\right)x = x$$

Step 2 $x = 13 \cdot 3$
$x = 39$ Solution

Check: $\frac{(39)}{3} = 13$
$13 = 13$

Now consider this equation.

$-\frac{x}{4} = 5$

Check your answer in **24**.

---

24 **Solve:** $-\frac{x}{4} = 5$

Step 1 $\left(-\frac{x}{4}\right)(-4) = 5\,(-4)$

Multiply both sides by -4.

Step 2 $\frac{(-x)(-4)}{4} = -20$

$\frac{4x}{4} = -20$

$x = -20$

$-\frac{x}{4} = \frac{-x}{4}$

$(-x)(-4) = +4x$

$\frac{4x}{4} = \left(\frac{4}{4}\right)x = 1x = x$

Check: $-\frac{(-20)}{4} = 5$
$-(-5) = 5$
$5 = 5$

Note ▶ When a negative sign precedes a fraction, we may move it either to the numerator or to the denominator, but not to both. In the previous problem, we placed the negative sign in the numerator in Step 2. If we had moved it to the denominator instead, we would have the following:

$\left(\frac{x}{-4}\right)(-4) = -20$ or $\frac{-4x}{-4} = -20$ or $x = -20$

which is exactly the same solution. ◀

Try the following problems for more practice with one-step multiplication and division equations.

(a) $-5x = -25$ (b) $\frac{x}{6} = 6$ (c) $-16 = 2x$

(d) $7x = 35$ (e) $-\frac{x}{2} = 14$ (f) $\frac{2x}{3} = 6$

(g) $-4x = 22$ (h) $\frac{x+3}{2} = 4$ (i) $2.4x = 0.972$

The step-by-step solutions are in **25**.

---

25 (a) **Solve:** $-5x = -25$ The variable x appears multiplied by -5.

Step 1 $\dfrac{-5x}{-5} = \dfrac{-25}{-5}$ Divide both sides by -5.

Step 2 $x = 5$ $\dfrac{-5x}{-5} = \left(\dfrac{-5}{-5}\right)x = 1x = x$

Check: $-5(5) = -25$
$-25 = -25$

(b) **Solve:** $\frac{x}{6} = 6$ The variable x appears divided by 6.

Step 1 $\left(\dfrac{x}{6}\right)(6) = 6(6)$ Multiply both sides by 6.

Step 2 $x = 36$ $\left(\dfrac{x}{6}\right)(6) = x\left(\dfrac{6}{6}\right) = x(1) = x$

Check: $\dfrac{(36)}{6} = 6$

$6 = 6$

(c) **Solve:** $-16 = 2x$

Step 1 $\dfrac{-16}{2} = \dfrac{2x}{2}$ Divide each side by 2.

Step 2 $-8 = x$ $\dfrac{2x}{2} = \left(\dfrac{2}{2}\right)x = 1x = x$

or $x = -8$

Check: $-16 = 2(-8)$
$-16 = -16$

(d) **Solve:** $7x = 35$

Step 1 $\dfrac{7x}{7} = \dfrac{35}{7}$ Divide both sides by 7.

Step 2 $x = \dfrac{35}{7}$ $\dfrac{7x}{7} = \left(\dfrac{7}{7}\right)x = x$

$x = 5$ Solution

Check: $7 \cdot (5) = 35$
$35 = 35$

(e) **Solve:** $-\dfrac{x}{2} = 14$

Step 1 $\left(-\dfrac{x}{2}\right)\boxed{-2} = (14)\boxed{-2}$ Multiply both sides by -2.

Step 2 $x\left(-\dfrac{1}{2}\right)(-2) = 14(-2)$ $-\dfrac{1}{2}x = x\left(-\dfrac{1}{2}\right)$

$\underbrace{x\left(-\dfrac{1}{2}\right)(-2)}_{1} = -28$ $\left(-\dfrac{1}{2}\right)(-2) = \dfrac{2}{2} = 1$

$x = -28$

Check: $-\dfrac{(-28)}{2} = 14$

$\dfrac{28}{2} = 14$

$14 = 14$

(f) **Solve:** $\dfrac{2x}{3} = 6$

Step 1 $\left(\dfrac{2x}{3}\right)\boxed{3} = (6)\boxed{3}$ Multiply both sides by 3.

$2x = 6 \cdot 3$ $\left(\dfrac{2x}{3}\right)3 = \dfrac{2 \cdot x \cdot 3}{3} = 2 \cdot x$

$2x = 18$

Step 2 $\dfrac{2x}{\boxed{2}} = \dfrac{18}{\boxed{2}}$ Divide both sides by 2. $\dfrac{2x}{2} = x$

$x = 9$ Solution

Check: $\dfrac{2 \cdot (9)}{3} = 6$

$\dfrac{18}{3} = 6$

$6 = 6$

(g) **Solve:** $-4x = 22$

$\dfrac{-4x}{\boxed{-4}} = \dfrac{22}{\boxed{-4}}$ Divide both sides by -4.

$x = -5.5$ Solution

Check: $-4(-5.5) = 22$

$22 = 22$

(h) **Solve:** $\dfrac{x+3}{2} = 4$ Multiply by 2.

$x + 3 = 8$ $\left(\dfrac{x+3}{2}\right)2 = \dfrac{(x+3) \cdot 2}{2} = x + 3$

$x + 3 \boxed{- 3} = 8 \boxed{- 3}$ Subtract 3.

$x = 5$ Solution

Check: $\dfrac{(5) + 3}{2} = 4$

$\dfrac{8}{2} = 4$

(i) **Solve:** $2.4x = 0.972$ Divide by 2.4.

$x = 0.405$

 .972 ÷ 2.4 = → | 0.405 |

We have now covered the four basic single-operation equations. Before moving on, try the following set of problems covering these four operations. For each problem, think carefully whether you must add, subtract, multiply, or divide in order to make the equation read $x =$ _____ or _____ $= x$.

(a) $14 = x - 8$

(b) $\dfrac{y}{6} = 3$

(c) $n + 11 = 4$

(d) $3x = 33$

(e) $7 = -\dfrac{a}{5}$

(f) $z - 2 = -10$

(g) $-8 = -16x$

(h) $26 = y + 5$

(i) $\dfrac{5x}{4} = -30$

After you work these problems, check your answers in **26**.

---

26   Each solution states the operation you should have performed on both sides and then gives the final answer.

(a)   Add 8. $x = 22$.

(b)   Multiply by 6. $y = 18$.

(c)   Subtract 11. $n = -7$.

(d)   Divide by 3. $x = 11$.

(e)   Multiply by -5. $a = -35$.

(f)   Add 2. $z = -8$.

(g)   Divide by -16. $x = \dfrac{1}{2}$.

(h)   Subtract 5. $y = 21$.

(i)   Multiply by $\dfrac{4}{5}$ (or multiply by 4, divide by 5). $x = -24$.

Now turn to **27** for practice on solving simple equations.

27

Exercises 7-3   **Solving Simple Equations**

A.   Solve the following equations.

1. $x + 4 = 13$

2. $23 = A + 6$

3. $17 - x = 41$

4. $z - 18 = 29$

5. $6 = a - 2\dfrac{1}{2}$

6. $73 + x = 11$

7. $y - 16.01 = 8.65$

8. $11.6 - R = 3.7$

9. $-39 = 3x$

10. $-9y = 117$

11. $13a = 0.078$

12. $\dfrac{x}{3} = 7$

13. $\dfrac{z}{1.3} = 0.5$

14. $\dfrac{N}{2} = \dfrac{3}{8}$

15. $m + 18 = 6$

16. $-34 = x - 7$

17. $-6y = -39$

18. $\dfrac{a}{-4} = 9$

19. $22 - T = 40$

20. $79.2 = 2.2y$

21. $-5.9 = -6.6 + Q$

22. $\dfrac{5}{4} = \dfrac{Z}{8}$

23. $12 + x = 37$

24. $66 - y = 42$

25. $\dfrac{K}{0.5} = 8.48$

26. $-12z = 3.6$

B. **Practical Problems**

1. **Electrician**   Ohm's law is often written in the form

$$I = \frac{E}{R}$$

where I is the current in amperes, E is the voltage, and R is the resistance in ohms. What is the voltage necessary to push a 0.8-ampere current through a resistance of 480 ohms?

2. **Airline Pilot**   The formula

$$D = RT$$

is used to calculate the distance D traveled by an object moving at a constant average speed R during an elapsed time T. How long would it take a pilot to fly 1240 miles at an average speed of 220 mi/hr?

3. **Wastewater Treatment Operator**   A wastewater treatment operator uses the formula

$$A = 8.34FC$$

to determine the amount A of chlorine in pounds to add to a basin. The flow F through the basin is in millions of gallons per day, and the desired concentration C of chlorine is in parts per million.

If 1800 pounds of chlorine was added to a basin with a flow rate of 7.5 million gallons per day, what would be the resulting concentration?

4. **Sheet Metal Technology**   The water pressure P in pounds per square foot is related to the depth of water D in feet by the formula

$$P = 62.4D$$

where 62.4 is the weight of 1 cu ft of water. If the material forming the bottom of a tank is made to withstand 800 pounds per square foot of pressure, what is the maximum safe height for the tank?

5. **Music Technology** The frequency f (in waves per second) and the wavelength w (in meters) of a sound traveling in air are related by the equation

$$fw = 343$$

where 343 m/s is the speed of sound in air. Find the wavelength of a musical note with a frequency of 200 waves per second.

6. At a constant temperature, the pressure P (in psi) and the volume V (in cubic feet) of a particular gas are related by the equation

$$PV = 1080$$

If the volume is 60 cu ft, what is the pressure?

7. **Electrical Technology** For a particular transformer, the voltage E in the circuits is related to the number of windings W of wire around the core by the equation

$$E = 40W$$

How many windings will produce a voltage of 840 V?

8. **Sheet Metal Technology** The surface speed S in fpm (feet per minute) of a rotating cylindrical object is

$$S = \frac{\pi dn}{12}$$

where d is the diameter of the object in inches and n is the speed of rotation in rpm (revolutions per minute). If an 8-in. grinder must have a surface speed of 6000 fpm, what should the speed of rotation be? Use $\pi = 3.14$, and round to the nearest hundred rpm.

When you have completed these exercises check your answers on p. 598, then continue in **28**.

7-4 SOLVING EQUATIONS INVOLVING TWO OPERATIONS

28 Solving many simple algebraic equations involves both kinds of operations: addition/subtraction and multiplication/division. For example,

Solve: $2x + 6 = 14$

Step 1 We want to change this equation to an equivalent equation with only x or terms that include x on the left, so subtract 6 from both sides.

$$2x + 6 \boxed{-6} = 14 \boxed{-6}$$

$$2x + \underbrace{6 \boxed{-6}}_{0} = 14 \boxed{-6}$$

Combine terms. (Be careful. Some careless students will add $2x$ and 6 to get $8x$! You may add only *like* terms.)

$$2x = 8$$

Now change this to an equivalent equation with only x on the left.

Step 2 $\dfrac{2x}{2} = \dfrac{8}{2}$

Divide both sides of the equation by 2.

$x = \dfrac{8}{2}$ $\qquad$ $\dfrac{2x}{2} = x$

$x = 4$ $\quad$ Solution

Check: $\qquad$ $2 \cdot (4) + 6 = 14$

$8 + 6 = 14$

$14 = 14$

Try this one to test your understanding of the process.

Solve: $3x - 7 = 11$

Check your work in **29**.

To solve $\frac{1}{2}x + 4 = 8$ can I multiply by 2 and get $x + 4 = 16$?

No. When you multiply by 2 you must multiply **all** of the expression on the left by 2:

$2(\frac{1}{2}x + 4) = 2(8)$ or $x + 8 = 16$

Careful.

29 **Solve:** $3x - 7 = 11$

Step 1 $3x - 7 \;\boxed{+\,7} = 11 \;\boxed{+\,7}$ $\qquad$ Add 7 to each side.

$3x \underbrace{-\,7 \;\boxed{+\,7}}_{0} = 11 \;\boxed{+\,7}$

$3x = 18$

Step 2 $\dfrac{3x}{} = \dfrac{18}{3}$ $\qquad$ Divide both sides of the equation by 3.

$x = 6$ $\quad$ Solution

Check: $\quad 3 \cdot (6) - 7 = 11$

$18 - 7 = 11$

$11 = 11$

Here is another two-operation equation.

Solve: $23 = 9 - \dfrac{y}{3}$

Give it a try and then check your work in **30**.

30 **Solve:** $23 = 9 - \dfrac{y}{3}$

Step 1 Notice that the variable y is on the right side of the equation. We must first eliminate the 9 by subtracting 9 from both sides. Be sure to keep the negative sign in front of the $\dfrac{y}{3}$ term.

$$23 - 9 = 9 - \dfrac{y}{3} - 9$$

$$\boxed{9 - 9 = 0}$$

$$14 = -\dfrac{y}{3}$$

Step 2 $\quad (14)\,(-3) = \left(-\dfrac{y}{3}\right)(-3)$ \qquad Multiply both sides by -3.

$$-42 = (-3)\left(-\dfrac{1}{3}\right)y \qquad\qquad -\dfrac{y}{3} = -\dfrac{1}{3}y$$

$$-42 = 1y \qquad\qquad\qquad (-3)\left(-\dfrac{1}{3}\right) = 1$$

$$\text{or} \quad y = -42$$

Check: $\quad 23 = 9 - \dfrac{(-42)}{3}$

$$23 = 9 - (-14)$$

$$23 = 23$$

Note ▶ In these two-step problems, we did the addition or subtraction in Step 1 and the multiplication or division in Step 2. We could have reversed this order and arrived at the correct solution, but the problem might have become more complicated to solve. Always add or subtract first, and when you multiply or divide, do so to *all* terms. ◀

If more than one variable term appears on the same side of an equation, combine these like terms before performing any operation to both sides. For example,

Solve: $\quad 2x + 5 + 4x = 17$

Step 1 $\quad (2x + 4x) + 5 = 17$ \qquad Combine the x-terms on the left side.

$$6x + 5 = 17$$

Step 2 $\quad 6x + 5 \boxed{-5} = 17 \boxed{-5}$ \qquad Subtract 5 from each side.

$$6x = 12$$

Step 3 $\quad \dfrac{6x}{6} = \dfrac{12}{6}$ \qquad\qquad Divide each side by 6.

$$x = 2$$

Check: $\quad 2(2) + 5 + 4(2) = 17$ \qquad Substitute 2 for x in each x-term.

$$4 + 5 + 8 = 17$$

$$17 = 17$$

Now you try one.

Solve: $\quad 32 = x - 12 - 5x$

Look for our solution in **31**.

31 **Step 1** $32 = (x - 5x) - 12$ Combine the x-terms on the right.

$32 = -4x - 12$

Step 2 $32 \boxed{+ 12} = -4x - 12 \boxed{+ 12}$ Add **12** to both sides.

$44 = -4x$

Step 3 $\dfrac{44}{-4} = \dfrac{-4x}{-4}$ Divide both sides by **−4**.

$-11 = x$ or $x = -11$

Check: $32 = (-11) - 12 - 5(-11)$

$32 = -11 - 12 + 55$

$32 = -23 + 55$

$32 = 32$

Here are a few more two-operation equations for practice.

(a) $7x + 2 = 51$ (b) $18 - 5x = 3$ (c) $15.3 = 4x - 1.5$

(d) $\dfrac{x}{5} - 4 = 6$ (e) $11 - x = 2$ (f) $5 = 7 - \dfrac{x}{4}$

(g) $2 + \dfrac{x}{2} = 3$ (h) $2x - 9.4 = 0$ (i) $2.75 = 14.25 - 0.20x$

(j) $3x + x = 18$ (k) $12 = 9x + 4 - 5x$ (l) $6 - 3x + x = 22$

Our complete step-by-step solutions are in **32**.

---

32 (a) **Solve:** $7x + 2 = 51$ Change this equation to an equivalent equation with only an x-term on the left.

Step 1 $7x + 2 \boxed{- 2} = 51 \boxed{- 2}$ Subtract **2** from each side.

$7x + \underbrace{2 \boxed{- 2}}_{0} = 51 \boxed{- 2}$ Combine terms.

$7x = 49$

Step 2 $\dfrac{7x}{7} = \dfrac{49}{7}$ Divide both sides by **7**.

$x = \dfrac{49}{7}$

$x = 7$ Solution

Check: $7 \cdot (7) + 2 = 51$

$49 + 2 = 51$

$51 = 51$

(b) **Solve:** $18 - 5x = 3$

Step 1 $18 - 5x \boxed{- 18} = 3 \boxed{- 18}$ Subtract **18** from each side.

$-5x + \underbrace{18 - 18}_{0} = 3 - 18$ Rearrange terms.

$-5x = -15$

Step 2 $\dfrac{-5x}{-5} = \dfrac{-15}{-5}$ Divide both sides by -5.

$x = 3$ Solution

Check $18 - 5 \cdot (3) = 3$
$18 - 15 = 3$
$3 = 3$

(c) **Solve:** $15.3 = 4x - 1.5$

Step 1 $15.3 \boxed{+\ 1.5} = 4x - \underbrace{1.5 \boxed{+\ 1.5}}_{0}$ Add 1.5 to each side.

$16.8 = 4x$ Combine terms.

$4x = 16.8$

Step 2 $\dfrac{4x}{4} = \dfrac{16.8}{4}$ Divide both sides by 4.

$x = 4.2$ Solution   Decimal number solutions are common in practical problems.

 15.3 $\boxed{+}$ **1.5** $\boxed{=}$ $\boxed{\div}$ **4** $\boxed{=}$ $\rightarrow$ $\boxed{4.2}$

Check: $15.3 = 4(4.2) - 1.5$
$15.3 = 16.8 - 1.5$
$15.3 = 15.3$

(d) **Solve:** $\dfrac{x}{5} - 4 = 6$

Step 1 $\dfrac{x}{5} - 4 \boxed{+\ 4} = 6 \boxed{+\ 4}$ Add 4 to both sides.

$\dfrac{x}{5} = 10$

Step 2 $\left(\dfrac{x}{5}\right) \cdot \boxed{5} = 10 \cdot \boxed{(5)}$ Multiply both sides by 5.

$x = 50$

Check: $\dfrac{(50)}{5} - 4 = 6$

$10 - 4 = 6$

$6 = 6$

(e) **Solve:** $11 - x = 2$

$11 - x \boxed{-\ 11} = 2 \boxed{-\ 11}$ Subtract 11 from each side.

$\underbrace{-x + 11 \boxed{-\ 11}}_{0} = 2 \boxed{-\ 11}$ Combine terms.

$-x = -9$ Multiply each side by -1.

$x = 9$ Solution

Check: $11 - (9) = 2$
$2 = 2$

(f) **Solve:** $5 = 7 - \frac{x}{4}$

Step 1 $5 \boxed{- 7} = 7 - \frac{x}{4} \boxed{- 7}$ Subtract **7** from both sides.

$$-2 = -\frac{x}{4} \qquad\qquad 7 - \frac{x}{4} - 7 = (7 - 7) - \frac{x}{4} = -\frac{x}{4}$$

Step 2 $-2 \boxed{(-4)} = \left(-\frac{x}{4}\right)(-4)$ Multiply both sides by **−4**.

$$8 = x \qquad\qquad \left(-\frac{x}{4}\right)(-4) = (-x)\left(\frac{-4}{4}\right) = (-x)(-1)$$

$$\text{or } x = 8 \qquad\qquad\qquad = x$$

Check: $5 = 7 - \frac{(8)}{4}$

$$5 = 7 - 2$$
$$5 = 5$$

(g) **Solve:** $2 + \frac{x}{2} = 3$

$$\frac{x}{2} + 2 \boxed{- 2} = 3 \boxed{- 2} \qquad\qquad\qquad \text{Subtract } \mathbf{2} \text{ from each side.}$$

$$\frac{x}{2} = 1$$

$$\left(\frac{x}{2}\right) \cdot \boxed{2} = 1 \cdot \boxed{2} \qquad\qquad\qquad \text{Multiple each side by } \mathbf{2}.$$

$$x = 2 \quad \text{Solution}$$

Check: $2 + \frac{(2)}{2} = 3$

$$2 + 1 = 3$$

(h) **Solve:** $2x - 9.4 = 0$

$$2x - 9.4 \boxed{+ 9.4} = 0 \boxed{+ 9.4} \qquad\qquad \text{Add } \mathbf{9.4} \text{ to each side.}$$

$$2x = 9.4 \qquad\qquad\qquad \text{Divide each side by } \mathbf{2}.$$

$$x = 4.7 \quad \text{Solution}$$

Check: $2(4.7) - 9.4 = 0$
$$9.4 - 9.4 = 0$$

(i) **Solve:** $2.75 = 14.25 - 0.20x$

Step 1 $2.75 - \mathbf{14.25} = 14.25 - 0.20x - \mathbf{14.25}$ Subtract **14.25** from both sides.

$$-11.5 = -0.20x$$

Step 2 $\dfrac{-11.5}{-0.20} = \dfrac{-0.20x}{-0.20}$ Divide both sides by **−0.20**.

$$57.5 = x \qquad \text{or} \qquad x = 57.5$$

 2.75 ⊟ **14.25** ⊜ ⨸ **.2** ⊞⁄− ⊜ → $\boxed{57.5}$

Check: $2.75 = 14.25 - 0.20(57.5)$
$$2.75 = 14.25 - 11.5$$
$$2.75 = 2.75$$

(j) **Solve:** $3x + x = 18$

Step 1 $\qquad 4x = 18$ Combine like terms on the left side.

Step 2 $\qquad \dfrac{4x}{4} = \dfrac{18}{4}$ Divide each side by 4.

$\qquad\qquad x = 4.5$

Check: $3(4.5) + (4.5) = 18$
$\qquad\qquad 13.5 + 4.5 = 18$
$\qquad\qquad\qquad\quad 18 = 18$

(k) **Solve:** $\qquad 12 = 9x + 4 - 5x$

Step 1 $\qquad 12 = 4x + 4$ Combine like terms.

Step 2 $\quad 12 - 4 = 4x + 4 - 4$ Subtract 4 from each side.

$\qquad\qquad 8 = 4x$

Step 3 $\qquad \dfrac{8}{4} = \dfrac{4x}{4}$ Divide each side by 4.

$\qquad\qquad 2 = x \quad$ or $\quad x = 2$

The check is left to you.

(l) **Solve:** $\qquad 6 - 3x + x = 22$

Step 1 $\qquad 6 - 2x = 22$ Combine like terms.

Careful: The $3x$-term is negative: $\quad -3x + x = -2x$

Step 2 $\quad 6 - 2x - 6 = 22 - 6$ Subtract 6 from both sides.

$\qquad\qquad -2x = 16$ (Keep the negative sign with the $2x$-term.)

Step 3 $\qquad \dfrac{-2x}{-2} = \dfrac{16}{-2}$ Divide both sides by -2.

$\qquad\qquad x = -8$

Be sure to check your solution.

Now turn to **33** for more practice on solving equations.

33

Exercises 7-4 **Solving Two-Step Equations**

A. Solve.

1. $2x - 3 = 17$
2. $4x + 6 = 2$

3. $\dfrac{x}{5} = 7$
4. $-8y + 12 = 32$

5. $4.4m - 1.2 = 9.8$
6. $\dfrac{3}{4} = 3x + \dfrac{1}{4}$

7. $14 - 7n = -56$
8. $38 = 58 - 4a$

9. $3z - 5z = 12$
10. $17 = 7q - 5q - 3$

11. $23 - \dfrac{x}{4} = 11$
12. $9m + 6 + 3m = -60$

13. $-15 = 12 - 2n + 5n$
14. $2.6y - 19 - 1.8y = 1$

15. $\frac{1}{2} + 2x = 1$ 16. $3x + 16 = 46$

17. $-4a + 45 = 17$ 18. $\frac{x}{2} + 1 = 8$

19. $-3Z + \frac{1}{2} = 17$ 20. $2x + 6 = 0$

21. $1 = 3 - 5x$ 22. $23 = 17 - \frac{x}{4}$

23. $-5P + 18 = 3$ 24. $5x - 2x = 24$

25. $6x + 2x = 80$ 26. $x + 12 - 6x = -18$

27. $27 = 2x - 5 + 4x$ 28. $-13 = 22 - 3x + 8x$

B. Practical Problems

1. **Office Services** A repair service charges $32 for a house call and an additional $24 per hour of work. The formula

   $T = 32 + 24H$

   represents the total charge T for H hours of work. If the total bill for a customer was $140, how many hours of actual labor were there?

2. **Meteorology** The air temperature T (in degrees Fahrenheit) at an altitude h (in feet) above a particular area can be approximated by the formula

   $T = -0.002h + G$

   where G is the temperature on the ground directly below. If the ground temperature is 76°F, at what altitude will the air temperature drop to freezing, 32°F?

3. **Sports Technology** Physical fitness experts sometimes use the following formula to approximate the maximum target heart rate R during exercise based on a person's age A:

   $R = -0.8A + 176$

   At what age should the heart rate during exercise not exceed 150 beats per minute?

4. The formula $A = p + prt$ is used to determine the total amount of money A in a bank account after an amount p is invested for t years at a rate of interest r. What rate of interest is needed for $8000 to grow to $12,000 after 5 years? (Be sure to convert your decimal answer to a percent and round to the nearest tenth of a percent.)

5. **Auto Mechanics** The formula

   $P + 2T = C$

   gives the overall diameter C of the crankshaft gear for a known pitch diameter P of the small gear and the height T of teeth above the pitch diameter circle. If $P = 2.875$ in. and $C = 3.125$ in., find T.

6. **Sheet Metal Technology** The allowance A for a Pittsburgh lock is given by

   $A = 2w + \frac{3}{16}$ in.

   where w is the width of the pocket. If the allowance for a Pittsburgh lock is $\frac{11}{16}$ in., what is the width of the pocket?

7. **Plumber**   A plumber's total bill A can be calculated using the formula

$$A = RT + M$$

where R is his hourly rate, T is the total labor time in hours, and M is the cost of materials. A plumber bids a particular job at $1850. If materials amount to $580, and his hourly rate is $40 per hour, how many hours should the job take for the estimate to be accurate?

8. **Machinist**   The formula

$$L = 2d + 3.26(r + R)$$

can be used to approximate the length L of belt needed to connect two pulleys of radii r and R if their centers are a distance d apart. How far apart can two pulleys be if their radii are 8 in. and 6 in., and the total length of the belt connecting them is 82 in.? (Round to the nearest inch.)

When you have completed these exercises, check your answers on page 598, then continue in **34**.

7-5   SOLVING MORE EQUATIONS AND FORMULAS

Parentheses in Equations

34   In Section 7-2 you learned how to deal with algebraic expressions involving parentheses. Now you will learn to solve equations containing parentheses by using these same skills. For example,

Solve: $2(x + 4) = 27$

Step 1   Use rule 3 on page 305. Multiply each term inside the parentheses by 2.

$$2x + 8 = 27$$

Step 2   Now solve this equation using the techniques of the previous section.

$$2x + 8 - 8 = 27 - 8 \qquad \text{Subtract } 8 \text{ from both sides.}$$
$$2x = 19$$
$$\frac{2x}{2} = \frac{19}{2} \qquad \text{Divide both sides by } 2.$$
$$x = 9.5$$

Check: $2[(9.5) + 4] = 27$
$$2(13.5) = 27$$
$$27 = 27$$

Try this similar example.

$$-3(y - 4) = 36$$

Look for our solution in **35**.

To solve this equation: $2x - 3 = 10$
Why can't I divide both sides by 2
to get $x - 3 = 5$ so that
$x = 8$. Isn't this ok?

No. When you divide both sides of an equation by a number, you must divide **all** of both sides. Divide **all** of the expresion on the left, not just part of it. $\frac{2x-3}{2} = \frac{10}{2}$ or $x - \frac{3}{2} = 5$.

35 **Solve:** $-3(y - 4) = 36$

Step 1 $-3y + 12 = 36$

Multiply each term inside parentheses by -3.
$-3(y - 4) = -3(y) + (-3)(-4)$

Step 2 $-3y + 12 - 12 = 36 - 12$

$-3y = 24$

Subtract 12 from each side.

Step 3 $\frac{-3y}{-3} = \frac{24}{-3}$

$y = -8$

Divide each side by -3.

Check: $-3((-8) - 4) = 36$

$-3(-12) = 36$

$36 = 36$

Note ▶ In each of the last two examples, you could have first divided both sides by the number in front of parentheses. Here is how each solution would have looked.

First Example

$\frac{2(x + 4)}{2} = \frac{27}{2}$

$x + 4 = 13.5$

$x + 4 - 4 = 13.5 - 4$

$x = 9.5$

Second Example

$\frac{-3(y - 4)}{-3} = \frac{36}{-3}$

$y - 4 = -12$

$y - 4 + 4 = -12 + 4$

$y = -8$

Some students may find this technique preferable, especially when the right side of the equation is exactly divisible by the number in front of the parentheses. ◀

Here is a more difficult equation involving parentheses.

Solve: $5x - (2x - 3) = 27$

Step 1 Use rule 2 on page 304. Change the sign of each term inside parentheses and then remove them.

$5x - 2x + 3 = 27$

$-(2x - 3) = -2x + 3$

Step 2 $3x + 3 = 27$

Combine the like terms on the left.
$5x - 2x = 3x$

Step 3 $3x + 3 \boxed{- 3} = 27 \boxed{- 3}$ Subtract 3 from both sides.

$$3x = 24$$

Step 4 $\dfrac{3x}{\boxed{3}} = \dfrac{24}{\boxed{3}}$ Divide both sides by 3.

$$x = 8$$

Check: $5(8) - [2(8) - 3] = 27$

$$40 - (16 - 3) = 27$$

$$40 - 13 = 27$$

$$27 = 27$$

Try one more.

Solve: $8 - 3(2 - 3x) = 34$

Step 1   Be very careful here. Some students are tempted to subtract the 3 from the 8. However, the order of operations rules on page 275 specify that multiplication must be performed before addition or subtraction. Therefore, your first step is to multiply the expression in parentheses by –3.

$8 - 6 + 9x = 34$ $8 \boxed{- 3}(2 - 3x) = 8 + \boxed{(-3)}(2) + \boxed{(-3)}(-3x)$

Step 2 $2 + 9x = 34$ Combine the like terms on the left.

Step 3 $2 + 9x \boxed{- 2} = 34 \boxed{- 2}$ Subtract 2 from each side.

$$9x = 32$$

Step 4 $\dfrac{9x}{9} = \dfrac{32}{9}$

$$x = 3\frac{5}{9}$$

Be sure to check your answer.

Here are more equations with parentheses for you to solve.

(a) $4(x - 2) = 26$ (b) $11 = -2(y + 5)$

(c) $23 = 6 - (3n + 4)$ (d) $4x - (6x - 9) = 41$

(e) $7 + 3(5x + 2) = 38$ (f) $20 = 2 - 5(9 - a)$

(g) $(3m - 2) - (5m - 3) = 19$ (h) $2(4x + 1) + 5(3x + 2) = 58$

The answers and some worked solutions are provided in **36**.

---

36   (a) $x = 8.5$ (b) $y = -10.5$ (c) $n = -7$ (d) $x = -16$

(e) $x = 1\frac{2}{3}$ (f) $a = 12.6$ (g) $m = -9$ (h) $x = 2$

Here are worked solutions to (b), (d), (f), and (h). The checks are left to you.

(b)   **Solve:** $11 = -2(y + 5)$

     Step 1 $11 = -2y - 10$ Multiply each term inside parentheses by –2.

Step 2 $11 \boxed{+ 10} = -2y - 10 \boxed{+ 10}$ Add 10 to each side.

$$21 = -2y$$

Step 3 $\dfrac{21}{-2} = \dfrac{-2y}{-2}$ Divide each side by –2.

$$-10.5 = y \quad \text{or} \quad y = -10.5$$

(d) **Solve:** $4x - (6x - 9) = 41$

Step 1 $\quad 4x - 6x + 9 = 41$ Change the sign of each term inside parentheses and then remove parentheses.

Step 2 $\quad\quad -2x + 9 = 41$ Combine like terms on the left side.

Step 3 $\quad -2x + 9 \boxed{- 9} = 41 \boxed{- 9}$ Subtract 9 from each side.

$$-2x = 32$$

Step 4 $\quad\quad \dfrac{-2x}{-2} = \dfrac{32}{-2}$ Divide each side by –2.

$$x = -16$$

(f) **Solve:** $\quad 20 = 2 - 5(9 - a)$

Step 1 Multiply both terms inside parentheses by –5 and then remove parentheses.

$$20 = 2 - 45 + 5a$$ $2 - 5(9 - a)$
$= 2 + (-5)(9) + (-5)(-a)$

Step 2 $\quad\quad 20 = -43 + 5a$ Combine like terms on the right side.

Step 3 $20 \boxed{+ 43} = -43 + 5a \boxed{+ 43}$ Add 43 to both sides.

$$63 = 5a$$

Step 4 $\quad\quad \dfrac{63}{5} = \dfrac{5a}{5}$ Divide both sides by 5.

$$12.6 = a \quad \text{or} \quad a = 12.6$$

(h) **Solve:** $\quad 2(4x + 1) + 5(3x + 2) = 58$

Step 1 Multiply each term inside the first parentheses by 2 and each term inside the second parentheses by 5.

$$8x + 2 + 15x + 10 = 58$$

Step 2 $\quad\quad 23x + 12 = 58$ Combine both pairs of like terms on the left side.

Step 3 $\quad 23x + 12 \boxed{- 12} = 58 \boxed{- 12}$ Subtract 12 from both sides.

$$23x = 46$$

Step 4 $\quad\quad \dfrac{23x}{23} = \dfrac{46}{23}$ Divide both sides by 23.

$$x = 2$$

Variable on Both Sides

In all the equations we have solved so far, the variable has been on only one side of the equation. Sometimes it is necessary to solve equations with variable terms on both sides.

For example, to solve

$$3x - 4 = 8 - x$$

First, move all variable terms to the left side by adding x to both sides.

$$3x - 4 + x = 8 - x + x$$

$$4x - 4 = 8 \qquad\qquad 3x + x = 4x \qquad -x + x = 0$$

Next, proceed as before.

$$4x - 4 + 4 = 8 + 4 \qquad\qquad \text{Add } 4 \text{ to both sides.}$$

$$4x = 12$$

$$\frac{4x}{4} = \frac{12}{4} \qquad\qquad \text{Divide both sides by } 4.$$

$$x = 3$$

Finally, check your answer.

$$3(3) - 4 = 8 - (3)$$

$$9 - 4 = 5$$

$$5 = 5$$

Ready to attempt one yourself? Give this one a try, then check it in **37**.

$$5y - 21 = 8y$$

---

37 Did you move the variable terms to the left? Here you can save a step by moving them to the right instead of to the left.

Solve: $\qquad 5y - 21 = 8y$

Step 1 $\quad 5y - 21 - 5y = 8y - 5y \qquad$ Subtract $5y$ from both sides.

$$-21 = 3y$$

Step 2 $\qquad\qquad \dfrac{-21}{3} = \dfrac{3y}{3} \qquad\qquad$ Divide both sides by 3.

$$-7 = y \qquad \text{or} \qquad y = -7$$

Check: $\qquad 5(-7) - 21 = 8(-7)$

$$-35 - 21 = -56$$

$$-56 = -56$$

Learning Hint ▶ If a variable-term is already by itself on one side of the equation, move all variable terms to this side. As in the last example, this will save a step. ◀

Now try these problems for practice in solving equations in which the variable appears on both sides.

(a) $\quad x - 6 = 3x$ $\qquad\qquad$ (b) $\quad 5(x - 2) = x + 4$

(c) $\quad 2(x - 1) = 3(x + 1)$ $\qquad$ (d) $\quad 4x + 9 = 7x - 15$

(e) $\quad 4n + 3 = 18 - 2n$ $\qquad\quad$ (f) $\quad 2A = 12 - A$

(g) $\quad 6y = 4(2y + 7)$ $\qquad\qquad$ (h) $\quad 8m - (2m - 3) = 3(m - 4)$

(i) $\quad 3 - (5x - 8) = 6x + 22$ $\quad$ (j) $\quad -2(3x - 5) = 7 - 5(2x + 3)$

Check your solutions in **38**.

---

38 (a) **Solve:** $\qquad x - 6 = 3x$

$$x - 6 \; -x = 3x \; -x$$

Subtract x from each side, so that x-terms will appear only on the right.

$$-6 = 2x$$

Combine terms.

or $\qquad\qquad 2x = -6$

Divide by 2.

$$x = -3 \quad \text{Solution}$$

Check $\quad (-3) - 6 = 3(-3)$

$$-9 = -9$$

(b) **Solve:** $\qquad 5(x - 2) = x + 4$

$$5x - 10 = x + 4$$

Multiply each term inside the parentheses by 5.

$$5x - 10 \; -x = x + 4 \; -x$$

Subtract x from each side.

$$4x - 10 = 4$$

$$4x - 10 \; +10 = 4 \; +10$$

Add 10 to each side.

$$4x = 14$$

Divide each side by 4.

$$x = 3\tfrac{1}{2} \quad \text{Solution}$$

Check: $\quad 5(3\tfrac{1}{2} - 2) = (3\tfrac{1}{2}) + 4$

$$5(1\tfrac{1}{2}) = 7\tfrac{1}{2}$$

$$7\tfrac{1}{2} = 7\tfrac{1}{2}$$

(c) **Solve:** $\quad 2(x - 1) = 3(x + 1)$

Remove parentheses by multiplying.

$$2x - 2 = 3x + 3$$

$$2x - 2 \; -3x = 3x + 3 \; -3x$$

Subtract $3x$ from each side.

$$-2 - x = 3$$

Combine terms.

$$-2 - x \; +2 = 3 \; +2$$

Add 2 to each side.

$$-x = 5$$

or $\qquad\qquad x = -5 \quad \text{Solution}$

Solve for x, not $-x$.

Do you see that $-x = 5$ is the same as $x = -5$? Multiply both sides of the equation $-x = 5$ by -1 to get $x = -5$.

Check: $\quad 2(-5 - 1) = 3(-5 + 1)$

$$2(-6) = 3(-4)$$

$$-12 = -12$$

(d) **Solve:** $\quad 4x + 9 = 7x - 15$

$$4x + 9 \; -7x = 7x - 15 \; -7x$$

Subtract $7x$ from each side.

$$-3x + 9 = -15$$

Combine terms.

$$-3x + 9 \; -9 = -15 \; -9$$

Subtract 9 from each side.

$$-3x = -24$$

Divide each side by -3.

$$x = 8 \quad \text{Solution}$$

Check: $\quad 4(8) + 9 = 7(8) - 15$

$$32 + 9 = 56 - 15$$

$$41 = 41$$

(e) **Solve:** $4n + 3 = 18 - 2n$

$$4n + 3 + 2n = 18 - 2n + 2n$$ Add $2n$ to both sides.

$$6n + 3 = 18$$ Now the n-terms appear only on the left side.

$$6n + 3 - 3 = 18 - 3$$ Subtract 3 from both sides.

$$6n = 15$$

$$\frac{6n}{6} = \frac{15}{6}$$ Divide by 6.

$$n = 2.5$$

Check: $4(2.5) + 3 = 18 - 2(2.5)$
$$10 + 3 = 18 - 5$$
$$13 = 13$$

(f) **Solve:** $2A = 12 - A$

$$2A + A = 12 - A + A$$ Add A to each side.

$$3A = 12$$ The A-terms now appear only on the left.

$$\frac{3A}{3} = \frac{12}{3}$$ Divide by 3.

$$A = 4$$

Check: $2(4) = 12 - (4)$
$$8 = 8$$

(g) **Solve:** $6y = 4(2y + 7)$

$$6y = 8y + 28$$ Multiply to remove parentheses.

$$6y - 8y = 8y + 28 - 8y$$ Subtract $8y$ from both sides.

$$-2y = 28$$

$$\frac{-2y}{-2} = \frac{28}{-2}$$ Divide by -2.

$$y = -14$$

Check the solution.

(h) **Solve:** $8m - (2m - 3) = 3(m - 4)$

$$8m - 2m + 3 = 3(m - 4)$$ Change signs to remove parentheses on the left.

$$8m - 2m + 3 = 3m - 12$$ Multiply by 3 to remove parentheses on the right.

$$6m + 3 = 3m - 12$$ Combine like terms.

$$6m + 3 - 3m = 3m - 12 - 3m$$ Subtract $3m$ from both sides.

$$3m + 3 = -12$$

$$3m + 3 - 3 = -12 - 3$$ Subtract 3 from both sides.

$$3m = -15$$

$$\frac{3m}{3} = \frac{-15}{3}$$ Divide by 3.

$$m = -5$$

Check the solution.

(i) **Solve:** $3 - (5x - 8) = 6x + 22$

Remove parentheses by changing the signs of both terms.

$$3 - 5x + 8 = 6x + 22$$

$$11 - 5x = 6x + 22$$

Combine terms.

$$11 - 5x - 6x = 6x + 22 - 6x$$

Subtract $6x$ from both sides.

$$11 - 11x = 22$$

Combine terms.

$$11 - 11x - 11 = 22 - 11$$

Subtract 11 from both sides.

$$-11x = 11$$

Divide by -11.

$$x = -1 \quad \text{Solution}$$

Check it.

(j) **Solve:** $-2(3x - 5) = 7 - 5(2x + 3)$

Remove parentheses. Multiply each term inside parentheses on the left by -2. Multiply each term inside parentheses on the right by -5 (leave the 7 alone!).

$$-6x + 10 = 7 - 10x - 15$$

$$-6x + 10 = -8 - 10x$$

Combine terms.

$$-6x + 10 + 10x = -8 - 10x + 10x$$

Add $10x$ to both sides.

$$4x + 10 = -8$$

Combine terms.

$$4x + 10 - 10 = -8 - 10$$

Subtract 10 from both sides.

$$4x = -18$$

Divide by 4.

$$x = -4.5 \quad \text{Solution}$$

Check it.

Remember:

1. Do only legal operations: add or subtract the same quantity from both sides of the equation; multiply or divide both sides of the equation by the same nonzero quantity.

2. Remove all parentheses carefully.

3. Combine like terms when they are on the same side of the equation.

4. Use legal operations to change the equation so that you have only x by itself on one side of the equation and a number on the other side of the equation.

5. Always check your answer.

Turn to **39** to learn to solve formulas.

Solving Formulas

39 To *solve a formula* for some letter means to rewrite the formula as an equivalent formula with that letter isolated on the left of the equals sign.

For example, the area of a triangle is given by the formula

$$A = \frac{BH}{2}$$ where A is the area, B is the length of the base, and H is the height

Solving for the base B gives the equivalent formula

$$B = \frac{2A}{H}$$

Solving for the height H gives the equivalent formula

$$H = \frac{2A}{B}$$

Solving formulas is a very important practical application of algebra. Very often a formula is not written in the form that is most useful. To use it you may need to rewrite the formula, solving it for the letter whose value you need to calculate.

To solve a formula, use the same balancing operations that you used to solve equations. You may add or subtract the same quantity on both sides of the formula and you may multiply or divide both sides of the formula by the same nonzero quantity.

For example, to solve the formula

$$S = \frac{R + P}{2} \qquad \text{for } R$$

First, multiply both sides of the equation by 2.

$$2 \cdot S = 2 \cdot \left(\frac{R + P}{2}\right)$$

$$2S = R + P$$

Second, subtract P from both sides of the equation.

$$2S - P = R + \underbrace{P - P}_{0}$$

$$2S - P = R$$

This formula can be reversed to read

$$R = 2S - P \qquad \text{We have solved the formula for } R.$$

Solve the following formulas for the variable indicated.

(a) $V = \dfrac{3K}{T}$ for K
(b) $Q = 1 - R + T$ for R

(c) $V = \pi R^2 H - AB$ for H
(d) $P = \dfrac{T}{A - B}$ for A

Check your solutions in **40**.

---

40 (a) $V = \dfrac{3K}{T}$

First, multiply both sides by T to get $\qquad VT = 3K$

Second, divide both sides by 3 to get $\qquad \dfrac{VT}{3} = K$

Solved for K, the formula is $\qquad K = \dfrac{VT}{3}$

(b) $Q = 1 - R + T$

    First, subtract T from both sides to get $Q - T = 1 - R$

    Second, subtract 1 from both sides to get $Q - T - 1 = -R$

    This is equivalent to $-R = Q - T - 1$

    or $R = -Q + T + 1$

> We have multiplied *all terms by −1.*

    or $R = 1 - Q + T$

(c) $V = \pi R^2 H - AB$

    First, add AB to both sides to get $V + AB = \pi R^2 H$

    Second, divide both sides by πR^2 to get $\dfrac{V + AB}{\pi R^2} = H$

    Notice that we divide *all* of the left side by πR^2

    Solved for H, the formula is $H = \dfrac{V + AB}{\pi R^2}$

(d) $P = \dfrac{T}{A - B}$

    First, multiply each side by $A - B$ to get $P(A - B) = T$

    Second, multiply to remove the parentheses, $PA - PB = T$

    Next, add PB to each side $PA = T + PB$

    Finally, divide each side by P $A = \dfrac{T + PB}{P}$

    Solved for A, the formula is $A = \dfrac{T + PB}{P}$

Careful ▶ Remember, when using the multiplication/division rule, you must multiply or divide *all* of both sides of the formula by the same quantity. ◀

Practice solving formulas with the following problems.

Solve:

(a) $P = 2A + 3B$ for A (b) $E = MC^2$ for M

(c) $S = \dfrac{A - RT}{1 - R}$ for A (d) $S = \dfrac{1}{2} gt^2$ for g

(e) $P = i^2 R$ for R (f) $I = \dfrac{V}{R + a}$ for R

(g) $A = \dfrac{2V - W}{R}$ for V (h) $F = \dfrac{9C}{5} + 32$ for C

(i) $A = \dfrac{\pi R^2 S}{360}$ for S (j) $P = \dfrac{t^2 dN}{3.78}$ for d

(k) $C = \dfrac{AD}{A + 12}$ for D (l) $V = \dfrac{\pi L T^2}{6} + 2$ for L

Check your answers in **41**.

41 (a) $A = \dfrac{P - 3B}{2}$

(b) $M = \dfrac{E}{C^2}$

(c) $A = S - SR + RT$

(d) $g = \dfrac{2S}{t^2}$

(e) $R = \dfrac{P}{i^2}$

(f) $R = \dfrac{V - aI}{I}$

(g) $V = \dfrac{AR + W}{2}$

(h) $C = \dfrac{5F - 160}{9}$

(i) $S = \dfrac{360A}{\pi R^2}$

(j) $d = \dfrac{3.78P}{t^2 N}$

(k) $D = \dfrac{CA + 12C}{A}$

(l) $L = \dfrac{6V - 12}{\pi T^2}$

USING SQUARE ROOTS IN SOLVING EQUATIONS

The equations you learned to solve in this chapter are all *linear* equations. The variable appears only to the first power—no x^2 or x^3 terms appear in the equations. You will learn how to solve more difficult algebraic equations in Chapter 11, but equations that look like

$x^2 = a$ where a is some positive number

can be solved easily.

To solve such an equation, simply take the square root of each side of the equation. The solution can be either

$x = \sqrt{a}$ or $x = -\sqrt{a}$

Example: Solve $x^2 = 36$.

Taking square roots, we have $x = +6$ or $x = -6$.
There are two possible solutions, one negative and one positive. Be careful, one of them, usually the negative one, may not be a reasonable answer to a practical problem.

Example: If the cross-sectional area of a square heating duct is 75 sq in., what must be the width of the duct?

Solve $x^2 = 75$.

Taking the square root of each side, we obtain the positive solution

$x = \sqrt{75}$

 $x \approx 8.7$ in.   rounded

75 $\boxed{\sqrt{}}$ $\rightarrow$ **8.6602540**

If you need to review the concept of square roots, return to Section 6-4 on page 276.

We will look at more problems like this in Chapter 11.

Now turn to **42** for a set of practice problems on solving equations and formulas.

Exercises 7-5 Solving More Equations and Formulas

A. Solve the following equations.

1. $5(x - 3) = 30$

2. $22 = -2(y + 6)$

3. $3(2n + 4) = 41$

4. $-6(3a - 7) = 21$

5. $2 - (x - 5) = 14$

6. $24 = 5 - (3 - 2m)$

7. $6 + 2(y - 4) = 13$

8. $7 - 11(2z + 3) = 18$

9. $8 = 5 - 3(3x - 4)$

10. $7 + 9(2w + 3) = 25$

11. $6c - (c - 4) = 29$

12. $9 = 4y - (y - 2)$

13. $5x - 3(2x - 8) = 31$

14. $6a + 2(a + 7) = 8$

15. $9t - 3 = 4t - 2$

16. $7y + 5 = 3y + 11$

17. $12x = 4x - 16$

18. $22n = 16n - 18$

19. $8y - 25 = 13 - 11y$

20. $6 - 2p = 14 - 4p$

21. $9x = 30 - 6x$

22. $12 - y = y$

23. $2(3t - 4) = 10t + 7$

24. $5A = 4(2 - A)$

25. $2 - (3x - 20) = 4(x - 2)$

26. $2(2x - 5) = 6x - (5 - x)$

27. $8 + 3(6 - 5x) = 11 - 10x$

28. $2(x - 5) - 3(2x - 8) = 16 - 6(4x - 3)$

B. Solve the following formulas for the variable shown.

1. $S = LW$ for L

2. $A = \frac{1}{2}BH$ for B

3. $V = IR$ for I

4. $H = \frac{D - R}{2}$ for D

5. $S = \frac{W}{2}(A + T)$ for T

6. $V = \pi R^2 H$ for H

7. $P = 2A + 2B$ for B

8. $H = \frac{R}{2} + 0.05$ for R

9. $T = \frac{RP}{R + 2}$ for P

10. $I = \frac{E + V}{R}$ for V

C. Practical Problems

Problem 1

1. **Sheet Metal Technology**   The length of arc of a sector of a circle is given by the formula

$$L = \frac{2\pi Ra}{360}$$

R is the radius of the circle and a is the central angle in degrees.

(a) Solve for a.   (b) Solve for R.

(c) Find L when $R = 10$ in. and $a = 30°$. Use $\pi \approx 3.14$.

2. **Sheet Metal Technology**   The area of the sector shown in problem 1 is $A = \pi R^2 a/360$.

(a) Solve this formula for a.

(b) Find A if $R = 12$ in., $a = 45°$, $\pi \approx 3.14$. Round to two significant digits.

3. **Electronics**   When two resistors R_1 and R_2 are put in series with a battery giving V volts, the current through the resistors is

$$i = \frac{V}{R_1 + R_2}$$

(a) Solve for R_1.

(b) Find R_2 if $V = 100$ volts, $i = 0.4$ ampere, $R_1 = 200$ ohms.

4. **Machine Technology**   Machinists use a formula known as Pomeroy's formula to determine roughly the power required by a metal punch machine.

$$P \approx \frac{t^2 dN}{3.78}$$

where P = power needed, in horsepower
t = thickness of the metal being punched
d = diameter of the hole being punched
N = number of holes to be punched at one time

(a) Solve this formula for N.

(b) Find the power needed to punch six 2-in.-diameter holes in a sheet $\frac{1}{8}$ in. thick. Round to one significant digit.

5. **Marine Technology**   When a gas is kept at constant temperature and the pressure on it is changed, its volume changes in accord with the pressure–volume relationship known as Boyle's law:

$$\frac{V_1}{V_2} = \frac{P_2}{P_1}$$

where P_1 and V_1 are the beginning volume and pressure, and P_2 and V_2 are the final volume and pressure.

(a) Solve for V_1.   (b) Solve for V_2.

(c) Solve for P_1.   (d) Solve for P_2.

(e) Find P_1 when $V_1 = 10$ cu ft, $V_2 = 25$ cu ft, and $P_2 = 120$ psi.

6. The volume of a football is roughly $V = \pi LT^2/6$, where L is its length and T is its thickness.

(a) Solve for L.

(b) Solve for T^2.

7. **Medical Technology** Nurses use a formula known as Young's rule to determine the amount of medicine to give a child under 12 years of age when the adult dosage is known.

$$C = \frac{AD}{A + 12}$$

C is the child's dose; A is the age of the child in years; D is the adult dose.

(a) Work backward and find the adult dose in terms of the child's dose. Solve for D.

(b) Find D if $C = 0.05$ gram and $A = 7$.

8. **Carpentry** The projection or width P of a protective overhang of a roof is determined by the height T of the window, the height H of the header above the window, and a factor F that depends on the latitude of the construction site.

$$P = \frac{T + H}{F} \qquad \text{Solve this equation for the header height } H.$$

9. **Electronics** For a current transformer, $\dfrac{i_L}{i_S} = \dfrac{T_P}{T_S}$

where i_L = line current
  i_S = secondary current
  T_P = number of turns of wire in the primary coil
  T_S = number of turns of wire in the secondary coil

(a) Solve for i_L.

(b) Solve for i_S.

(c) Find i_L when $i_S = 1.5$ amperes, $T_P = 1500$, $T_S = 100$.

10 **Electronics** The electrical power P dissipated in a circuit is equal to the product of the current I and the voltage V, where P is in watts, I is in amperes, and V is in volts.

(a) Write an equation giving P in terms of I and V.

(b) Solve for I.

(c) Find V when $P = 15,750$ watts, $I = 42$ amperes.

11. **Electronics** The inductance L in microhenrys of a coil constructed by a ham radio operator is given by the formula

$$P = \frac{R^2 N^2}{9R + 10D}$$

where R = radius of the coil
  D = length of the coil
  N = number of turns of wire in the coil

Find L if $R = 3$ in., $D = 6$ in., $N = 200$.

12. **Sheet Metal Technology** A sheet metal technician uses the following formula for calculating bend allowance, BA, in inches:

$$BA = N(0.01743R + 0.0078T)$$

where N = number of degrees in the bend
  R = inside radius of the bend, in inches
  T = thickness of the metal, in inches

Find BA for each of the following situations. (You'll want to use a calculator on this one.)

	N	R	T
(a)	50°	$1\frac{1}{4}$ in.	0.050 in.
(b)	65°	0.857 in.	0.035 in.
(c)	40°	1.025 in.	0.0856 in.

13. **Machinist** Suppose that, on the average, 3% of the parts produced by a particular machine have proven to be defective. Then the formula

$$N - 0.03N = P$$

will give the number of parts N that must be produced in order to manufacture a total of P nondefective ones. How many parts should be produced by this machine in order to end up with 7500 nondefective ones?

14. **Office Services** The formula

$$A = P(1 + rt)$$

is used to find the total amount A of money in an account when an original amount or principal P is invested at a rate of simple interest r for t years. How long would it take $8000 to grow to $10,000 at 8% simple interest?

15. **Civil Engineer** The formula

$$I = 0.000014L(T - t)$$

gives the expansion I of a particular highway of length L at a temperature of T degrees Fahrenheit. The variable t stands for the temperature at which the highway was built. If a 2-mile stretch of highway was built at an average temperature of 60°F, what is the maximum temperature it can withstand if expansion joints allow for 7.5 ft of expansion? (*Hint:* The units of L must be the same as the units of I.)

When you have completed these exercises check your answers on page 598, then turn to **43** to learn about word problems.

7-6 SOLVING WORD PROBLEMS

43

Translating English to Algebra

Algebra is a very useful tool for solving real problems. But in order to use it you may find it necessary to translate simple English sentences and phrases into mathematical expressions or equations. In technical work especially, the formulas to be used are often given in the form of English sentences, and they must be rewritten as algebraic formulas before they can be used. For example, the statement

> Horsepower required to overcome vehicle air resistance is equal to the cube of the vehicle speed in mph multiplied by the frontal area in square feet divided by 150,000

translates to the formula

$$hp = \frac{mph^3 \cdot area}{150,000}$$

or

$$P = \frac{v^3 A}{150,000} \qquad \text{in algebraic form, where } v \text{ is the vehicle speed}$$

In the next few pages of this chapter we show you how to translate English statements into algebraic formulas. To begin, try the following problem.

An automotive technician found the following statement in a manual:

The pitch diameter of a cam gear is twice the diameter of the crank gear.

Translate this sentence into an algebraic equation.

Check your work in **44**.

---

44   The equation is $P = 2C$ where P is the pitch diameter of the cam gear, and C is the diameter of the crank gear

You may use any letters you wish of course, but we have chosen letters that remind you of the quantities they represent: P for *pitch* and C for *crank*.

Notice that the phrase "twice C" means "two times C" and is written as $2C$ in algebra.

Certain words and phrases appear again and again in statements to be translated. They are signals alerting you to the mathematical operations to be used. Here is a handy list of the *signal words* and their mathematical translations.

SIGNAL WORDS

English Term	Math Translation	Example
Equals Is, is equal to, was, are, were The same as . . . What is left is . . . The result is . . . Gives, makes, leaves, having	=	$A = B$
Plus, sum of Increased by, more than	+	$A + B$
Minus B, subtract B Less B Decreased by B, take away B Reduced by B, diminished by B B less than A B subtracted from A Difference between A and B	−	$A - B$
Times, multiply, of Multiplied Product of	×	AB
Divide, divided by B Quotient of	÷	$A \div B$ or $\dfrac{A}{B}$
Twice, twice as much Double	×2	$2A$
Squared, square of A Cubed, cube of A		A^2 A^3

Translate the phrase "length plus 3 inches" into an algebraic expression. Try it, then turn to **45** to check your answer.

---

45 **First,** make a word equation by using parentheses.

(length)　　(plus)　　(3 inches)

Second, substitute mathematical symbols.

(length)　　(plus)　　(3 inches)

$\downarrow$　　　$\downarrow$　　　$\downarrow$

L　　　$+$　　　3　　　or $L + 3$

Notice that signal words, such as "plus," are translated directly into math symbols. Unknown quantities are represented by letters of the alphabet, chosen to remind you of their meaning.

Translate the following phrases into math expressions.

(a)　Weight divided by 13.6　　　＿＿＿＿＿＿＿

(b)　$6\frac{1}{4}$ in. more than the width　　　＿＿＿＿＿＿＿

(c)　One-half of the original torque　　　＿＿＿＿＿＿＿

(d)　The sum of the two lengths　　　＿＿＿＿＿＿＿

(e)　The voltage decreased by 5　　　＿＿＿＿＿＿＿

(f)　Five times the gear reduction　　　＿＿＿＿＿＿＿

(g)　8 in. less than twice the height　　　＿＿＿＿＿＿＿

Our step-by-step translations are given in **46**.

46　(a)　(weight)　(divided by)　(13.6)

$\downarrow$　　　　$\downarrow$　　　　$\downarrow$

W　　　　$\div$　　　13.6　or　$\dfrac{W}{13.6}$

(b)　$(6\frac{1}{4}$ in.)　(more than)　(the width)

$\downarrow$　　　　$\downarrow$　　　　$\downarrow$

$6\frac{1}{4}$　　　$+$　　　W　　or　$6\frac{1}{4} + W$

(c)　(one-half)　(of)　(the original torque)

$\downarrow$　　$\downarrow$　　　$\downarrow$

$\dfrac{1}{2}$　　$\times$　　　T　　or $\dfrac{1}{2} T$　or　$\dfrac{T}{2}$

(d)　(the sum of)　(the two lengths)

$\downarrow$　$\downarrow$

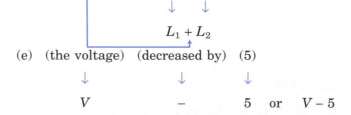

$L_1 + L_2$

(e)　(the voltage)　(decreased by)　(5)

$\downarrow$　　　　$\downarrow$　　　$\downarrow$

V　　　　$-$　　　5　or　$V - 5$

(f) (five) (times) (the gear reduction)

   5   × G or $5G$

(g) 8 less than twice the height means
     (twice the height) (less) (8)

       2   H –   8   or $2H - 8$

Of course, any letters could be used in place of the ones used above.

Note ▶   The phrase "less than" has a meaning very different from the word "less."

"8 ***less*** 5" means $8 - 5$

"8 ***less than*** 5" means $5 - 8$ ◀

Translating Sentences to Equations

So far we have translated only phrases, pieces of sentences, but complete sentences can also be translated. An English phrase translates into an algebraic expression, and an English sentence translates into an algebraic equation. For example, the sentence

The height of the duct is equal to its width

translates to H = W or $H = W$

Each word or phrase in the sentence becomes a mathematical term, letter, number, expression, or arithmetic operation sign.

Translate the following sentence into algebraic form as we did above.

    The size of a drill for a tap is equal to the tap diameter minus the depth.

Check your translation in **47**.

---

47 $S = T - D$

Follow these steps when you must translate an English sentence into an algebraic equation or formula.

Example:

| **Step 1** | Cross out all unnecessary words. | ~~The~~ size ~~of a drill for a tap~~ is equal to ~~the~~ tap diameter minus ~~the~~ depth. |
| **Step 2** | Make a word equation using parentheses. | (Size) (is equal to) (tap diameter) (minus) (depth) |

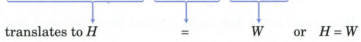

| **Step 3** | Substitute a letter or an arithmetic symbol for each parentheses. | S = T – D |
| **Step 4** | Combine and simplify | $S = T - D$ |

In most formulas the units for the quantities involved must be given. In the formula above, T and D are in inches.

Learning Help ▶   Translating English sentences or verbal rules into algebraic formulas requires that you read the sentences very differently from the way you read stories or newspaper articles. Very few people are able to write out the math formula after reading the problem only once. You should expect to read it several times, and you'll want to read it slowly. No speed reading here! ◀

The ideas in technical work and formulas are usually concentrated in a few key words, and you must find them. If you find a word you do not recognize, stop reading and look it up in a dictionary, textbook, or manual. It may be important. Translating and working with formulas is one of the skills you must have if you are to succeed at any technical occupation.

Here is another example of translating a verbal rule into an algebraic formula:

The electrical resistance of a length of wire is equal to the resistivity of the metal times the length of the wire divided by the square of the wire diameter.

Step 1	Eliminate all but the key words.	"~~The electrical~~ resistance ~~of a length of wire~~ is equal to ~~the~~ resistivity ~~of the metal~~ times ~~the~~ length ~~of the wire~~ divided by ~~the~~ square of ~~the wire~~ diameter."
Step 2	Make a word equation.	(Resistance) (is equal to) (resistivity) (times) (length) (divided by) (square of diameter)
Step 3	Substitute letters and symbols	$R = \dfrac{rL}{D^2}$

If the resistivity r has units of ohms times inches, L and D will be in inches.

The more translations you do, the easier it gets. Translate each of the following technical statements into algebraic formulas.

(a) A sheet metal worker measuring a duct cover finds that the width is $8\frac{1}{2}$ in. less than the height.

(b) One-quarter of a job takes $3\frac{1}{2}$ days.

(c) One half of a coil of wire weighs $16\frac{2}{3}$ lb.

(d) The volume of an elliptical tank is approximately 0.7854 times the product of its height, length, and width.

(e) The engine speed is equal to 168 times the overall gear reduction multiplied by the speed in mph and divided by the rolling radius of the tire.

(f) The air resistance force in pounds acting against a moving vehicle is equal to 0.0025 times the square of the speed in mph times the frontal area of the vehicle.

(g) Two pieces of wire have a combined length of 24 in. The longer piece is five times the length of the shorter piece. (*Hint:* Write two separate equations.)

(h) Two shims are to have a combined thickness of 0.090 in. The larger shim must be 3.5 times as thick as the smaller shim. (*Hint:* Write two separate equations.)

Work carefully. Check your formulas in **48**.

---

48 (a) $W = H - 8\dfrac{1}{2}$

(b) $\dfrac{1}{4}J = 3\dfrac{1}{2}$ or $\dfrac{J}{4} = 3\dfrac{1}{2}$

(c) $\dfrac{1}{2}C = 16\dfrac{2}{3}$ or $\dfrac{C}{2} = 16\dfrac{2}{3}$

(d) $V = 0.7854HLW$

(e) $S = \dfrac{168Gv}{R}$ where v is in mph.

(f) $R = 0.0025v^2A$ where v is in mph.

(g) $24 = L + S$ and $L = 5S$

(h) $S + L = 0.090$ and $L = 3.5S$

Notice in problems (g) and (h) that two equations can be written. These can be combined to form a single equation.

(g) $24 = L + S$ and $L = 5S$ give $24 = 5S + S$

(h) $S + L = 0.090$ and $L = 3.5S$ give $S + 3.5S = 0.090$

General Word Problems

Now that you can translate English phrases and sentences into algebraic expressions and equations, you should be able to solve many practical word problems. For example, consider this problem:

A machinist needs to use two shims with a combined thickness of 0.084 in. One shim is to be three times as thick as the other. What are the thicknesses of the two shims?

Let $x = $ thickness of the thinner shim

then $3x = $ thickness of the thicker shim

and the equation is

$3x + x = 0.084$ in.   Combine terms.

    $4x = 0.084$ in.   Divide each side by 4.

      $x = 0.021$ in. $3x = 3(0.021$ in.$) = 0.063$ in.

The thin shim is 0.021 in. and the thicker one is 0.063 in.

Check: 0.021 in. $+ 3(0.021$ in.$) = 0.084$ in.

             0.021 in. $+ 0.063$ in. $= 0.084$ in.   which is correct

Your turn. Use your knowledge of algebra to translate the following problem to an algebraic equation and then solve it.

Two carpenters, Al and Bill, produced 42 assembly frames in one day. Al worked faster and produced 8 more than Bill. How many frames did each build?

Try it. Check your work in **49**.

---

49   Let $B = $ number of frames built by Bill
then $B + 8 = $ number of frames built by Al

and

   $B + (B + 8) = 42$ Combine terms, $B + B = 2B$.

      $2B + 8 = 42$ Subtract 8 from each side.

         $2B = 34$ Divide both sides by 2.

          $B = 17$ Bill built 17 frames.

then $B + 8 = 25$ Al built 25 frames.

Check: $17 + 25 = 42$

Some of the more difficult percent problems can be simplified using algebraic equations. One such problem, backing the tax out of a total, is commonly encountered by all those who own their own businesses. As an example, suppose that an auto mechanic has collected $1468.63 for parts, including 6% tax. For accounting purposes he must determine exactly how much of the total is sales tax. If we let x stand for the dollar amount of the parts before tax was added, then the tax is 6% of that or $0.06x$, and we have

$$x \quad + \quad 0.06x \quad = \quad \$1468.63$$

| Cost of the parts | Tax on the parts | Total amount |

Solving for x yields

$$x + 0.06x = \$1468.63$$
$$1.00x + 0.06x = \$1468.63 \qquad (x = 1x = 1.00x). \qquad \text{Combine like terms.}$$
$$1.06x = \$1468.63 \qquad \text{Now divide by 1.06.}$$
$$x = \$1385.50 \qquad x \text{ is the cost of the parts.}$$
$$0.06x = 0.06(\$1385.50) = \$83.13 \quad\Big\} \quad \text{Multiply by 0.06 or}$$
$$\$1468.63 - \$1385.50 = \$83.13 \qquad \text{subtract from } \$1468.63 \text{ to get the tax.}$$

Practice makes perfect. Work the following set of word problems.

(a) The area of a rectangular shop floor is 400 sq ft. If the width is 16 ft, what is the length? (*Hint:* Area = length · width.)

(b) A 14-ft-long steel rod is cut into two pieces. The longer piece is $2\frac{1}{2}$ times the length of the shorter piece. Find the length of each piece. (Ignore waste.)

(c) A carpenter wants to cut a 12-ft board into three pieces. The longest piece must have three times the length of the shortest piece, and the medium-sized piece is 2 ft longer than the shortest piece. Find the actual length of each piece. (Ignore waste in cutting.)

(d) Find the dimensions of a rectangular cover plate if its length is 6 in. longer than its width and if its perimeter is 68 in. (*Hint:* Perimeter = 2 · length + 2 · width.)

(e) Mike and Jeff are partners in a small manufacturing firm. Because Mike provided more of the initial capital for the business, they have agreed that Mike's share of the profits should be $\frac{1}{4}$ greater than Jeff's. The total profit for the first quarter of this year was $17,550. How should they divide it?

(f) A plumber collected $2784.84 during the day, $1650 for labor, which is not taxed, and the rest for parts, which includes 5% sales tax. Determine the total amount of tax that was collected.

(g) A printer knows from past experience that about 2% of a particular run of posters will be spoiled. How many should she print in order to end up with 1500 usable posters?

When you have solved these problems, turn to **50** to check your work.

---

50 (a) Width = 16 ft   Length = L
Area = 400 sq ft = 16 ft · L
$16L = 400$ Divide by 16.
$L = 25$ ft

(b) x = shorter piece

$(2\frac{1}{2})x$ = longer piece

$x + (2\frac{1}{2})x = 14$ ft

$(3\frac{1}{2})x = 14$ ft

or $\dfrac{7x}{2} = 14$ Multiply by 2.

$7x = 28$ Divide by 7.

$x = 4$ ft   Larger piece $= 2\frac{1}{2}x = 2\frac{1}{2}(4) = 10$ ft.

The shorter piece is 4 ft long and the longer piece is 10 ft long.

(c) Let x = the shortest piece
then $3x$ = longest piece
and $x + 2$ = medium-sized piece

$x + (x + 2) + (3x) = 12$ ft   Combine terms.

$5x + 2 = 12$ Subtract 2 from each side.

$5x = 10$ Divide by 5.

$x = 2$ ft   The shortest piece is 2 ft long.

$3x = 6$ ft   The longest piece is 6 ft long.

$x + 2 = 4$ ft   The third piece is 4 ft long.

(d) Let width = W
then length = $W + 6$
and perimeter = 68 in. = $2W + 2(W + 6)$ Remove parentheses,
$2(W + 6) = 2W + 12.$

$68 = 2W + 2W + 12$ Combine terms.

or $4W + 12 = 68$ Subtract 12 from each side.

$4W = 56$ Divide by 4.

$W = 14$ in.

length = $W + 6 = 20$ in.

(e) Let Jeff's share = J
then Mike's share = $J + \frac{1}{4}J$

$\$17{,}550 = J + \left(J + \frac{1}{4}J\right)$ Combine terms; remember $J = 1 \cdot J$.

$= 2\frac{1}{4}J$ Write the mixed number as a fraction.

$= \dfrac{9J}{4}$

or $\dfrac{9J}{4} = 17{,}550$ Multiply by 4.

$9J = 70{,}200$ Divide by 9.

$J = \$7800$ Joe's share

Mike's share $= \$7800 + \dfrac{\$7800}{4}$

$= \$7800 + \$1950$

$= \$9750$

(f) Let x = cost of the parts
then $0.05x$ = amount of the tax

$$x + 0.05x = \$2784.84 - \$1650 \qquad \text{Subtract out the labor to get the}$$
$$1.05x = \$1134.84 \qquad \text{dollar amount of the parts.}$$
$$x = \$1080.80 \qquad \text{This represents the cost of the parts.}$$
$$0.05(\$1080.80) = \$54.04 \qquad \text{The total tax is } \$54.04.$$

(g) Let x = number of posters she needs to print
then $0.02x$ = number that will be spoiled

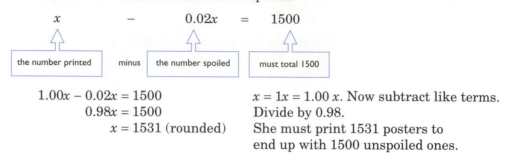

$$1.00x - 0.02x = 1500 \qquad x = 1x = 1.00\,x. \text{ Now subtract like terms.}$$
$$0.98x = 1500 \qquad \text{Divide by 0.98.}$$
$$x = 1531 \text{ (rounded)} \qquad \text{She must print 1531 posters to}$$
$$\text{end up with 1500 unspoiled ones.}$$

Now turn to **51** for a set of exercises on solving word problems.

51

Exercises 7-6   **Solving Word Problems**

A. Translate the following into algebraic equations:

1. The height of the tank is 1.4 times its width.

2. The sum of the two weights is 167 lb.

3. The volume of a cylinder is equal to $\frac{1}{4}$ of its height times π times its diameter squared.

4. The weight in kilograms is equal to 0.454 times the weight in pounds.

5. **Machine Technology**   The volume of a solid bar is equal to the product of the cross-sectional area and the length of the bar.

6. The volume of a cone is $\frac{1}{3}$ times π times the height times the square of the radius of the base.

7. **Machine Technology**   The pitch diameter D of a spur gear is equal to the number of teeth on the gear divided by the pitch.

8. **Machine Technology**   The cutting time for a lathe operation is equal to the length of the cut divided by the product of the tool feed rate and the revolution rate of the workpiece.

9. **Machine Technology**   The weight of a metal cylinder is approximately equal to 0.785 times the height of the cylinder times the density of the metal times the square of the diameter of the cylinder.

10. The voltage across a simple circuit is equal to the product of the resistance of the circuit and the current flowing in the circuit.

B. Set up equations and solve.

1. **Wastewater Technology**   Two tanks must have a total capacity of 400 gallons. If one tank needs to be twice the size of the other, how many gallons should each tank hold?

2. **Machine Technology**   Two metal castings weigh a total of 84 lb. One weighs 12 lb more than the other. How much does each one weigh?

3. **Plumbing** The plumber wants to cut a 24-ft length of pipe into three sections. The largest piece should be twice the length of the middle piece, and the middle piece should be twice the length of the smallest piece. How long should each piece be?

4. **Electrical Technology** The sum of three voltages in a circuit is 38 volts. The middle-sized one is 2 volts more than the smallest. The largest is 6 volts more than the smallest. What is the value of each voltage?

5. **Masonry** There are 156 concrete blocks available to make a retaining wall. The bottom three rows will all have the same number of blocks. The next six rows will each have two blocks fewer than the row below it. How many blocks are in each row?

6. **Construction Technology** A concrete mix is in volume proportions of 1 part cement, 2 parts water, 2 parts aggregate, and 3 parts sand. How many cubic feet of each ingredient are needed to make 54 cu ft of concrete?

7. **Office Services** Jo and Ellen are partners in a painting business. Because Jo is the office manager, she is to receive one-third more than Ellen when the profits are distributed. If their profit is $85,400, how much should each of them receive?

8. **Construction Technology** A total of $560,000 is budgeted for constructing a roadway. The rule of thumb for this type of project is that the pavement costs twice the amount of the base material, and the sidewalk costs one-fourth the amount of the pavement. Using these figures, how much should each item cost?

9. **Photography** A photographer has enough liquid toner for fifty 5-in. by 7-in. prints. How many 8-in. by 10-in. prints will this amount cover? (*Hint:* The toner covers the *area* of the prints: Area = length × width.)

10. **Construction Technology** A building foundation has a length of 82 ft and a perimeter of 292 ft. What is the width? (*Hint:* Perimeter = 2 × length + 2 × width.)

11. **Roofing** A roofer earns $24 per hour plus $36 per hour overtime. If he puts in 40 hours of regular time during a certain week and he wishes to earn $1200, how many hours of overtime should he work?

12. **Agricultural Technology** An empty avocado crate weighs 4.2 kg. How many avocados weighing 0.3 kg each can be added before the total weight of the crate reaches 15.0 kg?

13. **Machine Technology** One-tenth of the parts tooled by a machine are rejects. How many parts must be tooled to assure 4500 acceptable ones?

14. **Office Services** An auto mechanic has total receipts of $1861.16 for a certain day. He determines that $1240 of this is labor. The remaining amount is for parts plus 6% sales tax on the parts only. Find the amount of the sales tax he collected.

15. **Office Services** A travel agent receives a 7% commission on the base fare of a customer—that is, the fare *before* sales tax is added. If the total fare charged to a customer comes to $753.84 *including* 5.5% sales tax, how much commission should the agent receive?

16. **Office Services** A newly established carpenter wishes to mail out letters advertising her services to the families of a small town. She has a choice of mailing the letters first class or obtaining a bulk-rate permit and mailing them at the cheaper bulk rate. The bulk-rate permit costs $80, and each piece of bulk mail then costs 16.7 cents. If the first-class rate is 32 cents, how many pieces would she have to mail in order to make bulk rate the cheaper way to go?

Check your answers on pages 599, then turn to **52** to learn about proportions.

7-7 MORE RATIO AND PROPORTION

52 In Chapter 4 we used the concepts of ratio and proportion to solve percent problems. In this section we shall learn about other applications of ratio and proportion in trade and technical areas. First, try these review problems to refresh your memory.

1. Solve the following proportions.

(a) $\dfrac{4}{9} = \dfrac{6}{x}$

(b) $\dfrac{25}{x} = \dfrac{15}{24}$

(c) $\dfrac{16.4}{20.6} = \dfrac{x}{51.5}$

(d) $\dfrac{x}{4\frac{1}{2}} = \dfrac{3\frac{3}{4}}{5\frac{5}{8}}$

2. The pulley system of an assembly belt has a pulley diameter ratio of 7 to 2. If the smaller pulley has a diameter of 9 inches, what is the diameter of the larger pulley?

3. If the gear ratio of a machine is 4:1, and the driving gear has 48 teeth, how many teeth are on the driven gear?

Look for the solutions in **53**.

---

53 1. (a) $\dfrac{4}{9} = \dfrac{6}{x}$

Step 1: Use the cross-product rule. $\qquad\qquad$ $4 \cdot x = 9 \cdot 6 = 54$

Step 2: Divide both sides of the equation by 4 to solve for x. $\quad$ $\dfrac{4x}{4} = \dfrac{54}{4}$

$$x = 13.5$$

(b) $\dfrac{25}{x} = \dfrac{15}{24}$

Step 1: Take the cross-products. $\qquad$ $15 \cdot x = 25 \cdot 24 = 600$

Step 2: Divide both sides by 15. $\qquad$ $x = \dfrac{600}{15} = 40$

(c) $\dfrac{16.4}{20.6} = \dfrac{x}{51.5}$

$$20.6x = (16.4)(51.5)$$

$$20.6x = 844.6$$

$$x = \dfrac{844.6}{20.6} = 41$$

16.4 ☒ **51.5** ÷ **20.6** = → ▮▮▮▮ *41.*

(d) $\dfrac{x}{4\frac{1}{2}} = \dfrac{3\frac{3}{4}}{5\frac{5}{8}}$

$$5\frac{5}{8}\, x = \left(4\frac{1}{2}\right)\left(3\frac{3}{4}\right)$$

$$\dfrac{45}{8}x = \dfrac{135}{8}$$

$$x = \dfrac{135}{8} \div \dfrac{45}{8}$$

$$x = \dfrac{135}{8} \cdot \dfrac{8}{45} = 3$$

2. Pulley ratio = $\dfrac{\text{diameter of larger pulley}}{\text{diameter of smaller pulley}} = \dfrac{7}{2}$

We know that the actual diameter of the smaller pulley is 9 inches. Let L represent the actual diameter of the larger pulley. We can now write the proportion

$\dfrac{7}{2} = \dfrac{L}{9}$ Notice that the L appears in the top of the fraction opposite the larger number in the pulley ratio, and the 9 goes in the bottom of the fraction opposite the smaller number in the pulley ratio.

Solve in the usual manner.

$2L = 63$

$L = 31.5$ The diameter of the larger pulley is 31.5 inches.

3. Gear ratio = $\dfrac{\text{number of teeth on driving gear}}{\text{number of teeth on driven gear}} = \dfrac{4}{1}$

We know that the driving gear actually has 48 teeth. Let N represent the actual number of teeth on the driven gear. We can now write the proportion

$\dfrac{4}{1} = \dfrac{48}{N}$ The gear ratio specifies that the driving gear is represented by the number in the top of the fraction. Therefore, 48 is in the numerator and N is in the denominator.

Solving, we have

$4N = 48$

$N = 12$ There are 12 teeth on the driven gear.

Scale Drawings   Proportion equations are found in a wide variety of practical situations. For example, when a drafter makes a drawing of a machine part, building layout, or other large structure, he or she must *scale it down*. The drawing must represent the object accurately, but it must be small enough to fit on the paper. The draftsperson reduces every dimension by some fixed ratio.

Drawings that are larger than life involve an expanded scale.

For the automobile shown, the ratio of the actual length to the scale-drawing length is equal to the ratio of the actual width to the scale-drawing width.

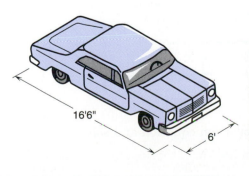

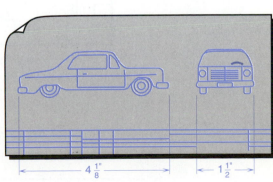

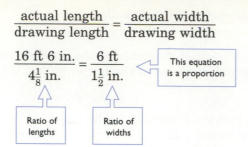

$$\frac{\text{actual length}}{\text{drawing length}} = \frac{\text{actual width}}{\text{drawing width}}$$

$$\frac{16 \text{ ft } 6 \text{ in.}}{4\frac{1}{8} \text{ in.}} = \frac{6 \text{ ft}}{1\frac{1}{2} \text{ in.}}$$

This equation is a proportion

Ratio of lengths

Ratio of widths

Rewrite all quantities in the same units:

$$\frac{198 \text{ in.}}{4\frac{1}{8} \text{ in.}} = \frac{72 \text{ in.}}{1\frac{1}{2} \text{ in.}}$$

You should notice first of all that each side of this equation is a ratio. Each side is a ratio of *like* quantities: lengths on the left and widths on the right.

Second, notice that the ratio $\dfrac{198 \text{ in.}}{4\frac{1}{8} \text{ in.}}$ is equal to $\dfrac{48}{1}$.

Divide it out: $198 \div 4\frac{1}{8} = 198 \div \dfrac{33}{8}$

$$= 198 \times \frac{8}{33}$$

$$= 48$$

198 ⊗ **8** ÷ **33** ⊜ → 48.

Notice also that the ratio $\dfrac{72 \text{ in.}}{1\frac{1}{2} \text{ in.}}$ is equal to $\dfrac{48}{1}$.

The common ratio $\dfrac{48}{1}$ is called the *scale* factor of the drawing.

In the following problem one of the dimensions is unknown:

> Suppose that a rectangular room has a length of 18 ft and a width of 12 ft. An architectural scale drawing of this room is made so that on the drawing the length of the room is 9 in. What will be the width of the room on the drawing?

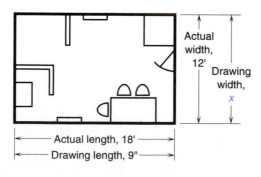

Actual width, 12'

Drawing width, *x*

Actual length, 18'

Drawing length, 9"

Try it. Set up a proportion equation and solve.

Check your work in **54**.

54 **First,** set up a ratio of lengths and a ratio of widths.

Length ratio = $\dfrac{\text{actual length}}{\text{drawing length}}$

$\qquad = \dfrac{18 \text{ ft}}{9 \text{ in.}} = \dfrac{216 \text{ in.}}{9 \text{ in.}}$ ←— Convert 18 ft to inches so that the top and the bottom of the fraction have the same units.

Width ratio = $\dfrac{\text{actual width}}{\text{drawing width}}$

$\qquad = \dfrac{12 \text{ ft}}{x \text{ in.}} = \dfrac{144 \text{ in.}}{x \text{ in.}}$ ←— Change 12 ft to 144 in.

Second, write a proportion equation

$$\dfrac{216}{9} = \dfrac{144}{x}$$

Third, solve this proportion. Cross-multiply to get

$216x = 9 \cdot 144$
$216x = 1296$
$\quad x = 6 \text{ in.}$

For the room drawing shown, the scale factor is

Scale factor = $\dfrac{18 \text{ ft}}{9 \text{ in.}} = \dfrac{216 \text{ in.}}{9 \text{ in.}}$

$\qquad = 24 \qquad \text{or} \qquad 24 \text{ to } 1$

One inch on the drawing corresponds to 24 in. or 2 ft on the actual object. A draftsperson would write this as $\frac{1}{2}$ in. = 1 ft.

The triangular plate shown has a height of 18 in. and a base length of 14 in.

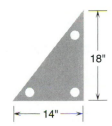

(a) What will be the corresponding height of a blueprint drawing of this plate if the base length of the drawn figure is $2\frac{3}{16}$ in.?

(b) Find the scale factor of the blueprint.

Check your work in **55**.

---

55 (a) $\dfrac{18 \text{ in.}}{x} = \dfrac{14 \text{ in.}}{2\frac{3}{16} \text{ in.}}$

Cross-multiply $\quad (18)\left(2\frac{3}{16}\right) = 14x$

$14x = 39\frac{3}{8}$

$x = 2\frac{13}{16} \text{ in.}$

 2 $\boxed{a\frac{b}{c}}$ **3** $\boxed{a\frac{b}{c}}$ **16** $\boxed{\times}$ **18** $\boxed{=}$ $\boxed{\div}$ **14** $\boxed{=}$ → | 2⎦13⎦16. | or | 2.8125 |

(b) Scale factor = $\dfrac{18 \text{ in.}}{2\frac{13}{16} \text{ in.}}$

$\qquad = 6\frac{2}{5} \qquad \text{or} \qquad 6.4 \text{ to } 1$

 18 $\boxed{\div}$ **2** $\boxed{a\frac{b}{c}}$ **13** $\boxed{a\frac{b}{c}}$ **16** $\boxed{=}$ → | 6⎦2⎦5. | or | 6.4 |

Similar Figures   In general, two geometric figures that have the same shape but are not the same size are said to be *similar* figures. The blueprint drawing and the actual object are a pair of similar figures. An enlarged photograph and the smaller original are similar.

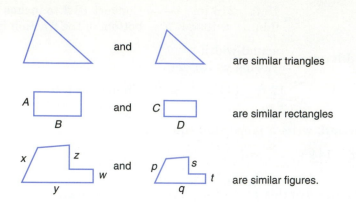

are similar triangles

A ▭   and   *C* ▭   are similar rectangles
B   *D*

and   are similar figures.

In any similar figures, all parts of corresponding dimensions have the same scale ratio. For example, in the rectangles above

$$\frac{A}{C} = \frac{B}{D}$$

In the irregular figure above,

$$\frac{x}{p} = \frac{y}{q} = \frac{z}{s} = \frac{w}{t} \qquad \text{and so on}$$

The triangles shown here are *not* similar:

Find the missing dimension in each of the following pairs of similar figures.

(a)

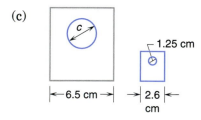

(b)   *B* ▭   with $5\frac{1}{2}''$ and $11''$, $4\frac{1}{8}''$

(c)   6.5 cm, 1.25 cm, 2.6 cm, *c*

(d)   6.3", *D*, $3\frac{1}{4}''$, $1\frac{1}{2}''$

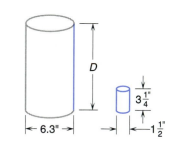

Check your answers in **56**.

_____

56   (a) $\dfrac{A}{144} = \dfrac{4}{120}$ Convert ft to in.

$$A = \frac{4(144)}{120} = 4.8 \text{ in.}$$

(b) $\dfrac{B}{4\frac{1}{8}} = \dfrac{5\frac{1}{2}}{11}$

$$B = \frac{5\frac{1}{2}\left(4\frac{1}{8}\right)}{11} = 2\frac{1}{16} \text{ in.}$$

(c) $\dfrac{c}{1.25} = \dfrac{6.5}{2.6}$

(d) $\dfrac{D}{3\frac{1}{4}} = \dfrac{6.3}{1\frac{1}{2}}$

$$c = \dfrac{6.5(1.25)}{2.6} = 3.125 \text{ cm}$$

$$D = \dfrac{6.3\left(3\frac{1}{4}\right)}{1\frac{1}{2}} = 13.65 \text{ in.}$$

Work problem (d) on a calculator this way:

3 $\boxed{a\frac{b}{c}}$ 1 $\boxed{a\frac{b}{c}}$ 4 $\boxed{\times}$ 6.3 $\boxed{=}$ $\boxed{\div}$ 1.5 $\boxed{=}$ → 13.65

Direct and Inverse Proportion

Many trade problems can be solved by setting up a proportion involving four related quantities. But it is important that you recognize that there are *two* kinds of proportions—direct and inverse. Two quantities are said to be *directly proportional* if an increase in one quantity leads to a proportional increase in the other quantity, or if a decrease in one leads to a decrease in the other.

> **Direct proportion:** increase → increase
>
> or decrease → decrease

For example, the electrical resistance of a wire is directly proportional to its length—the longer the wire, the greater the resistance. If 1 ft of Nichrome heater element wire has a resistance of 1.65 ohms, what length of wire is needed to provide a resistance of 19.8 ohms?

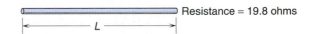

Resistance = 1.65 ohms | 1 ft | Resistance = 19.8 ohms | L |

First, recognize that this problem involves a *direct* proportion. As the length of wire increases, the resistance increases proportionally.

As L increases . . . R increases

$$\dfrac{L}{1 \text{ ft}} = \dfrac{R}{1.65 \text{ ohms}}$$

Both ratios increase in size when L increases.

Second, set up a direct proportion and solve.

$$\dfrac{L}{1 \text{ ft}} = \dfrac{19.8 \text{ ohms}}{1.65 \text{ ohms}}$$

$$L = 12 \text{ ft}$$

Solve each of the following problems by setting up a direct proportion.

(a) If a widget machine produces 88 widgets in 2 hours, how many will it produce in $3\frac{1}{2}$ hours?

(b) If one gallon of paint covers 825 sq ft, how much paint is needed to cover 2640 sq ft?

(c) What is the cost of six filters if eight filters cost $39.92?

(d) A diesel truck was driven 273 miles on 42 gallons of fuel. How much fuel is needed for a trip of 600 miles? (Round to the nearest gallon.)

(e) A cylindrical oil tank holds 450 gallons when it is filled to its full height of 8 ft. When it contains oil to a height of 2 ft 4 in., how many gallons of oil are in the tank?

Check your solution in **57**.

___

57 (a) A direct proportion—the more time spent, the more widgets produced.

$$\frac{88}{x} = \frac{2 \text{ hr}}{3\frac{1}{2} \text{ hr}}$$

Cross-multiply:

$$2x = 308$$
$$x = 154 \text{ widgets}$$

 88 ⊠ **3.5** ÷ **2** = → $\boxed{154.}$

(b) A direct proportion—the more paint, the greater the area that can be covered:

$$\frac{1 \text{ gal}}{x \text{ gal}} = \frac{825 \text{ sq ft}}{2640 \text{ sq ft}}$$

$$x = 3.2 \text{ gallons}$$

(c) A direct proportion—the more you pay, the more you get:

$$\frac{6 \text{ filters}}{8 \text{ filters}} = \frac{x}{\$39.92}$$

$$x = \$29.94$$

(d) A direct proportion—the more miles you drive, the more fuel it takes:

$$\frac{273 \text{ mi}}{600 \text{ mi}} = \frac{42 \text{ gal}}{x \text{ gal}}$$

$$x \approx 92 \text{ gallons}$$

(e) A direct proportion—the volume is directly proportional to the height:

$$\frac{450 \text{ gal}}{x \text{ gal}} = \frac{8 \text{ ft}}{2\frac{1}{3} \text{ ft}} \quad \Leftarrow \boxed{2' \ 4'' = 2\frac{4}{12}' = 2\frac{1}{3}'}$$

$$x = 131\frac{1}{4} \text{ gallons}$$

Two quantities are said to be *inversely proportional* if an increase in one quantity leads to a proportional decrease in the other quantity, or if a decrease in one leads to an increase in the other.

Inverse proportion: increase → decrease

or   decrease → increase

For example, the time required for a trip of a certain length is *inversely* proportional to the speed of travel. If a certain trip takes 2 hours at 50 mph, how long will it take at 60 mph?

The correct proportion equation is

$$\frac{50 \text{ mph}}{60 \text{ mph}} = \frac{x \text{ hr}}{2 \text{ hr}}$$

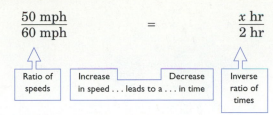

By inverting the time ratio, we have set it up so that both sides of the equation are in balance—both ratios decrease as speed increases.

Before attempting to solve the problem, make an estimate of the answer. We expect that the time to make the trip at 60 mph will be *less* than the time at 50 mph. The correct answer should be less than 2 hours.

Now solve it by cross-multiplying:

$$60x = 2 \cdot 50$$
$$x = 1\tfrac{2}{3} \text{ hr}$$

Learning Help ▶ Remember, in a *direct* proportion

$$\frac{A}{B} = \frac{C}{D}$$ These terms to together. If B and D are fixed, then an increase in A goes with an increase in C.

In an *inverse* proportion

$$\frac{X}{Y} = \frac{P}{Q}$$ These terms go together. If P and Y are fixed, then an increase in X goes with a decrease in Q. ◀

Try this problem. In an automobile cylinder, the pressure is inversely proportional to the volume if the temperature does not change. If the volume of gas in the cylinder is 300 cu cm when the pressure is 20 psi, what is the volume when the pressure is increased to 80 psi?

Set this up as an inverse proportion and solve.

Check your work in **58**.

---

58 Pressure is inversely proportional to volume. If the pressure increases, we expect the volume to decrease.

The answer should be less than 300 cu cm.

$$\frac{P_1}{P_2} = \frac{V_2}{V_1}$$

$$\frac{20 \text{ psi}}{80 \text{ psi}} = \frac{V}{300 \text{ cu cm}}$$

Cross-multiply

$$80V = 20 \cdot 300$$
$$V = 75 \text{ cu cm}$$

Gears and Pulleys A particularly useful kind of inverse proportion involves the relationship between the size of a gear or pulley and the speed with which it rotates.

In the following diagram, *A* is the driving gear.

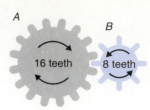

Because gear *A* has twice as many teeth as gear *B*, when *A* turns one turn, *B* will make two turns. If gear *A* turns at 10 turns per second, gear *B* will turn at 20 turns per second. The speed of the gear is inversely proportional to the number of teeth.

$$\frac{\text{speed of gear } A}{\text{speed of gear } B} = \frac{\text{teeth in gear } B}{\text{teeth in gear } A}$$

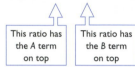

| This ratio has the A term on top | This ratio has the B term on top |

In this proportion, gear speed is measured in revolutions per minute, abbreviated rpm.

For the gear assembly shown, if gear *A* is turned by a shaft at 40 rpm, what will be the speed of gear *B*?

Check your work in **59**.

---

59   Because the relation is an inverse proportion, the smaller gear moves with the greater speed. We expect the speed of gear *B* to be faster than 40 rpm.

$$\frac{40 \text{ rpm}}{B} = \frac{8 \text{ teeth}}{16 \text{ teeth}}$$

$$B = 80 \text{ rpm}$$

On an automobile the speed of the drive shaft is converted to rear axle motion by the ring and pinion gear system.

$$\frac{\text{drive shaft speed}}{\text{rear axle speed}} = \frac{\text{teeth in ring gear on axle}}{\text{teeth in pinion gear on drive shaft}}$$

| This ratio has the drive shaft term on top | This ratio has the drive shaft term on the bottom |

Again, gear speed is inversely proportional to the number of teeth on the gear.

If the pinion gear has 9 teeth and the ring gear has 40 teeth, what is the rear axle speed when the drive shaft turns at 1200 rpm?

Check your work in **60**.

---

60 $$\frac{1200 \text{ rpm}}{R} = \frac{40 \text{ teeth}}{9 \text{ teeth}}$$

Cross-multiply: $40R = 9 \cdot 1200$

$$R = 270 \text{ rpm}$$

Pulleys transfer power in much the same way as gears. For the pulley system shown, the speed of a pulley is inversely proportional to its diameter

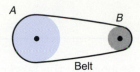

If pulley A has a diameter twice that of pulley B, then when pulley A makes one turn, pulley B will make two turns, assuming of course that there is no slippage of the belt.

$$\frac{\text{speed of pulley } A}{\text{speed of pulley } B} = \frac{\text{diameter of pulley } B}{\text{diameter of pulley } A}$$

This ratio has the A term on top	This ratio has the A term on the bottom

If pulley B is 16 in. in diameter and is rotating at 240 rpm, what is the speed of pulley A if its diameter is 20 in.?

Check your work in **61**.

---

61 $\dfrac{A}{240 \text{ rpm}} = \dfrac{16 \text{ in}}{20 \text{ in.}}$

Cross-multiply $20A = 16 \cdot 240$

$A = 192 \text{ rpm}$

Solve each of the following problems by setting up an inverse proportion.

(a)   A 9-in. pulley on a drill press rotates at 1260 rpm. It is belted to a 5-in. pulley on an electric motor. Find the speed of the motor shaft.

(b)   A 12-tooth gear mounted on a motor shaft drives a larger gear. The motor shaft rotates at 1450 rpm. If the speed of the large gear is to be 425 rpm, how many teeth must be on the large gear?

(c)   For gases, pressure is inversely proportional to volume if the temperature does not change. If 30 cu ft of air at 15 psi is compressed to 6 cu ft, what is the new pressure?

(d)   If five assembly machines can complete a given job in 3 hours, how many hours will it take for two assembly machines to do the same job?

(e)   The forces and lever arm distances for a lever obey an inverse proportion.

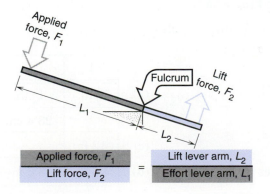

Applied force, F_1		Lift lever arm, L_2
Lift force, F_2	$=$	Effort lever arm, L_1

If a 100-lb force is applied to a 22-in. crowbar pivoted 2 in. from the end, what lift force is exerted?

Check your work in **62**.

---

62   (a) $\dfrac{9 \text{ in.}}{5 \text{ in.}} = \dfrac{x}{1260 \text{ rpm}}$ An inverse proportion: the larger pulley turns more slowly.

$$5x = 9 \cdot 1260$$

$$x = 2268 \text{ rpm}$$

(b) $\dfrac{12 \text{ teeth}}{x \text{ teeth}} = \dfrac{425 \text{ rpm}}{1450 \text{ rpm}}$ The larger gear turns more slowly.

$$425x = 12 \cdot 1450$$

$$x = 40.941\ldots \qquad \text{or 41 teeth, rounding to the nearest whole number. We}$$
can't have a part of a gear tooth!

$$x \approx 41$$

(c) $\dfrac{30 \text{ cu ft}}{6 \text{ cu ft}} = \dfrac{P}{15 \text{ psi}}$ An inverse proportion: the higher the pressure, the smaller the volume.

$$6P = 30 \cdot 15$$

$$P = 75 \text{ psi}$$

(c)   Careful on this one! An inverse proportion should be used. The *more* machines used, the *fewer* the hours needed to do the job.

$$\frac{5 \text{ machines}}{2 \text{ machines}} = \frac{x \text{ hr}}{3 \text{ hr}}$$

$$2x = 15$$

$$x = 7\tfrac{1}{2} \text{ hr} \qquad \text{Two machines will take much longer to do the job than will five machines.}$$

(e) $\dfrac{100 \text{ lb}}{F} = \dfrac{2 \text{ in.}}{20 \text{ in.}}$ If the entire bar is 22 in. long, and $L_2 = 2$ in., then $L_1 = 20$ in.

$$2F = 20 \cdot 100$$

$$F = 1000 \text{ lb}$$

Now turn to **63** for a set of practice problems on ratio and proportion.

63
Exercises 7-7   **More Ratio and Proportion**

A.   Complete the following tables.

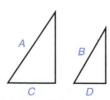

Similar triangles

1.

	A	B	C	D
(a)	$5\tfrac{1}{2}$ in.	$1\tfrac{1}{4}$ in.	$2\tfrac{3}{4}$ in.	
(b)		23.4 cm	20.8 cm	15.6 cm
(c)	12 ft		9 ft	6 ft
(d)	4.5 m	3.6 m		2.4 m

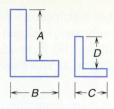

Similar figures

2.

	A	B	C	D
(a)	4 in.	5 in.	$3\frac{1}{2}$ in.	
(b)	5 ft		5 ft	2 ft
(c)	6.4 m	5.6 m		1.6 m
(d)		36 m	12 cm	14 cm

3.

	Number of Teeth on Gear 1	Number of Teeth on Gear 2	RPM of Gear 1	RPM of Gear 2
(a)	20	48	240	
(b)	25		150	420
(c)		40	160	100
(d)	32	40		1200

4.

	Diameter of Pulley 1	Diameter of Pulley 2	RPM of Pulley 1	RPM of Pulley 2
(a)	18 in.	24 in.	200	
(b)	12 in.		300	240
(c)		5 in.	400	640
(d)	14 in.	6 in.	300	

B. Practical Problems

1. **Metalworking** A line shaft rotating at 250 rpm is connected to a grinding wheel by the pulley assembly shown. If the grinder shaft must turn at 1200 rpm, what size pulley should be attached to the line shaft?

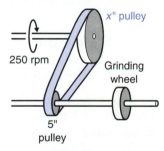

Problem 1

2. **Architectural Technology** The architectural drawing of an outside deck is $3\frac{1}{2}$ in. wide by $10\frac{7}{8}$ in. long. If the deck will actually be 14 ft wide, calculate the following:
   (a) The actual length of the deck.
   (b) The scale factor

3. **Auto Mechanics** Horsepower developed by an engine varies directly as its displacement. How many horsepower will be developed by an engine with a displacement of 240 cu in. if a 380-cu in. engine of the same kind develops 220 hp?

4. **Transportation Technology** The distance necessary to stop a subway train at a given speed is inversely proportional to its deceleration. If a train traveling at 30 mph requires 180 ft to stop when decelerating at 0.18 g, what is the stopping distance at the same speed when it is decelerating at 0.15 g?

5. A crowbar 28 in. long is pivoted 6 in. from the end. What force must be applied at the long end in order to lift a 400-lb object at the short end?

6. If 60 gal of oil flow through a certain pipe in 16 minutes, how long will it take to fill a 450-gal tank using this pipe?

7. **Auto Mechanics** If the alternator-to-engine drive ratio is 2.45 to 1, what rpm will the alternator have when the engine is idling at 400 rpm? (*Hint:* Use a direct proportion.)

8. The length of a wrench is inversely proportional to the amount of force needed to loosen a bolt. A wrench 6 in. long requires a force of 240 lb to loosen a rusty bolt. How much force would be required to loosen the same bolt using a 10-in. wrench?

9. **Metalworking** A bearing alloy is made up of 74% copper, 16% lead, and 10% tin. If 17 lb of copper is used to obtain a batch of this alloy, what weight of alloy is obtained? (Round to the nearest pound.)

10. **Machine Technology** A pair of belted pulleys have diameters of 20 in. and 16 in., respectively. If the larger pulley turns at 2000 rpm, how fast will the smaller pulley turn?

11. **Manufacturing** A 15-tooth gear on a motor shaft drives a larger gear having 36 teeth. If the motor shaft rotates at 1200 rpm, what is the speed of the larger gear?

12. **Electronics** The power gain of an amplifier circuit is defined as

$$\text{Power gain} = \frac{\text{output power}}{\text{input power}}$$

If the audio power amplifier circuit has an input power of 0.72 watt and a power gain of 30, what output power will be available at the speaker?

13. **Industrial Technology** If 12 assemblers can complete a certain job in 4 hours, how long will the same job take if the number of assemblers is cut back to 8?

14. **Electronics** A 115-volt power transformer has 320 turns on the primary. If it delivers a secondary voltage of 12 volts, how many turns are on the secondary? (*Hint:* Use a direct proportion.)

15. **Auto Mechanics** The headlights of a car are mounted at a height of 3.5 ft. If the light beam drops 1 in. per 35 ft, how far ahead in the road will the headlights illuminate?

16. A truck driver covers a certain stretch of the interstate in $4\frac{1}{2}$ hours traveling at the posted speed limit of 55 mph. If the speed limit is raised to 65 mph, how much time will the same trip require?

17. **Machine Technology** It is known that a $\frac{7}{8}$-in. thickness of a particular kind of cable has a capacity to hold 2500 lb. If the capacity of the cable is proportional to its thickness, what size cable is needed to hold 4000 lb?

18. **Masonry** Cement, sand, and gravel are mixed to a proportion of 1:3:6 for a particular batch of concrete. How many cubic yards of each should be used to mix 125 cubic yards of concrete?

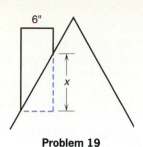

Problem 19

19. **Plumbing**   A cylindrical vent 6 in. in diameter must be cut at an angle to fit on a gable roof with a $\frac{1}{3}$ pitch. This means that for the vent itself the ratio of rise to run will be 2:3. Find the height x of the cut that must be made on the cylinder to make it fit the slope of the roof. (See the figure.)

20. **Architectural Technology**   A wrought-iron gate for a new house will be 18 ft 9 in. long and 9 ft 6 in. high. An architect makes a drawing of the gate using a scale factor of $\frac{1}{2}$ in. = 1 ft. What will be the dimensions of the gate on the drawing?

21. **Auto Mechanics**   When a tire is inflated, the air pressure is inversely proportional to the volume of the air. If the pressure of a certain tire is 28 psi when the volume is 120 cu in., what is the pressure when the volume is 150 cu in.?

22. **Electrical Technology**   The electrical resistance of a given length of wire is inversely proportional to the square of the diameter of the wire:

$$\frac{\text{resistance of wire } A}{\text{resistance of wire } B} = \frac{(\text{diameter of } B)^2}{(\text{diameter of } A)^2}$$

If a certain length of wire with a diameter of 34.852 mils has a resistance of 8.125 ohms, what is the resistance of the same length of the same composition wire with a diameter of 45.507 mils?

23. If you are paid \$238.74 for $21\frac{1}{2}$ hr of work, what amount should you be paid for 34 hours of work at this same rate of pay?

24. **Sheet Metal Technology**   If the triangular plate shown is cut into eight pieces along equally spaced dashed lines, find the height of each cut. (Round to two decimal digits.)

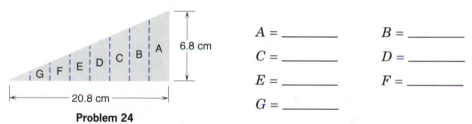

$A =$ _____	$B =$ _____
$C =$ _____	$D =$ _____
$E =$ _____	$F =$ _____
$G =$ _____	

Problem 24

25. **Wastewater Technology**   The time required for an outlet pipe to empty a tank is inversely proportional to the cross-sectional area of the pipe. A pipe with a cross-sectional area of 113.0 sq in. requires 6.4 hours to empty a certain tank. If the pipe was replaced with one with a cross-sectional area of 50.25 sq in., how long would it take this pipe to empty the same tank?

26. A gas has a volume of 2480 cm^3 at a pressure of 63.5 psi. What is the pressure when the gas is compressed to 1830 cm^3? (Round to the nearest tenth.)

When you have completed these exercises, check your answers on page 599, then turn to **64** to learn about multiplying and dividing algebraic expressions.

7-8 MULTIPLYING
AND DIVIDING
ALGEBRAIC
EXPRESSIONS

**Multiplying Simple
Factors**

64 Earlier we learned that only like algebraic terms, those with the same literal part, can be added and subtracted. However, any two terms, like or unlike, can be multiplied.

In order to multiply two terms such as $2x$ and $3xy$, first remember that $2x$ means 2 times x. Second, recall from arithmetic that the order in which you do multiplications does not make a difference. For example, in arithmetic

$$2 \cdot 3 \cdot 4 = (2 \cdot 4) \cdot 3 = (3 \cdot 4) \cdot 2$$

and in algebra

$$a \cdot b \cdot c = (a \cdot c) \cdot b = (c \cdot b) \cdot a \qquad \text{or} \qquad 2 \cdot x \cdot 3 \cdot x \cdot y = 2 \cdot 3 \cdot x \cdot x \cdot y$$

Remember that $x \cdot x = x^2$, $x \cdot x \cdot x = x^3$, and so on. Therefore,

$$2x \cdot 3xy = 2 \cdot 3 \cdot x \cdot x \cdot y = 6x^2y$$

The following examples show how to multiply two terms.

Example 1: $a \cdot 2a = a \cdot 2 \cdot a$

$\qquad\qquad = 2 \cdot \underline{a \cdot a} \qquad$ Group like factors together.

$\qquad\qquad = 2 \cdot a^2$

$\qquad\qquad = 2a^2$

Example 2: $2x^2 \cdot 3xy = 2 \cdot x \cdot x \cdot 3 \cdot x \cdot y$

$\qquad\qquad = \underline{2 \cdot 3} \cdot \underline{x \cdot x \cdot x} \cdot y \qquad$ Group like factors together.

$\qquad\qquad = 6 \cdot x^3 \cdot y$

$\qquad\qquad = 6x^3y$

Example 3: $3x^2yz \cdot 2xy = 3 \cdot x^2 \cdot y \cdot z \cdot 2 \cdot x \cdot y$

$\qquad\qquad = \underline{3 \cdot 2} \cdot \underline{x^2 \cdot x} \cdot \underline{y \cdot y} \cdot z$

$\qquad\qquad = 6 \cdot x^3 \cdot y^2 \cdot z$

$\qquad\qquad = 6x^3y^2z$

Remember to group like factors together before multiplying.

If you need to review exponents, return to frame **15** on page 272.

Now try the following problems.

(a) $x \cdot y$ $\qquad = \underline{\qquad}$

(b) $2x \cdot 3x$ $\qquad = \underline{\qquad}$

(c) $2x \cdot 5xy$ $\qquad = \underline{\qquad}$

(d) $4a^2b \cdot 2a$ $\qquad = \underline{\qquad}$

(e) $3x^2y \cdot 4xy^2$ $\qquad = \underline{\qquad}$

(f) $5xyz \cdot 2ax^2$ $\qquad = \underline{\qquad}$

(g) $3x \cdot 2y^2 \cdot 2x^2y = \underline{\qquad}$

(h) $x^2y^2 \cdot 2x \cdot y^2$ $\qquad = \underline{\qquad}$

(i) $2x^2(x + 3x^2)$ $\qquad = \underline{\qquad}$

(j) $-2a^2b(a^2 - 3b^2) = \underline{\qquad}$

The answers are in **65**.

65 (a) xy (b) $6x^2$

(c) $10x^2y$ (d) $8a^3b$

(e) $12x^3y^3$ (f) $10ax^3yz$

(g) $12x^3y^3$ (h) $2x^3y^4$

(i) $2x^3 + 6x^4$ (j) $-2a^4b + 6a^2b^3$

Were the last two problems tricky for you? Recall from Rule 3 on page 305 that when there is a multiplier in front of parentheses, every term inside must be multiplied by this factor. Try it this way:

(i) $2x^2(x + 3x^2) = (2x^2)(x) + (2x^2)(3x^2)$

$$= 2 \cdot \underline{x^2 \cdot x} + \underline{2 \cdot 3} \cdot \underline{x^2 \cdot x^2}$$

$$= 2 \cdot x^3 + 6 \cdot x^4$$

$$= 2x^3 + 6x^4$$

(j) $-2a^2b(a^2 - 3b^2) = (-2a^2b)(a^2) + (-2a^2b)(-3b^2)$

$$= (-2) \cdot \underline{a^2 \cdot a^2} \cdot b + \underline{(-2) \cdot (-3)} \, a^2 \cdot \underline{b \cdot b^2}$$

$$= -2 \cdot a^4 \cdot b + 6 \cdot a^2 \cdot b^3$$

$$= -2a^4b + 6a^2b^3$$

Rule for Multiplication   As you did the preceding problems, you may have noticed that when you multiply like factors together, you actually add their exponents. For example,

$$x^2 \cdot x^3 = x \cdot x \cdot x \cdot x \cdot x = x^5$$

In general, we can state the following rule:

Rule 1   To multiply numbers written in exponential form having the same base, add the exponents.

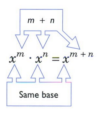

Example: $x^2 \cdot x^3 = x^{2+3} = x^5$

Example: $3^5 \times 3^2 = 3^{5+2} = 3^7$

Careful ▶

1. Notice in the second example that the like bases remain unchanged when you multiply—that is, $3^5 \times 3^2 = 3^7$, not 9^7.

2. Exponents are added during multiplication, but numerical factors are multiplied as usual. For example,

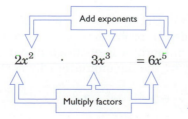

Dividing Simple Factors

To divide a^5 by a^2, think of it this way:

$$\frac{a^5}{a^2} = \frac{\cancel{a} \cdot \cancel{a} \cdot a \cdot a \cdot a}{\cancel{a} \cdot \cancel{a}} = a^3$$

Cancel common factors

Notice that the final exponent, 3, is the difference between the original two exponents, 5 and 2. Remember that multiplication and division are reverse operations. It makes sense that if you add exponents when multiplying, you would subtract exponents when dividing. We can now state the following rule:

Rule 2   To divide numbers written in exponential form having the same base, subtract the exponents.

$$\frac{x^m}{x^n} = x^{m-n}$$

Example: $\dfrac{x^7}{x^3} = x^{7-3} = x^4$

Example: $\dfrac{4^5}{4^2} = 4^{5-2} = 4^3$

Careful ▶

1. As with multiplication, the like bases remain unchanged when dividing with exponents. This is especially important to keep in mind when the base is a number, as in the second preceding example.

2. When dividing expressions that contain both exponential expressions and numerical factors, *subtract* the exponents but *divide* the numerical factors. For example,

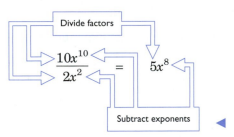

Divide factors

$$\frac{10x^{10}}{2x^2} = 5x^8$$

Subtract exponents ◀

Try out the division rule on these problems:

(a) $\dfrac{x^6}{x^3}$ (b) $\dfrac{6a^4}{2a}$ (c) $\dfrac{-12m^3n^5}{4m^2n^2}$ (d) $\dfrac{5^8}{5^2}$ (e) $\dfrac{4x^5 - 8x^4 + 6x^3}{2x^2}$

Check your answers in **66**.

---

66   (a) x^3 (b) $3a^3$ (c) $-3mn^3$ (d) 5^6 (e) $2x^3 - 4x^2 + 3x$

Did you have trouble with (e)? Unlike the other problems, which have only one term in the numerator (dividend), problem (e) has *three* terms in the numerator. Just imagine that the entire numerator is enclosed in parentheses and divide each of the three terms separately by the denominator (divisor). Write it out this way:

$$\frac{4x^5 - 8x^4 + 6x^3}{2x^2} = \frac{4x^5}{2x^2} - \frac{8x^4}{2x^2} + \frac{6x^3}{2x^2} = 2x^3 - 4x^2 + 3x$$

Note ▶ The rule for division confirms the fact stated in Section 6-4 that $a^0 = 1$. We know from arithmetic that any quantity divided by itself is equal to 1, so

$$\frac{a^n}{a^n} = 1 \qquad \text{(for } a \neq 0\text{)}$$

But according to the division rule,

$$\frac{a^n}{a^n} = a^{n-n} = a^0 \qquad \text{Therefore, } a^0 \text{ must be equal to 1. ◀}$$

Negative Exponents If we apply the division rule to a problem in which the divisor has a larger exponent than the dividend, the answer will contain a negative exponent. For example,

$$\frac{x^4}{x^6} = x^{4-6} = x^{-2}$$

If we use the "cancellation method" to do this problem, we have

$$\frac{x^4}{x^6} = \frac{x \cdot x \cdot x \cdot x}{x \cdot x \cdot x \cdot x \cdot x \cdot x} = \frac{1}{x^2} \qquad \text{Therefore, } x^{-2} = \frac{1}{x^2}.$$

In general, we define negative exponents as follows:

$$x^{-n} = \frac{1}{x^n}$$

Practice using the definition of negative exponents by giving two answers for each of the following problems—one with negative exponents, and one using fractions to eliminate the negative exponents.

(a) $\dfrac{a^2}{a^6}$ (b) $\dfrac{8x^2}{-2x^3}$ (c) $\dfrac{6^3}{6^5}$ (d) $\dfrac{5x^4y^2}{15xy^7}$ (e) $\dfrac{12x^6 - 9x^3 + 6x}{3x^2}$

The answers are given in **67**.

---

67 (a) a^{-4} or $\dfrac{1}{a^4}$ (b) $-4x^{-1}$ or $\dfrac{-4}{x}$

(c) 6^{-2} or $\dfrac{1}{6^2}$ (d) $\dfrac{1}{3}x^3y^{-5}$ or $\dfrac{x^3}{3y^5}$

(e) $4x^4 - 3x + 2x^{-1}$ or $4x^4 - 3x + \dfrac{2}{x}$

Now turn to **68** for a set of exercises on multiplication and division of algebraic expressions.

68

Exercises 7-8 **Multiplying and Dividing Algebraic Expressions**

A. Simplify by multiplying.

1. $5 \cdot 4x$ 2. $(a^2)(a^3)$

3. $-3R(-2R)$ 4. $3x \cdot 3x$

5. $(4x^2y)(-2xy^3)$ 6. $2x \cdot 2x \cdot 2x$

7. $0.4a \cdot 1.5a$

8. $x \cdot x \cdot A \cdot x \cdot 2x \cdot A^2$

9. $\frac{1}{2}Q \cdot \frac{1}{2}Q \cdot \frac{1}{2}Q \cdot \frac{1}{2}Q$

10. $(pq^2)(\frac{1}{2}pq)(2.4p^2q)$

11. $(-2M)(3M^2)(-4M^3)$

12. $2x(3x - 1)$

13. $-2(1 - 2y)$

14. $ab(a^2 - b^2)$

15. $p^2 \cdot p^4$

16. $3 \cdot 2x^2$

17. $5x^3 \cdot x^4$

18. $4x \cdot 2x^2 \cdot x^3$

19. $2y^5 \cdot 5y^2$

20. $2a^2b \cdot 2ab^2 \cdot a$

21. $xy \cdot x^2 \cdot xy^3$

22. $x(x + 2)$

23. $p^2(p + 2p)$

24. $xy(x + y)$

25. $5^6 \cdot 5^4$

26. $3^2 \cdot 3^7$

27. $10^2 \cdot 10^{-5}$

28. $2^{-6} \cdot 2^9$

29. $-3x^2(2x^2 - 5x^3)$

30. $6x^3(4x^4 + 3x^2)$

31. $2ab^2(3a^2 - 5ab + 7b^2)$

32. $-5xy^3(4x^2 - 6xy + 8y^2)$

B. Divide as indicated. Express all answers using positive exponents.

1. $\dfrac{4^5}{4^3}$

2. $\dfrac{10^9}{10^3}$

3. $\dfrac{x^4}{x}$

4. $\dfrac{y^6}{y^2}$

5. $\dfrac{10^4}{10^6}$

6. $\dfrac{m^5}{m^{10}}$

7. $\dfrac{8a^8}{4a^4}$

8. $\dfrac{-15m^{12}}{5m^6}$

9. $\dfrac{16y^2}{-24y^3}$

10. $\dfrac{-20t}{-5t^3}$

11. $\dfrac{-6a^2b^4}{-2ab}$

12. $\dfrac{36x^4y^5}{-9x^2y}$

13. $\dfrac{48m^2n^3}{-16m^6n^4}$

14. $\dfrac{15c^6d}{20c^2d^3}$

15. $\dfrac{-12a^2b^2}{20a^2b^5}$

16. $\dfrac{100xy}{-10x^2y^4}$

17. $\dfrac{6x^4 - 8x^2}{2x^2}$

18. $\dfrac{9y^3 + 6y^2}{3y}$

19. $\dfrac{12a^7 - 6a^5 + 18a^3}{-6a^2}$

20. $\dfrac{-15m^5 + 10m^2 + 5m}{5m}$

When you have completed these exercises, check your answers on page 600, then turn to **69** to learn about scientific notation.

7-9 SCIENTIFIC NOTATION

69 In technical work you will often deal with very small and very large numbers which may require a lot of space to write. For example, a certain computer function may take 5 nanoseconds (0.000000005 s), and in electricity, 1 kilowatt-hour (kWh) is equal to 3,600,000 joules (J) of work. Rather than take the time and space to write all the zeros, we use scientific notation to express such numbers.

Definition of Scientific Notation

In scientific notation 5 nanoseconds would be written as 5×10^{-9} s, and 1 kWh would be 3.6×10^6 J. As you can see, a number written in **scientific notation** is a product of a number between 1 and 10 and a power of 10. Stated formally, a number is expressed in scientific notation when it is in the form

$$P \times 10^k$$

where P is a number less than 10 and greater than or equal to 1, and k is an integer.

Numbers that are powers of ten are easy to write in scientific notation:

$$1000 = 1 \times 10^3$$

$$100 = 1 \times 10^2$$

$$10 = 1 \times 10^1 \qquad \text{and so on}$$

In general, to write a positive decimal number in scientific notation, first write it as a number between 1 and 10 times a power of 10, then write the power using an exponent.

For example,

$$3,600,000 = 3.6 \times 1,000,000$$

— A multiple of 10

— A number between 1 and 10

$$= 3.6 \times 10^6$$

$$0.0004 = 4 \times 0.0001 = 4 \times \frac{1}{10,000} = 4 \times \frac{1}{10^4} = 4 \times 10^{-4}$$

Note ▶ If you need to review negative exponents, see page 367. ◀

Converting to Scientific Notation

For a shorthand way of converting to scientific notation, follow these steps.

Step 1 If the number is given without a decimal point, rewrite it with a decimal point.

Example: 57,400 = 57,400.   0.0038

Decimal point   Decimal point already shown

Step 2 Place a mark ∧ after the first nonzero digit.

Example: 5 ∧ 7400.   0.003 ∧ 8

Step 3 Count the number of digits from the mark ∧ to the decimal point.

Example: 5 ∧ 7400.   0.003 ∧ 8

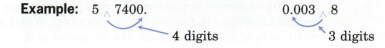

4 digits   3 digits

Step 4 Place the decimal point in the marked position and use the resulting number as the multiplier P in the scientific notation form. Discard any right end zeros on this number. Use the number of digits from the mark $_\wedge$ to the original decimal point as the exponent. If the shift from the $_\wedge$ to the original decimal point is to the right, the exponent is positive; if the shift is to the left, the exponent is negative.

Example: $5_\wedge 7400. = 5.74 \times 10^4$ $\qquad$ $0.003_\wedge 8 = 3.8 \times 10^{-3}$

Shift 4 digits right. Exponent is + 4. $\qquad$ Shift 3 digits left. Exponent is −3.

More Examples:

$150{,}000 = 1_\wedge 50000. = 1.5 \times 10^5$ $\qquad$ $0.00000205 = 0.000002_\wedge 5 = 2.05 \times 10^{-6}$

Shift 5 digits right. Exponent is + 5. $\qquad$ Shift 6 digits left. Exponent is −6.

$47 = 4_\wedge 7. = 4.7 \times 10^1$

Shift 1 digit right.

For practice, write the following numbers in scientific notation.

(a) 2900 $\qquad$ (b) 0.006 $\qquad$ (c) 1,960,100

(d) 0.0000028 $\qquad$ (e) 600 $\qquad$ (f) 0.0001005

Check your answers in **70**.

---

70 (a) $2{,}900 = 2_\wedge 900. = 2.9 \times 10^3$

Shift 3 digits right.

(b) $0.006 = 0.006_\wedge = 6 \times 10^{-3}$

Shift 3 digits left.

(c) $1{,}960{,}100 = 1_\wedge 960100. = 1.9601 \times 10^6$

Shift 6 digits right.

(d) $0.0000028 = 0.000002_\wedge 8 = 2.8 \times 10^{-6}$

Shift 6 digits left.

(e) $600 = 6_\wedge 00. = 6 \times 10^2$

Shift 2 digits right.

(f) $0.0001005 = 0.0001_\wedge 005 = 1.005 \times 10^{-4}$

Shift 4 digits left.

Learning Help ▶ When a positive number greater than or equal to 10 is written in scientific notation, the exponent is positive. When a positive number less than 1 is written in scientific notation, the exponent is negative. When a number between 1 and 10 is written in scientific notation, the exponent is 0. ◀

Converting from Scientific Notation to Decimal Form

In order to convert a number from scientific notation to decimal form, shift the decimal point as indicated by the power of 10—to the right for a positive exponent, to the left for a negative exponent. Attach additional zeros as needed.

Example:

Attach additional zeros.

$$3.2 \times 10^4 = 3\ 2000. = 32,000$$

Shift the decimal point 4 places right.

For a positive exponent, shift the decimal point to the right.

Example:

Attach additional zeros.

$$2.7 \times 10^{-3} = 0.002\ 7 = 0.0027$$

Shift the decimal point 3 places left.

For a negative exponent, shift the decimal point to the left.

Write each of the following numbers in decimal form.

(a) 8.2×10^3 (b) 1.25×10^{-6}

(c) 2×10^{-4} (d) 5.301×10^5

Check your answers in **71**.

---

71   (a) $8.2 \times 10^3 = 8\ 200\ . = 8200$

      (b) $1.25 \times 10^{-6} = 0.000001\ 25 = 0.00000125$

      (c) $2 \times 10^{-4} = 0.0002 = 0.0002$

      (d) $5.301 \times 10^5 = 5\ 30100. = 530,100$

Multiplying and Dividing in Scientific Notation

Scientific notation is especially useful when we must multiply or divide very large or very small numbers. Although most calculators have a means of converting decimal numbers to scientific notation and can perform arithmetic with scientific notation, it is important that you be able to do simple calculations of this kind quickly and accurately without a calculator.

To multiply or divide numbers given in exponential form, use the rules given in Section 7-8. (See frame **65**, page 365.)

$$a^m \times a^n = a^{m+n}$$

$$a^m \div a^n = \frac{a^m}{a^n} = a^{m-n}$$

To multiply numbers written in scientific notation, work with the decimal and exponential parts separately.

$$(A \times 10^B) \times (C \times 10^D) = (A \times C) \times 10^{B+D}$$

Example: $26{,}000 \times 3{,}500{,}000 = ?$

Step 1   Rewrite each number in scientific notation.

$$= (2.6 \times 10^4) \times (3.5 \times 10^6)$$

Step 2   Regroup to work with the decimal and exponential parts separately.

$$= (2.6 \times 3.5) \times (10^4 \times 10^6)$$

Step 3   Multiply using the rule for multiplying exponential numbers.

$$= 9.1 \times 10^{4+6}$$
$$= 9.1 \times 10^{10}$$

When dividing numbers written in scientific notation, it may help to think of the division as a fraction.

$$(A \times 10^B) \div (C \times 10^D) = \frac{A \times 10^B}{C \times 10^D} = \frac{A}{C} \times \frac{10^B}{10^D}$$

Therefore,

$$(A \times 10^B) \div (C \times 10^D) = (A \div C) \times 10^{B-D}$$

Example: $45{,}000 \div 0.0018 = ?$

Step 1   Rewrite. $= (4.5 \times 10^4) \div (1.8 \times 10^{-3})$

Step 2   Regroup. $= (4.5 \div 1.8) \times (10^4 \div 10^{-3})$

Step 3   Divide using the rules for exponential division.

$$= 2.5 \times 10^{4-(-3)}$$
$$= 2.5 \times 10^7 \qquad \boxed{4 - (-3) = 4 + 3 = 7}$$

If the result of the calculation is not in scientific notation, that is, if the decimal part is greater than 10 or less than 1, rewrite the decimal part so that it is in scientific notation.

Example: $0.0000072 \div 0.0009 = (7.2 \times 10^{-6}) \div (9 \times 10^{-4})$
$$= (7.2 \div 9) \times (10^{-6} \div 10^{-4})$$
$$= 0.8 \times 10^{-6-(-4)} \qquad \boxed{-6 - (-4) = -6 + 4 = -2}$$
$$= 0.8 \times 10^{-2}$$

But 0.8 is not a number between 1 and 10, so write it as

$0.8 = 8 \times 10^{-1}$. Then, $0.8 \times 10^{-2} = 8 \times 10^{-1} \times 10^{-2}$
$$= 8 \times 10^{-3} \quad \text{in scientific notation}$$

Perform the following calculations using scientific notation, and write the answer in scientific notation.

(a) $1600 \times 350{,}000$ (b) $64{,}000 \times 250{,}000$

(c) 2700×0.0000045 (d) $15{,}600 \div 0.0013$

(e) $0.000348 \div 0.087$ (f) $0.00378 \div 540{,}000{,}000$

Check your work in **72**.

72 (a) $(1.6 \times 10^3) \times (3.5 \times 10^5) = (1.6 \times 3.5) \times (10^3 \times 10^5)$
$$= 5.6 \times 10^{3+5}$$
$$= 5.6 \times 10^8$$

(b) $(6.4 \times 10^4) \times (2.5 \times 10^5) = (6.4 \times 2.5) \times (10^4 \times 10^5)$
$$= 16 \times 10^{4+5}$$
$$= 16 \times 10^9$$
$$= 1.6 \times 10^1 \times 10^9 \quad \Longleftarrow \boxed{16 = 1.6 \times 10^1}$$
$$= 1.6 \times 10^{10}$$

(c) $(2.7 \times 10^3) \times (4.5 \times 10^{-6}) = (2.7 \times 4.5) \times (10^3 \times 10^{-6})$
$$= 12.15 \times 10^{3 + (-6)} \quad \Longleftarrow \boxed{3 + (-6) = -3}$$
$$= 12.15 \times 10^{-3}$$
$$= 1.215 \times 10^1 \times 10^{-3}$$
$$= 1.215 \times 10^{-2}$$

(d) $(1.56 \times 10^4) \div (1.3 \times 10^{-3}) = (1.56 \div 1.3) \times (10^4 \div 10^{-3})$
$$= 1.2 \times 10^{4-(-3)} \quad \Longleftarrow \boxed{4 - (-3) = 4 + 3 = 7}$$
$$= 1.2 \times 10^7$$

(e) $(3.48 \times 10^{-4}) \div (8.7 \times 10^{-2}) = (3.48 \div 8.7) \times (10^{-4} \div 10^{-2})$
$$= 0.4 \times 10^{-4-(-2)} \quad \Longleftarrow \boxed{-4 - (-2) = -4 + 2 = -2}$$
$$= 0.4 \times 10^{-2}$$
$$= 4 \times 10^{-1} \times 10^{-2} \quad \Longleftarrow \boxed{0.4 = 4 \times 10^{-1}}$$
$$= 4 \times 10^{-3}$$

(f) $(3.78 \times 10^{-3}) \div (5.4 \times 10^8) = (3.78 \div 5.4) \times (10^{-3} \div 10^8)$
$$= 0.7 \times 10^{-3-8}$$
$$= 0.7 \times 10^{-11}$$
$$= 7 \times 10^{-1} \times 10^{-11}$$
$$= 7 \times 10^{-12}$$

CALCULATORS AND SCIENTIFIC NOTATION

If a very large or very small number contains too many digits to be shown on the display of a scientific calculator, it will be converted to scientific notation. For example, if you enter the product

6480000 $\boxed{\times}$ **75000** $\boxed{=}$

on a calculator, the answer will be displayed as

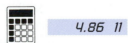 `4.86 11`

The space between the 4.86 and the 11 indicates that the number is in scientific notation, and 11 is the power of 10. Interpret this as

$6{,}480{,}000 \times 75{,}000 = 4.86 \times 10^{11}$

Similarly, the division

.000006 $\boxed{\div}$ **48000000** $\boxed{=}$

gives the display `1.25 -13`.

So $0.000006 \div 48{,}000{,}000 = 1.25 \times 10^{-13}$.

You may also enter numbers in scientific notation directly into your calculator using a key labeled EXP or EE . To enter

$0.000006 \div 48,000,000$ or $(6 \times 10^{-6}) \div (4.8 \times 10^7)$

enter

6 EXP **6** +/– ÷ **4.8** EXP **7** =

and your calculator will again display $\boxed{1.25 \quad -13}$.

Now turn to frame **73** for a set of problems on scientific notation.

73

Exercises 7-9 **Scientific Notation**

A.   Rewrite each number in scientific notation.

1. 5000	2. 450	3. 90	5. 40,700
5. 0.003	6. 0.071	7. 0.0004	8. 0.0059
9. 6,770,000	10. 38,200	11. 0.0292	12. 0.009901
13. 1001	14. 0.0020	15. 0.000107	16. 810,000
17. 31.4	18. 0.6	19. 125	20. 0.74

21.   Young's modulus for the elasticity of steel: 29,000,000 lb/in.^{2}

22.   Thermal conductivity of wood: 0.00024 cal/cm·s

23.   Power output: 95,500,000 watts

24.   Speed of light: 658,800,000 mph

B.   Rewrite each number in decimal form.

1. 2×10^5	2. 7×10^6	3. 9×10^{-5}	4. 3×10^{-4}
5. 1.7×10^{-3}	6. 3.7×10^{-2}	7. 5.1×10^4	8. 8.7×10^2
9. 4.05×10^4	10. 7.01×10^{-6}	11. 3.205×10^{-3}	12. 1.007×10^3
13. 2.45×10^6	14. 3.19×10^{-4}	15. 6.47×10^5	16. 8.26×10^{-7}

C.   Rewrite each of the following in scientific notation and calculate the answer in scientific notation. If necessary, convert your answer to scientific notation. Round to one decimal place if necessary.

1. $2000 \times 40,000$	2. 0.0037×0.0000024
3. $460,000 \times 0.0017$	4. $0.0018 \times 550,000$
5. $0.0000089 \div 3200$	6. $0.000125 \div 5000$
7. $45,500 \div 0.0091$	8. $12,450 \div 0.0083$
9. $2,240,000 \div 16,000$	10. $25,500 \div 1,700,000$
11. $0.000045 \div 0.00071$	12. $0.000086 \div 0.000901$

13. $9,501,000 \times 2410$

14. 9800×0.000066

15. $0.0000064 \div 80,000$

16. 1070×0.0000055

17. $\dfrac{0.000056}{0.0020}$

18. $\dfrac{0.0507}{43,000}$

19. $\dfrac{0.00602 \times 0.000070}{72,000}$

20. $\dfrac{2,780,000 \times 512,000}{0.000721}$

21. $64,000 \times 2800 \times 370,000$

22. $0.00075 \times 0.000062 \times 0.014$

23. $\dfrac{0.0517}{0.0273 \times 0.00469}$

24. $\dfrac{893,000}{5620 \times 387,000}$

D. Solve.

1. A brick wall 15 m by 25 m is 0.48 m thick. Under particular temperature conditions, the rate of heat flow through the wall, in calories per second, is given by the expression

$$(1.7 \times 10^{-4}) \times \left(\frac{1500 \times 2500}{48}\right)$$ Calculate the value of this quantity to the nearest tenth.

2. If an atom has a diameter of 4×10^{-8} cm and its nucleus has a diameter of 8×10^{-13} cm, find the ratio of the diameter of the nucleus to the diameter of the atom.

3. **Electrical Technology** The capacitance of a certain capacitor (in microfarads) can be found using the following expression. Calculate this capacitance.

$$(5.75 \times 10^{-8}) \times \left(\frac{8.00}{3.00 \times 10^{-3}}\right)$$

4. The energy (in joules) of a photon of visible light with a wavelength of 5.00×10^{-7} m is given by the following expression. Calculate this energy.

$$\frac{(6.63 \times 10^{-34}) \times (3.00 \times 10^{8})}{5.00 \times 10^{-7}}$$

Check your answers on page 600 then turn to 74 for a set of practice problems on the work of this chapter.

74   Answers are given on pages 600.

A.   Simplify.

1. $4x + 6x$

2. $6y + y + 5y$

3. $3xy + 9xy$

4. $6xy^3 + 9xy^3$

5. $3\frac{1}{3}x - 1\frac{1}{4}x - \frac{5}{8}x$

6. $8v - 8v$

7. $0.27G + 0.78G - 0.65G$

8. $7y - 8y^2 + 9y$

9. $3x \cdot 7x$

10. $(4m)(2m^2)$

11. $(2xy)(-5xyz)$

12. $3x \cdot 3x \cdot 3x$

13. $3(4x - 7)$

14. $2ab(3a^2 - 5b^2)$

15. $(4x + 3) + (5x + 8)$

16. $(7y - 4) - (2y + 3)$

17. $3(x + y) + 6(x - y)$

18. $(x^2 - 6) - 3(2x^2 - 5)$

19. $\dfrac{6^8}{6^4}$

20. $\dfrac{x^3}{x^8}$

21. $\dfrac{-12m^{10}}{4m^2}$

22. $\dfrac{6y}{-10y^3}$

23. $\dfrac{-16a^2b^2}{-4ab^3}$

24. $\dfrac{8x^4 - 4x^3 + 12x^2}{4x^2}$

Name

Date

Course/Section

B.   Find the value of each of the following. Round to two decimal places when necessary.

1. $L = 2W - 3$ for $W = 8$

2. $M = 3x - 5y + 4z$ for $x = 3, y = 5, z = 6$

3. $I = PRt$ for $P = 800, R = 0.06, t = 3$

4. $I = \dfrac{V}{R}$ for $V = 220, R = 0.0012$

5. $V = LWH$ for $L = 3\frac{1}{2}, W = 2\frac{1}{4}, H = 5\frac{3}{8}$

6. $N = (a + b)(a - b)$ for $a = 7, b = 12$

7. $L = \dfrac{s(P + p)}{2}$ for $s = 3.6, P = 38, p = 26$

8. $f = \dfrac{1}{8N}$ for $N = 6$

9. $t = \dfrac{D - d}{L}$ for $D = 12, d = 4, L = 2$

10. $V = \dfrac{gt^2}{2}$ for $g = 32.2, t = 4.1$

C. **Solve the following equations and proportions. Round your answer to two decimal places if necessary.**

1. $x + 5 = 17$

2. $m + 0.4 = 0.75$

3. $e - 12 = 32$

4. $a - 2.1 = -1.2$

5. $12 = 7 - x$

6. $4\frac{1}{2} - x = -6$

7. $5x = 20$

8. $-4m = 24$

9. $\frac{1}{2}y = 16$

10. $\dfrac{M}{5} = 12$

11. $\dfrac{B}{2} = \dfrac{7}{4}$

12. $\dfrac{x}{11} = \dfrac{17}{30}$

13. $\dfrac{75}{2} = \dfrac{1500}{y}$

14. $\dfrac{32}{20} = \dfrac{E}{30}$

15. $\dfrac{F}{2.4} = \dfrac{3.6}{12}$

16. $2x + 7 = 13$

17. $-4x + 11 = 35$

18. $0.75 - 5f = 6\frac{1}{2}$

19. $0.5x - 16 = -18$

20. $5y + 8 + 3y = 24$

21. $3g - 12 = g + 8$

22. $7m - 4 = 11 - 3m$

23. $3(x - 5) = 33$

24. $2 - (2x + 3) = 7$

25. $5x - 2(x - 1) = 11$

26. $2(x + 1) - x = 8$

Solve the following formulas for the variable shown.

27. $A = bH$ for b 　　　　　　　**28.** $R = S + P$ for P

29. $P = 2L + 2W$ for L 　　　　**30.** $P = \dfrac{w}{F}$ for F

31. $S = \frac{1}{2}gt - 4$ for g 　　　　**32.** $V = \pi R^2 H - AB$ for A

D.　Rewrite in scientific notation.

1. 7500 　　　**2.** 12,800 　　　**3.** 0.041 　　　**4.** 0.000236

5. 0.00572 　**6.** 0.00000482 　**7.** 447,000 　**8.** 2,127,000

9. 80,200,000 　**10.** 46,710 　**11.** 0.00000705 　**12.** 0.001006

Rewrite as a decimal.

13. 9.3×10^5 　**14.** 6.02×10^3 　**15.** 2.9×10^{-4} 　**16.** 3.05×10^{-6}

17. 5.146×10^{-7} 　**18.** 6.203×10^{-3} 　**19.** 9.071×10^4 　**20.** 4.006×10^6

Calculate using scientific notation, and write your answer in standard scientific notation. Round to one decimal place if necessary.

21. $45,000 \times 1,260,000$ 　　　　**22.** $625,000 \times 12,000,000$

23. 0.0007×0.0043 　　　　　**24.** 0.0000065×0.032

25. $56,000 \times 0.0000075$ 　　　　**26.** $1,020 \times 0.00055$

27. $0.0074 \div 0.00006$ 　　　　　**28.** $0.000063 \div 0.0078$

29. $96,000 \div 3,400,000$ 　　　　**30.** $26,500,000 \div 12,000$

31. $0.00089 \div 37,000$ 　　　　　**32.** $123,500 \div 0.00077$

E.　Practical Problems

1.　Aeronautical Mechanics In weight and balance calculations, airplane mechanics are concerned with the center of gravity of an airplane. The center of gravity may be calculated from the formula

$$CG = \frac{100(H - x)}{L}$$

where H is the distance from the datum to the empty CG, x is the distance from the datum to the leading edge of the mean aerodynamic chord (MAC), and L is the length of the MAC. CG is expressed as a percent of the MAC.

(All lengths are in inches.)
(a)　Find the center of gravity if $H = 180$, $x = 155$, and $L = 80$.
(b)　Solve the formula for L, and find L when CG = 30%, $H = 200$, and $x = 150$.

(c) Solve the formula for H, and find H when CG = 25%, $x = 125$, and $L = 60$.

(d) Solve the formula for x, and find x when CG = 28%, $H = 170$, and $L = 50$.

2. **Manufacturing**   Two belted pulleys have diameters of 24 in. and 10 in.

   (a) Find the pulley ratio.

   (b) If the larger pulley turns at 1500 rpm, how fast will the smaller pulley turn?

3. **Electrical Technology**   Electricians use a formula which states that the level of light in a room, in foot-candles, is equal to the product of the fixture rating, in lumens, the coefficient of depreciation, and the coefficient of utilization, all divided by the area, in square feet.

   (a) State this as an algebraic equation.

   (b) Find the level of illumination for four fixtures rated at 2800 lumens each if the coefficient of depreciation is 0.75, the coefficient of utilization is 0.6, and the area of the room is 120 sq ft.

   (c) Solve for area.

   (d) Use your answer to (c) to determine the size of the room in which a level of 60 foot-candles can be achieved with 10,000 lumens, given that the coefficient of depreciation is 0.8 and the coefficient of utilization is 0.5.

4. **Printing**   If six printing presses can do a certain job in $2\frac{1}{2}$ hours, how long will it take four presses to do the same job? (*Hint:* Use an inverse proportion.)

5. **Sheet Metal Technology**   The length of a piece of sheet metal is four times its width. The perimeter of the sheet is 80 cm. Set up an equation relating these measurements and solve the equation to find the dimensions of the sheet.

6. **Drafting**   Drafters usually use a scale of $\frac{1}{4}$ in. = 1 ft on their drawings. Find the actual length of each of the following items if their blueprint length is given.

   (a) A printing press $2\frac{1}{4}$ in. long.   (b) A building $7\frac{1}{2}$ in. long.

   (c) A bolt $\frac{1}{32}$ in. long.   (d) A car $1\frac{7}{8}$ in. long.

7. **Construction Technology**   A certain concrete mix requires one sack of cement for every 650 lb of concrete. How many sacks of cement are needed for 3785 lb of concrete? (*Remember:* You cannot buy a fraction of a sack.)

8. **Construction Technology**   Structural engineers have found that a good estimate of the crushing load for a square wooden pillar is given by the formula

$$L = \frac{25T^4}{H^2}$$ where L is the crushing load in tons, T is the thickness of the wood in inches, and H is the height of the post in feet

   Find the crushing load for a 6-in.-thick post 12 ft high.

9. **Auto Mechanics**   The rear axle of a car turns at 320 rpm when the drive shaft turns at 1600 rpm. If the pinion gear has 12 teeth, how many teeth are on the ring gear?

10. If a 120-lb force is applied to a 36-in. crowbar pivoted 4 in. from the end, what lift force is exerted?

11. **Machine Technology**   How many turns will a pinion gear having 16 teeth make if a ring gear having 48 teeth makes 120 turns?

12. **Machine Technology**   Two shims must have a combined thickness of 0.048 in. One shim must be twice as thick as the other. Set up an equation and solve for the thickness of each shim.

13. **Electrical Technology**   The current in a certain electrical circuit is inversely proportional to the line voltage. If a current of 0.40 ampere is delivered at 420 volts, what is the current of a similar system operating at 120 volts?

14. **Printing**   A printer needs 12,000 good copies of a flyer for a particular job. Previous experience has shown that she can expect no more than a 5% spoilage rate on this type of job. How many should she print to assure 12,000 clean copies?

15. **Office Services**   Sal, Martin, and Gloria are partners in a welding business. Because of the differences in their contributions, it is decided that Gloria's share of the profits will be one-and-a-half times as much as Sal's share, and

that Martin's share will be twice as much as Gloria's share. How should a profit of $48,200 be divided?

16. **Electrical Engineering**   The calculation shown is used to find the velocity of an electron at an anode. Calculate this velocity (units are in meters per second).

$$\sqrt{\frac{2(1.6 \times 10^{-19})(2.7 \times 10^2)}{9.1 \times 10^{-31}}}$$

17. **Plumbing**   The formula

$$N = \sqrt{\left(\frac{D}{d}\right)^5}$$

gives the approximate number N of smaller pipes of diameter d necessary to supply the same total flow as one larger pipe of diameter D. Unlike the formula in problem 22 on p. 300, this formula takes into account the extra friction caused by the smaller pipes. Use this formula to determine the number of $\frac{1}{2}$-in. pipes that will provide the same flow as one $1\frac{1}{2}$-in. pipe.

18. **Sheet Metal Technology**   The length L of the stretchout for a square pipe with a grooved seam is given by

$$L = 4s + 3w$$

where s is the side length of the square end and w is the width of the lock. Find the length of stretchout for a square pipe $1\frac{3}{4}$ in. on a side with a lock width of $\frac{3}{16}$ in.

19. **Auto Mechanics**   If a mechanic's helper is paid $456.28 for $38\frac{1}{4}$ hours of work, how much should he receive for $56\frac{1}{2}$ hours at the same rate?

20. **Manufacturing**   A pair of belted pulleys have diameters of $8\frac{3}{4}$ in. and $5\frac{5}{8}$ in. If the smaller pulley turns at 850 rpm, how fast will the large one turn?

21. **Refrigeration and Air Conditioning**   In planning a solar energy heating system for a house, a contractor uses the formula

$$Q = 8.33GDT \qquad \text{to determine the energy necessary for heating water.}$$

In this formula Q is the energy in Btu, G is the number of gallons heated per day, D is the number of days, and T is the temperature difference between tap water and the desired temperature of hot water. Find Q when G is 50 gallons, D is 30 days, and the water must be heated from 60° to 140°.

22. **Drafting**   Each production employee in a plant requires an average of 125 sq ft of work area. How many employees will be able to work in an area that measures $2\frac{1}{16}$ in. by $8\frac{5}{8}$ in. on a blueprint if the scale of the drawing is $\frac{1}{32}$ in. = 1 ft?

23.   Find the missing dimension in the following pair of similar figures.

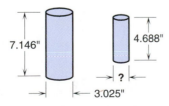

24. **Machine Technology**   In a closed container, the pressure is inversely proportional to the volume when the temperature is held constant. Find the pressure of a gas compressed to 0.386 cu ft if the pressure is 12.86 psi at 2.52 cu ft.

25. **Sheet Metal Technology**   The bend allowance for sheet metal is given by the formula

$$BA = N(0.01743R + 0.0078T)$$ where N is the angle of the bend in degrees, R is the inside radius of the bend in inches, and T is the thickness of the metal in inches.

Find BA if N is 47°, R is 0.725 in., and T is 0.0625 in. Round to three significant digits.

26. **Machine Technology** To find the taper per inch of a piece of work, a machinist uses the formula

$$T = \frac{D - d}{L}$$ where D is the diameter of the large end, d is the diameter of the small end, and L is the length.

Find T if $D = 4.1625$ in., $d = 3.2513$ in., and $L = 8$ in.

27. **Construction Technology** The modulus of elasticity of a beam is 2,650,000 at a deflection limit of 360. If the modulus is directly proportional to the deflection limit, find the modulus at a deflection limit of 240. Round to three significant digits.

Practical
Plane Geometry

Objective	Sample Problems		Where to Go for Help	
			Page	Frame

When you finish this unit you will be able to:

1. Measure angles with a protractor.

$\sphericalangle ABC =$ _____   385   **1**

2. Use simple geometric relationships involving intersecting lines and triangles.

(a)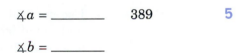

Assume that the two horizontal lines are parallel.

$\sphericalangle a =$ _____   389   **5**

$\sphericalangle b =$ _____

$\sphericalangle c =$ _____

$\sphericalangle d =$ _____

$\sphericalangle e =$ _____

(b)

$\sphericalangle x =$ _____   391   **9**

3. Identify polygons, including triangles (right, isosceles, equilateral), squares, rectangles, parallelograms, trapezoids, and hexagons.

(a) _____   409   **27**

(b) _____   398   **15**

(c) _____

(d) _____

(e) _____   418   **36**

Name _____

Date _____

Course/Section _____

4. Use the Pythago-
rean theorem.

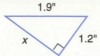

Round to one decimal
place.

$x =$ _____ 411 **28**

5. Find the area and
perimeter of geo-
metric figures

(a)

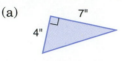

Area = _____ 414 **31**

(b)

Area = _____ 403 **22**

(c)

Area = _____
(Round to two decimal places.)

Perimeter = _____ 418 **36**

(d)

Area = _____ 431 **48**

$r = 2.70"$

Circumference = _____ 427 **44**

Use $\pi \approx 3.14$
and round both
answers to the
nearest tenth.

(e) Find the area of a
ring with outside
diameter 6.5 in.
and inside diame-
ter 4.2 in. Round
to the nearest
tenth.   _____ 433 **49**

(Answers to these preview problems are given on page 578.)

If you are certain that you can work *all* these problems correctly, turn to page 439
for a set of practice problems. If you cannot work one or more of the preview problems, turn to the page indicated to the right of the problem. Those who wish to
master this material with the greatest success should turn to frame **1** and begin
work there.

8 Practical Plane Geometry

"NO! NO! I SAID BUILD AN ARK!"

© 1974 Reprinted by permission of *Saturday Review* and Orlando Busino and Omni International.

1 Geometry is one of the oldest branches of mathematics. It involves study of the properties of points, lines, plane surfaces, and solid figures. Ancient Egyptian engineers used these properties when they built the Pyramids, Noah used them to build the Ark, and modern trades workers use them when cutting sheet metal, installing plumbing, building cabinets, and performing countless other tasks. When they study formal geometry, students concentrate on theory and actually prove the truth of various geometric theorems. In this chapter we concentrate on the applications of the ideas of plane geometry to do technical work and the trades.

8-1 ANGLE MEASUREMENT

In geometry, the word *plane* is used to refer to a mathematical concept describing an infinite set of points in space. For practical purposes, a plane surface can be thought of as a perfectly flat, smooth surface such as a tabletop or a window pane.

Labeling Angles

An *angle* is a measure of the size of the opening between two intersecting lines. The point of intersection is called the *vertex,* and the lines forming the opening are called the *sides*.

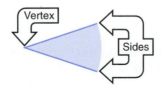

An angle may be identified in any one of the following ways:

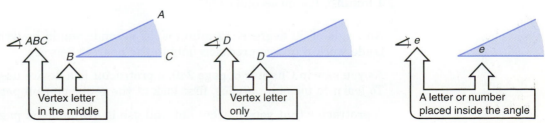

The angle symbol ∡ is simply a shorthand way to write the word "angle."

For the first angle, the middle letter is the vertex letter of the angle, while the other two letters represent points on each side. Notice that capital letters are used. The second angle is identified by the letter *D* near the vertex. The third angle is named by a small letter or number placed inside the angle.

The first and third methods of naming angles are most useful when several angles are drawn with the same vertex. For example,

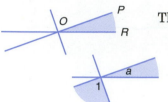

The interior of angle *POR* (or *ROP*) is shaded.

The interiors of angles *a* and 1 are shaded.

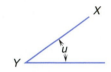

Name the angle shown in the margin using the three different methods.

See frame **2** to check your answers.

---

2   ∡*XYZ* or ∡*ZYX* or ∡*Y* or ∡*u*

Measuring Angles   Naming or labeling an angle is the first step. Measuring it may be even more important. Angles are measured according to the size of their opening. The length of the sides of the angle is not related to the size of the angle.

and   are the same size angle.

The basic unit of measurement of angle size is the *degree*. Because of traditions going back thousands of years, we define a degree as an angle that is $\frac{1}{360}$ of a full circle, or $\frac{1}{90}$ of a quarter circle.

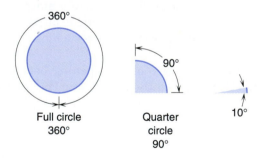

Full circle
360°

Quarter circle
90°

10°

The degree can be further subdivided into finer angle units known as *minutes* and *seconds* (which are not related to the time units).

1 degree, 1° = 60 minutes = 60′

1 minute, 1′ = 60 seconds = 60″

An angle of 62 degrees, 12 minutes, 37 seconds would be written 62°12′37″. Most trade work requires precision only to the nearest degree.

As you saw in Chapter 5, page 240, a protractor is a device used to measure angles. To learn to measure angles, first look at the protractor on page 387.

A protractor that you may cut out and use is available on page 610.

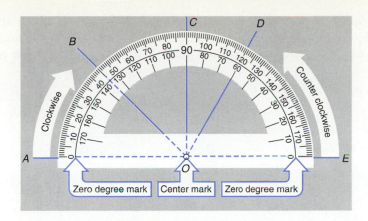

To measure an angle, place the protractor over it so that the zero-degree mark is lined up with one side of the angle, and the center mark is on the vertex. Read the measure in degrees where the other side of the angle intersects the scale of the protractor. When reading an angle clockwise, use the upper scale and, when reading counterclockwise, use the lower scale.

In the drawing, ∡*EOB* should be read counterclockwise from the 0° mark on the right. ∡*EOB* = 135°.

∡*AOD* should be read clockwise from the 0° mark on the left. ∡*AOD* = 120°.

Find (a)   ∡*EOD*   (b)   ∡*AOB*

Check your answers in **3**.

---

3   (a)   ∡*EOD* = 60°   Measure counterclockwise from the right 0° mark and read the lower scale.

(b)   ∡*AOB* = 45°   Measure clockwise from the left 0° mark and read the upper scale.

Classifying Angles

An *acute* angle is an angle *less than* 90°. Angles *AOB* and *EOD* are both less than 90°, so both are acute angles.

Acute angles:

An *obtuse* angle is an angle *greater than* 90° and less than 180°. Angles *AOD* and *EOB* are both greater than 90°, so both are obtuse angles.

Obtuse angles:

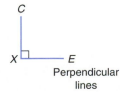

Perpendicular lines

A *right* angle is an angle exactly equal to 90°. Angle *EXC* in the drawing at the left is a right angle.

Two straight lines that meet in a right or 90° angle are said to be *perpendicular*. In this drawing, *CX* is perpendicular to *XE*.

Notice that a small square is placed at the vertex of a right angle to show that the sides are perpendicular.

A *straight* angle is an angle equal to 180°.

Angle *BOA* shown here is a straight angle.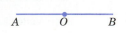

Use your protractor to measure each of the angles given and tell whether each is acute, obtuse, right, or straight. Estimate the size of the angle before you measure. (You will need to extend the sides of the angles to fit your protractor.)

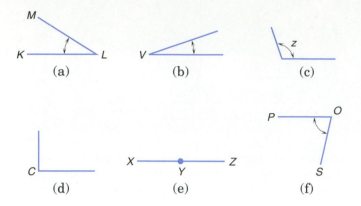

(a)   (b)   (c)

C   X ——●—— Z   P ——— O
                Y   S

(d)   (e)   (f)

Check your answers in **4**.

---

4   (a)   ⊰$KLM = 33°$, an acute angle

(b)   ⊰$V = 20°$, an acute angle

(c)   ⊰$z = 113°$, an obtuse angle

(d)   ⊰$C = 90°$, a right angle

(e)   ⊰$XYZ = 180°$, a straight angle

(f)   ⊰$POS = 80°$, an acute angle

Two angles that add up to 90° are called *complementary* angles. Since 40° + 50° = 90°, the 40° angle is called the *complement* of 50°, and the 50° angle is the complement of the 40° angle.

Two angles that add up to 180° are called *supplementary* angles. Since 120° + 60° = 180°, the 120° angle is called the *supplement* of 60°, and the 60° angle is the supplement of the 120° angle.

Drawing Angles   In addition to measuring angles that already exist, trades workers will also need to draw their own angles. To draw an angle of a given size follow these steps:

Step 1   Draw a line representing one side of the angle.

Step 2   Place the protractor over the line so that the center mark is on the vertex, and the 0° mark coincides with the other end of the line. (**Note:** We could have chosen either end of the line for the vertex.)

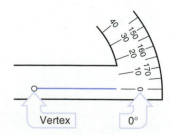

Vertex   0°

Step 3 Now place a small dot above the degree mark corresponding to the size of your angle. In this case the angle is 30°.

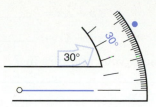

Step 4 Remove the protractor and connect the dot to the vertex.

Now you try it. Use your protractor to draw the following angles.

$\angle A = 15°$ $\angle B = 115°$ $\angle C = 90°$

Check your work in **5**.

_____

5 Here are our drawings:

Angle Facts There are several important geometric relationships involving angles that you should know. For example, measure the four angles created by the intersecting lines shown.

$a =$ _____ $b =$ _____ $c =$ _____ $d =$ _____

Measure them now, then check your answers in **6**.

_____

6 $\angle a = 30°$ $\angle b = 150°$ $\angle c = 30°$ $\angle d = 150°$

Did you notice that opposite angles are equal?

For any pair of intersecting lines the opposite angles are called *vertical angles*. In the preceding drawing, angles a and c are a pair of vertical angles. Angles b and d are also a pair of vertical angles.

 are pairs of vertical angles.

An important geometric rule is that

┌───┐
│ When two straight lines intersect, the vertical angles are always equal. │
└───┘

$\angle p = \angle r$
and $\angle q = \angle s$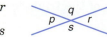

If you remember that a straight angle = 180°, you can complete the following statements for the preceding drawing.

$\angle p + \angle q = $ _____ $\angle p + \angle s = $ _____

$\angle s + \angle r = $ _____ $\angle q + \angle r = $ _____

Complete these, then check your answers in **7**.

---

7 $\angle p + \angle q = 180°$ $\angle p + \angle s = 180°$

$\angle s + \angle r = 180°$ $\angle q + \angle r = 180°$

This leads to a second important geometry relation:

> When two lines intersect, each pair of adjacent angles sums to 180°.

Use these relationships to answer the following questions.

(a) $\angle a = $ _____   (b) $\angle b = $ _____   (c) $\angle c = $ _____

Check your answers in **8**.

---

8 (a) $\angle a = 40°$ since   vertical angles are equal

(b) $\angle b = 140°$ since $\angle a + \angle b = 180°$

(c) $\angle c = 140°$ since   vertical angles are equal, $\angle b = \angle c$

A triangle is formed from three distinct line segments. The line segments are called the *sides* of the triangle, and the points where they meet are called its *vertices*. Every triangle forms three angles, and these are related by an important geometric rule.

Use your protractor to measure each angle in the following triangle, *ABC*.

$\angle A = $ _____

$\angle B = $ _____

$\angle C = $ _____

Check your measurements in **9**.

---

9 $\angle A = 65°$ $\angle B = 40°$ $\angle C = 75°$

Notice that the three angles of the triangle sum to 180°.

$65° + 40° + 75° = 180°$

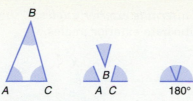

Measuring angles on many triangles of every possible size and shape leads to the third geometry relationship:

> The interior angles of a triangle always add to 180°.

Use this fact to solve the following problems.

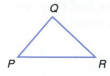

(a) If $\angle P = 50°$ and $\angle Q = 85°$, find $\angle R$.

(b) If $\angle P = 27° \, 42'$ and $\angle Q = 76° \, 25'$, find $\angle R$.

Check your answers in **10**.

---

10   (a)   By the relationship given, $\angle P + \angle Q + \angle R = 180°$, so

$\angle R = 180° - \angle P - \angle Q$
$\angle R = 180° - 50° - 85°$
$\angle R = 45°$

(b) $\angle R = 180° - 27° \, 42' - 76° \, 25'$
$= 180° - 103° \, 67'$
$= 180° - 104° \, 7'$ Use $180° = 179° \, 60'$.
$= 75° \, 53'$

A fourth and final geometry fact you will find useful involves parallel lines. Two straight lines are said to be *parallel* if they always are the same distance apart and never meet. If a pair of parallel lines is cut by a third line, several angles are formed.

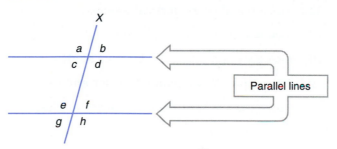

Angles c and f are called *alternate interior angles*. Angles d and e are also alternate interior angles.

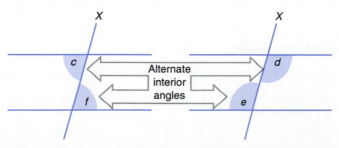

The angles are *interior* (or inside) the original parallel lines and they are on *alternate* sides of the line XY. The alternate interior angles are equal.

Angles *a* and *h* are called *alternate exterior angles*.
Angles *b* and *g* are also alternate exterior angles.

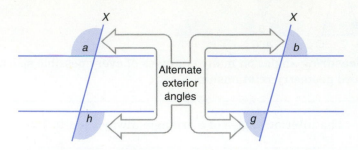

The angles are *exterior* (or outside) the original parallel lines and they are on *alternate* sides of the line *XY*. The alternate exterior angles are equal.

The geometric rule is:

> When parallel lines are cut by a third line, the alternate exterior angles are equal, and the alternate interior angles are equal.

Use this rule and the previous ones to answer the following questions. Assume that the two horizontal lines are parallel.

∢*p* = _____   ∢*q* = _____

∢*r* = _____   ∢*s* = _____

∢*t* = _____   ∢*u* = _____

∢*w* = _____

Check your work in **11**.

---

11   ∢*p* = 110° since 70° + ∢*p* = 180°.

∢*q* = 110° since *p* and *q* are vertical angles.

∢*r* = 70° since it is a vertical angle to the 70° angle.

∢*s* = 70° since ∢*s* = ∢*r*, alternate interior angles.

∢*t* = 110° since ∢*t* = ∢*q*, alternate interior angles.

∢*u* = 110° since ∢*u* = ∢*p*, alternate exterior angles.

∢*w* = 70° since ∢*u* + ∢*w* = 180°.

Now, ready for some problems on angle measure? Turn to **12** for practice in measuring and working with angles.

12

Exercises 8-1   **Angle Measurement**

A. Solve the following. If you need a protractor, cut out and use the one on page 610.

1. Name each of the following angles using the three kinds of notation shown in the text.

(a)   (b)   (c)

(d) T   (e)   (f)  R   S   T

2. For this figure,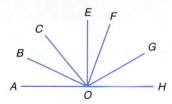

   (a) Name an acute angle that has *AO* as a side.

   (b) Name an acute angle that has *HO* as a side.

   (c) Name an obtuse angle that has *AO* as a side.

   (d) Name a right angle.

   (e) Use your protractor to measure ∡*AOB*, ∡*GOF*, ∡*HOC*, ∡*FOH*, ∡*BOF*, ∡*COG*.

3. Use your protractor to draw the following angles.

   (a) ∡*LMN* = 29°   (b) ∡3 = 100°

   (c) ∡*Y* = 152°   (d) ∡*PQR* = 137°

   (e) ∡*t* = 68°   (f) ∡*F* = 48°

4. Measure the indicated angles on the following shapes. (You will need to extend the sides of the angles to improve the accuracy of your measurement.)

   (a)    (b)   (c)

   (d)   (e)   (f)

   (g)   (h)   (i)

B. Use the geometry relationships to answer the following questions.

   1. In each problem one angle measurement is given. Determine the others for each figure without using a protractor.

      (a) 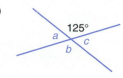   (b) 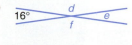   (c)

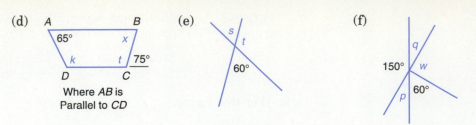

(d) *A*   *B*
65°   *x*
k   *t* 75°
D   *C*
Where *AB* is
Parallel to *CD*

(e)
s
t
60°

(f)
q
150° *w*
p 60°

2. For each triangle shown, two angles are given. Find the third angle without using a protractor.

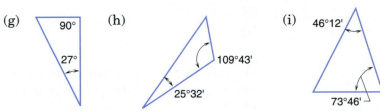

(a) *A*
60°   52°
B   *C*

(b) *D*
23°
20°
E
F

(c) *G*
90° 48°
H   *K*

(d)
27°
20°

(e)   (The angles
labelled *a*
40° are equal.)
a   *a*

(f)
100°
27°

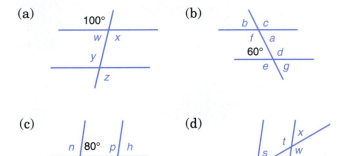

(g)
90°
27°

(h)
109°43'
25°32'

(i) 46°12'
73°46'

3. In each figure, two parallel lines are cut by a third line. Find the angles marked.

(a)
100°
w *x*
y
z

(b)
b *c*
f *a*
60° *d*
e *g*

(c)
n 80° *p* *h*
q *m*   *k*

(d)
x
t
s *w*
z *u*
y
40°

C. Applied Problems

1. **Machine Technology** A machinist must punch holes *A* and *B* in the piece of steel shown by rotating the piece through a certain angle from vertex *C*. What is angle *ACB*?

Problem 1

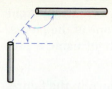

Problem 2

2. **Plumbing** A plumber must connect the two pipes shown by first selecting the proper elbows. What angle elbows does he need?

3. **Carpentry** A carpenter needs to build a triangular hutch that will fit into the corner of a room as shown. Measure the three angles of the hutch.

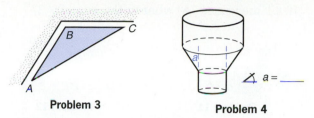

Problem 3

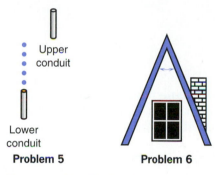

∡ *a* = _____

Problem 4

4. **Sheet Metal Technology** A sheet metal worker must connect the two vent openings shown. At what angle must the sides of the connecting pieces flare out?

5. **Electrical Technology** An electrician wants to connect the two conduits shown with a 45° elbow. How far up must she extend the lower conduit before they will connect at that angle? Give the answer in inches. (*Hint:* Draw the 45° connection on the upper conduit first.) Each dot represents a 1-in. extension.

Upper conduit

Lower conduit

Problem 5   **Problem 6**

6. **Construction Technology** At what angle must the rafters be set to create the roof gable shown in the drawing?

7. **Machine Technology** A machinist receives a sketch of a part he must make as shown. If the indicated angle *BAC* is precisely half of angle *BAD*, find ∡*BAD*.

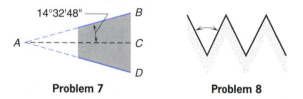

14°32'48"

Problem 7   **Problem 8**

8. **Machine Technology** The drawing illustrates a cross section of a V-thread. Find the indicated angle.

9. **Carpentry** At what angles must drywall board be cut to create the wall shown in the figure?

Problem 9

Problem 10

10. **Carpentry** A carpenter wishes to make a semicircular deck by cutting eight equal angular pieces of wood as shown. Measure the angle of one of the pieces. Was there another way to determine the angle measure without using a protractor?

11. **Carpentry** A carpenter's square is placed over a board as shown in the figure. If ∡1 measures 74° what is the size of ∡2?

Problem 11   **Problem 12**

12. Board 1 must be joined to board 2 at a right angle. If ∡x measures 42°, what must ∡y measure?

When you have finished these problems check your answers on pages 602 then continue in **13** with the study of plane figures.

___

8-2 AREA AND PERIMETER OF POLYGONS

13

Polygons   A *polygon* is a closed plane figure containing three or more angles and bounded by three or more straight sides. The word "polygon" itself means "many sides." The following figures are all polygons.

A figure with a curved side is *not* a polygon.

 and [curved figure] are not polygons

Every trades worker will find that an understanding of polygons is important. In this section you will learn how to identify the parts of a polygon, recognize some different kinds of polygons, and compute the perimeter and area of certain polygons.

In the general polygon shown, each *vertex* or corner is labeled with a letter: *A, B, C, D,* and *E.* The polygon is named simply "polygon *ABCDE*"—not a very fancy name, but it will do.

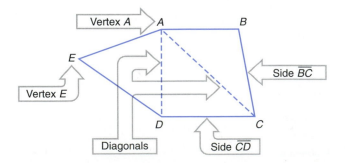

The *sides* of the polygon are named after the line segments that form each side: $\overline{AB}$, $\overline{BC}$, $\overline{CD}$, and so on. Notice that we place a bar over the letters to indicate that we are talking about a line segment. A *line segment* is a finite portion of a straight line; it has two endpoints.

The *diagonals* of the polygon are the line segments connecting nonconsecutive vertices such as $\overline{AC}$ and $\overline{AD}$. These are shown as dashed lines in the figure.

In the following polygon, identify all sides, vertices, and diagonals.

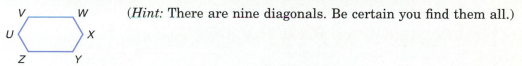

(*Hint:* There are nine diagonals. Be certain you find them all.)

See **14** to check your answers.

---

14   The sides are $\overline{UV}$, $\overline{VW}$, $\overline{WX}$, $\overline{XY}$, $\overline{YZ}$, and $\overline{ZU}$.

Note ▶   The letters naming a side can be reversed: side $\overline{VU}$ is the same as side $\overline{UV}$. ◀

The vertices are *U, V, W, X, Y,* and *Z.*

The diagonals are $\overline{UW}$, $\overline{UX}$, $\overline{UY}$, $\overline{VX}$, $\overline{VY}$, $\overline{VZ}$, $\overline{WY}$, $\overline{WZ}$, and $\overline{XZ}$.

Perimeter   For those who do practical work with polygons, the most important measurements are the lengths of the sides, the *perimeter,* and the *area.* The perimeter of any polygon is simply the sum of the lengths of its sides. It is the distance around the outside of the polygon.

In the polygon *KLMN* the perimeter is

3 in. + 5 in. + 6 in. + 4 in. = 18 in.

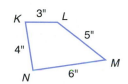

The perimeter is a length; therefore, it has length units, in this case inches.

Find the perimeter of each of the following polygons.

(a)
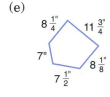

(b)

(c)
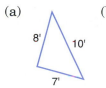

(d)
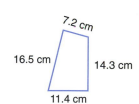

(e)
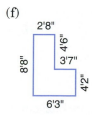

(f)
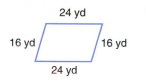

(g)
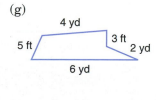

Check your answers in **15**.

---

15   (a)   25 ft   (b)   20 in.   (c)   80 yd   (d)   49.4 cm

(e) $42\frac{5}{8}$ in.   (f)   29 ft 10 in.   (g)   14 yd 2 ft or 44 ft

Adding up the lengths of the sides will always give you the perimeter of a polygon, but handy formulas are needed to enable you to calculate the area of a polygon. Before you can use these formulas you must learn to identify the different types of polygons. Let's look at a few. In this section we examine the *quadrilaterals* or four-sided polygons.

Quadrilaterals

Figure *ABCD* is a *parallelogram*. In a parallelogram opposite sides are parallel and equal in length.

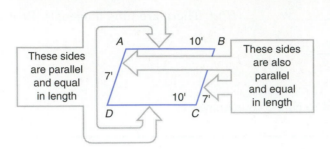

Figure *EFGH* is a *rectangle,* a parallelogram in which the four corner angles are right angles. The ∟ symbols at the vertices indicate right angles. Since a rectangle is a parallelogram, opposite sides are parallel and equal.

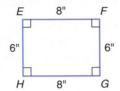

IJKL is a *square,* a rectangle in which *all* sides are the same length.

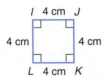

MNOP is called a *trapezoid*. A trapezoid contains two parallel sides and two non-parallel sides. $\overline{MN}$ and $\overline{OP}$ are parallel; $\overline{MP}$ and $\overline{NO}$ are not.

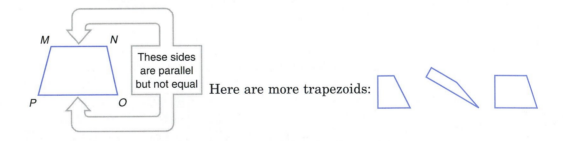

Here are more trapezoids:

If a four-sided polygon has none of these special features—no parallel sides—we call it a *quadrilateral*. *QRST* is a quadrilateral.

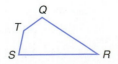

Name each of the following quadrilaterals. If more than one name fits, give the most specific name.

(a)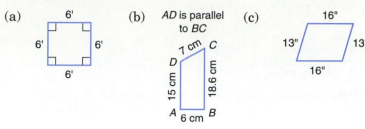

(b) *AD* is parallel to *BC*

(c)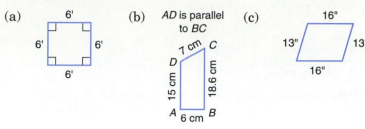

(d)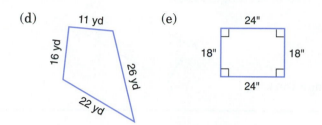

(e)

Look in **16** for the correct answers.

---

16 (a) Square. (It is also a rectangle and a parallelogram, but "square" is a more specific name.)

(b) Trapezoid.

(c) Parallelogram.

(d) Quadrilateral.

(e) Rectangle. (It is also a parallelogram, but "rectangle" is a more specific name.)

Once you can identify a polygon, you can use a formula to find its area. Next let's examine and use some area formulas for these quadrilaterals.

Rectangles *The area of any plane figure is the number of square units of surface within the figure.*

Example: In this rectangle

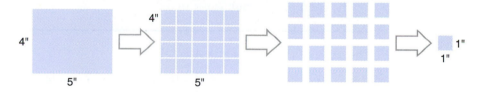

we can divide the surface into exactly 20 small squares each one inch on a side. By counting squares we can see that the area of this 4-in. by 5-in. rectangle contains 20 square inches. We abbreviate these area units as 20 square inches, 20 sq in., or 20 in.^{2}.

Of course there is no need to draw lines in this messy way. We can find the area by multiplying the two dimensions of the rectangle.

AREA OF A RECTANGLE

$A = \text{length} \times \text{width}$

$A = LW$

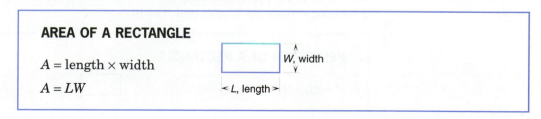

Another Example: Find the area of rectangle *EFGH*.

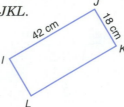

$L = 9$ in.

$W = 5$ in.

A, area $= LW = (9$ in.$)(5$ in.$)$

$A = 45$ sq in.

Careful ▶ The formula is easy to use, but be careful. It is valid only for rectangles. If you use this formula with a different polygon, you will not get the correct answer. ◀

Find the area of rectangle *IJKL*.

Check your work in **17**.

---

17 First, write the formula: $A = LW$

Second, substitute *L* and *W*: $A = (42$ cm$)(18$ cm$)$

Third, calculate the answer $A = 756$ sq cm
    including the units.

42 ⊠ 18 ⊜ → **756.**

Of course the formula $A = LW$ may also be used to find *L* or *W* when the other quantities are known. As you learned in your study of basic algebra in Chapter 7, the following formulas are all equivalent.

$$A = LW \qquad L = \frac{A}{W} \qquad \text{or} \qquad W = \frac{A}{L}$$

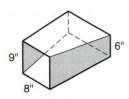

For example, a sheet metal worker must build a heating duct in which a rectangular vent 6 in. high must have the same area as a rectangular opening 8 in. by 9 in. Find the length of the vent.

Try it, then check your work in **18**.

---

18 The area of the opening is $A = LW$

$$A = (8 \text{ in.})(9 \text{ in.}) = 72 \text{ sq in.}$$

The length of the vent is $L = \dfrac{A}{W}$

$$L = \frac{72 \text{ sq in.}}{6 \text{ in.}}$$

or $L = 12$ in.

Another formula that you may find useful allows you to calculate the perimeter of a rectangle from its length and width.

PERIMETER OF A RECTANGLE

$P = 2L + 2W$

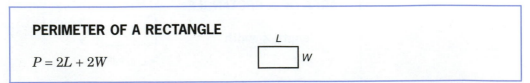

Use this formula to find the perimeter of the rectangle shown here.

Check your work in **19**.

 3"

7"

19 Perimeter = $2L + 2W$
$= 2\,(7\text{ in.}) + 2\,(3\text{ in.})$
$= 14\text{ in.} + 6\text{ in.}$
$= 20\text{ in.}$

 2 ⊗ **7** ⊕ **2** ⊗ **3** ⊜ → [**20.**]

For practice on using the area and perimeter formulas for rectangles, work the following problems.

(a) Find the area and perimeter of each of the following rectangles.

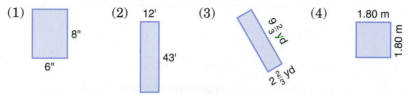

(1) 8" 6"   (2) 12' 43'   (3) $9\frac{2}{3}$ yd $2\frac{2}{3}$ yd   (4) 1.80 m 1.80 m

(b) Find the area of a rectangle whose length is 4 ft 6 in. and whose width is 3 ft 3 in. (Give your answer in square feet in decimal form. Round to one decimal place.)

(c) A rectangular opening 12 in. wide must have the same total area as two smaller rectangular vents 6 in. by 4 in. and 8 in. by 5 in. What must be the height of the opening?

(d) What is the cost of refinishing a wood floor in a room 35 ft by 20 ft at a cost of $2.25 per square foot?

(e) Find the area of a rectangular opening 20 in. wide and 3 ft long.

(f) What is the cost of fencing a rectangular yard 45 ft by 58 ft at $4.65 per foot?

Check your answers in **20**.

---

20 (a) (1) $A = 48$ sq in., $P = 28$ in. (2) $A = 516$ sq ft, $P = 110$ ft

(3) $A = 25\frac{7}{9}$ sq yd, $P = 24\frac{2}{3}$ yd (4) $A = 3.24$ sq m, $P = 7.20$ m

(b) 4 ft 6 in. $= 4\frac{6}{12}$ ft $= 4.5$ ft
3 ft 3 in. $= 3\frac{3}{12}$ ft $= 3.25$ ft
4.5 ft $\times$ 3.25 ft ≈ 14.6 sq ft

(c) Total area of smaller vents $= (6 \times 4) + (8 \times 5) = 64$ sq in.
Height of opening $= A / W = 64$ sq in. $\div$ 12 in. $= 5\frac{1}{3}$ in.

(d) Area $= 35$ ft $\times 20$ ft $= 700$ sq ft
Cost $= 700$ sq ft $\times$ 2.25 per sq ft $= 1575

(e) 720 sq in. or 5 sq ft

(f) $957.90

 2 ⊗ **45** ⊕ **2** ⊗ **58** ⊜ ⊗ **4.65** ⊜ → [**957.9**]

Squares   Did you notice that the rectangle in problem (a)(4) in **19** is actually a square? All sides are the same length. To save time when calculating the area or perimeter of a square, use the following formulas.

AREA OF A SQUARE

$A = s^2$

PERIMETER OF A SQUARE

$P = 4s$

Use these formulas to find the area and perimeter of the square shown. Check your work in **21**.

9'

---

21   Area, $A = s^2$
$$A = (9 \text{ ft})^2$$
$$A = (9 \text{ ft})(9 \text{ ft}) = 81 \text{ sq ft}$$

Remember s^2 means s times s and not $2s$.

Perimeter, $P = 4s$
$$P = 4(9 \text{ ft})$$
$$P = 36 \text{ ft}$$

A calculation that is very useful in carpentry, sheet metal work, and many other trades involves finding the length of one side of a square given its area. The following formula will help.

SIDE OF A SQUARE

$s = \sqrt{A}$

For example, if a square opening is to have an area of 64 sq in. what must be its side length?

$s = \sqrt{A}$

$s = \sqrt{64 \text{ sq in.}}$

$s = 8$ in.

Ready for a few problems involving squares? Try these.

(a)   Find the area and perimeter of each of the following squares.

(1) 6'   (2) $17\frac{1}{2}$"   (3) 2.50 cm   (4) 3'4"

(b)   Find the area of a square whose side length is $2\frac{1}{3}$ yd. (Round to two decimal places.)

(c)   Find the length of the side of a square whose area is (1) 144 sq m, (2) 927 sq in. (Use a calculator and round to one decimal place.)

(d)   At 96 cents per square foot, how much will it cost to sod a square lawn 14 ft 6 in. on a side? (Round to the nearest cent.)

(e) A square hot-air duct must have the same area as two rectangular ones 5 in. × 8 in. and 4 in. × 6 in. How long are the sides of the square duct?

Check your answers in **22**.

---

22 (a) (1) $A = 36$ sq ft; $P = 24$ ft

(2) $A = 306.25$ sq in.; $P = 70$ in.

(3) $A = 6.25$ sq cm; $P = 10$ cm

(4) $A = 11.11$ sq ft (rounded) or $11\frac{1}{9}$ sq ft, using fractions; $P = 13$ ft 4 in.

(b) 5.44 sq yd

(c) (1) 12 m   (2) 30.4 in.

(d) $201.84

(e) $(5 \times 8) + (4 \times 6) = 40$ sq in. $+ 24$ sq in. $= 64$ sq in.
$\sqrt{64}$ sq in. $= 8$ in.

For problem (c)(2),

927 $\boxed{\sqrt{}}$ → $\boxed{30.446675}$

For problem (d),

14.5 $\boxed{x^2}$ $\boxed{\times}$ **.96** $\boxed{=}$ → $\boxed{201.84}$

Parallelograms You should recall that a parallelogram is a four-sided figure whose opposite pairs of sides are equal and parallel. Here are a few parallelograms:

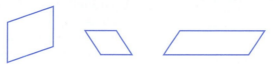

To find the area of a parallelogram use the following formula:

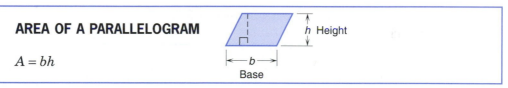

Notice that we do not use only the lengths of the sides to find the area of a parallelogram. The height h is the perpendicular distance between the base and the side opposite the base. The height h is perpendicular to the base b.

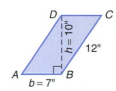

Example: In parallelogram *ABCD*, the base is 7 in. and the height is 10 in. Find its area.

First, write the formula: $A = bh$.

Second, substitute the given values: $A = (7 \text{ in.})(10 \text{ in.})$

Third, calculate the area including the units: $A = 70$ sq in.

Notice that we ignore the length of the slant side 12 in.

Find the area of this parallelogram:

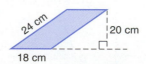

Check your work in **23**.

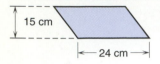

23 Area = bh
$\quad\quad\quad = (18 \text{ cm})(20 \text{ cm})$
$\quad\quad\quad = 360 \text{ cm}^2$

Be careful to use the correct dimensions.

This problem could also be worked using the 24-cm side as the base.

In this case the new height is 15 cm and the area is

$A = (24 \text{ cm})(15 \text{ cm})$
$A = 360 \text{ cm}^2$

To find the perimeter of a parallelogram, simply add the lengths of the four sides.

$P = 2a + 2b$
$P = 2(24 \text{ cm}) + 2(18 \text{ cm})$
$P = 48 \text{ cm} + 36 \text{ cm} = 84 \text{ cm}$

Now try the following problems.

Find the perimeter and area for each parallelogram shown.

(a)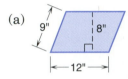

(b)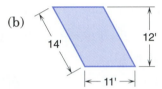

(c)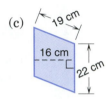

(d)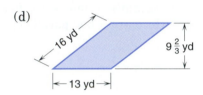

(e) A parallelogram-shaped opening has a base of 2 ft 9 in. and a height of 3 ft 3 in. Find its area. (Your answer should be in square feet rounded to two decimal places.)

Check your answers in **24**.

AREA OF A PARALLELOGRAM

You may be interested in where the formula in frame **22** came from. If so, follow this explanation.

Here is a typical parallelogram. As you can see, opposite sides $\overline{AB}$ and $\overline{DC}$ are parallel, and opposite sides $\overline{AD}$ and $\overline{BC}$ are also parallel.

Now let's cut off a small triangle from one side of the parallelogram by drawing the perpendicular line AE.

Next, reattach the triangular section to the other side of the figure, forming a rectangle. Notice that $EF = DC = b$.

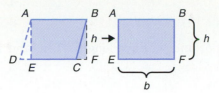

The area of the rectangle $ABFE$ is $A = bh$. This is exactly the same as the area of the original parallelogram $ABCD$.

---

24 (a) $P = 42$ in., $\qquad$ $A = 96$ sq in.

(b) $P = 50$ ft, $\qquad$ $A = 132$ sq ft

(c) $P = 82$ cm, $\qquad$ $A = 352$ cm^2

(d) $P = 58$ yd, $\qquad$ $A = 125\frac{2}{3}$ sq yd

(e) $A = 8.94$ sq ft

2.75 $\boxed{\times}$ **3.25** $\boxed{=}$ $\rightarrow$ $\boxed{8.9375}$ ≈ 8.94

Trapezoids A trapezoid is a four-sided figure with only one pair of sides parallel. Here are a few trapezoids:

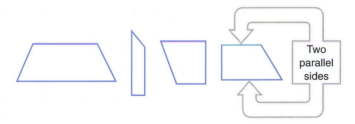

To find the area of a trapezoid use the following formula:

AREA OF A TRAPEZOID

$$A = \left(\frac{b_1 + b_2}{2}\right)h$$

or $\dfrac{h}{2}(b_1 + b_2)$

b_1

h

b_2

The factor $\left(\dfrac{b_1 + b_2}{2}\right)$ is the average length of the two parallel sides b_1 and b_2. The height h is the perpendicular distance between the two parallel sides.

Follow the example on p. 406. Find the area of this trapezoid-shaped metal plate.

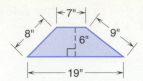

The parallel sides, b_1 and b_2, have lengths 7 in. and 19 in. The height h is 6 in.

$$A = \left(\frac{b_1 + b_2}{2}\right)h$$

$A = \left(\dfrac{7 \text{ in.} + 19 \text{ in.}}{2}\right)(6 \text{ in.})$ First, evaluate the quantity $\dfrac{7 + 19}{2}$.

$A = \left(\dfrac{26}{2} \text{ in.}\right)(6 \text{ in.})$

$A = (13 \text{ in.})(6 \text{ in.})$ Now multiply 13×6.

$A = 78$ sq in.

7 $+$ **19** $=$ $\div$ **2** $\times$ **6** $=$ → *78.*

The **perimeter** of a trapezoid is simply the sum of the lengths of all four sides.

Find the area and perimeter of each of the following trapezoids.

(a)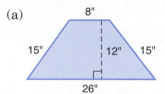

(b)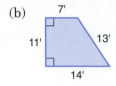

(c)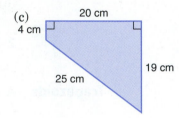

(d)

(e)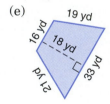

(f)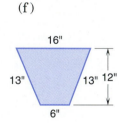

Check your answers in **25**.

---

25 (a) $A = \left(\dfrac{8 \text{ in.} + 26 \text{ in.}}{2}\right)(12 \text{ in.})$

$A = (17 \text{ in.})(12 \text{ in.})$

$A = 204$ sq in.

$P = 8 \text{ in.} + 15 \text{ in.} + 26 \text{ in.} + 15 \text{ in.}$

$P = 64$ in.

(b) $A = \left(\dfrac{7 \text{ ft} + 14 \text{ ft}}{2}\right)(11 \text{ ft})$

$A = (10.5 \text{ ft})(11 \text{ ft})$

$A = 115.5$ sq ft

$P = 11 \text{ ft} + 7 \text{ ft} + 13 \text{ ft} + 14 \text{ ft}$

$P = 45$ ft

(c) $A = \left(\dfrac{4 \text{ cm} + 19 \text{ cm}}{2}\right)(20 \text{ cm})$

$A = (11.5 \text{ cm})(20 \text{ cm})$

$A = 230 \text{ cm}^2$

$P = 4 \text{ cm} + 20 \text{ cm} + 19 \text{ cm} + 25 \text{ cm}$

$P = 68$ cm

(d) $A = \left(\dfrac{9\frac{1}{2} \text{ in.} + 6\frac{1}{2} \text{ in.}}{2}\right)(2\frac{1}{4} \text{ in.})$ $P = 3\frac{3}{4} \text{ in.} + 6\frac{1}{2} \text{ in.} + 2\frac{1}{4} \text{ in.} + 9\frac{1}{2} \text{ in.}$

$P = 22 \text{ in.}$

$A = (8 \text{ in.})(2\frac{1}{4} \text{ in.})$

$A = (8 \text{ in.})(2.25 \text{ in.})$

$A = 18 \text{ sq in.}$

(e) $A = \left(\dfrac{16 \text{ yd} + 33 \text{ yd}}{2}\right)(18 \text{ yd})$ $P = 16 \text{ yd} + 19 \text{ yd} + 33 \text{ yd} + 21 \text{ yd}$

$P = 89 \text{ yd}$

$A = (24.5 \text{ yd})(18 \text{ yd})$

$A = 441 \text{ sq yd}$

(f) $A = \left(\dfrac{16 \text{ in.} + 6 \text{ in.}}{2}\right)(12 \text{ in.})$ $P = 13 \text{ in.} + 16 \text{ in.} + 13 \text{ in.} + 6 \text{ in.}$

$P = 48 \text{ in.}$

$A = (11 \text{ in.})(12 \text{ in.})$

$A = 132 \text{ sq in.}$

Note ▶ In a real job situation you will usually need to measure the height of the figure. Remember, the bases of a trapezoid are the parallel sides and h is the distance between them. **◀**

Turn to **26** for practice in finding the area and perimeter of polygons.

26

Exercises 8-2 **Area and Perimeter of Polygons**

A. Find the perimeter of each polygon.

1.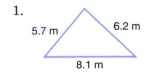
5.7 m   6.2 m   8.1 m

2.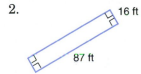
16 ft   87 ft

3.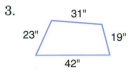
31"   23"   19"   42"

4.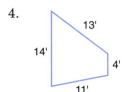
13'   14'   4'   11'

5.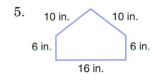
10 in.   10 in.   6 in.   6 in.   16 in.

6.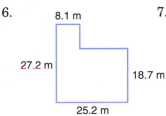
8.1 m   27.2 m   25.2 m

7.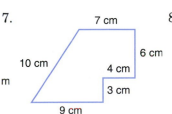
7 cm   10 cm   6 cm   4 cm   3 cm   9 cm   18.7 m

8.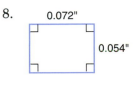
0.072"   0.054"

9.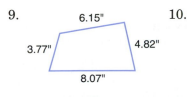
6.15"   3.77"   4.82"   8.07"

10.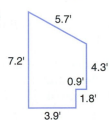
5.7'   7.2'   4.3'   0.9'   1.8'   3.9'

11.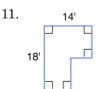
14'   18'

12.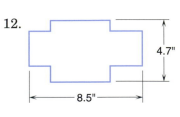
4.7"   8.5"

B. Find the perimeter and area of each figure. Round to the nearest tenth if necessary. (Assume right angles and parallel sides except where obviously otherwise.)

1.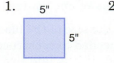
5"   5"

2.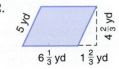
5 yd   4$\frac{2}{3}$ yd   6$\frac{1}{3}$ yd   1$\frac{2}{3}$ yd

3.
12.5"   22.8"

4.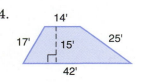
14'   17'   15'   25'   42'

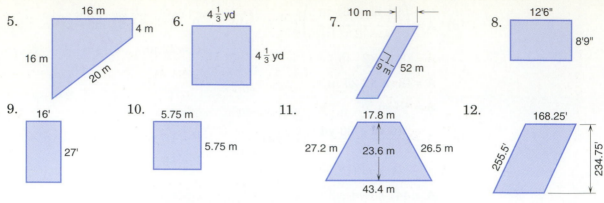

5. 16 m, 4 m, 16 m, 20 m

6. $4\frac{1}{3}$ yd, $4\frac{1}{3}$ yd

7. 10 m, 9 m, 52 m

8. 12'6", 8'9"

9. 16', 27'

10. 5.75 m, 5.75 m

11. 17.8 m, 27.2 m, 23.6 m, 26.5 m, 43.4 m

12. 168.25', 255.5', 234.75'

C. Practical Problems

1. **Masonry** How many bricks will it take to build the wall shown if each brick measures $4\frac{1}{4}$ in. × $8\frac{3}{4}$ in. including mortar? (*Hint:* Change the wall dimensions to inches.)

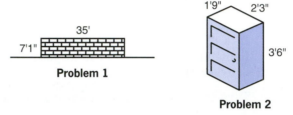

1'9"   2'3"

35'

7'1"

3'6"

Problem 1

Problem 2

2. **Carpentry** How many square feet of wood are needed to build the cabinet in the figure? (Assume that wood is needed for all six surfaces.)

12'

19'

Problem 3

3. **Interior Decorating** The floor plan of a room is shown in the figure. If the entire room is to be carpeted, how many square *yards* of carpet are needed? At $25.95 per square yard, how much will it cost?

4. **Building Construction** The amount of horizontal wood siding (in sq) needed to cover a given wall area can be calculated from the formula

   Area of siding (sq) = K × total area to be covered (sq ft)

   where K is a constant that depends on the type and size of the siding used. How much 1 × 6 rustic shiplapped drop siding is needed to cover a rectangular wall 60 ft by 14 ft? Use $K = 2.19$ and round to the nearest 10 sq.

5. **Building Construction** How many bundles of strip flooring are needed to cover a rectangular floor area 16 ft by 12 ft 6 in. in size if a bundle of strip flooring contains 24 sq ft, and if 30% extra must be allowed for side and end matching?

6. **Masonry** A large rectangular window opening measures $72\frac{5}{8}$ in. by $60\frac{3}{8}$ in. Calculate the area of this opening. Round to the nearest square inch.

7. **Masonry** Find the area of one stretcher course of 20 concrete blocks if the blocks are $7\frac{5}{8}$ in. long and 4 in. high, and if the mortar joints are $\frac{3}{8}$ in. thick. Round to the nearest 10 sq in. (*Hint:* Twenty blocks require 19 mortar joints.)

8. **Masonry** To determine the approximate number of concrete blocks needed to construct a wall, the following formula is often used.

   Number of blocks per course = N × area of one course (sq ft)

   where N is a number that depends on the size of the block. How many 8-in. by 8-in. by 16-in. concrete blocks are needed to construct a foundation wall 8 in. thick with a total length of 120 ft, if it is laid five courses high? Use

$N = \frac{9}{8}$ for this block size. (*Hint:* The blocks are set lengthwise. Translate all dimensions to inches.)

9. **Auto Mechanics**   A roll of gasket material is 18 in. wide. What length is needed to obtain 16 sq ft of the material? (*Careful:* The numbers are not expressed in compatible units.)

10. **Welding**   A type of steel sheet weighs 3.125 lb per square foot. What is the weight of a piece measuring 24 in. by 108 in.? (*Hint:* Be careful of units.)

11. **Printing**   A rule of thumb in printing says that the area of the typed page should be half the area of the paper page. If the paper page is 5 in. by 8 in., what should the length of the typed page be if the width is 3 in.?

12. **Printing**   A ream of 17-in. by 22-in. paper weighs 16 lb. Find the weight of a ream of 19-in. by 24-in. sheets of the same density of paper.

13. **Printing**   A sheet of 25-in. by 38-in. paper must be cut into 6-in. by 8-in. cards. The cuts can be made in either direction but they must be consistent. Which plan will result in the least amount of waste, cutting the 6-in. side of the card along the 25-in. side of the sheet or along the 38-in. side of the sheet? State the amount of waste created by each possibility. (*Hint:* Make a cutting diagram for each possibility.)

14. **Refrigeration and Air Conditioning**   To determine the size of a heating system needed for a home, an installer needs to calculate the heat loss through surfaces such as walls, windows, and ceilings exposed to the outside temperatures. The formula

$$L = kDA$$

gives the heat loss L (in BTU per hour) if D is the temperature difference between the inside and outside, A is the area of the surface in sq ft, and k is the insulation rating of the surface. Find the heat loss per hour for a 7-ft by 16-ft single-pane glass window ($k = 1.13$) if the outside temperature is 40° and the desired inside temperature is 68°. Round to the nearest hundred BTU.

When you have finished these problems, check your answers on page 602, then continue in **27** with the study of other polygons.

8-3   TRIANGLES, HEXAGONS, AND IRREGULAR POLYGONS

27   A *triangle* is a polygon having three sides and therefore three angles.

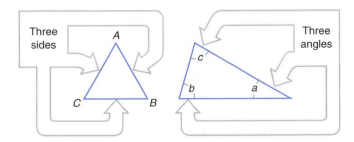

Just as there are several varieties of four-sided figures—squares, rectangles, parallelograms, and trapezoids—there are also several varieties of triangles. Fortunately, one area formula can be used with all triangles. First, you must learn to identify the many kinds of triangles that appear in practical work.

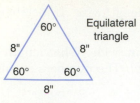

Equilateral triangle

An *equilateral triangle* is one in which all three sides have the same length. An equilateral triangle is also said to be *equiangular* since, if the three sides are equal, the three angles will also be equal. In fact, since the interior angles of any triangle always add up to 180°, each angle of an equilateral triangle must equal 60°.

An *isosceles triangle* is one in which at least two of the three sides are equal. It is always true that the two angles opposite the equal sides are also equal.

A *scalene triangle* is one in which no sides are equal.

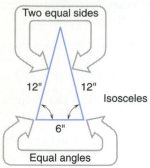

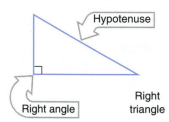

A *right triangle* contains a 90° angle. Two of the sides, called the *legs,* are perpendicular to each other. The longest side of a right triangle is always the side opposite to the right angle. This side is called the *hypotenuse* of the triangle. An isosceles or scalene triangle can be a right triangle, but an equilateral triangle can never be a right triangle.

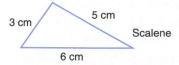

Note ▶ Right triangles are by far the most important in trades work. If a triangle is not a right triangle, it is called **acute** if all three angles are acute, and it is called **obtuse** if one angle is an obtuse angle. ◀

Identify the following triangles as being equilateral, isosceles, or scalene. Also name the ones that are right triangles.

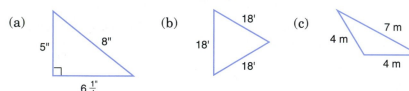

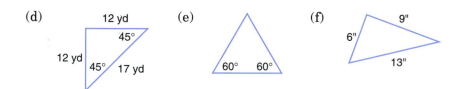

Check your answers in **28**.

---

28   (a) Scalene   (b) Equilateral   (c) Isosceles

(d) Isosceles   (e) Equilateral   (f) Scalene

(a) and (d) are also right triangles.

Did you have trouble with (d) or (e)? The answer to each depends on the fact that the interior angles of a triangle add up to 180°. In (d) the two given angles add up to 90°. Therefore, the missing angle must be 90° if the three angles sum to 180°. This makes (d) a right triangle.

In (e) the two angles shown sum to 120°, so the third angle must be 180° − 120° or 60°. Because the three angles are equal we know the triangle is an equiangular triangle, and this means it is also equilateral.

Pythagorean Theorem

The Pythagorean theorem is a rule or formula that allows us to calculate the length of one side of a right triangle when we are given the lengths of the other two sides. Although the formula is named after the ancient Greek mathematician Pythagoras, it was known to Babylonian engineers and surveyors more than a thousand years before Pythagoras lived.

PYTHAGOREAN THEOREM

For any right triangle, the square of the hypotenuse is equal to the sum of the squares of the other two sides.

$$c^2 = a^2 + b^2$$

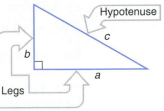

Geometrically, this means that if the squares are built on the sides of the triangle, the area of the larger square is equal to the sum of the areas of the two smaller squares. For the triangle shown

$$3^2 + 4^2 = 5^2$$

$$9 + 16 = 25$$

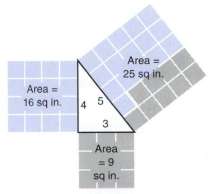

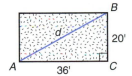

Use this formula to solve this practical problem. Find the distance d between points A and B for this rectangular plot of land.

Try it. Check your work in **29**.

---

29   The Pythagorean theorem tells us that for the right triangle ABC,

$$d^2 = 36^2 + 20^2$$
$$d^2 = 1296 + 400$$
$$d^2 = 1696$$

Taking the square root of both sides of the equation, we have

$$d \approx 41 \text{ ft}   \text{ rounded to the nearest foot}$$

36 $\boxed{x^2}$ $\boxed{+}$ **20** $\boxed{x^2}$ $\boxed{=}$ $\boxed{\sqrt{}}$ $\rightarrow$ | 41.182521 |

This rule may be used to find any one of the three sides of a right triangle if the other two sides are given. The formula $c^2 = a^2 + b^2$ can be rewritten as

$$c = \sqrt{a^2 + b^2}   \text{ or }   a = \sqrt{c^2 - b^2}   \text{ or }   b = \sqrt{c^2 - a^2}$$

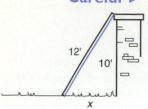

Careful ▶ This is true only for a *right* triangle—a triangle with a right or 90° angle. ◀

Use these formulas to solve the problem.

A carpenter wants to use a 12-ft ladder to reach the top of a 10-ft wall. How far must the base of the ladder be from the base of the wall?

Try it, then check your work in **30.**

---

30 Use the formula $a = \sqrt{c^2 - b^2}$ to get

$x = \sqrt{12^2 - 10^2}$

$x = \sqrt{144 - 100}$

$x = \sqrt{44}$

$x \approx 6.63$ ft or 6 ft 8 in., rounded to the nearest inch

If you need help with square roots, pause here and return to page 276 for a review.

For some right triangles, the three side lengths are all whole numbers rather than fractions or decimals. Such special triangles have been used in technical work since Egyptian surveyors used them 2000 years ago to lay out rectangular fields for farming.

Here are a few "Pythagorean triple" right triangles:

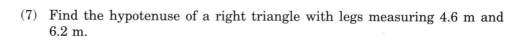

Solve the following problems, using the Pythagorean theorem. (If your answer does not come out exactly, round to the nearest tenth.)

(a) Find the missing side for each of the following right triangles.

(1)   (2)   (3)

(4)   (5)   (6)

(7) Find the hypotenuse of a right triangle with legs measuring 4.6 m and 6.2 m.

(b) A rectangular table measures 4 ft by 6 ft. Find the length of a diagonal brace placed beneath the table top.

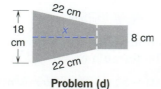

Problem (c)

(c) What is the distance between the centers of two pulleys if one is placed 9 in. to the left and 6 in. above the other?

(d) Find the length of the missing dimension x in the part shown in the figure.

Problem (d)

412

Chap. 8 Practical Plane Geometry

(e) Find the diagonal length of the stairway shown in the drawing. Round to the nearest tenth of a foot.

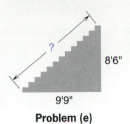

Problem (e)

Check your answers in **31**.

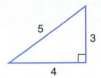
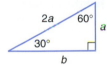

SOME SPECIAL TRIANGLES

3-4-5 TRIANGLE

If you draw a triangle with legs 3, 4, and 5 in. long, or in the ratio 3:4:5, it will always be a right triangle.

30-60-90 TRIANGLE

In a triangle with angles 30°, 60°, and 90° the shortest side will be exactly half as long as the longest side—the hypotenuse.

Side b is about $1.7a$.

45-45-90 TRIANGLE

If the two legs of a right triangle are equal, the angles will be 45°, 45°, and 90°. The hypotenuse will be $\sqrt{2}$ times the length of the other sides.

This kind of triangle is called a right-isosceles triangle.

Hypotenuse = $a\sqrt{2}$ or about $1.4a$.

---

31 (a) (1) 20 ft (2) 14 in. (3) 12.0 cm (4) 15.5 yd

 (5) 51 ft (6) 2.4 in. (7) 7.7 m

 (b) 7.2 ft or 7 ft $2\frac{1}{2}$ in.

 (c) 10.8 in.

 (d) 21.4 cm

 (e) 12.9 ft or 12 ft 11 in., rounded to the nearest inch.

In feet: **8.5** x^2 + **9.75** x^2 = $\sqrt{}$ → *12.934933* ≈ 12.9 ft

In inches: − **12** = × **12** = → *11.219200* ≈ 12 ft 11 in.

In problem (d), the first difficulty is to locate the correct triangle.

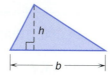

$$x = \sqrt{22^2 - 5^2}$$
$$x = \sqrt{459}$$

Area of a Triangle The area of any triangle, no matter what its shape or size, can be found using the same simple formula.

> **AREA OF A TRIANGLE**
>
> Area, $A = \frac{1}{2}bh$ or $A = \frac{bh}{2}$
>
> b = base
> h = height

Any side of a triangle can be used as the base. For a given base, the height is the perpendicular distance from the opposite vertex to the base. The following three identical triangles show the three different base-height combinations.

Notice that for the last two cases the height falls outside the triangle, and the base must be extended to meet the height at a right angle. For any such triangle, the three area calculations are equal:

$$A = \tfrac{1}{2}b_1h_1 = \tfrac{1}{2}b_2h_2 = \tfrac{1}{2}b_3h_3$$

For example, in this triangle the base b is 13 in., and the height h is 8 in. Applying the formula, we have

Area, $A = \frac{1}{2}bh$
$A = \frac{1}{2}(13\text{ in.})(8\text{ in.})$
$A = 52$ sq in.

In this triangle

$b = 25$ cm

$h = 16$ cm

Area, $A = \frac{1}{2}(25\text{ cm})(16\text{ cm})$
$A = 200$ sq cm

And in this triangle,

$b = 20$ cm

$h = 12$ cm

Area, $A = \frac{1}{2}(20\text{ cm})(12\text{ cm}) = 120$ sq cm

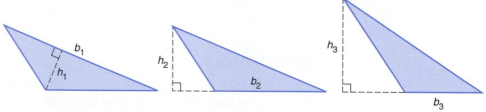

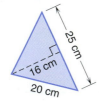

Find the area of triangle ABC.

Check your work in **32**.

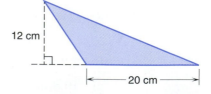

AREA OF A TRIANGLE

The area of any triangle is equal to one-half of its base times its height, $A = \frac{1}{2}bh$.

To learn where this formula comes from, follow this explanation:

First, for any triangle *ABC*,

draw a second identical triangle *DEF*.

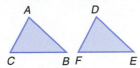

Second, flip the second triangle over and attach it to the first,

to get a parallelogram.

Finally, if the base and height of the original triangle are *b* and *h*, then these are also the base and height for the parallelogram.

The area of the parallelogram is

$A = bh$

so the area of the original triangle must be

$A = \frac{1}{2}bh$

---

32 Base, $b = 6.5$ ft

height, $h = 8$ ft

Area, $A = \frac{1}{2}bh$

$A = \frac{1}{2}(6.5 \text{ ft})(8 \text{ ft})$

$A = 26$ sq ft

Notice that for a right triangle, the two sides meeting at the right angle can be used as the base and height.

Solve this slightly tricky problem. Find the area of the isosceles triangle shown.

(*Hint:* First use the Pythagorean theorem to find the height, then find the area. The perpendicular line from *C* to $\overline{AB}$ will divide side $\overline{AB}$ in half.)

Check your work in **33**.

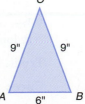

33 Draw the height perpendicular to the 6-in. base. Then in the triangle below

$$h = \sqrt{9^2 - 3^2}$$

$$h = \sqrt{81 - 9}$$

$$h = \sqrt{72} \approx 8.485 \text{ in.}$$

Now find the area:

$$A = \frac{1}{2}bh$$

$$A = \frac{1}{2}(6 \text{ in.})(8.485 \text{ in.}) = 25.455 \text{ sq in.}$$

$$A \approx 25 \text{ sq in.} \quad \text{rounded}$$

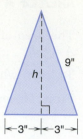

Using a calculator the entire calculation looks like this:

 9 x^2 $-$ **3** x^2 $=$ $\sqrt{}$ $\times$ **6** $\div$ **2** $=$ → 25.455844

The height   The area

In general, to find the area of an isosceles triangle use the following formula.

AREA OF AN ISOSCELES TRIANGLE

$$\text{Area, } A = \frac{1}{2}b\sqrt{a^2 - \left(\frac{b}{2}\right)^2}$$

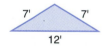

An attic wall has the shape shown. How many square feet of insulation are needed to cover the wall? (Round to the nearest square foot.)

Check your work in **34**.

---

34 The triangle is isosceles since two sides are equal. Substituting $a = 7$ ft and $b = 12$ ft into the formula, we have

$$\text{Area, } A = \frac{1}{2}(12 \text{ ft})\sqrt{(7 \text{ ft})^2 - \left(\frac{12 \text{ ft}}{2}\right)^2}$$

$$A = 6\sqrt{49 - 36}$$

$$A = 6\sqrt{13} \approx 21.63 \text{ sq ft, or } 22 \text{ sq ft} \quad \text{rounded}$$

 7 x^2 $-$ **6** x^2 $=$ $\sqrt{}$ $\times$ **12** $\div$ **2** $=$ → 21.633308

For an equilateral triangle an even simpler formula may be used.

AREA OF AN EQUILATERAL TRIANGLE

$$\text{Area, } A \approx 0.433a^2, \quad \text{approximately}$$

What area of sheet steel is needed to make a triangular pattern where each side is 3 ft long?

Try it. Use the given formula and round to one decimal place.

Check your work in **35**.

---

35 Area, $A \approx 0.433a^2$, where $a = 3$ ft
$A \approx 0.433(3 \text{ ft})^2$
$A \approx 0.433 \cdot 9$ sq ft
$A \approx 3.9$ sq ft   rounded

Work these problems for practice.

(a) In the following problems find the area of each triangle. (Round your answer to one decimal place if it does not come out exactly.)

(1)

15"
18"

(2)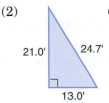

21.0'   24.7'
13.0'

(3)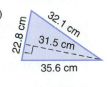

22.8 cm   32.1 cm
31.5 cm
35.6 cm

(4)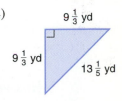

$9\frac{1}{3}$ yd
$9\frac{1}{3}$ yd
$13\frac{1}{5}$ yd

(5)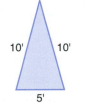

15 m   30 m
17 m
8 m   18 m

(6)

10'   10'
5'

(7)

16"
16"   22.6"

(8)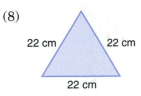

22 cm   22 cm
22 cm

13'3"
9'6"

Problem (b)

(b) Find the area, to the nearest square foot, of the triangular patio deck shown.

(c) At a cost of 97 cents per square foot, what would it cost to tile the triangular work space shown?

10'   8'
16'

Problem (c)

Check your answers in **36**.

HERO'S FORMULA

It is possible to find the area of any triangle from the lengths of its sides. The formula that enables us to do this was first devised almost 2000 years ago by Hero or Heron, a Greek mathematician. It is a complicated formula, and you may feel like a hero yourself if you learn how to use it.

Area, $A = \sqrt{s(s - a)(s - b)(s - c)}$ For any triangle

where $s = \dfrac{a + b + c}{2}$

a   c
b

(s stands for "semiperimeter," or half of the perimeter.)

Example:

$a = 6$ in., $b = 5$ in., $c = 7$ in.

First, find s. $s = \dfrac{6 + 5 + 7}{2} = \dfrac{18}{2} = 9$

5"   7"
6"

Next, substitute to find the area:

$A = \sqrt{9(9 - 6)(9 - 5)(9 - 7)}$

$A = \sqrt{9 \cdot 3 \cdot 4 \cdot 2}$

$A = \sqrt{216} \approx 14.7$ sq in.   rounded

36 (a) (1) 135 sq in.   (2) 136.5 sq ft   (3) 359.1 cm²

       (4) 43.6 sq yd   (5) 135 m²   (6) 24.2 sq ft

       (7) 128 sq in.   (8) 209.6 cm²

 For (a)(8): **.433** ⊠ **22** $\boxed{x^2}$ $\boxed{=}$ → `209.572`

(b) 63 sq ft   (c) $62.08

In problem (a)(5) you should have used 18 m as the base and *not* 26 m. b = 18 m, h = 15 m.

In problem (c) you should have noticed that the height was given as 8 ft.

Hexagons   A polygon is a plane geometric figure with three or more sides. A *regular polygon* is one in which all sides are the same length and all angles are equal. The equilateral triangle and the square are both regular polygons.

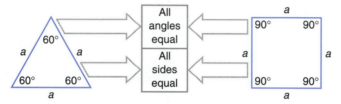

A *pentagon* is a five-sided polygon. A *regular pentagon* is one whose sides are all the same length and whose angles are equal.

The regular *hexagon* is a six-sided polygon in which each interior angle is 120° and all sides are the same length.

Regular hexagon:
All sides equal
All angles 120°

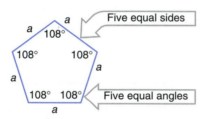

If we draw the diagonals, six equilateral triangles are formed.

Since you already know that the area of one equilateral triangle is approximately $A \approx 0.433a^2$, the area of the complete hexagon will be 6 times $0.433a^2$ or $2.598a^2$.

AREA OF A REGULAR HEXAGON

Area, $A \approx 2.598a^2$ approximately

For example, if $a = 6.00$ cm, $\qquad$ $A \approx 2.598(6 \text{ cm})^2$

First, square the number in parentheses. $\qquad A \approx 2.598(36 \text{ cm}^2)$

$\qquad\qquad\qquad\qquad\qquad\qquad\qquad\qquad A \approx 93.5 \text{ cm}^2$, rounded

 2.598 $\boxed{\times}$ **6** $\boxed{x^2}$ $\boxed{=}$ → $\boxed{93.528}$

Use this formula to find the area of a hexagonal plate 4.0 in. on each side.

4"

Check your work in **37**.

What does isosceles mean?

The prefix *iso* comes from a Greek word that means *same*, so an isosceles triangle is one that has two sides the same length.

---

37 Area, $A \approx 2.598(a^2)$ $\quad a = 4$ in.

$\qquad A \approx 2.598(4 \text{ in.})^2$

$\qquad A \approx 2.598(16 \text{ sq in.})$

$\qquad A \approx 42$ sq in., rounded

Because the hexagon is used so often in practical and technical work, several other formulas may be helpful to you.

A hexagon is often measured by specifying the distance across the corners or the distance across the flats.

Distance across
the corners

Distance across
the flats

DIMENSIONS OF A REGULAR HEXAGON

Distance across the corners, $d = 2a$ $\quad$ or $\quad a = 0.5d$

Distance across the flats, $f \approx 1.732a$ $\quad$ or $\quad a \approx 0.577f$

For example, for a piece of hexagonal bar stock where $a = \frac{1}{2}$ in., the distance across the corners would be

$d = 2a \qquad$ or $\qquad d = 1$ in.

The distance across the flats would be

$f \approx 1.732a \qquad$ or $\qquad f \approx 0.866$ in.

If the cross section of some hexagonal stock measures $\frac{3}{4}$ in. across the flats, find

(a) The side length, a.

(b) The distance across the corners, d.

(c) The cross-sectional area.

Round to three decimal places.

Try it, then check your work in **38**.

---

38 $f = \frac{3}{4}$ in.   or   0.75 in.

(a) $a \approx 0.577f$
     $a \approx 0.577(0.75$ in.$)$
     $a \approx 0.433$ in.

(b) $d = 2a$
     $d \approx 2(0.433$ in.$)$
     $d \approx 0.866$ in.

(c)   Area, $A \approx 2.598a^2$
          $A \approx 2.598(0.433$ in.$)^2$
          $A \approx 2.598(0.1875$ sq in.$)$
          $A \approx 0.487$ sq in.

Work these problems for practice.

(a) Fill in the blanks with the missing dimensions of each regular hexagon. (Round to two decimal places if necessary.)

	a	d	f	A
(1)	2 in.			
(2)		$\frac{3}{4}$ in.		
(3)			6 mm	

(b) A hex nut has a side length of $\frac{1}{4}$ in. From the following list, pick the smallest size wrench that will fit it.

   (1) $\frac{1}{4}$ in.   (2) $\frac{3}{8}$ in.   (3) $\frac{7}{16}$ in.   (4) $\frac{1}{2}$ in.   (5) $\frac{5}{8}$ in.

Check your answers in **39**.

39

		a	d	f	A
(a)	(1)	2 in.	4 in.	3.46 in.	10.39 sq in.
	(2)	$\frac{3}{8}$ in.	$\frac{3}{4}$ in.	0.65 in.	0.37 sq in.
	(3)	3.46 mm	6.92 mm	6 mm	31.14 mm^2

For problem (a)(3),

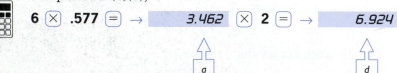

6 $\times$ **.577** $=$ $\rightarrow$ 3.462 $\times$ **2** $=$ $\rightarrow$ 6.924

                           a d

$$2.598 \;\boxed{\times}\; 3.462 \;\boxed{x^2}\; \boxed{=}\; \boxed{31.138184} \;\Longleftarrow\; \boxed{A}$$

(b) $\frac{7}{16}$ in.

Let's take a closer look at problem (b).

By drawing a picture we can see that we must find f. Since $a = \frac{1}{4}$ in., we have

$f \approx 1.732a$
$f \approx 1.732(0.25)$
$f \approx 0.433$ in.

By checking a list of wrench sizes in their decimal equivalents or converting the fractions given, we find that $\frac{7}{16}$ in. (0.4375 in.) is the smallest size that will fit the nut.

CLASSIFICATION OF POLYGONS

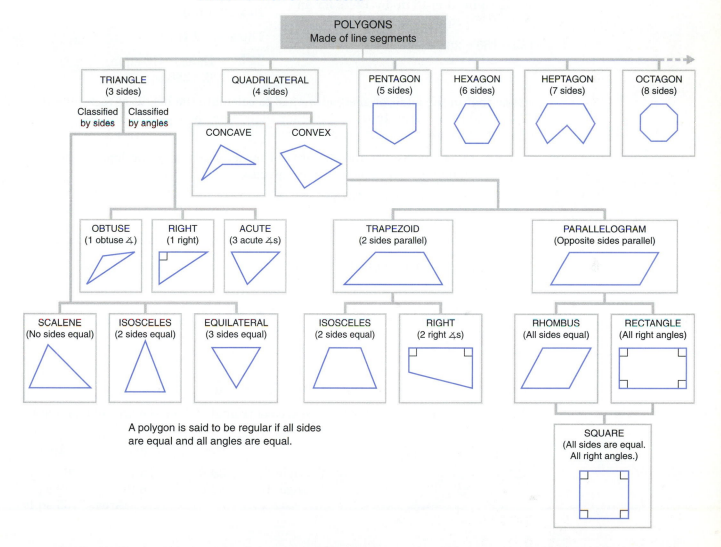

Irregular Polygons   Often, the shapes of polygons that appear in practical work are not the simple geometric figures that you have seen so far. The easiest way to work with irregular polygon shapes is to divide them into simpler, more familiar figures.

For example, look at this L-shaped figure:

There are two ways to find the area of this shape.

Method 1: Addition

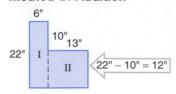

By drawing the dashed line, we have divided this figure into two rectangles, I and II. I is 22 in. by 6 in. for an area of 132 sq in. II is 13 in. by 12 in. for an area of 156 sq in. The total area is

132 + 156 = 288 sq in.

Method 2: Subtraction

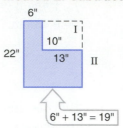

In this method we subtract the area of the small rectangle (I) from the area of the large rectangle (II). The area of I is

$10 \times 13 = 130$ sq in.

The area of II is

22 in. $\times$ 19 in. = 418 sq in.
418 − 130 = 288 sq in.

Notice that in the addition method we had to calculate the height of rectangle II as 22 in. − 10 in. or 12 in. In the subtraction method we had to calculate the width of rectangle II as 6 in. + 13 in. or 19 in.

Which method you use to find the area of an irregular polygon depends on your ingenuity and on which dimensions are given.

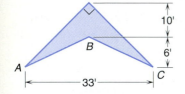

Find the area of the shape *ABCD*.

Check your work in **40**.

---

40 You may at first wish to divide this into two triangles and add their areas.

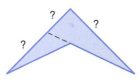

However, the dimensions provided do not allow you to do this.

The only option here is to subtract the area of triangle I from the area of triangle II.

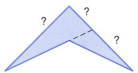

	Base	Height	Area
Triangle II	33 ft	16 ft	264 sq ft
Triangle I	33 ft	6 ft	− 99 sq ft
			Area = 165 sq ft

If both methods will work, pick the one that looks easier and requires the simpler arithmetic.

Find the area of this shape. (Round to the nearest square inch.)

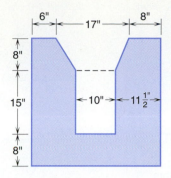

Check your work in **41**.

---

41 You might have been tempted to split the shape up into the five regions indicated in the figure. This will work, but there is an easier way.

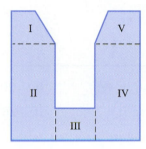

For the alternative shown in the next figure we need to compute only three areas, then we subtract II and III from I.

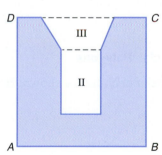

The outside rectangle *ABCD* is area I.

	Base	Height	Area	
Figure I	6 in. + 17 in. + 8 in. = 31 in.	8 in. + 15 in. + 8 in. = 31 in.	31 in. × 31 in.	= 961 sq in.
Figure II	10 in.	15 in.	10 in. × 15 in.	= 150 sq in.
Figure III (trapezoid)	17 in. and 10 in.	8 in.	$\left(\dfrac{17\text{ in.} + 10\text{ in.}}{2}\right)(8\text{ in.})$	= 108 sq in.

The total area is 961 sq in. − 150 sq in. − 108 sq in. = 703 sq in.

Learning Help ▶ When working with complex figures, organize your work carefully. Neatness will help eliminate careless mistakes. ◀

Now, for practice, find the areas of the following shapes. (Round to the nearest tenth.)

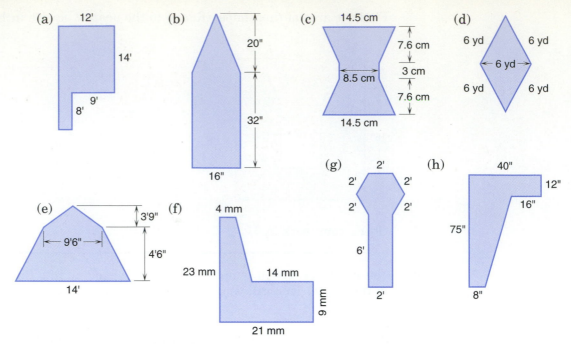

Check your answers in **42**.

42
(a) 192 sq ft
(b) 672 sq in.
(c) 200.3 cm²

(d) 31.2 sq yd
(e) 70.7 sq ft
(f) 266 mm²

(g) 22.4 sq ft
(h) 1488 sq in.

In problem (d) divide the figure into two equilateral triangles, then use the area formula $A \approx 0.433a^2$. Figure (g) is a hexagon plus a rectangle.

Now turn to **43** for a set of problems on the work of this section.

43

Exercises 8-3   **Triangles, Hexagons, and Irregular Polygons**

A. Find the missing dimensions of each figure shown. (Round to the nearest tenth if necessary.)

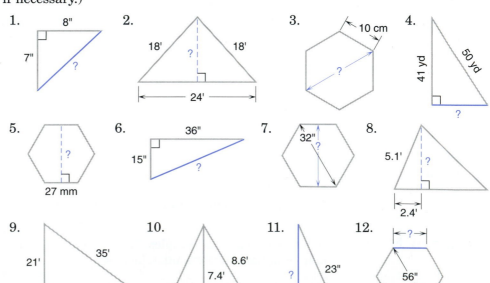

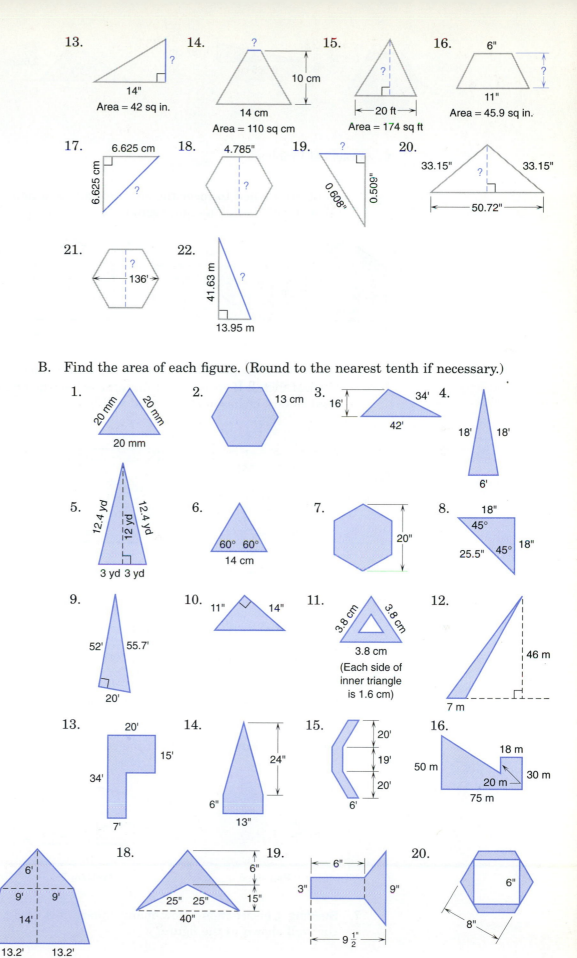

13.
14"
?
Area = 42 sq in.

14.
?
10 cm
14 cm
Area = 110 sq cm

15.
?
20 ft
Area = 174 sq ft

16.
6"
?
11"
Area = 45.9 sq in.

17.
6.625 cm
6.625 cm
?

18.
4.785"
?

19.
?
0.608"
0.509"

20.
33.15"   33.15"
?
50.72"

21.
?
136'

22.
41.63 m
?
13.95 m

B. Find the area of each figure. (Round to the nearest tenth if necessary.)

1.
20 mm   20 mm
20 mm

2.
13 cm

3.
16'   34'
42'

4.
18'   18'
6'

5.
12.4 yd   12.4 yd
12 yd
3 yd 3 yd

6.
60° 60°
14 cm

7.
20"

8.
18"
45°
25.5"   45°   18"

9.
52'   55.7'
20'

10.
11"   14"

11.
3.8 cm   3.8 cm
3.8 cm
(Each side of inner triangle is 1.6 cm)

12.
46 m
7 m

13.
20'
15'
34'
7'

14.
24"
6"
13"

15.
20'
19'
20'
6'

16.
18 m
50 m
20 m   30 m
75 m

17.
6'
9'   9'
14'
13.2'   13.2'

18.
6"
15"
25" 25"
40"

19.
6"
3"   9"
9 1/2"

20.
6"
8"

21.  22. 0.137 cm
0.565 cm 23. 2.55 m
6.23 m 24. 0.367"

C. Practical Problems

1. **Painting and Decorating**   At 460 sq ft per gallon, how many gallons of paint are needed to cover the outside walls of a house as pictured below? (Do not count windows and doors.)

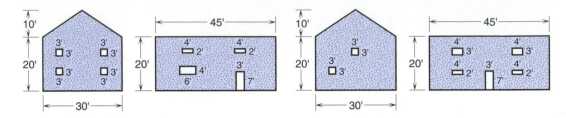

2. **Sheet Metal Technology**   A four-sided vent connection must be made out of sheet metal. If each side of the vent is like the one pictured, how many total square inches of sheet metal will be used?

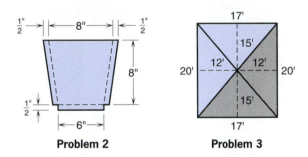

Problem 2   Problem 3

3. **Roofing**   The aerial view of a roof is shown in the figure. How many squares (100 sq ft) of shingles are needed for the roof?

4. **Roofing**   In problem 3 how many feet of gutters are needed?

5. **Carpentry**   Allowing for a 3-ft overhang, how long a rafter is needed for the gable of the house in the figure? Round to the nearest tenth.

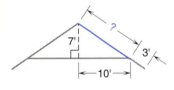

Problem 5

6. **Metalworking**   A cut is to be made in a piece of metal as indicated by the dashed line. Find the length of the cut to the nearest tenth of an inch.

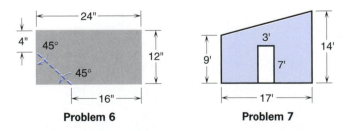

Problem 6   Problem 7

7. **Building Construction**   How many square feet of drywall are needed for the wall shown in the figure?

8. **Building Construction** Find the area of the gable end of the house shown, not counting windows. Each window opening is 4 ft $3\frac{1}{4}$ in. by 2 ft $9\frac{1}{4}$ in. Round to the nearest square foot.

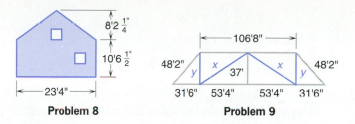

Problem 8   Problem 9

9. **Construction Technology** Find the missing dimensions x and y on the bridge truss shown.

When you have finished these problems, check your answers on page 602, then continue in **44** with the study of circles.

8-4 CIRCLES

44   A circle is probably the most familiar and the simplest plane figure. Certainly, it is the geometric figure that is most often used in practical and technical work. Mathematically, a circle is a closed curve representing the set of points some fixed distance from a given point called the *center*.

In this circle the center point is labeled O. The *radius r* is the distance from the center of the circle to the circle itself. The *diameter d* is the straight-line distance across the circle through the center point. It should be obvious from the drawing that the diameter is twice the radius.

Circumference   The *circumference* of a circle is the distance around it. It is a distance similar to the perimeter measure for a polygon. For all circles, the ratio of the circumference to the diameter, the distance around the circle to the distance across, is the same number.

$$\frac{\text{circumference}}{\text{diameter}} = 3.14159\ldots = \pi$$

The value of this ratio has been given the label π, the Greek letter "pi." The value of π is approximately 3.14, but the number has no simple decimal form—the digits continue without ending or repeating. For most practical purposes use 3.1416 for π.

The relationships between radius, diameter, and circumference can be summarized as follows:

CIRCLE FORMULAS

Diameter of a circle, $d = 2r$ or $r = \dfrac{d}{2}$

Circumference of a circle, $C = \pi d$ or $C = 2\pi r$

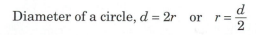

In the problems that follow, use 3.1416 for π unless you are told otherwise.

Use these formulas to find the radius and circumference of a hole with a diameter of 4.00 in. Round to one decimal place.

Check your work in **45**.

45 Radius, $r = \dfrac{d}{2}$

$$r = \frac{4\text{ in.}}{2} = 2\text{ in.}$$

Circumference, $C = \pi d$

$$\approx (3.1416)(4\text{ in.})$$

$$\approx 12.5664\text{ in.} \approx 12.6\text{ in.} \quad \text{rounded}$$

Use these formulas to solve the following practical problems. (Round to two decimal places.)

(a) If a hole is cut using a $\frac{3}{4}$-in.-diameter wood bit, what will be the radius and circumference of the hole?

(b) An 8-ft strip of sheet metal is bent into a circular tube. What will be the diameter of this tube?

(c) A circular redwood hot tub has a diameter of 5 ft. What length of steel band is needed to encircle the tub to hold the individual boards in position? Allow 4 in. for fastening.

(d) If eight bolts are to be equally spaced around a circular steel plate at a radius of 9 in., what must be the spacing between the centers of the bolts along the curve?

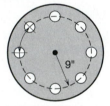

Problem (d)

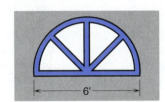

Problem (e)

(e) At $1.85 per foot, what will be the cost of the molding around the semicircular window shown? (Compute for the curved portion only.)

Check your work in **46**.

---

46 (a) $r = \dfrac{d}{2}$

$$r = \frac{\frac{3}{4}\text{ in.}}{2} = \frac{3\text{ in.}}{8} \qquad \frac{\frac{3}{4}}{2} = \frac{3}{4} \cdot \frac{1}{2} = \frac{3}{8}$$

$C = \pi d$
$C \approx (3.1416)(\frac{3}{4}\text{ in.})$
$C \approx (3.1416)(0.75\text{ in.})$ Change the fraction to a decimal.
$C \approx 2.36\text{ in.}$

(b) If $C = \pi d$ then $d = \dfrac{C}{\pi}$ Divide both sides by π.

and $d \approx \dfrac{8\text{ ft}}{3.1416}$

$$d \approx 2.55\text{ ft} \qquad \text{rounded}$$

(c) $C = \pi d$
$C \approx (3.1416)(5 \text{ ft})$
$C \approx 15.71 \text{ ft} \approx 15 \text{ ft } 8\frac{1}{2} \text{ in.}$

Adding 4 in., the length needed is $16 \text{ ft } \frac{1}{2} \text{ in.}$

Notice that 0.7 ft is (0.7)(12 in.) = 8.4 in. or about $8\frac{1}{2}$ in.

Some calculators have a π key	Decimal feet

0.708′ changed to inches

(d) $C = 2\pi r$
$C \approx 2(3.1416)(9 \text{ in.})$
$C \approx 56.55 \text{ in.}$

Since eight bolts must be spaced evenly around the 9-in. circle, divide by 8.
Spacing = $56.55 \div 8 \approx 7.07$ in.

(e) $C = \pi d$ for a complete circle
$C \approx (3.1416)(6 \text{ ft})$
$C \approx 18.8496 \text{ ft}$

The length of molding is half of this or 9.4248 ft. The cost is ($1.85)(9.4248 ft)
$\approx$ $17.44, rounded to the nearest cent.

Parts of Circles

Carpenters, plumbers, sheet metal workers, and other trades people often work with parts of circles. A common problem is to determine the length of a piece of material that contains both curved and straight segments.

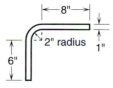

Suppose that we need to determine the total length of steel rod needed to form the curved piece shown. The two straight segments are no problem: We will need a total of 8 in. + 6 in. or 14 in. of rod for these. When a rod is bent, the material on the outside of the curve is stretched, and the material on the inside is compressed. We must calculate the circumference for the curved section from a *neutral line* midway between the inside and outside radius.

The inside radius is given as 2 in. The stock is 1 in. thick, so the midline or neutral line is at a radius of $2\frac{1}{2}$ in. The curved section is one-quarter of a full circle, so the length of bar needed for the curved arc is

$C = \dfrac{2\pi r}{4}$ Divide the circumference by 4 to find the length of the quarter-circle arc.

$C = \dfrac{\pi r}{2}$

$C \approx \dfrac{(3.1416)(2.5 \text{ in.})}{2}$

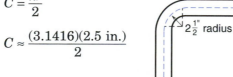

$2\frac{1}{2}″$ radius

$C \approx 3.93$ in.   rounded to two decimal places.

The total length of bar needed for the piece is

$L = 3.93$ in. + 14 in.
$L = 17.93$ in.

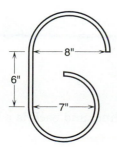

Your turn. Find the total length of ornamental iron $\frac{1}{2}$ in. thick needed to bend into this shape. Round to two decimal places.

Check your work in **47**.

47 Measuring to the midline we find that the upper curve has a radius of $4\frac{1}{4}$ in. and the lower curve has a radius of $3\frac{3}{4}$ in. Calculate the lengths of the two curved pieces this way:

Upper Curve

(Half circle)

$C = \dfrac{2\pi r}{2} = \pi r$ Multiply by $\frac{1}{2}$ for half circle.

$C \approx (3.1416)(4\frac{1}{4}\text{ in.})$

$C \approx 13.35\text{ in.}$

Lower Curve

(Three-fourths circle)

$C = \dfrac{3}{4} \cdot 2\pi r = \dfrac{3\pi r}{2}$ Multiply by $\frac{3}{4}$.

$C \approx \dfrac{(3)(3.1416)(3\frac{3}{4}\text{ in.})}{2}$

$C \approx 17.67\text{ in.}$

The total length of straight stock needed to create this shape is

13.35 in. + 17.67 in. + 6 in. = 37.02 in.

Once you have your work organized, you can perform the entire calculation on a calculator as follows:

 3.1416 ☒ **4.25** ⊞ **3** ☒ **3.1416** ☒ **3.75** ⊡ **2** ⊜ ⊞ **6** ⊜ →   *37.0233*

Find the length of stock needed to create each of the following shapes. Use 3.14 for π and round to one decimal place.

(a)

(b)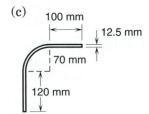

(c)

(d)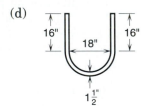

(e)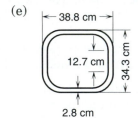

Check your answers in **48**.

---

48 (a) 31.4 in. (b) 31.8 in. (c) 339.7 mm

(d) 62.6 in. (e) 118.8 cm

On problem (e) a bit of careful reasoning will convince you that the dimensions are as follows:

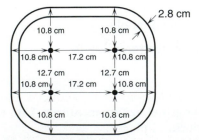

Each of the four corner arcs has radius 9.4 cm to the midline, and each is a quarter circle.

430

Radius = 10.8 cm − 1.4 cm

= 9.4 cm

Area of a Circle To find the *area* of a circle use one of the following formulas.

AREA OF A CIRCLE

$$A = \pi r^2$$

or $A = \dfrac{\pi d^2}{4} \approx 0.7854 d^2$

If you are curious about how these formulas are obtained, don't miss the box on page 432.

Use the formula to find the area of a circle with a diameter of 8 in.

$$A = \frac{\pi d^2}{4}$$

$$A \approx \frac{(3.1416)(8 \text{ in.})^2}{4}$$

8"

$A \approx 50.2656$ sq in.   or   50.3 sq in.   rounded

 3.1416 ⊗ **8** $\boxed{x^2}$ ⊘ **4** = → 50.2656

Of course you would find exactly the same area if you used the radius $r = 4$ in. in the first formula. Most people who need the area formula in their work memorize the formula $A = \pi r^2$ and calculate the radius r if they are given the diameter d.

Now for practice try these problems.

(a) Find the areas of the following circles. (Round as indicated.)

(1)

10"

(nearest tenth)

(2)

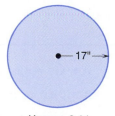

17"

Use $\pi = 3.14$

(nearest whole number)

(3) Find the area of the head of a piston with a radius of 6.5 cm. (Round to the nearest tenth.)

(4)

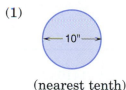

19 yd

Use $\pi = 3.14$

(nearest whole number)

(5)

2.2 mm

(nearest thousandth)

(b) Find the number of square feet of wood needed to make a circular table 6 ft in diameter. Use $\pi \approx 3.14$ and round to one decimal place.

(c) Find the number of square inches of stained glass used in a semicircular window with a radius of 9 in. Use $\pi \approx 3.14$ and round to the nearest whole number.

(d) What is the area of the largest circle that can be cut out of a square piece of sheet metal 4 ft 6 in. on a side? Round to the nearest hundredth.

(e) A pressure of 860 lb per square foot is exerted on the bottom of a cylindrical water tank. If the bottom has a diameter of 15 ft, what is the total force on the bottom? Round the final answer to the nearest thousand pounds.

Check your answers in **49**.

AREA OF A CIRCLE

To find the formula for the area of a circle, first divide the circle into many pie-shaped sectors, then rearrange them like this:

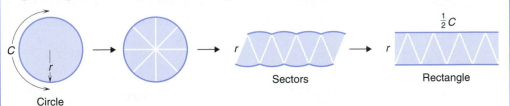

When the sectors are rearranged, they fit into a rectangle whose height is r, the radius of the original circle, and whose base length is roughly half the circumference of the circle. The more sectors the circle is divided into, the better the fit. The area of the sectors is approximately equal to the area of the rectangle.

Area of circle, $A = r \cdot \frac{1}{2}C$

$$A = r \cdot \frac{1}{2}(2\pi r)$$

$$A = \pi r^2$$

---

49 (a) (1) $A = \pi r^2 \approx (3.1416)(5 \text{ ft})^2 \approx (3.1416)(25 \text{ sq ft})$

$A \approx 78.5$ sq ft   rounded

(2) $A = \pi r^2 \approx (3.14)(17 \text{ in.})^2$

$A \approx 907$ sq in.

(3) $A = \pi r^2 \approx (3.1416)(6.5 \text{ cm})^2$

$A \approx 132.7$ sq cm   rounded

(4) $A = \frac{1}{2}(\pi r^2) \approx \frac{1}{2}(3.14)(9.5 \text{ yd})^2$

$A \approx 142$ sq yd   rounded

(5) $A = \frac{1}{4}(\pi r^2) \approx \frac{1}{4}(3.1416)(2.2 \text{ mm})^2$

$A \approx 3.801 \text{ mm}^2$ rounded

(b) $A = \pi r^2 \approx (3.14)(3 \text{ ft})^2$

$A \approx 28.3$ sq ft

(c) $A = \frac{1}{2}(\pi r^2) \approx \frac{1}{2}(3.14)(9 \text{ in.})^2$

$A \approx 127$ sq in.

(d) $A = \pi r^2 \approx (3.1416)(2 \text{ ft } 3 \text{ in.})^2$

$A \approx (3.1416)(2.25 \text{ ft})^2 \approx 15.90 \text{ sq ft}$

(e) $A = \pi r^2 \approx (3.1416)(7.5 \text{ ft})^2$

$A \approx 176.715 \text{ sq ft}$

$\text{Force} = 860 \dfrac{\text{lb}}{\text{sq ft}} \times 176.715 \text{ sq ft}$

$\text{Force} = 151{,}974.9 \text{ or } 152{,}000 \text{ lb}$ rounded

 3.1416 $\boxed{\times}$ **7.5** $\boxed{x^2}$ $\boxed{\times}$ **860** $\boxed{=}$ → $\boxed{151974.9}$

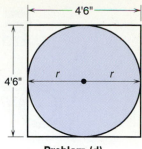

Problem (d)

Rings A circular *ring* is defined as the area between two concentric circles—two circles having the same center point. A washer is a ring; so is the cross section of a pipe, a collar, or a cylinder.

To find the area of a ring, subtract the area of the inner circle from the area of the outer circle. Use the following formula.

AREA OF A RING

Shaded area = area of outer circle − area of inner circle

$$A = \pi(R^2 - r^2) \quad \text{or} \quad A = \tfrac{1}{4}\pi(D^2 - d^2)$$

$$A \approx 0.7854(D^2 - d^2)$$

For example, to find the cross-sectional area of the wall of a ceramic pipe whose i.d. (inside diameter) is 8 in. and whose o.d. (outside diameter) is 10 in., substitute $d = 8$ in. and $D = 10$ in. into the second formula.

$A \approx \tfrac{1}{4}(3.1416)[(10 \text{ in.})^2 - (8 \text{ in.})^2]$

$A \approx (0.7854)(100 \text{ sq in.} - 64 \text{ sq in.})$

$A \approx (0.7854)(36 \text{ sq in.})$

$A \approx 28.3 \text{ sq in.}$ rounded

Note ▶ Notice that $\dfrac{\pi}{4}$ is equal to 0.7854 rounded to four decimal places. Also notice that the diameters are *squared first* and *then* the results are *subtracted*. ◀

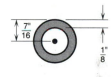

Try it. Find the area of a washer with an outside radius of $\frac{7}{16}$ in. and a thickness of $\frac{1}{8}$ in.

Round to the nearest thousandth.

Check your work in **50**.

---

50 $R = \frac{7}{16}$ in. $r = \frac{7}{16}$ in. $- \frac{1}{8}$ in.   or $r = \frac{5}{16}$ in.

then

$A = \pi(R^2 - r^2)$

$A \approx (3.1416)\left[\left(\frac{7}{16} \text{ in.}\right)^2 - \left(\frac{5}{16} \text{ in.}\right)^2\right]$

$A \approx (3.1416)[(0.4375 \text{ in.})^2 - (0.3125 \text{ in.})^2]$

$A \approx (3.1416)[0.1914 \text{ sq in.} - 0.0977 \text{ sq in.}]$

$A \approx (3.1416)[0.0937 \text{ sq in.}]$

$A \approx 0.295 \text{ sq in.}$ rounded

 → $\boxed{0.294525}$

$5 \div 16 = \text{[STO]} \ 7 \div 16 = \ x^2 \ - \ \text{[RCL]} \ x^2 \ = \ \times \ 3.1416 = $

Most people find it easiest to convert the fractions to decimals before squaring, especially if they are working with a calculator.

Solve the following problems involving rings. (Round to two decimal places.)

Problem (a)

(a) Find the cross-sectional area of the pipe shown in the figure.

(b) Find the area of a washer with an inside diameter of $\frac{1}{2}$ in. and an outside diameter of $\frac{7}{8}$ in.

(c) What is the cross-sectional area of a steel collar 0.6 cm thick with an inside diameter of 9.4 cm?

(d) At 95¢ per square foot find the cost of sodding a circular lawn 6 ft wide surrounding a flower garden with a radius of 20 ft.

(e) For a cross section of pipe with an inside diameter of 8 in., how much more material is needed to make the pipe $\frac{1}{2}$ in. thick than to make it $\frac{3}{8}$ in. thick?

The correct answers are given in **51**.

COMBINING CIRCULAR AREAS

In many practical situations it is necessary to determine what size circle contains the same area as two or more smaller circles.

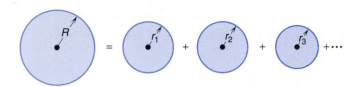

Area of larger circle = combined areas of the smaller circles

$$\pi R^2 = \pi r_1^2 + \pi r_2^2 + \cdots$$

Dividing all terms by π, we have

$$R^2 = r_1^2 + r_2^2 + \cdots$$

Taking the square root of both sides of the equation yields

$$R = \sqrt{r_1^2 + r_2^2 + \cdots}$$

As an example of how to apply this formula, suppose the flow from a 4-in.-radius pipe and a 6-in.-radius pipe must be combined into one large pipe of area equal to the total of the smaller two. To find the radius of the larger pipe, substitute into the formula

$$R = \sqrt{r_1^2 + r_2^2}$$

$$R = \sqrt{(4 \text{ in.})^2 + (6 \text{ in.})^2}$$

$$R = \sqrt{16 \text{ sq in.} + 36 \text{ sq in.}}$$

$$R = \sqrt{52 \text{ sq in.}}$$

$$R \approx 7.2 \text{ in.} \quad \text{rounded}$$

A similar relationship holds true for the diameters of circles.

$$D = \sqrt{d_1^2 + d_2^2 + \cdots}$$

Try these problems for practice.

1. Find the radius of a steel rod whose area must equal that of four rods each of radius 2 in.

2. What diameter gas line would be needed to replace two 6-in.-diameter lines?

3. Find the diameter of a water main with the same total area as two mains of diameter 3 in. and 4 in.

4. What radius vent will have the same cross-sectional area as three 6-in.-radius vents?

5. Find the diameter of a water pipe that has the same area as four pipes, two with 2-in. diameters and two with 3-in. diameters.

All answers are given on page 603.

---

51 (a) 351.86 sq in. (b) 0.40 sq in. (c) 18.85 cm^2

(d) $823.73 (e) 3.49 sq in.

In problem (c),

Radius of inner hole $= \dfrac{9.4 \text{ cm}}{2} = 4.7$ cm

Radius of outer edge $= 4.7$ cm $+ 0.6$ cm $= 5.3$ cm

Then,

area $= 3.1416 \times (5.3^2 - 4.7^2) = 18.849$ cm^2

To solve problem (e) using a calculator, first calculate the cross-sectional area of the pipe with the smaller thickness:

 3 ÷ **8** × **2** + **8** = x^2 − **8** x^2 = × **.7854** = STO → *9.8665875*

Then calculate the cross-sectional area of the thicker pipe and subtract the value in memory:

 9 x^2 − **8** x^2 = × **.7854** − RCL = → *3.4852125*

Now turn to **52** for a set of review exercises on circles.

52

Exercises 8-4   **Circles**

A. Find the circumference and area of each circle. Round to three significant digits.

1.   2.   3.   4.

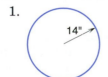   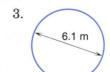

5. Circle of radius 12 ft.

6. Circle of diameter 0.50 m.

7. Circle of diameter 1.2 cm.

8. Circle of radius 0.4 in.

B. Find the area of each figure. Round to the nearest tenth.

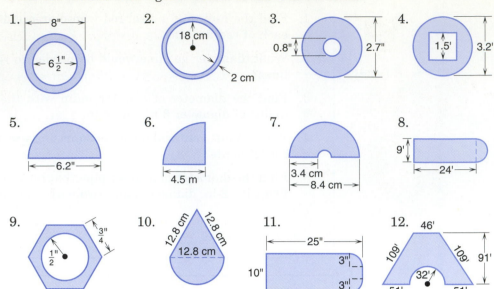

1. 8" 6 1/2"

2. 18 cm 2 cm

3. 0.8" 2.7"

4. 1.5' 3.2'

5. 6.2"

6. 4.5 m

7. 3.4 cm 8.4 cm

8. 9' 24'

9. 3/4" 1/2"

10. 12.8 cm 12.8 cm 12.8 cm

11. 25" 3" 3" 10"

12. 46' 109' 109' 91' 32' 51' 51'

Find the length of stock needed to create each shape. Round to the nearest tenth.

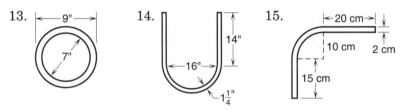

13. 9" 7"

14. 14" 16" 1 1/4"

15. 20 cm 10 cm 2 cm 15 cm

Problem 1

8" 4" 8"

Problem 3

0.3"

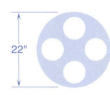

22"

Problem 6

C. Practical Problems

1. **Metalworking** What length of 1-in. stock is needed to bend a piece of steel into this shape? Use π ≈ 3.14 and round to the nearest tenth.

2. **Landscaping** (a) Calculate, to the nearest square foot, the cultivable area of a circular garden 38 ft in diameter if the outer 2 ft are to be used for a path. (b) Find the area of the path.

3. **Machine Technology** What diameter must a circular piece of stock be to mill a hexagonal shape with a side length of 0.3 in.?

4. **Machine Technology** In problem 3 how many square inches of stock will be wasted? Round to the nearest hundredth.

5. **Plumbing** What is the cross-sectional area of a cement pipe 2 in. thick with an inside diameter of 2 ft? Use π ≈ 3.14 and round to the nearest whole number.

6. **Machine Technology** In the piece of steel shown, $4\frac{1}{2}$-in.-diameter holes are drilled in a 22-in.-diameter circular plate. Find the area remaining. Use π ≈ 3.14 and round to the nearest sq in.

7. **Plumbing** How much additional cross-sectional area is there in a 30-in.-i.d. pipe 2 in. thick than one 1 in. thick? Round to one decimal place.

8. **Machine Technology** What diameter round stock is needed to mill a hexagonal nut $\frac{7}{8}$ in. on a side?

9. **Landscaping** How many plants spaced every 6 in. are needed to surround a circular walkway with a 25-ft radius? Use π ≈ 3.14.

Problem 10

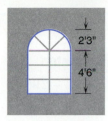

Problem 11

Problem 14

10. **Machine Technology** What size-diameter circular stock is needed to mill a square end 3 cm on a side? Round to the nearest hundredth.

11. **Carpentry** At $1.25 per foot, how much will it cost to put molding around the window pictured? Use π ≈ 3.14.

12. Given the radius of the semicircular ends of the track is 32 m, how long must each straightaway be to make a 400-m track? Use π ≈ 3.14.

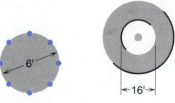

Problem 12

Problem 13

13. **Manufacturing** What is the total length of belting needed for the pulley shown? Round to the nearest whole number.

14. **Electrical Technology** An electrician bends a ½-in. conduit as shown in the figure to follow the bend in a wall. Find the total length of conduit needed between *A* and *B*. Use π ≈ 3.14.

15. **Metalworking** How many inches apart will eight bolts be along the circumference of the metal plate shown in the figure? Round to the nearest tenth.

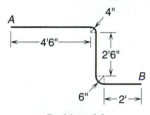

Problem 15   **Problem 16**

16. **Landscaping** How many square *yards* of concrete surface are there in a walkway 8 ft wide surrounding a tree if the inside diameter is 16 ft? Round to one decimal place.

17. **Carpentry** The circular saw shown has a diameter of 18 cm and 22 teeth. What is the spacing between teeth?

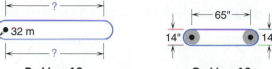

Problem 17

18. **Auto Mechanics** A hose must fit over a cylindrical opening 57.2 mm in diameter. If the hose wall is 0.3 mm thick, what length of clamp strap is needed to go around the hose connection? Round to the nearest whole number.

19. **Auto Mechanics** What length on the circumference would a 20° angle span on a 16-in. flywheel? Round to the nearest hundredth.

20. **Auto Mechanics** In making a certain turn on a road, the outside wheels of a car must travel 8.75 ft farther than the inside wheels. If the tires are 23 in. in diameter, how many additional turns do the outside wheels make during this turn? (*Hint:* In one "turn," a tire travels a distance equal to its circumference. Convert 8.75 ft to inches.)

21. **Wastewater Technology** To find the velocity of flow (in feet per second, or fps) through a pipe, divide the rate of flow (in cubic feet per second, or cfs) by the cross-sectional area of the pipe (in square feet). If the rate of flow through an 8-in.-diameter pipe is 0.0075 cfs, find the velocity of flow. Round to the nearest thousandth.

22. **Plumbing** The pressure exerted by a column of water amounts to 0.434 psi of surface for every foot of height or *head*. What is the total force (pressure times area) exerted on the bottom surface of a cylindrical container 16 in. in diameter if the head is 4.5 ft? Round to the nearest pound.

Check your answers on page 603, then turn to **53** for a set of practice problems over the work of Chapter 8.

A summary of all the formulas you have learned for plane figures appears in the box. Be certain you know these before you continue.

SUMMARY OF FORMULAS FOR PLANE FIGURES

Figure		Area	Perimeter
Rectangle		$A = LW$	$P = 2L + 2W$
Square		$A = s^2$	$P = 4s$
Parallelogram		$A = bh$	$P = 2a + 2b$
Triangle		$A = \frac{1}{2}bh \quad$ or $\quad \frac{bh}{2}$	$P = a + b + c$
Equilateral triangle		$A \approx 0.433a^2$	$P = 3a$
Isosceles triangle		$A = \frac{b}{2}\sqrt{a^2 - \left(\frac{b}{2}\right)^2}$	$P = 2a + b$
Right triangle		$A = \frac{1}{2}ab$	$P = a + b + c$
Trapezoid		$A = \left(\frac{b_1 + b_2}{2}\right)h$ or $\frac{h}{2}(b_1 + b_2)$	$P = a + c + b_1 + b_2$
Regular hexagon		$A \approx 2.598a^2$	$P = 6a$
Circle		$A = \pi r^2$ $A \approx 0.7854d^2$	$C = \pi d$ $C = 2\pi r$
Ring		$A = \pi(R^2 - r^2)$ $A \approx 0.7854(D^2 - d^2)$	$C = 2\pi R$

PROBLEM SET 8

Practical Plane Geometry

53 Answers are given on page 603.

A. Solve the following problems involving angles.

Name each angle and tell whether it is acute, obtuse, or right.

1.

1

2. A

B   C

3.
m

4.

E □

5. P

Q   R

6.
G ◇ 90°

Measure the indicated angles using a protractor.

7.

8.

9.

10.

11. •——————•

12.

13.
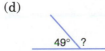

14.

15.

16. Use a protractor to draw angles of the following measures.
   (a) 65°   (b) 138°   (c) 12°   (d) 90°

17. Find the size of each indicated angle without using a protractor.
   (a)
   77°
   48°   ?

   (b)
   85°
   a \ c
   b

   (c)
   a / b
   50° / c
   e / d
   f / g

   The horizontal lines are parallel.

   (d)
   49° ?

   (e)
   ?
   40°   40°

   (f)
   a
   5"   5"
   b   c
   5"

Name _____

Date _____

Course/Section _____

(g)

(h)

(i)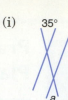

B. Solve the following problems involving plane figures.

Find the perimeter (or circumference) and area of each figure. (Round to the nearest tenth.)

1.

2.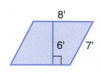

3.

4.

5.

6.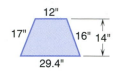

7.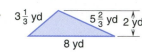

8.

9.

10.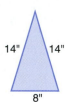

11.

12.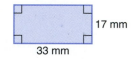

13.

14.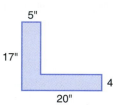

15.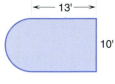

16.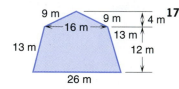

17.

18.

Find the missing dimensions of the following figures. (Round to the nearest hundredth if necessary.)

19.

20.

21.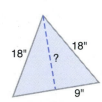

22.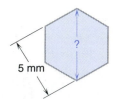

23.

24.

C. Practical Problems

1. **Building Construction**   What will it cost to pave a rectangular parking lot 220 ft by 85 ft at $4.75 per square foot?

2. **Metalworking**   A steel brace is used to strengthen the table leg as shown. What is the length of the brace? (Round to the nearest hundredth of an inch.)

Problem 2

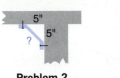

Problem 3

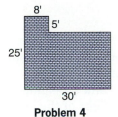

Problem 4

3. **Machine Technology**   Holes are punched in a steel plate as indicated. Through what angle must the plate be rotated to locate the second hole?

4. **Masonry**   How many square feet of brick are needed to lay a patio in the shape shown?

5. **Electrical Technology**   A coil of wire has an average diameter of 8 in. How many feet of wire are there if it contains 120 turns? (Round to the nearest foot.)

6. **Sheet Metal Technology**   A sheet metal worker needs to make a vent connection in the shape of the trapezoidal prism shown in the drawing. Use a protractor to find the indicated angle.

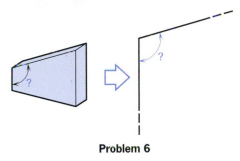

Problem 6

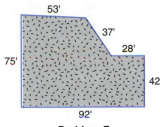

Problem 7

7. **Landscaping**   At $3.12 per foot, how much does it cost to fence the yard shown in the figure?

8. **Metalworking**   Find the length of straight stock needed to bend $\frac{1}{2}$-in. steel into the shape indicated. Use $\pi \approx 3.14$.

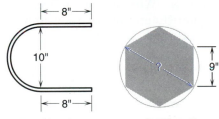

Problem 8   **Problem 9**

9. **Machine Technology**   A hexagonal piece of steel 9 in. on a side must be milled by a machinist. What diameter round stock does he need? (Round to the nearest tenth of an inch.)

10. How much guy wire is needed to anchor a 30-ft pole to a spot 20 ft from its base? Allow 4 in. for fastening. (Round to the nearest inch.)

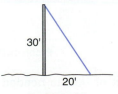

Problem 10

11. **Carpentry**   How many 4 ft by 8 ft sheets of exterior plywood must be ordered for the 4 walls of a building 20 ft long, 32 ft wide, and 12 ft high? Assume that there are 120 sq ft of window and door space. (*Note:* You can cut the sheets to fit, but you may not order a fraction of a sheet.)

12. **Metalworking**   Find the perimeter of the steel plate shown in the figure. (Round to the nearest tenth of an inch.)

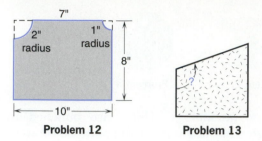

Problem 12   Problem 13

13. **Building Construction**   At what angle must sheet rock be cut to conform to the shape of the wall shown? Measure with a protractor.

14. **Metalworking**   A triangular shape with a base of 6 in. and a height of 9 in. is cut from a steel plate weighing 5.1 lb/sq ft. Find the weight of the shape (1 sq ft = 144 sq in.) (Round to the nearest ounce.)

15. **Mining Technology**   An air shaft is drilled from the indicated spot on the hill to the mine tunnel below. How long is the shaft to the nearest foot?

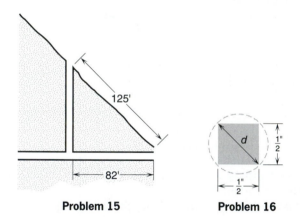

Problem 15   Problem 16

16. **Machine Technology**   What diameter circular stock is needed to mill a square bolt $\frac{1}{2}$ in. on a side? (Round to one-thousandth of an inch.)

17. **Interior Design**   At 73 cents a foot how much will it cost to weatherstrip the following: 6 windows measuring 4 ft by 6 ft; 8 windows measuring 3 ft by 2 ft; and all sides except the bottom of two doors 3 ft wide and 7 ft high?

18. **Building Construction**   How many square feet must be plastered to surface the wall shown? (Round to the nearest square foot.)

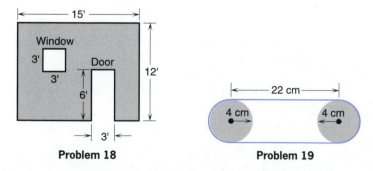

Problem 18   Problem 19

19. **Manufacturing**   What length of belt is needed for the pulley shown? (Round to the nearest tenth.)

20. **Electrical Technology** Electrical conduit must conform to the shape and dimensions shown. Find the total length of conduit in inches. (Use π ≈ 3.14 and round to the nearest tenth.)

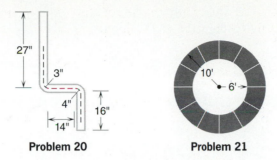

Problem 20   Problem 21

21. **Building Construction** At $4.75 per square foot, how much will it cost to pave the circular pathway shown? (Use π ≈ 3.14 and round to the nearest cent.)

22. **Carpentry** Find the size of ∡2 when the carpenter's square is placed over the board as shown.

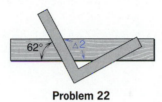

Problem 22

23. **Welding** The top to a container is made of steel weighing 4.375 lb/sq ft. If the top is a 30 in. by 30 in. square, find its weight to the nearest 0.1 lb.

24. **Machine Technology** A square bolt must be machined from round stock. If the bolt must be $2\frac{1}{8}$ in. on each side, what diameter stock is needed? (See the figure.) Round to the nearest thousandth.

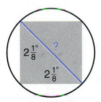

Problem 24

25. **Auto Mechanics** A fan pulley has a diameter of $3\frac{7}{8}$ in. What is its circumference? Round to the nearest 0.01 in.

Objective	Sample Problems		Where to Go for Help	
			Page	Frame

When you finish this unit you will be able to:

1. Identify solid figures, including prisms, cubes, cones, cylinders, pyramids, spheres, and frustums.

(a) _____ 447 **1**

(b) _____

(c) _____

2. Find the surface area and volume of solid objects. (Round to one decimal place.)

(a) Volume = _____ 460 **12**

(b) Sphere $r = 2.1$ cm

Volume = _____ 466 **17**

Surface area = _____

(c) Volume = _____ 449 **2**

(d) Lateral surface area 462 **14**

= _____

Volume = _____

Name _____

(e) Lateral surface area 455 **9**

= _____

Date _____

$h = 4$ cm
$s = 5$ cm

Volume = _____

Course/Section _____

3. Do geometric constructions.

Construct a perpendicular to line AB at point P and a parallel to line AB through point Q.

$Q\bullet$ 471 **21**

A ———————P———— B

(Answers to these preview problems are given on page 579.)

If you are certain that you can work *all* these problems correctly, turn to page 487 for a set of practice problems. If you cannot work one or more of the preview problems, turn to the page indicated to the right of the problem. Those who wish to master this material with the greatest success should turn to frame **1** and begin work there.

9 Solid Figures and Geometric Constructions

By permission of Johnny Hart and Creators Syndicate, Inc.

Plane figures have only two dimensions: length and width or base and height. A plane figure can be drawn exactly on a flat plane surface. Solid figures are three-dimensional: they have length, width, and height. Making an exact model of a solid figure requires shaping it in clay, wood, or paper. A square is a two-dimensional or plane figure; a cube is a three-dimensional or solid figure.

Trades workers encounter solid figures in the form of tanks, pipes, ducts, boxes, and buildings. They must be able to identify the solid and its component parts and compute its surface area and volume.

In many trades, workers must be able to use geometric concepts to construct perpendicular lines, draw parallel lines, construct a circle of given diameter, or perform similar geometric constructions without using a measuring device or other special equipment.

9-1 PRISMS AND PYRAMIDS

Prisms

1   A *prism* is a solid figure having at least one pair of parallel surfaces that create a uniform cross section. The figure shows a hexagonal prism.

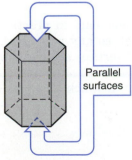

Parallel surfaces

Hexagonal prism

Cutting the prism anywhere parallel to the hexagonal surfaces produces the same hexagonal cross section.

All the polygons that form the prism are called *faces*. The faces that create the uniform cross section are called the *bases*. They give the prism its name. The others are called lateral faces.

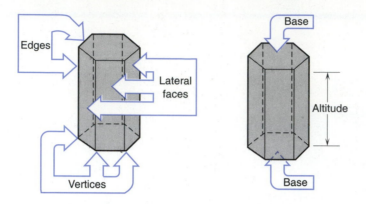

The sides of the polygons are the *edges* of the prism, and the corners are still referred to as *vertices*. The perpendicular distance between the bases is called the *altitude* of the prism.

In a *right prism* the lateral edges are perpendicular to the bases, and the lateral faces are rectangles. In this section you will learn about right prisms only.

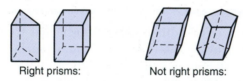

The components of a prism are its faces, vertices, and edges. Count the number of faces, vertices, and edges in this prism:

Number of faces = _____

Number of vertices = _____

Number of edges = _____

Check your answer in **2**.

---

2   There are 7 faces (2 bases and 5 lateral faces), 15 edges, and 10 vertices on this prism.

Here are some examples of simple prisms.

Rectangular Prism   The most common solid in practical work is the *rectangular prism*. All opposite faces are parallel, so any pair can be chosen as bases. Every face is a rectangle. The angles at all vertices are right angles.

Cube   A *cube* is a rectangular prism in which all edges are the same length.

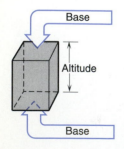

Triangular Prism In a *triangular prism* the bases are triangular.

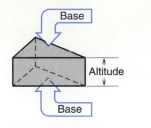

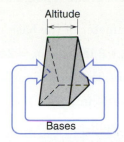

Trapezoidal Prism In a *trapezoidal prism* the bases are identical trapezoids.

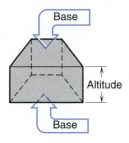

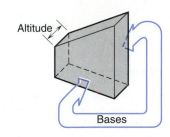

We know that the trapezoids and not the other faces are the bases because cutting the prism anywhere parallel to the trapezoid faces produces another trapezoidal prism.

Three important quantities may be calculated for any prism: the lateral surface area, the total surface area, and the volume. The following formulas apply to all right prisms.

RIGHT PRISMS

Lateral surface area	$L = ph$		h = altitude
Total surface area	$S = L + 2A$		p = perimeter of the base
Volume	$V = Ah$		A = area of the base

The lateral surface area L is the area of all surfaces *excluding* the two bases.

The total surface area S is the lateral surface area *plus* the area of the two bases.

The *volume* or capacity of a prism is the total amount of space inside it. Volume is measured in cubic units. (You may want to review these units in Chapter 5, page 210.

Let's look at an example. Find the (a) lateral surface area, (b) total surface area, and (c) volume of the triangular cross-section duct shown.

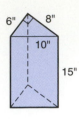

(a) The perimeter of the base is

$p = 6$ in. $+ 8$ in. $+ 10$ in. $= 24$ in. and $h = 15$ in.

Therefore, the lateral surface area is

$L = ph = (24$ in.$)(15$ in.$)$

$L = 360$ sq in.

(b) In this prism the bases are right triangles, so that the area of each base is

$A = \frac{1}{2}bh$

$A = \frac{1}{2}(6$ in.$)(8$ in.$)$

$A = 24$ sq in.

Therefore, the total surface area is

$S = L + 2A$

$S = 360$ sq in. $+ 2(24$ sq in.$)$

$S = 408$ sq in.

(c) The volume of the prism is

$V = Ah$

$V = (24$ sq in.$)(15$ in.$)$

$V = 360$ cu in.

Your turn. Find L, S, and V for this trapezoidal right prism.

Check your work in **3**.

---

3 $h = 10$ ft   and $p = 4$ ft $+ 15$ ft $+ 12$ ft $+ 17$ ft $= 48$ ft

The area of each trapezoidal base is

$A = \left(\dfrac{b_1 + b_2}{2}\right)h$

$A = \left(\dfrac{4 \text{ ft} + 12 \text{ ft}}{2}\right)(15 \text{ ft})$

$A = 120$ sq ft

Therefore,

$L = ph$	$S = L + 2A$	$V = Ah$
$L = (48$ ft$)(10$ ft$)$	$S = 480$ sq ft $+ 2(120$ sq ft$)$	$V = (120$ sq ft$)(10$ ft$)$
$L = 480$ sq ft	$S = 720$ sq ft	$V = 1200$ cu ft

Find the lateral surface area, total surface area, and volume for each of the following prisms.

(a) (b) (c) (d)

$4\frac{1}{2}''$ · 2'' · 3''

18' · 6' · 16' · 18' · 20'

9'' · 5'' · 15'' · $6\frac{3}{4}''$ · 12''

7 cm · 7 cm · 7 cm

Check your work in **4**.

---

4 (a) $h = 4\frac{1}{2}$ in.

$p = 3$ in. $+ 2$ in. $+ 3$ in. $+ 2$ in. $= 10$ in. $A = (3$ in.$)(2$ in.$) = 6$ sq in.

$L = ph = (10$ in.$)(4.5$ in.$)$ $S = L + 2A$

$L = 45$ sq in. $S = 45$ sq in. $+ 2(6$ sq in.$)$

                                                   $S = 57$ sq in.

$V = Ah$

$V = (6$ sq in.$)(4.5$ in.$)$

$V = 27$ cu in.

(b) $L = 1080$ sq ft $S = 1464$ sq ft $V = 3456$ cu ft

(c) $L = 416.25$ sq in. $S = 476.25$ sq in. $V = 450$ cu in.

(d) For a cube, $L = 4s^2 = 196$ sq cm

$S = 6s^2 = 294$ sq cm

$V = s^3 = 343$ cu cm

In practical problems it is often necessary to convert volume units from cu in. or cu ft to gallons or similar units. Use the following conversion factors and set up unity fractions as shown in Chapter 5.

---

1 cu ft = 1728 cu in.   1 cu yd = 27 cu ft
  ≈ 7.48 gallons
  ≈ 28.3 liters   1 cu in. ≈ 16.38 cm^3
  ≈ 0.0283 m^3
  ≈ 0.806 bushel   1 gal ≈ 231 cu in.

For example, calculate the volume of water needed to fill this swimming pool in gallons. Round to the nearest hundred gallons.

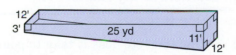

12' · 3' · 25 yd · 11' · 12'

Check your calculation in **5**.

---

5 The pool is a prism with trapezoidal sides. The height of the trapezoid is 25 yd or 75 ft and the altitude of the prism is 12 ft.

$$V = Ah$$

$$V = \left(\frac{3 \text{ ft} + 11 \text{ ft}}{2}\right)(75 \text{ ft})(12 \text{ ft})$$

$$V = (7 \text{ ft})(75 \text{ ft})(12 \text{ ft})$$

$$V = 6300 \text{ cu ft}$$

In gallons,

$$V = 6300 \text{ cu ft} \times \frac{7.48 \text{ gal}}{1 \text{ cu ft}}$$

$$V = 47{,}124 \text{ gal} \qquad \text{or} \qquad 47{,}100 \text{ gal, rounded}$$

3 ⊞ **11** ⊜ ⊙ **2** ⊠ **75** ⊠ **12** ⊠ **7.48** ⊜ → ░ 47124. ░

In many practical problems the prism encountered has an irregular polygon for its base. For example, find the volume of the slotted bar shown. If steel weighs 0.283 lb/cu in., what would be the weight of this bar? Round to one decimal place.

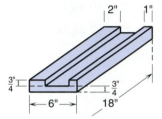

Try it. We will work it out in detail in **6**.

---

6 First, we must calculate the volume of this prism. To find this, we first calculate the cross-sectional area by subtracting.

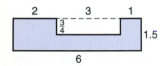

$$A = (6'' \times 1.5'') - \left(3'' \times \frac{3}{4}''\right)$$

$$A = 9 \text{ sq in.} - 2.25 \text{ sq in.}$$

$$A = 6.75 \text{ sq in.}$$

and the volume of the bar is

$$V = Ah$$
$$V = (6.75 \text{ sq in.})(18 \text{ in.})$$
$$V = 121.5 \text{ cu in.}$$

The weight of the bar is

$$W = 121.5 \text{ cu in.} \times 0.283 \, \frac{\text{lb}}{\text{cu in.}}$$

$$W = 34.3845 \text{ lb} \qquad \text{or} \qquad 34.4 \text{ lb} \quad \text{rounded}$$

Learning Help ▶ The most common source of errors in most calculations of this kind is carelessness. Organize your work neatly. Work slowly and carefully. ◀

Now try these practice problems.

(a) How many cubic *yards* of concrete does it take to pour a square pillar 2 ft on a side if it is 12 ft high? (Round to one decimal place.)

(b) What is the capacity to the nearest gallon of a septic tank in the shape of a rectangular prism 12 ft by 16 ft by 6 ft?

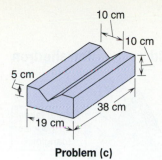

Problem (c)

(c) What is the weight of the steel V-block in the figure at 0.0173 lb/cm³? (Round to the nearest tenth.)

(d) Find the weight of the piece of brass pictured at 0.2963 lb/cu in. (Round to the nearest tenth.)

(e) Find the weight (in ounces) of the steel pin shown if its density is 0.0173 lb/cm³. (Round to the nearest tenth.)

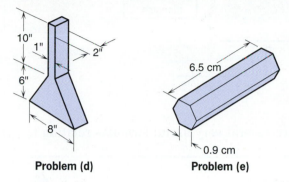

Problem (d)   **Problem (e)**

The correct answers are in **7**.

---

7 (a) $V = Ah = (2 \text{ ft})(2 \text{ ft})(12 \text{ ft})$

$= 48 \text{ cu ft}$ But 1 cu yd = 27 cu ft

$= 48 \text{ cu ft} \times \dfrac{1 \text{ cu yd}}{27 \text{ cu ft}}$

$= \dfrac{48}{27} \text{ cu yd} \approx 1.8 \text{ cu yd}$

(b) 8617 gal   (c) 108.5 lb   (d) 21.9 lb   (e) 3.8 oz

Pyramids A *pyramid* is a solid object with one base and three or more *lateral faces* that taper to a single point opposite the base. This single point is called the *apex*.

A *right pyramid* is one whose base is a regular polygon and whose apex is centered over the base. We will examine only right pyramids here.

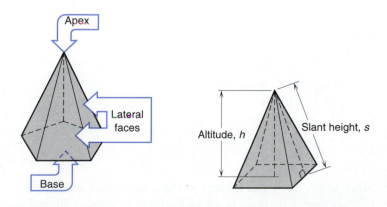

The *altitude* of a pyramid is the perpendicular distance from the apex to the base. The *slant height* of the pyramid is the height of one of the lateral faces.

Several kinds of right pyramids can be constructed:

Square Pyramid   **Triangular Pyramid**   or   **Tetrahedron**

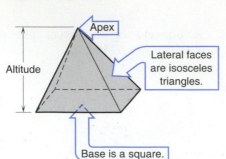

Apex

Altitude

Lateral faces are isosceles triangles.

Base is a square.

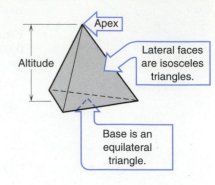

Apex

Altitude

Lateral faces are isosceles triangles.

Base is an equilateral triangle.

There are several very useful formulas relating to pyramids.

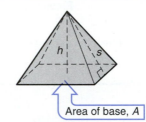

PYRAMIDS

Lateral surface area $L = \frac{1}{2}ps$

Volume $V = \frac{1}{3}Ah$

h s

Area of base, A

where p = perimeter of the base
s = slant height of a lateral face
A = area of the base
h = altitude

For example, find the lateral surface area and volume of this square pyramid:

$s \approx 10.77$ ft, $h = 10$ ft   Round to the nearest tenth.

$p = 8$ ft + 8 ft + 8 ft + 8 ft = 32 ft $A = (8$ ft$)(8$ ft$) = 64$ sq ft

The lateral surface area is

$L = \frac{1}{2}ps$

$L \approx \frac{1}{2}(32$ ft$)(10.77$ ft$)$

$L \approx 172.3$ sq ft   This is the area of the four triangular lateral faces.

The volume is

$V = \frac{1}{3}Ah$

$V = \frac{1}{3}(64$ sq ft$)(10$ ft$)$

$V \approx 213.3$ cu ft   rounded

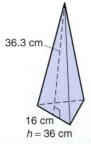

36.3 cm

16 cm

$h = 36$ cm

Your turn. Find the lateral surface area and volume of this triangular pyramid. (*Remember:* The base is an equilateral triangle.) Round to the nearest whole unit.

Our solution is in **8**.

8 $p = 16 \text{ cm} + 16 \text{ cm} + 16 \text{ cm}$

$p = 48 \text{ cm}$ perimeter of the base

$s \approx 36.3 \text{ cm}$

Therefore, $L = \frac{1}{2}ps$

$$L \approx \frac{1}{2}(48 \text{ cm})(36.3 \text{ cm})$$

$$L \approx 871.2 \text{ cm}^2 \text{ or } 871 \text{ cm}^2 \quad \text{rounded}$$

$A \approx 0.433s^2$ for an equilateral triangle

$A \approx 0.433(16 \text{ cm})^2$

$A \approx 110.8 \text{ cm}^2$

The volume is

$$V = \frac{1}{3}Ah$$

$$V \approx \frac{1}{3}(110.8 \text{ cm}^2)(36 \text{ cm})$$

$$V \approx 1330 \text{ cm}^3 \quad \text{rounded}$$

The entire calculation for volume can be done as follows using a calculator:

 .433 $\boxed{\times}$ **16** $\boxed{x^2}$ $\boxed{\times}$ **36** $\boxed{\div}$ **3** $\boxed{=}$ $\rightarrow$ *1330.176*

A

Multiplying by $\frac{1}{3}$ is the same as dividing by 3

More practice.

(a) Find the lateral surface area and volume of each pyramid. (Round to one decimal place.)

(1)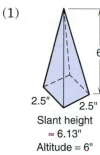

6"

2.5"   2.5"

Slant height ≈ 6.13"

Altitude = 6"

(2)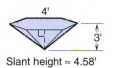

4'

3'

Slant height ≈ 4.58'

Altitude = 3'

(3)

13 cm

Altitude = 16 cm

Slant height ≈ 16.4 cm

(b) Find the volume of an octagonal pyramid if the area of the base is 245 sq in. and the altitude is 16 in. (Round to the nearest cubic inch.)

The answers are in **9**.

---

9 (a) (1) 30.7 sq in.; 12.5 cu in.   (2) 55.0 sq ft; 41.6 cu ft

   (3) 319.8 cm²; 390.3 cm³

(b) 1307 cu in.

Frustum of a Pyramid

A *frustum* of a pyramid is the solid figure remaining after the top of the pyramid is cut off parallel to the base. Frustum shapes appear as containers, as building foundations, and as transition sections in ducts.

Most frustums used in practical work, such as sheet metal construction, are frustums of square pyramids or cones.

Every frustum has two bases, upper and lower, that are parallel and different in size. The altitude is the perpendicular distance between the bases.

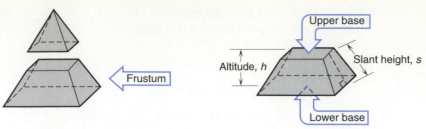

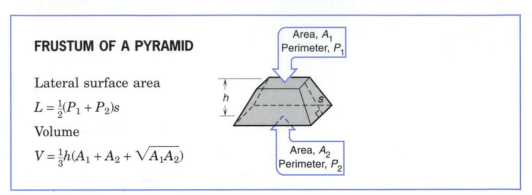

The following formulas enable you to find the lateral area and volume of any pyramid frustum.

FRUSTUM OF A PYRAMID

Lateral surface area

$L = \frac{1}{2}(P_1 + P_2)s$

Volume

$V = \frac{1}{3}h(A_1 + A_2 + \sqrt{A_1 A_2})$

where P_1 and P_2 = upper and lower perimeters for a pyramid frustum

A_1 and A_2 = upper and lower base areas for any frustum

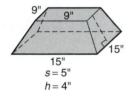

For example, find the lateral surface area and volume of the pyramid frustum shown.

Upper perimeter, $P_1 = 9$ in. + 9 in. + 9 in. + 9 in. = 36 in.

Lower perimeter, $P_2 = 15$ in. + 15 in. + 15 in. + 15 in. = 60 in.

$L = \frac{1}{2}(36 \text{ in.} + 60 \text{ in.})(5 \text{ in.})$

$L = 240$ sq in.

Upper area, $A_1 = (9 \text{ in.})(9 \text{ in.}) = 81$ sq in.

Lower area, $A_2 = (15 \text{ in.})(15 \text{ in.}) = 225$ sq in.

$V = \frac{1}{3}(4 \text{ in.})(81 \text{ sq in.} + 225 \text{ sq in.} + \sqrt{81 \cdot 225})$

$V = \frac{1}{3}(4 \text{ in.})(81 + 225 + 135)$ sq in.

$V = 588$ cu in.

Use the following calculator sequence to find the volume:

 81 $\times$ **225** $=$ $\sqrt{}$ $+$ **225** $+$ **81** $=$ $\times$ **4** $\div$ **3** $=$ $\rightarrow$ *588.*

To find the total surface area of the frustum, add the areas A_1 and A_2 to the lateral surface area.

Practice with these problems.

(a) Find the lateral surface area and volume of each of the following frustums. (Round to the nearest whole unit.)

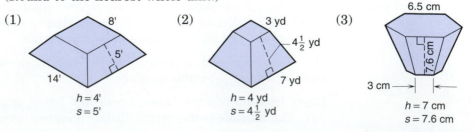

(1) 8' 5' 14' h = 4' s = 5'

(2) 3 yd $4\frac{1}{2}$ yd 7 yd h = 4 yd s = $4\frac{1}{2}$ yd

(3) 6.5 cm 7.6 cm 3 cm h = 7 cm s = 7.6 cm

(b) A connection must be made between two square vent openings in an air conditioning system. Find the total amount of sheet metal needed using the dimensions shown in the figure.

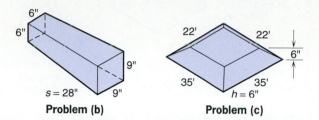

Problem (b) **Problem (c)**

(c) How many cubic *yards* of concrete are needed to pour the foundation shown in the figure? (Watch your units!) Round to the nearest tenth.

Check your answers in **10**.

---

10 (a) (1) 220 sq ft, 496 cu ft (2) 90 sq yd, 105 cu yd
(3) 217 cm², 429 cm³

(b) 840 sq in. (c) 15.3 cu yd

Now turn to **11** for a set of exercises on prisms and pyramids.

11

Exercises 9-1 **Prisms and Pyramids**

A. Find the volume of each of the following shapes. (Round to the nearest tenth if necessary.)

1.

2.

3.
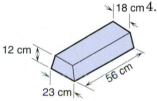

4.
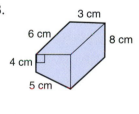

Round to nearest cu in.

Round to nearest hundred

5.

Round to the nearest tenth.

6.
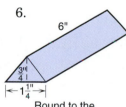

Round to the nearest hundredth.

7.
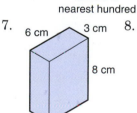

8.

B. Find the lateral surface area and the volume of each of the following solids. (Round to the nearest tenth.)

1.
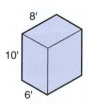

2.
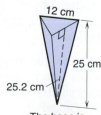

The base is equilateral.
s = 25.2 cm
h = 25 cm

3.

4.

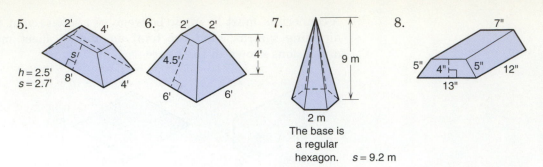

5.   2'
      4'
      s
h = 2.5'   8'
s = 2.7'

6.   2' 2'
      4.5'   4'
      6'   6'

7.   9 m
      2 m
      The base is
      a regular
      hexagon.   s = 9.2 m

8.   7"
      5" 4" 5" 12"
      13"

C. Find the total outside surface area and the volume of each of the following
    solids. (Round to the nearest tenth.)

1.   5 yd   5 yd
      5 yd

2.   9.3 m
      7.4 m
      5.2 m

3.   9.4"   9"
      9" 9"
      The base is
      equilateral.

4. $\frac{1}{2}$"
      $3\frac{1}{2}$"
      The base
      is a
      regular
      hexagon.

5.   3'
      8'
      4' 4'
      The pyramid cap
      has s = 3.6'
      and h = 3'.

6.   13 cm
      35 cm
      25 cm
      37 cm
      36 cm

7.   18.5"
      10"   s = 19.2"
      The base
      is a square.

8.   6 yd
      2 yd   8 yd   12 yd
      15 yd

D. Practical Problems (Round to the nearest tenth.)

1. **Manufacturing** How many cubic *feet* of warehouse space are needed for
    450 boxes 16 in. by 8 in. by 10 in.?

2. **Metalworking** Find the weight of the piece of steel pictured. (Steel
    weighs 0.2833 lb/cu in.)

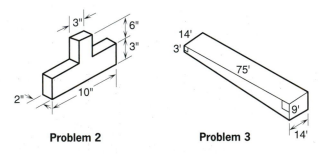

3"
6"
3"
2"   10"

14'
3'
75'
9'
14'

Problem 2   **Problem 3**

3. **Building Construction** How many gallons of water are needed to fill a
    swimming pool that approximates the shape in the figure? (Round to the
    nearest gallon.)

4. **Painting and Decorating** How many gallons of paint are needed for the
    outside walls of a building 26 ft high by 42 ft by 28 ft if there are 480 sq
    ft of windows? One gallon covers 400 sq ft. (Assume that you cannot buy
    a fraction of a gallon.)

5. **Building Construction** Dirt must be excavated for the foundation of a
    building 30 yards by 15 yards to a depth of 3 yards. How many trips will
    it take to haul the dirt away if a truck with a capacity of 3 cu yd is used?

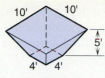

Problem 6

6. **Manufacturing** How many bushels will the bin in the figure hold? (1 cu ft ≈ 1.24 bushels.) Round to the nearest bushel.

7. **Masonry** If brick has a density of 103 lb/cu ft, what will be the weight of 500 bricks, each, $3\frac{3}{4}$ in. by $2\frac{1}{4}$ in. by 8 in.? Round to the nearest pound.

8. **Landscaping** Dirt cut 2 ft deep from a section of land 38 ft by 50 ft is used to fill a section 25 ft by 25 ft to a depth of 3 ft. How many cubic *yards* of dirt are left after the fill?

9. **Sheet Metal Technology** How many square inches of sheet metal are used to make the vent transition shown? (The ends are open.)

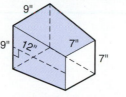

Problem 9   **Problem 10**

10. **Building Construction** How many cubic yards of concrete are needed to pour the building foundation shown in the figure?

11. **Metalworking** Find the weight of the cast iron shape in the figure. (Cast iron weighs 0.2607 lb/cu in. The base is an equilateral triangle.)

Problem 11

12. **Metalworking** Find the weight of the piece of brass shown. Brass weighs 0.296 lb/cu in.

13. **Manufacturing** How high must a 400-gallon rectangular tank be if the base is a square 3 ft 9 in. on a side? (1 cu ft ≈ 7.48 gallons.)

14. **Landscaping** How many pounds of rock will you need to fill an area 25 ft by 35 ft to a depth of 2 in. if the rock weighs 1050 lb/cu yd? Round to the nearest ten lb.

15. **Wastewater Technology** A rectangular sludge bed measuring 80 ft by 120 ft is filled to a depth of 2 ft. If a truck with a capacity of 3 cu yd is used to haul away the sludge, how many trips must the truck make to empty the sludge bed?

16. **Wastewater Technology** A reservoir with a surface area of 2.5 acres loses 0.3 in. per day to evaporation. How many gallons does it lose in a week? (1 acre = 43,560 sq ft, 1 sq ft = 144 sq in., 1 gal = 231 cu in.) (Round to the nearest thousand.)

17. **Carpentry** A carpenter needs to pour a rectangular slab 28 ft long, 16 ft wide, and 6 in. thick. How many cubic yards of concrete does the carpenter need? (Round to the nearest tenth.)

18. **Construction Technology** Concrete weighs 150 lb/cu ft. What is the weight of a concrete wall 18 ft long, 4 ft 6 in. high, and 8 in. thick?

19. **Masonry** The concrete footings for piers for a raised foundation are cubes measuring 18 in. by 18 in. by 18 in. How many cubic yards of concrete are needed for 28 footings?

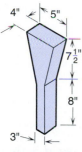

Problem 12

20. **Plumbing** A flush tank 21 in. by 6.5 in. contains water to a depth of 12 in. How many gallons of water will be saved per flush if a conservation device reduces the capacity to two-thirds of this amount? (Round to the nearest tenth of a gallon.)

21. **Roofing** A flat roof 28 ft long and 16 ft wide has a 3-in. depth of water sitting on it. What is the weight of the water on the roof? (Water weighs 62.5 lb/cu ft.)

When you have completed these exercises, check your answers on page 604 then turn to **12** to study cylinders, cones, and spheres.

9-2 CYLINDERS, CONES, AND SPHERES

Cylinders

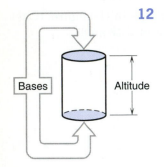
Bases   Altitude

12 A *cylinder* is a solid object with two identical circular bases. The *altitude* of a cylinder is the perpendicular distance between the bases.

A *right cylinder* is one whose curved side walls are perpendicular to its circular base. Whenever we mention the radius, diameter, or circumference of a cylinder, we are referring to those dimensions of its circular base.

Two important formulas enable us to find the lateral surface and volume of a cylinder.

CYLINDERS

Lateral surface area

$L = Ch$ or $L = 2\pi rh = \pi dh$

Volume

$V = \pi r^2 h$ or $V \approx 0.7854 d^2 h$

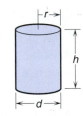

---

where C is the circumference of the base, r is the radius of the base, d is the diameter of the base, and h is the altitude of the cylinder. To find the total surface area of a cylinder, simply add the areas of the two circular bases to the lateral area. The formula is

Total surface area $= 2\pi r^2 + 2\pi rh = 2\pi r(r + h)$

Learning Help ▶

You can visualize the lateral surface area by imagining the cylinder wall flattened into a rectangle as shown.

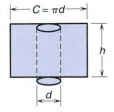

Lateral surface area = area of rectangle
$$= L \times w$$
$$= \pi d \times h$$
$$= \pi dh \quad ◀$$

Use these formulas to find the lateral surface area and volume of a cylindrical container whose altitude is 24 in. and whose diameter is 22 in.

$d = 22$ in. $r = 11$ in. $h = 24$ in.

Then the lateral surface area is

$L = \pi dh$
$L \approx (3.1416)(22 \text{ in.})(24 \text{ in.})$
$L \approx 1659 \text{ sq in.}$ rounded

and the volume is

$V = \pi r^2 h$
$V \approx (3.1416)(11 \text{ in.})^2(24 \text{ in.})$
$V \approx (3.1416)(121 \text{ sq in.})(24 \text{ in.})$
$V \approx 9123 \text{ cu in.}$ rounded

Find the lateral surface area and the volume in gallons of the cylindrical storage tank shown. Round answers to the nearest unit.

Check your work in **13**.

---

13 $L = \pi dh$
$L \approx (3.1416)(14 \text{ ft})(30 \text{ ft})$
$L \approx 1319.472 \text{ sq ft}$ or 1319 sq ft rounded

$V \approx 0.7854 d^2 h$ approximately
$V \approx (0.7854)(14 \text{ ft})^2(30 \text{ ft})$
$V \approx (0.7854)(196 \text{ sq ft})(30 \text{ ft})$
$V \approx 4618.152 \text{ cu ft}$ or 4618 cu ft rounded

In gallons

$$V \approx 4618.152 \text{ cu ft} \times \frac{7.48 \text{ gal}}{1 \text{ cu ft}}$$

$V \approx 34,544 \text{ gal}$ rounded

Careful ▶   When using an intermediate value in a later calculation, do not round it prematurely. For example, if you had rounded V to 4618 and used this value to calculate V in gallons, the result would be 34,543, which is not correct to the nearest gallon. ◀

After organizing your work, you can key it into your calculator this way:

.7854 ⊠ **14** $\boxed{x^2}$ ⊠ **30** ⊠ **7.48** ⊟ → ‎ *34543.777*

Try these problems for practice in working with cylinders.

(a) Find the lateral surface area and volume of each cylinder. (Round to the nearest whole number.)

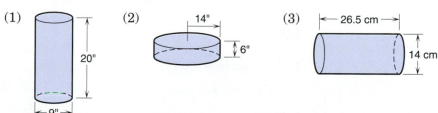

(b) Find the total surface area of a cylinder 9 yd high and 16 yd in diameter. (Round to the nearest whole number.)

(c) How many cubic *yards* of concrete will it take to pour a cylindrical column 14 ft high with a diameter of 2 ft? (Round to the nearest tenth.)

(d) What is the capacity in gallons of a cylindrical tank with a radius of 8 ft and an altitude of 20 ft? (Round to the nearest 100 gallons.)

(e) A pipe 3 in. in diameter and 40 ft high is filled to the top with water. If 1 cu ft of water weighs 62.4 lb, what is the weight at the base of the pipe? (Be careful of your units.) (Round to the nearest pound.)

(f) The *bottom* and the *inside walls* of the cylindrical tank shown in the figure must be lined with sheet copper. How many square feet of sheet copper are needed? (Round to the nearest tenth.)

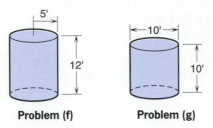

Problem (f)   **Problem (g)**

(g) How many quarts of paint are needed to cover the outside wall (not including the top and bottom) of a cylindrical water tank 10 ft in diameter and 10 ft in height if one quart covers 100 sq ft? (Assume that you cannot buy a fraction of a quart.)

Check your answers in **14**.

---

14 (a) (1) 565 sq in., 1272 cu in.   (2) 528 sq in., 3695 cu in.
       (3) 1166 sq cm, 4079 cu cm

(b) 855 sq yd   (c) 1.6 cu yd

(d) 30,100 gallons, rounded   (e) 123 lb, rounded

(f) 455.5 sq ft   (g) 4 qt

In problem (e), you must change the diameter from 3 in. to 0.25 ft before substituting into the formula.

Cones   A *cone* is a pyramid-like solid figure with a circular base. The radius and diameter of a cone refer to its circular base. The *altitude* is the perpendicular distance from the apex to the base. The *slant height* is the apex-to-base distance along the surface of the cone.

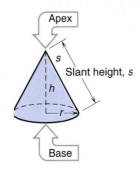

The following formulas enable us to find the lateral surface area and volume of a cone.

CONES

Lateral surface area

$L = \pi r s$ or $L = \frac{1}{2}\pi d s$

Volume

$V = \frac{1}{3}\pi r^2 h$ or $V \approx 0.2618 d^2 h$

$d = 2r$

---

where r = radius of the base
         d = diameter of the base
         h = altitude
         s = slant height

To find the total surface area of a cone, simply add the area of the circular base to the lateral area. The formula would be

Total surface area = $\pi r^2 + \pi rs = \pi r(r + s)$

For example, find the lateral surface area and volume of the cone shown here. (Round to the nearest whole unit.) $h = 15$ in., $s = 18$ in., $r = 10$ in.

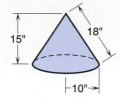

Then the lateral surface area is

$L = \pi rs$
$L \approx (3.1416)(10 \text{ in.})(18 \text{ in.})$
$L \approx 565.488$ sq in.   or   565 sq in.   rounded

and the volume is

$V = \frac{1}{3}\pi r^2 h$

$V \approx \frac{1}{3}(3.1416)(10 \text{ in.})^2(15 \text{ in.})$

$V \approx 1570.8$ cu in.   or   1571 cu in.   rounded

 Using a calculator to find the volume, we obtain

3.1416 ✕ **10** x^2 ✕ **15** ÷ **3** = → ▓ *1570.8*

Find the lateral surface area, total surface area, and volume of this cone. (Round to the nearest whole unit.)

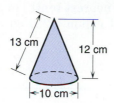

Check your work in **15**.

---

15 $h = 12$ cm $r = 5$ cm $s = 13$ cm

Lateral surface area $= L = \pi rs$
$\qquad\qquad L \approx (3.1416)(5 \text{ cm})(13 \text{ cm})$
$\qquad\qquad L \approx 204.204$ cm^2 or   204 cm^2 rounded

Total surface area $= \pi r^2 + \pi rs$
$\qquad\qquad \approx (3.1416)(5 \text{ cm})^2 + 204.204 \text{ cm}^2$
$\qquad\qquad \approx 78.54 + 204.204$
$\qquad\qquad \approx 282.744$ or 283 cm^2 rounded

Volume $= V = \frac{1}{3}\pi r^2 h$

$\qquad\qquad V \approx \frac{1}{3}(3.1416)(5 \text{ cm})^2(12 \text{ cm})$

$\qquad\qquad V \approx 314.16$ cm^3 or   314 cm^3 rounded

You should realize that the altitude, radius, and slant height for any cone are always related by the Pythagorean theorem.

$h \quad s \qquad s^2 = h^2 + r^2$
$\quad r$

If you are given any two of these quantities, you can use this formula to find the third.

Notice that the volume of a cone is exactly one-third the volume of the cylinder that just encloses it.

Solve the following practice problems.

(a) Find the lateral surface area and volume of each of the following cones. (Round to the nearest whole number.)

(1)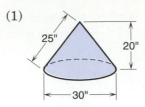

(2)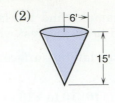

(3)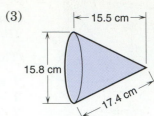

(b) Find the total surface area of a cone with radius $\frac{3}{4}$ in., altitude 1 in., and slant height $1\frac{1}{4}$ in. (Round to two decimal places.)

(c) A pile of sand dumped by a hopper is cone-shaped. If the diameter of the base is 18 ft 3 in. and the altitude is 8 ft 6 in., how many cubic feet of sand are in the pile? (Round to the nearest tenth.)

(d) What is the capacity in gallons of a conical drum 4 ft high with a radius of 3 ft? (1 cubic foot ≈ 7.48 gallons.) (Round to the nearest gallon.)

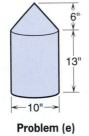

Problem (e)

(e) Find the weight of the cast iron shape shown at 0.26 lb/cu in. (Round to the nearest pound.)

(f) Find the capacity in gallons (to the nearest tenth) of a conical oil container 15 in. high with a diameter of 16 in. (1 gallon ≈ 231 cu in.).

Check your answers in **16**.

---

16 (a) (1) 1178 sq in., 4712 cu in.   (2) 305 sq ft, 565 cu ft

(3) 432 cm², 1013 cm³

(b) 4.71 sq in.   (c) 741.2 cu ft   (d) 282 gal   (e) 306 lb   (f) 4.4 gal

On (a)(2) use $s^2 = h^2 + r^2$ or $s = \sqrt{15^2 + 6^2} \approx 16.1555$ ft

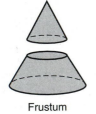

Here is the calculator sequence for problem (f):

.2618 ⊠ 16 x^2 ⊠ 15 ÷ 231 = → 4.352

Frustum of a Cone   The frustum of a cone, like the frustum of a pyramid, is the figure remaining when the top of the cone is cut off parallel to its base.

Frustum shapes appear as containers, as funnels, and as transition sections in ducts.

Every frustum has two bases, upper and lower, that are parallel and different in size. The altitude is the perpendicular distance between bases.

Frustum

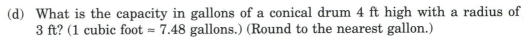

The following formulas enable you to find the lateral area and volume of any frustum of a cone.

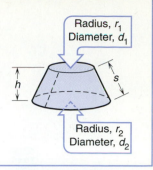

FRUSTUM OF A CONE

Lateral surface area

$$L = \pi s(r_1 + r_2) \quad \text{or} \quad \tfrac{1}{2}\pi s(d_1 + d_2)$$

Volume

$$V = \tfrac{1}{3}\pi h(r_1^2 + r_2^2 + r_1 r_2) \quad \text{or} \quad \tfrac{1}{12}\pi h(d_1^2 + d_2^2 + d_1 d_2)$$

These formulas can be obtained from the equation for the lateral surface area and volume of the frustum of a pyramid, using πr^2 or $\tfrac{1}{4}\pi d^2$ for the area of a circle.

To find the *total surface area* of the frustum, add the areas of the circular ends to the lateral surface area.

For example, to find the lateral surface area, total surface area, and volume of the frustum shown, use $s = 9.0$ ft, $r_1 = 2$ ft, $r_2 = 5$ ft, and $h = 8.5$ ft.

Lateral surface area, $\quad L \approx 3.1416(9)(2 + 5)$
$$\approx 197.9208 \text{ sq ft} \quad \text{or} \quad 198 \text{ sq ft} \quad \text{rounded}$$

Total surface area, $\quad A \approx L + \pi r_1^2 + \pi r_2^2$
$$A \approx 197.9208 + 3.1416(2^2) + 3.1416(5^2)$$
$$A \approx 289.0272 \text{ sq ft} \quad \text{or} \quad 289 \text{ sq ft} \quad \text{rounded}$$

Volume, $\quad V \approx \tfrac{1}{3}(3.1416)(8.5)(2^2 + 5^2 + 2 \cdot 5)$
$$V \approx 347.1468 \text{ cu ft} \quad \text{or} \quad 347 \text{ cu ft} \quad \text{rounded}$$

Using a calculator to find the volume, we have

3.1416 ÷ **3** × **8.5** × (**2** x^2 + **5** x^2 + **2** × **5**) = → *347.1468*

Practice with these problems.

(a) Find the lateral surface area and volume of each of the following frustums. (Round to the nearest whole unit and use $\pi \approx 3.14$.)

(1)
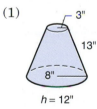

3"
13"
8"
$h = 12"$

(2)
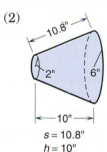

10.8"
2" 6"
10"
$s = 10.8"$
$h = 10"$

(3)
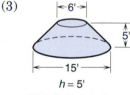

6'
5'
15'
$h = 5'$

Hint: Calculate the slant height s using the Pythagorean theorem.

(b) What is the capacity in gallons of the oil can shown in the figure? (1 gallon $\approx$ 231 in.³) Round to the nearest tenth.

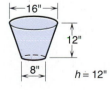

16"
12"
8" $h = 12"$

Check your answers in **17**.

---

17 (a) (1) 449 sq in., 1218 cu in.   (2) 271 sq in., 544 cu in.
      (3) 222 sq ft, 459 cu ft

(b) 6.1 gal

Find the slant height in problem (a)(3) this way:

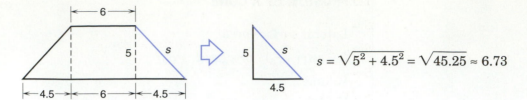

$$s = \sqrt{5^2 + 4.5^2} = \sqrt{45.25} \approx 6.73$$

 Using a calculator on problem (b),

3.1416 ÷ **12** × **12** × **(** **16** x^2 **+** **8** x^2 **+** **8** × **16** **)** ÷ **231** **=** → ‎ *6.0928*

Spheres

The *sphere* is the simplest of all solid geometric figures. Geometrically, it is defined as the surface whose points are all equidistant from a given point called the *center*. The *radius* is the distance from the center to the surface. The *diameter* is the straight-line distance across the sphere on a line through its center.

The following formulas enable you to find the surface area and volume of any sphere.

SPHERE

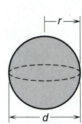

Surface area

$S = 4\pi r^2$ or $S = \pi d^2$

Volume

$V = \dfrac{4\pi r^3}{3}$ or $V = \dfrac{\pi d^3}{6}$

The volume formulas can be written approximately as

$V \approx 4.1888 r^3$ and $V \approx 0.5236 d^3$

For example, find the surface area and volume of a spherical basketball of diameter $9\frac{1}{2}$ in. Round to the nearest tenth.

The surface area is

$S = \pi d^2$
$S \approx (3.1416)(9.5 \text{ in.})^2$
$S \approx (3.1416)(90.25 \text{ sq in.})$
$S \approx 283.5 \text{ sq in.}$

The volume is

$V = \dfrac{\pi d^3}{6}$

$V \approx \dfrac{(3.1416)(9.5 \text{ in.})^3}{6}$

$V \approx 448.9 \text{ cu in.}$ rounded

Using a calculator to find the volume, we have

 3.1416 × **9.5** y^x **3** ÷ **6** **=** → *448.92155*

Your turn. Find the surface area and the volume in gallons of a spherical tank 8 ft in radius. Round to the nearest whole unit.

Check your solution in **18**.

18 The surface area is

$$S = 4\pi r^2$$

$$S \approx 4(3.1416)(8 \text{ ft})^2 \quad \longleftarrow \boxed{\text{Calculate } 8^2 \text{ before multiplying}}$$

$$S \approx 4(3.1416)(64 \text{ sq ft}) = 804.2496$$

$$S \approx 804 \text{ sq ft}$$

The volume is

$$V \approx \frac{4\pi r^3}{3} \qquad \boxed{\text{Calculate } 8^3 \text{ before multiplying}}$$

$$V \approx \frac{4(3.1416)(8 \text{ ft})^3}{3} \approx 2144.6656$$

$$V \approx 2145 \text{ cu ft, rounded}$$

$$V \approx 2144.7 \text{ cu ft} \times \frac{7.48 \text{ gallons}}{1 \text{ cu ft}} \qquad \text{Convert to gallons using a unity fraction.}$$

$$V \approx 16{,}042 \text{ gallons} \quad \text{rounded to the nearest gallon}$$

For practice, try the following problems.

(a) Find the surface area and volume of each of the following spheres. (Round to the nearest whole unit.)

    (1)   radius = 15 in.   (2)   diameter = 22 ft

    (3)   radius = 6.5 cm   (4)   diameter = 8 ft 6 in.

(b) How many gallons of water can be stored in a spherical tank 50 in. in diameter? (Use 1 gallon ≈ 231 cu in.   Round to the nearest gallon.)

(c) What will be the weight of a spherical steel tank of radius 9 ft when it is full of water? The steel used weighs approximately 127 lb/sq ft and water weighs 62.4 lb/cu ft. (Round to the nearest 100 lb.)

Check your answers in **19**.

---

19 (a) (1)   2827 sq in., 14,137 cu in.
          (2)   1521 sq ft, 5575 cu ft
          (3)   531 cm^2, 1150 cm^3
          (4)   227 sq ft, 322 cu ft

    (b) 283 gallons   (c)   319,800 lb

In problem (b), the volume is 65,450 cu in. Multiply by a unity fraction to convert to gallons.

$$65{,}450 \text{ cu in.} \times \frac{1 \text{ gal}}{231 \text{ cu in.}} = \frac{65{,}450}{231} \text{ gal or about 283 gal}$$

In problem (c) the area is 1017.9 sq ft. To find its weight, multiply by a unity fraction:

$$1017.9 \text{ sq ft} \times \frac{127 \text{ lb}}{1 \text{ sq ft}} = 129{,}300 \text{ lb} \quad \text{rounded}$$

The volume is 3053.6 cu ft or the weight of water is

$$3053.6 \text{ cu ft} \times \frac{62.4 \text{ lb}}{1 \text{ cu ft}} = 190{,}500 \text{ lb} \quad \text{rounded}$$

We can perform the entire calculation on a calculator as follows.

First, calculate the weight of the tank and store it in memory.

4 ⊗ **3.1416** ⊗ **9** [x²] ⊗ **127** [=] [STO] → *129270.56*

Then, calculate the weight of the water and add it to the tank weight in memory.

4 ⊗ **3.1416** ⊗ **9** [yˣ] **3** ÷ **3** ⊗ **62.4** [+] [RCL] [=] → *319817.39*

Now turn to **20** for a set of exercises on solid geometric figures.

20

Exercises 9-2 **Cylinders, Cones, and Spheres**

A. Find the lateral surface area and volume of each of the following solids. In addition, find the total surface area for problems 3, 8, and 9. (Round to three significant digits and use π ≈ 3.14.)

1.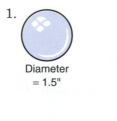
Diameter = 1.5"

2.
r = 18'

3.
28 mm
17 mm

4.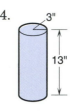
3"
13"

5.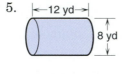
←12 yd→
8 yd

6.
8"
17"
15"

7.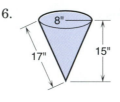
15'
33.5'
30'
s = 33.5'
h = 30'

8.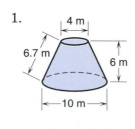
18'
20.6'
←20'→

9.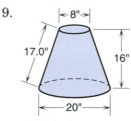
←8"→
17.0"
16"
←20"→

10.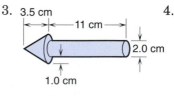
4.5'
5.0'
4.0'
←7.5'→
s = 5.0'
h = 4.0'

B. Find the volume of each figure. (Use π ≈ 3.14 and round to two significant digits.)

1.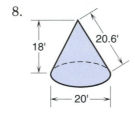
4 m
6.7 m
6 m
←10 m→

2.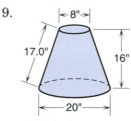
6"
16.3"
16"
3"
s = 16.3"
h = 16"

3.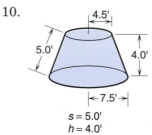
3.5 cm
←11 cm→
2.0 cm
1.0 cm

4.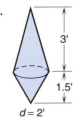
3'
1.5'
d = 2'

C. Practical Problems (Round to the nearest tenth if necessary.)

1. **Landscaping** How many cubic meters of dirt are there in a pile, conical in shape, 10 m in diameter and 4 m high?

2. **Plumbing** A 4-ft-high cylindrical pipe has a diameter of 3 in. If water weighs 62.4 lb/cu ft, what is the weight of water in the pipe when filled to the top? (*Hint:* Change the diameter to feet.)

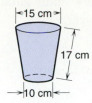

Problem 4

3. **Landscaping**   A layer of crushed rock must be spread over a circular area 23 ft in diameter. How deep a layer will be obtained using 100 cu ft of rock? (Round to the nearest hundredth of a foot.)

4. **Sheet Metal Technology**   How many square centimeters of sheet metal are needed for the sides and bottom of the pail shown in the figure?

5. How many liters of water can be stored in a cylindrical tank 8 m high with a radius of 0.5 m? (1 m^3 contains 1000 liters.)

6. **Metalworking**   What is the weight of the piece of aluminum shown at 0.0975 lb/cu in.? (*Hint*: Subtract the volume of the cylindrical hole.)

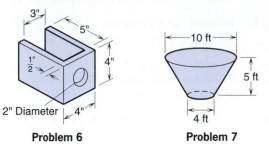

Problem 6   **Problem 7**

7. **Manufacturing**   How many bushels will the bin in the figure hold? (1 cu ft ≈ 1.24 bushels.)

8. How high should a 50-gallon tank be if it must fit into an area allowing a 16-in. diameter? (1 gallon ≈ 231 cu in.)

9. **Building Construction**   In a 4-hour percolation test, 8 ft of water seeped out of a cylindrical pit 4 ft in diameter. The plumbing code says the soil must be able to absorb 5000 gallons in 24 hours. At this rate will the soil be up to code?

10. **Sheet Metal Technology**   How many square inches of sheet copper are needed to cover the inside wall of a cylindrical tank 6 ft high with a 9-in. radius? (*Hint*: Change the altitude to inches.)

11. **Sheet Metal Technology**   How many square centimeters of sheet metal will it take to make the open-top container shown in the figure? All measurements are in centimeters. (*Hint*: No metal is needed for the small opening at the top.)

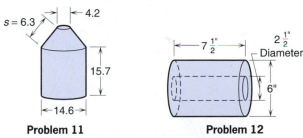

Problem 11   **Problem 12**

12. **Machine Technology**   What is the weight of the bushing shown if it is made of steel weighing 0.2833 lb/cu in.? (The inner cylinder is hollow.)

13. **Manufacturing**   Find the capacity in gallons of the oil can in the figure. (1 gallon ≈ 231 cu in.   Round to the nearest tenth.)

Problem 13

14. **Plumbing** A septic tank has the shape shown. How many gallons does it hold? (1 cu ft ≈ 7.48 gallons.)

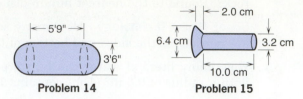

Problem 14   **Problem 15**

15. **Metalworking** Find the weight of the steel rivet shown in the figure. (Steel weighs 0.0172 lb/cu cm.)

16. **Building Construction** How many cubic yards of concrete are needed to pour eight cylindrical pillars 12 ft 6 in. high, each with a diameter of 1 ft 9 in.?

17. **Machine Technology** At a density of 0.0925 lb/cu in., calculate the weight of the aluminum piece shown.

Diameter of four outside holes: $\frac{1}{4}$ in.

Diameter of inside hole: $\frac{1}{3}$ in.

Thickness of piece: $\frac{1}{2}$ in.

18. **Industrial Technology** A spherical tank has a diameter of 16.5 ft. If a gallon of paint will cover 80 sq ft, how many gallons are needed to cover the tank? Round to the nearest gallon.

19. **Plumbing** A cylindrical tank 28 in. in diameter must have a capacity of 75 gallons. How high must it be?

20. **Plumbing** The water tower shown consists of a 16-ft-high cylinder and a 10-ft-radius hemisphere. How many gallons of water does it contain when it is filled to the top of the cylinder? Round to the nearest hundred.

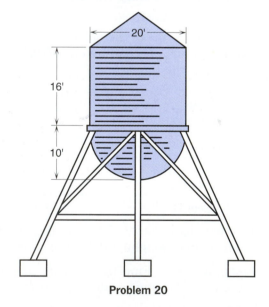

Problem 20

When you have completed these exercises, check your answers on page 604 then turn to **21** to learn about some important geometric constructions.

The following table provides a handy summary of the formulas for solid figures presented in this chapter.

SUMMARY OF FORMULAS FOR SOLID FIGURES

Figure	Lateral Surface Area	Volume
Prism	$L = ph$	$V = Ah$
Pyramid	$L = \frac{1}{2} ps$	$V = \frac{1}{3} Ah$
Cylinder	$L = \pi \, dh$ or $L = 2\pi rh$	$V = \pi r^2 h$ or $V \approx 0.7854 d^2 h$
Cone	$L = \frac{1}{2} \pi ds$ or $L = \pi rs$	$V = \frac{1}{3} \pi r^2 h$ or $V \approx 0.2618 d^2 h$
Frustum of pyramid	$L = \frac{1}{2}(P_1 + P_2)s$	$V = \frac{1}{3} h (A_1 + A_2 + \sqrt{A_1 A_2})$
Frustum of cone	$L = \frac{1}{2} \pi s (d_1 + d_2)$ or $L = \pi s (r_1 + r_2)$	$V = \frac{1}{12} \pi h (d_1^2 + d_2^2 + d_1 d_2)$ or $V = \frac{1}{3} \pi h (r_1^2 + r_2^2 + r_1 r_2)$
Sphere	$S = 4\pi r^2$ or $S = \pi d^2$	$V = \frac{4}{3} \pi r^3$ or $V \approx 4.1888 r^3$ $V = \frac{1}{6} \pi d^3$ or $V \approx 0.5236 d^3$

9-3 GEOMETRIC CONSTRUCTIONS

21 Long before the invention of protractors, ancient Greek geometers knew how to construct angles of a given size or lines that were exactly perpendicular. They found it possible to make such geometric constructions using only a straightedge or unmarked rule and a compass for drawing circles or arcs of circles. Many of the techniques they discovered are still used today, particularly in sheet metal work, drafting, carpentry, machine shops, and the construction trades. In this section we take a step-by-step look at a few of the most useful constructions.

1. Bisecting a Line Segment To bisect a line segment means to divide it into two exactly equal parts—to cut it in half.

Example: Bisect the line segment AB.

$A \text{———} B$

Step 1 Using endpoint A as a center and any radius greater than half the length of the line segment, draw arcs above and below the line.

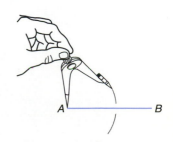

Step 2   Repeat the process using point B as the center. Do not change the radius, and be sure to intersect the arcs made in step 1. Label the points of intersection C and D.

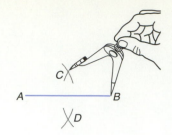

Step 3   Connect points C and D. The segment CD bisects line segment AB at E. Point E is the center point of line segment AB.

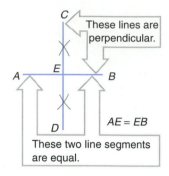

It is also true that the line segment CD is perpendicular to the original line segment AB.

Ready for a little practice? Draw three line segments of any lengths, and then bisect them using only a compass and a straightedge. Check your work with a ruler to see if the line has actually been divided into two pieces of equal length. If you have any doubts about the procedure, ask your instructor or a tutor to check your work.

Turn to **22**.

---

22   **2. Bisecting an Angle**   To bisect an angle means to construct a line that divides the angle exactly in half.

Example: Bisect angle X.

Step 1   Using X as the center and any convenient radius, draw an arc that cuts both sides of the angle at Y and Z.

Step 2   Using Y as the center and any radius greater than half the distance between Y and Z, strike an arc above the first arc.

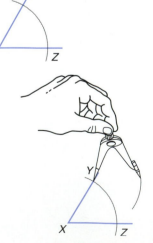

Step 3   Using Z as the center and the same radius from step 2, draw another arc intersecting the first arc at W.

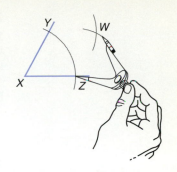

Step 4   Connect point W to point X. The line WX bisects angle X. Angles YXW and ZXW are exactly equal, and each of these angles is half of angle ZXY.

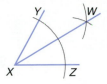

For practice, bisect the following angles.

(a)

(b)

(c)   Use the first construction method to draw a perpendicular to line PQ. Use the second construction to bisect one of the four right angles formed.

Check your work in **23**.

---

23   (a)   Angle ABD = angle CBD.

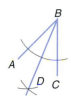

Check this result by measuring the angles with a protractor.

(b)   Angle RSW = angle TSW.

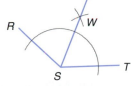

Check by measuring with a protractor.

(c)   Line MN is perpendicular to line PQ. Angle MOK equals angle QOK.

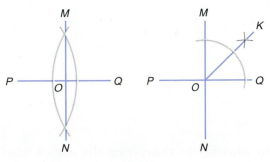

Check with a protractor.

3. Constructing a Perpendicular from a Point to a Line   Construction 1 enabled you to draw a perpendicular to any line segment at the midpoint of that line segment. This third construction enables you to draw a perpendicular to any line from any point not on the line.

Example: Construct a perpendicular from point F to line segment GH.

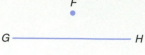

Step 1   Using F as the center, draw any arc that cuts line segment GH in two places, I and J.

Step 2   Using I as the center, and a radius greater than half the line segment IJ, draw an arc below GH.

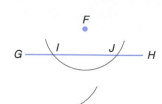

Step 3   Using J as the center and the *same* radius from step 2, draw another arc below GH that intersects the first arc at K.

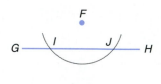

Step 4   Connect F to K. Line FK is exactly perpendicular to line segment GH.

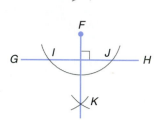

For practice, draw a line from point P perpendicular to the line segment AB.

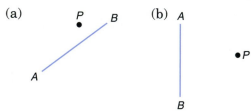

(a)

(b)

Check your work in **24**.

---

24 (a)   (b)

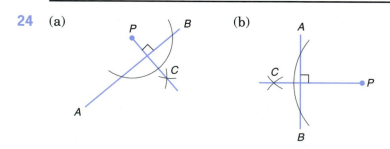

Check your drawing by measuring the angles where PC and AB intersect. Each angle should be 90°, of course.

4. Constructing a Perpendicular at a Given Point on a Line

This construction enables you to draw a perpendicular to a line at any given point on that line.

Example: Construct a perpendicular to line segment *LM* at point *N*.

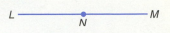

Step 1 Using *N* as the center and a convenient radius, draw equal arcs cutting the line segment *LM* on both sides of point *N*. The points where the arcs cut the line are labeled *P* and *Q*.

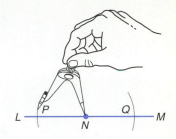

Step 2 Using *P* as the center and a radius greater than *PN*, draw an arc above (or below) the line.

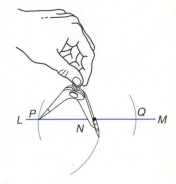

Step 3 Using *Q* as the center and the same radius, draw another arc that intersects the one from step 2. The point where these arcs intersect is labeled *R*.

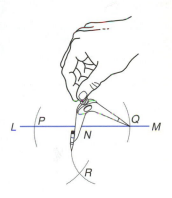

Step 4 Connect point *R* to point *N*. The line *RN* is perpendicular to the original line segment *LM*.

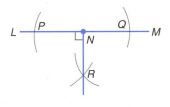

If the point to which the perpendicular must be constructed is at one end of the line segment, extend the line segment and use the same procedure.

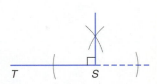

If for some reason you cannot extend the line, try this construction instead.

5. Constructing a Perpendicular to a Line at an End Point

Example: Construct a perpendicular to line segment *XY* at point *X*.

Step 1 Pick any point P off the line. Draw an arc as shown with radius PX. The arc should cut line segment XY at some point W.

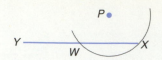

Step 2 From point W draw a line through P intersecting the arc at Z.

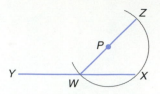

Step 3 Draw a line segment between points Z and X.

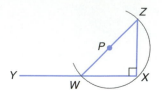

Line ZX is perpendicular to line XY.

Now use the constructions you have learned so far to do the following:

(a) Construct a square with side 2 in.

(b) Construct a rectangle whose length is twice its width.

(c) Construct a 3–4–5 right triangle.

(d) Construct an isosceles right triangle.

Check your work in **25**.

---

25 (a) **First,** draw a line segment 2 in. long (Our drawing is not full scale.)

Second, draw a perpendicular to the line segment at one end using construction 5.

Third, using B as a center and with radius AB, draw an arc at C.

Fourth, from A strike an arc of radius AB above A.

Fifth, with C as a center and radius AB, draw a second arc intersecting the last one. Label this point D.

Finally, connect AD and DC.

(b) Follow the procedure in problem (a) but make line segment *BC* equal to twice *AB*. Similarly, make *AD* equal to *BC*.

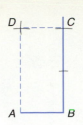

(c) **First,** use construction 3, 4, or 5 to draw a perpendicular *CD* to some line segment *AB*.

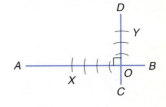

Second, using some convenient radius, mark three arcs along *OD* and four arcs along *OA*.

Finally, connect points *X* and *Y*. Triangle *OXY* is a 3-4-5 right triangle.

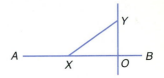

(d) **First,** use construction 3, 4, or 5 to draw a perpendicular to some given line.

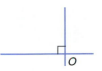

Second, using the intersection *O* as a center, cut both perpendicular lines with the same arc at *A* and *B*.

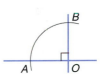

Triangle *AOB* is isosceles. Sides *AO* and *BO* are equal.

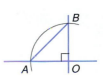

6. Duplicating an Angle   This construction enables you to draw an angle exactly equal to some given angle, without using a protractor, of course.

Example: Construct an angle *PQR* equal to the given angle *MNO*.

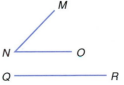

Step 1 Draw one side of angle *PQR* and label it. It can be of any convenient length.

Step 2 Using *N* as the center and with any convenient radius, draw an arc cutting both sides of angle *MNO*. Label these points *S* and *T*.

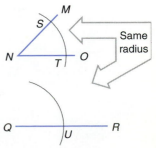

Step 3 Using the *same* radius and *Q* as the center, strike a wide arc that cuts line *QR* at point *U*.

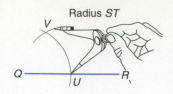

Step 4 Using *ST* as the radius, draw an arc from *U* that cuts the existing arc at *V*.

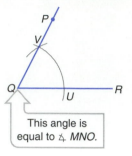

Step 5 Draw a line from point *Q* through point *V*. Then angle *PQR* is equal to the original angle *MNO*.

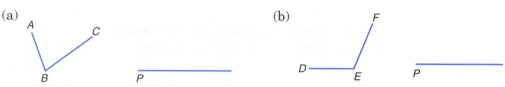

This angle is equal to ∡ *MNO*.

Practice this construction by duplicating the following angles. (The vertex of the duplicated angle should be at point *P*.)

(a)   (b)

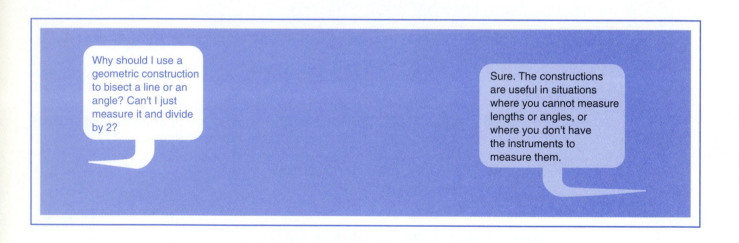

Check your work against ours in **26**.

Why should I use a geometric construction to bisect a line or an angle? Can't I just measure it and divide by 2?

Sure. The constructions are useful in situations where you cannot measure lengths or angles, or where you don't have the instruments to measure them.

---

26 (a)

(b)

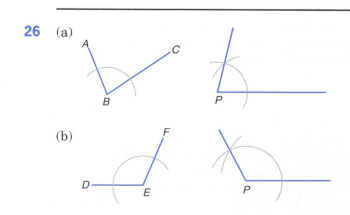

7. Duplicating a Triangle This construction enables you to duplicate a given triangle exactly, drawing a second triangle of the same size and shape.

Example: Duplicate triangle *XYZ*.

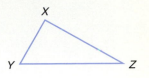

Step 1 Draw a line segment longer than *YZ*. From one endpoint, *B*, draw an arc equal to *YZ* cutting the line at *C*.

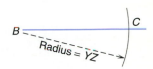

Step 2 Using *B* as the center and radius equal to *YX*, draw an arc in the direction of the third vertex of the triangle.

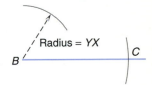

Step 3 Using *C* as the center and radius equal to *XZ*, draw an arc that intersects the last arc at *A*.

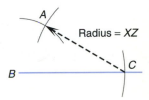

Step 4 Connect point *A* to point *B* and point *C* to point *A*. Triangle *ABC* is identical in both size and shape with triangle *XYZ*. All three angles are the same as the corresponding angles in the original triangle.

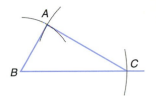

For practice, duplicate each of the following triangles, using construction 7.

(a) 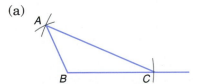   (b)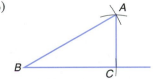

Check your work in **27**.

---

27 (a) (b)

8. Constructing an Equilateral Triangle Each angle in an equilateral triangle measures 60°. Constructing an equilateral triangle enables you to draw a 60° angle without using a protractor.

Example: Construct an equilateral triangle.

Step 1 Draw a line segment of whatever length you choose for the side of the triangle. Label it *AB*.

A ———————— B

Step 2 Using AB as a radius, draw an arc from A and a second arc from B. These will intersect at point C.

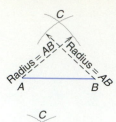

Step 3 Draw line segments AC and BC. Triangle ABC is equilateral. All sides are equal in length. Each angle measures $60°$.

That was an easy one. For practice construct an equilateral triangle with side equal to 2 in. Check your work by measuring all sides with a ruler to see if they are equal.

When this is completed, turn to **28** for another construction.

28   **9. Constructing a Parallel to a Given Line**   Many geometric figures involve parallel lines. This construction enables you to draw a line that is parallel to some given line and that goes through some given point.

Example: Construct a parallel to MN that goes through point O.

Step 1 Using the distance MN as a radius and using point O as the center, draw an arc as shown.

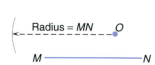

Step 2 Next, draw a second arc with radius ON using point M as the center. Label the intersection of the arcs P.

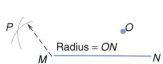

Step 3 Connect points O and P. The line OP is parallel to the line MN. If we connect points M and P and points O and N, we will have a parallelogram $MNOP$.

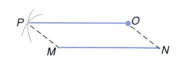

Practice this construction by doing the following problems.

(a) Draw a line through A parallel to the base of the triangle.

(b) Draw a line parallel to line XY and $1\frac{1}{2}$ in. from it.

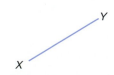

Check your work in **29**.

29 (a)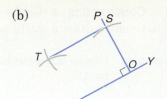

(b)

Not drawn to scale.

Construct *PO* perpendicular to *XY*. Measure *OS* equal to $1\frac{1}{2}$ in. Construct *ST* parallel to *XY* through point *S*.

10. Dividing a Line Segment into a Number of Equal Parts This construction is one of the simplest and most useful of all. It enables a sheet metal worker or carpenter to divide *any* length into *any* number of equal parts without making a measurement. For example, it will allow you to divide a 1-in. length into thirds or fifths or even tenths rather than only the fourths, eighths, or sixteenths shown on a measuring rule.

Example: Divide line segment *AB* into five equal parts.

Step 1 Draw line segment *AC* of any length and at any convenient angle to *AB*.

Step 2 Using any radius that will fit, mark off five arcs of equal length along *AC*, starting at *A*. Label the resulting points *D, E, F, G,* and *H* and connect the last point *H* to *B*.

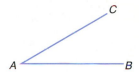

Step 3 Construct a line parallel to *BH* through point *G*. This line intersects line *AB* at point *I*. Use construction 9 to draw the parallel line.

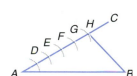

Step 4 The distance *BI* is one-fifth of the distance *AB*. To cut *AB* into five equal parts use *BI* as a radius and draw arcs at *J, K,* and *L* as shown.

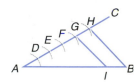

Distances *AL, LK, KJ, JI,* and *IB* are equal.

For practice,

(a) Divide the following line segment into three equal parts.

(b) Construct a line one-fifth of an inch long.

Check your work by measuring with a ruler.

Continue in **30**.

11. Constructing a Circle of a Given Diameter This very simple procedure allows you to draw a circle with any given diameter.

Example: Construct a circle with a diameter equal to *AB*.

Step 1 Bisect *AB*. Label the point of bisection *C*.

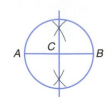

Step 2 Using radius equal to *AC* (or *CB*) construct a circle using *C* as the center.

The construction is a snap, but can you reverse the procedure?

12. Finding the Center of a Circle Given an Arc In this construction you begin with a circle or arc of a circle and locate its center point. Knowing the center point, you can find its radius or diameter.

Example: Find the center of the circle with arc *AB*.

Step 1 Pick a third point *C* anywhere on the arc between *A* and *B*. Connect *A* to *C* with a straight line and connect *B* to *C* with a straight line.

Step 2 Use construction 3 to bisect line segments *AC* and *BC*. Extend the bisectors inside the arc until they meet. They meet at point *D*, the center of the circle of which arc *AB* is a part.

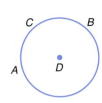

Step 3 Now you can construct the rest of the circle using *DA* (or *DB* or *DC*) as the radius.

Find the center of the following arcs.

(a)    (b)

Check your work by using the center and radius you find to draw the complete circle. Do the arcs fit?

Continue in **31**.

31 Two more constructions involving circles will be useful to you.

13. Constructing a Tangent to a Circle A *tangent* is a line that touches a circle at exactly one point.

Example: Construct a tangent to point A on the given circle with center O.

Step 1 Extend a line from O through A to X.

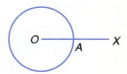

Step 2 Using construction 4, draw a perpendicular to line OX at point A. This perpendicular is tangent to the circle.

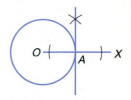

But suppose that the line must be tangent and also go through a point outside the circle. Do it this way.

14. Construction of a Tangent to a Circle from a Point Off the Circle

Example: Construct a tangent to the circle with center X from point M.

Step 1 Draw the line segment MX and bisect it using construction 1. Label the midpoint P.

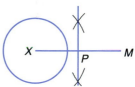

Step 2 Using P as the center and a radius equal to PX, strike an arc that cuts the circle at two points A and B.

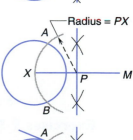

Step 3 Draw lines joining M to A and M to B. The lines MA and MB are tangent to the circle.

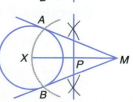

Notice that there are two lines tangent to any circle from some point outside the circle, and $MA = MB$.

Practice these last two constructions by doing the following problems.

(a) Draw a tangent to this circle at point P. (The center of the circle is at O.)

(b) Draw the two tangents to this circle from point Q. (The center of the circle is at O.)

(c) Draw a tangent to this arc at point B. (*Hint:* First use construction 12 to find the center of the arc.)

Check your constructions in **32**.

---

32 (a)   (b)   (c)

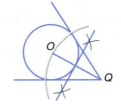
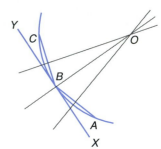

In problem (c), draw line segments AB and BC. Use these to find the center O. Draw line OB. Construct line XY perpendicular to OB. Then XY is tangent to the arc.

Turn to **33** for a set of exercises designed to help you become more expert at using these constructions.

33

Exercises 9-3   **Geometric Constructions**

Perform the following constructions. Check your own work with a ruler or protractor.

1. Draw a line segment of any length and bisect it.

2. Draw a line segment of any length and divide it into four equal parts.

3. Draw a line segment of any length and divide it into seven equal parts.

4. Draw an acute angle and bisect it.

5. Draw an obtuse angle and bisect it.

6. Construct a 90° angle using any method.

7. Construct a 45° angle.

8. Draw an acute angle and duplicate it.

9. Draw an obtuse angle and duplicate it.

10. Construct a square with side length 4 in.

11. Construct a rectangle such that one pair of opposite sides has three times the length of the other pair.

12. Draw a line and a point off the line. Construct a perpendicular from the point to the line.

13. Draw a scalene triangle and copy it.

14. Construct an equilateral triangle with side length $3\frac{1}{2}$ in.

15. Construct a 30° angle.

16. Construct a 120° angle.

17. Construct a 135° angle.

18. Draw a line and pick a point on the line. Construct a perpendicular to the line at that point.

19. Construct a line parallel to AB through C.

Problem 19   **Problem 20**

20. Construct a line parallel to DE 3 in. away.

21. Divide this line into five equal parts.

Problem 21   **Problem 22**

22. Use construction methods to find the center *and* complete the circle for the arc shown.

23. Construct a tangent to the circle at A.

Problems 23 and 24

24. For the circle shown, construct the two tangent lines from B to the circle.

Now go to **34** for a set of problems on practical geometry.

34   Answers are given on page 604.

A.   Solve the following problems involving solid figures.

Find the lateral surface area and volume of each of the following. (Round to the nearest tenth.)

1.

2.

3.

4.

5.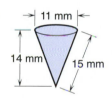

6.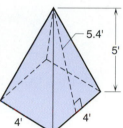

7.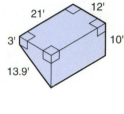

8.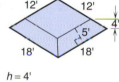

h = 4'
s = 5'

Find the total surface area and volume of each of the following. (Round to the nearest tenth.)

9.

10.

11.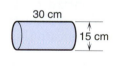

12.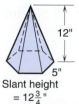

Slant height
= 12¾"

13.

14.

Name

Date

Course/Section

15. A sphere with a radius of 10 cm.

16.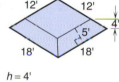

17. A sphere with a diameter of 13 in.

18.

19.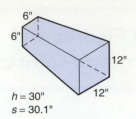

20.

21.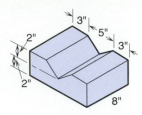

B. Perform the following constructions.

1. Copy this acute angle, and then bisect the duplicate. Check your accuracy by measuring with a protractor.

2. Bisect this line segment.

3. Construct a 45° angle.
4. Construct a perpendicular from the point *A* to the line *XY*.

5. Construct a square with side length $2\frac{1}{2}$ in.
6. Construct an equilateral triangle with side length 3 in.
7. Construct a 15° angle.
8. Construct a 150° angle.
9. Duplicate this triangle.

10. Construct a line parallel to the line *AB* passing through the point *P*.

11. Construct a line parallel to this line and $2\frac{1}{2}$ in. from it.

12. Divide this line segment into seven equal parts.

13. Find the center of the arc AB by construction and then complete the circle.

A $\qquad$ B

14. Construct a line tangent to the circle at B.

15. Construct the two tangent lines from C to the circle.

C. Practical Problems

1. Manufacturing How many boxes 16 in. by 12 in. by 10 in. will fit in 1250 cu ft of warehouse space? (1 cu ft = 1728 cu in.)

2. Construction Technology How many cubic *yards* of concrete are needed to pour the highway support shown in the figure? (1 cu yd = 27 cu ft.) (Round to nearest cubic yard.)

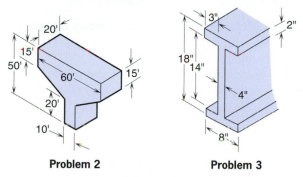

Problem 2   **Problem 3**

3. Metalworking What is the weight of the 10-ft-long steel I-beam if the density of steel is 0.2833 lb/cu in.? (Round to the nearest pound.)

4. Plumbing What is the capacity in gallons of a cylindrical water tank 3 ft in diameter and 8 ft high? (1 cu ft ≈ 7.48 gal.) (Round to one decimal place.)

5. Painting and Decorating How many quarts of paint will it take to cover a spherical water tank 25 ft in diameter if one quart covers 50 sq ft?

6. Construction Technology A hole must be excavated for a swimming pool in the shape shown in the figure. How many trips will be needed to haul the dirt away if the truck has a capacity of 9 m³? (*Remember:* A fraction of a load constitutes a trip.)

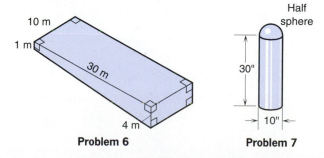

Problem 6   **Problem 7**

7. Manufacturing How many cubic feet of propane will the tank in the figure contain? (1 cu ft = 1728 cu in.) (Round to the nearest tenth of a cubic foot.)

8. **Manufacturing** What must be the height of a cylindrical 750-gallon tank if it is 4 ft in diameter? (1 cu ft ≈ 7.48 gallons.) (Round to the nearest inch.)

9. At a density of 42 lb/cu ft, what is the weight of fuel in the rectangular gas tank pictured?

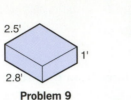

Problem 9   **Problem 10**

10. **Manufacturing** How many square inches of paper are needed to produce 2000 conical cups like the one in the figure? (Round to the nearest 100 sq in.)

11. **Building Construction** How many cubic yards of concrete are needed to pour the foundation shown in the figure? (Round to the nearest cubic yard.)

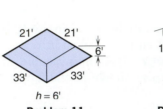

Problem 11   **Problem 12**

12. **Sheet Metal Technology** How many square inches of sheet metal are needed to make the bottom and sides of the pail in the figure? (Round to the nearest square inch and use π ≈ 3.14.)

13. Find the capacity in gallons of the oil can in the figure. It is a cylinder with a cone on top. (1 gallon ≈ 231 cu in.) (Round to the nearest tenth of a gallon.)

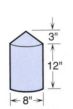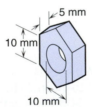

Problem 13   **Problem 14**

14. **Machine Technology** At 0.000017 lb/mm^3, find the weight of 1200 hex nuts, each 10 mm on a side and 5 mm thick with a hole 10 mm in diameter. (Round to the nearest tenth of a pound.)

15. A cylindrical tank can be no more than 6 ft high. What diameter must it have in order to hold 1400 gallons?

PREVIEW
10

Triangle Trigonometry

Objective	Sample Problems		Where to Go for Help

When you finish this unit you will be able to:

1. Determine whether or not a figure is a right triangle, given certain information.

(a)

(b)

(c)

(d)

(e)

2. Find the values of trig ratios.

(a) sin 26°

(b) cos 84°

(c) tan 43°20′

(Round to three decimal places.)

3. Find the angle when given the value of a trig ratio.

Find ∡x to the nearest minute:

(a) $\sin x = 0.242$ ∡$x =$ _____

(b) $\cos x = 0.549$ ∡$x =$ _____

Name _____

Find ∡x to the nearest tenth of a degree.

Date _____

(c) $\tan x = 3.821$ ∡$x =$ _____

Course/Section _____

(d) $\sin x = 0.750$ ∡$x =$ _____

4. Work with angles in radian measure.

Express each angle in radians.

(a) 46° _____ 497 **4**

(b) 8.4° _____

(c) 80°15′ _____

(d) Calculate the arc length and area of the circular sector shown. (Round to one decimal place.) _____ 498 **5**

(e) What is the angular velocity in radians per second of a large flywheel that rotates through an angle of 200° in 2 seconds? (Round to the nearest tenth.) _____ 499 **6**

5. Solve problems involving triangles. (Round sides to the nearest tenth and angles to the nearest minute.)

(a) $X =$ _____ 515 **29**

$Y =$ _____

(b) $\angle m =$ _____ 518 **33**

$X =$ _____

(c) In a pipe-fitting job, what is the run if the offset is $22\frac{1}{2}°$ and the length of set is $16\frac{1}{2}$ in.? run = _____

(d) Find the angle of taper on the figure. angle = _____

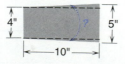

6. Solve oblique triangles. (Round sides to the nearest tenth and angles to the nearest degree.)

Solve these triangles.

(a) $a = 6.5$ in., $b = 6.0$ in., $c = 3.5$ in.

_____   523   **35**

(b) $A = 68°$, $B = 79°$, $b = 8.0$ ft

_____

(Answers to these preview problems are given on page 579.)

If you are certain that you can work all of these problems correctly, turn to page 535 for a set of practice problems. If you cannot work one or more of the preview problems, turn to the page indicated to the right of the problem. Those who wish to master this material with the greatest success should turn to frame **1** and begin work there.

10 Triangle Trigonometry

"Not the Bermuda triangle?"

Drawing by Richter © 1977. The New Yorker Magazine, Inc

10-1 ANGLES AND TRIANGLES

The word *trigonometry* means simply "triangle measurement." We know that the ancient Egyptian engineers and architects of 4000 years ago used practical trigonometry in building the pyramids. By 140 B.C. the Greek mathematician Hipparchus had made trigonometry a part of formal mathematics and taught it as astronomy. In this chapter we look at only the simple practical trigonometry used in electronics, drafting, machine tool technology, aviation mechanics, and similar technical work.

1 In Chapter 8 you learned some of the vocabulary of angles and triangles. To be certain you remember this information, try this short quiz.

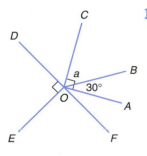

(a) Find each of the angles shown.

    (1) $\sphericalangle AOB =$ _____   (2) $\sphericalangle DOF =$ _____

    (3) $\sphericalangle a$ $=$ _____   (4) $\sphericalangle DOE =$ _____

(b) Label each of these angles with the correct name: acute angle, obtuse angle, straight angle, right angle.

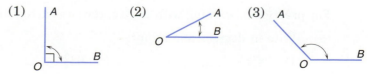

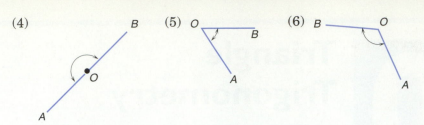

(4)   (5)   (6)

Check your answers in **2**.

---

2   (a)   (1) $\angle AOB = 30°$ (2) $\angle DOF = 180°$ (The symbol ∟ means "right angle" or 90°.)

   (3) $\angle a = 60°$ (4) $\angle DOE = 90°$

   (b)   (1)   right angle   (2)   acute angle   (3)   obtuse angle

   (4)   straight angle   (5)   acute angle   (6)   obtuse angle

If you missed any of these, you should return to **1** on page 385 for a quick review; otherwise, continue here.

Angle Measurement   In most practical work angles are measured in degrees and fractions of a degree. Smaller units have been defined as follows:

60 minutes = 1 degree   abbreviated, $60' = 1°$
60 seconds = 1 minute   abbreviated, $60'' = 1'$

For most technical purposes angles can be rounded to the nearest minute.

Often in your work—especially if you use a hand calculator—you will need to convert an angle measured in decimal degrees to its equivalent in degrees and minutes. For example,

$17\frac{1}{2}° = 17° + \frac{1}{2}°$

$\qquad = 17° + \frac{1}{2}(60')$ Multiply the fractional part by 60', since $1° = 60'$.

$\qquad = 17°30'$

Write this angle to the nearest minute:

$36.25° = $ _____

Check your answer in **3**.

---

3 $36.25° = 36° + 0.25°$
$\qquad = 36° + (0.25)(60')$ Multiply the decimal part by 60'.
$\qquad = 36°15'$

The reverse procedure is also useful. For example, to convert the angle $72°6'$ to decimal form,

$72°6' = 72° + 6'$

$\qquad = 72° + \left(\frac{6'}{60'}\right)$ Divide the minutes part by 60', since $1° = 60'$.

$\qquad = 72° + 0.1°$

$\qquad = 72.1°$

For practice in working with angles, rewrite each of the following angles as shown.

(a)   Write in degrees and minutes.

   (1) $24\frac{1}{3}° = $ _____   (2) $64\frac{3}{4}° = $ _____

(3) $15.3° = $ _____   (4) $38.6° = $ _____

(5) $46\frac{3}{8}° = $ _____   (6) $124.8° = $ _____

(b) Write in decimal form.

(1) $10°48' = $ _____   (2) $96°9' = $ _____

(3) $57°36' = $ _____   (4) $168°54' = $ _____

(5) $33' = $ _____   (6) $1°12' = $ _____

Check your answers in **4**.

---

4 (a) (1) 24°20'   (2) 64°45'   (3) 15°18'
            (4) 38°36'   (5) $46°22\frac{1}{2}'$ (6) 124°48'

   (b) (1) 10.8°   (2) 96.15°   (3) 57.6°
            (4) 168.9°   (5) 0.55°   (6) 1.2°

Radian Measure In some scientific and technical work angles are measured or described in an angle unit called the *radian*. By definition,

$$1 \text{ radian} = \frac{180°}{\pi} \text{ or about } 57.296°$$

One radian is the angle at the center of a circle that corresponds to an arc exactly one radius in length.

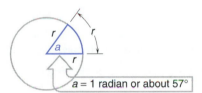

a = 1 radian or about 57°

To understand where the definition comes from, set up a proportion:

$$\frac{\text{angle } a}{\text{total angle in the circle}} = \frac{\text{arc length}}{\text{circumference of the circle}}$$

$$\frac{a}{360°} = \frac{r}{2\pi r}$$

$$a = \frac{360° \; r}{2\pi r}$$

$$a = \frac{180°}{\pi} \approx 57.295779°$$

Degrees to Radians To convert angle measurements from degrees to radians (rad), simply multiply by $\pi/180$ and round as needed. Use $\pi = 3.1416$ or the π button on your calculator unless you are told otherwise. For example,

$$43° = \frac{43 \cdot \pi}{180}$$

$$\approx 0.75 \text{ rad}\quad \text{rounded}$$

and

$$60° = \frac{60 \cdot \pi}{180}$$

$$= \frac{\pi}{3} \text{ rad} \approx 1.0472 \text{ rad}$$

Radians to Degrees To convert angles from radians to degrees, multiply by $180/\pi$ and round. For example,

$$1.3 \text{ rad} = \frac{1.3 \cdot 180}{\pi}$$

$$\approx 74.5°$$

and

$$0.5 \text{ rad} = \frac{0.5 \cdot 180}{\pi}$$

$$= \frac{90}{\pi} \text{ degrees or about } 28.6°$$

Try these problems for practice in using radian measure.

(a) Convert to radians. (Round to two decimal places.)

    (1) 10°   (2) 35°   (3) 90°   (4) 120°

(b) Convert to degrees. (Round to the nearest tenth.)

    (1) 0.3 rad   (2) 1.5 rad   (3) 0.8 rad   (4) 0.05 rad

Check your answers in **5**.

---

5 (a) (1) $\dfrac{10 \cdot \pi}{180} \approx 0.17 \text{ rad}$ (2) 0.61 rad   (3) 1.57 rad   (4) 2.09 rad

    (b) (1) $\dfrac{0.3 \cdot 180}{\pi} \approx 17.2°$ (2) 85.9°   (3) 45.8°   (4) 2.9°

Sectors The wedge-shaped portion of a circle shown is called a **sector.** Both its area and the length of arc S can be expressed most directly using radian measure.

SECTORS

Arc length $S = ra$
Area $A = \frac{1}{2}r^2a$ where a is the central
                         angle given in radians.

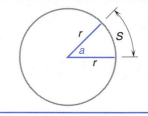

For example, a sheet metal worker wants to know the arc length and area of a sector with central angle 150° cut from a circular sheet of metal with radius 16.0 in.

First, calculate angle a in radians: $a = \dfrac{150 \times \pi}{180} \approx 2.618 \text{ rad}$

Second, calculate the arc length S: $S \approx 16 \text{ in.} \times 2.618$
                                       $\approx 41.9 \text{ in.}$

Finally, calculate the area A: $A \approx \frac{1}{2}(16^2)(2.618)$
                               $\approx 335 \text{ sq in.}$

Using a calculator, we have

S: 16 $\times$ 150 $\times$ 3.1416 $\div$ 180 $=$ →   *41.888*

A: .5 $\times$ 16 x^2 $\times$ 2.618 $=$ →   *335.104*

Calculate the arc length and area of a sector with central angle 80° and radius 25.0 ft.

Check your work in **6**.

---

6 The central angle is $a = \dfrac{80 \times \pi}{180} \approx 1.396$ rad.

The arc length is $S \approx 25 \times 1.396 \approx 34.9$ ft.

The area is $A \approx \frac{1}{2} \times 25^2 \times 1.396 \approx 436$ sq ft.

Linear and Angular Speed Radian units are also used in science and technology to describe the rotation of an object. When an object moves along a straight line for a distance d in time t, its **average linear speed** v is defined as

$$v = \frac{d}{t}$$

For example, if a car travels 1800 ft in 24 seconds, its average linear speed for the trip is

$$v = \frac{1800 \text{ ft}}{24 \text{ sec}} = 75 \text{ ft/sec}$$

When an object moves along a circular arc for a distance S in time t, it goes through an angle a in radians, and its **average angular speed** w is defined as

$$w = \frac{a}{t}$$

If a 24-in.-diameter flywheel rotates through an angle of 140° in 1.1 seconds, (a) what distance does a point on the rim travel? (b) What is the average angular speed of the flywheel?

Check your work in **7**.

---

7 (a) The angle traveled is $a = \dfrac{140 \times \pi}{180} \approx 2.44$ rad.

The distance traveled is $S \approx 12 \times 2.44 \approx 29$ in.

(b) The average angular speed is $w \approx \dfrac{2.44}{1.1} \approx 2.2$ rad/sec.

Right Triangles Trigonometry is basically triangle measurement, and, to keep it as simple as possible, we can begin by studying only *right* triangles. You should remember from Chapter 8 that a right triangle is one that contains a right angle, a 90° angle. All of the following are right triangles. In each triangle the right angle is marked.

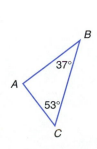

Is triangle *ABC* a right triangle?

The answer is given in **8**.

---

8 You should remember that the sum of the three angles of any triangle is 180°. For triangle *ABC* two of the angles total 37° + 53° or 90°; therefore, angle *A* must equal 180° − 90° or 90°. The triangle *is* a right triangle.

To make your study of triangle trigonometry even easier, we will always work with triangles that are in a *standard position*. For this collection of right triangles

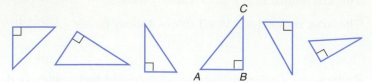

only triangle *ABC* is in standard position.

Place the triangle so that the right angle is on the right side, one side (*AB*) is horizontal, and the other side (*BC*) is vertical. The longest side (*AC*) will slope up from left to right.

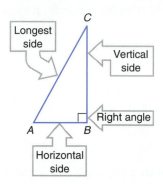

Try it. Rotate the following triangle so that it is in standard position.

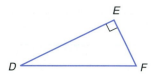

Check it in **9**.

9

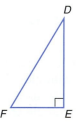

Once we have placed the right triangle in this position, we can name the sides in a way that has also become standard.

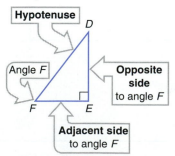

The longest side *DF*, the side opposite the right angle, is called the *hypotenuse*.

You should recall from Chapter 8 that the perpendicular sides are also called the *legs* of the right triangle.

The side or leg *DE* is called the *opposite side* for angle *F*. The side or leg *EF* is called the *adjacent side* for angle *F*.

For angle *D*, side *DE* is its adjacent side, and side *EF* is its opposite side.

Note ▶ The hypotenuse is never considered to be either the adjacent side or the opposite side. It is not a leg of the triangle. ◀

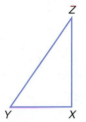

Check your understanding of this vocabulary by putting triangle *XYZ* in standard position and by completing the following statements.

  (a) The hypotenuse is _____.

  (b) The opposite side for angle *Y* is _____.

  (c) The adjacent side for angle *Y* is _____.

  (d) The opposite side for angle *Z* is _____.

  (e) The adjacent side for angle *Z* is _____.

Check your answers in **10**.

10

  (a) *YZ*   (b) *XZ*   (c) *XY*

  (d) *XY*   (e) *XZ*

Of course, the phrases "adjacent side," "opposite side," and "hypotenuse" are used only with right triangles.

For each of the following right triangles, place the triangle in standard position, mark the hypotenuse with the letter *H*, mark the side adjacent to the shaded angle with the letter *A*, and mark the side opposite the shaded angle with the letter *B*.

  (a)   (b)   (c)   (d)   (e)

 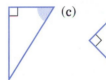

Check your work in **11**.

---

11 (a)   (b)   (c)   (d)   (e)

  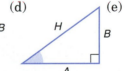

Pythagorean Theorem You should also recall from Chapter 8 that the *Pythagorean theorem* is an equation relating the lengths of the sides of any right triangle.

PYTHAGOREAN THEOREM

$$H^2 = A^2 + B^2$$

This equation is true for every right triangle. For example, the triangle

gives $39^2 = 36^2 + 15^2$
or $1521 = 1296 + 225$ which is correct

For practice, use this rule to find the missing side for each of the following triangles.

(a)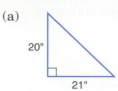

(b)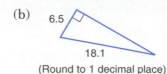
(Round to 1 decimal place)

(c)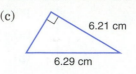

Check your answers in **12**.

---

12 (a) $H^2 = 21^2 + 20^2$
$H^2 = 441 + 400$
$H^2 = 841$
$H = 29$ in.

(b) $18.1^2 = B^2 + 6.5^2$
$327.61 = B^2 + 42.25$
$B^2 = 327.61 - 42.25$
$B^2 = 285.36$
$B \approx 16.9$

(c) $6.29^2 = 6.21^2 + A^2$
$39.5641 = 38.5641 + A^2$
$A^2 = 39.5641 - 38.5641$
$A^2 = 1$
$A = 1$ cm

 Here's how to work problem (b) on a calculator:

18.1 $\boxed{x^2}$ $\boxed{-}$ **6.5** $\boxed{x^2}$ $\boxed{=}$ $\boxed{\surd}$ → `16.892602`

Special Right Triangles

Three very special right triangles are used very often in practical work. Here they are:

  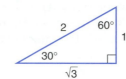 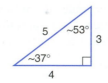

 45°–45°–90°△ The first triangle is a *45° right triangle*. Its angles are 45°, 45°, and 90°, and it is formed by the diagonal and two sides of a square. Notice that it is an isosceles triangle—the two shorter sides or legs are the same length.

Our triangle has legs one unit long, but we can draw a 45° right triangle with any length sides. If we increase the length of one side, the others increase in proportion while the angles stay the same size.

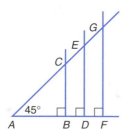

Triangle *ABC* is a 45° right triangle.
Triangle *ADE* is also a 45° right triangle.
Triangle *AFG* is also a 45° right triangle, . . . and so on.

If *AD* = 2 in.

(a) How long is *DE*?

(b) How long is *AE*?

(c) If *AC* = 5 in., how long is *AB*?

Look for the answers in **13**.

---

13 (a) DE = 2 in.　　　　The legs are always equal in length. If AD is 2 in. long, DE is also 2 in. long.

(b) AE = 2 $\sqrt{2}$ in. or about 2.8 in.　The hypotenuse of a 45° right triangle is $\sqrt{2}$ times the length of each leg.

(c) AC = AB $\sqrt{2}$, so $AB = \dfrac{AC}{\sqrt{2}} = \dfrac{5}{\sqrt{2}}$

Using a calculator gives

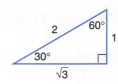 **5** ÷ **2** √ = → *3.5355339* ≈ 3.5 in.

No matter what the size of the triangle, if it is a right triangle and if one angle is 45°, the third angle is also 45°. The two legs are equal, and the hypotenuse is $\sqrt{2} \approx 1.4$ times the length of a leg.

30°–60°–90° △　The second of our special triangles is the 30°–60° right triangle. In a 30°–60° right triangle the length of the side opposite the 30° angle, the smallest side, is exactly one-half the length of the hypotenuse. The length of the third side is $\sqrt{3}$ times the length of the smaller side.

To see where this relationship comes from, start with an equilateral triangle with sides 2 units long.

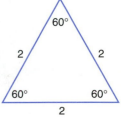

Draw a perpendicular from one corner to the opposite side.

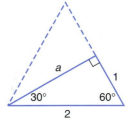

Then in the 30°–60° right triangle the Pythagorean theorem gives

$$a^2 + 1^2 = 2^2$$
$$a^2 + 1 = 4$$
$$a^2 = 3$$
$$a = \sqrt{3}$$

In this triangle, if a = 30° and A = 2 in., calculate

(a) H =

(b) B =

(c) b =

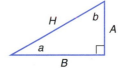

(d) If a = 30°, H = 3 in., then B = ?, A = ?

Compare your work with ours in **14**.

---

14 (a) H = 4 in.　　　(b) B = 2 $\sqrt{3}$ in. or about 3.5 in.　　　(c) b = 60°

(d) In a 30°–60° right triangle, $A = \frac{1}{2}H = \frac{1}{2}(3) = 1.5$ in.

　　and $B = A \sqrt{3} = 1.5 \times \sqrt{3} \approx 2.6$ in.

Again, we may form any number of 30°–60° right triangles, and in each triangle the hypotenuse is twice the length of the smallest side. The third side is $\sqrt{3}$ or about 1.7 times the length of the smallest side. When you find a right triangle with one angle equal to 30° or 60°, you automatically know a lot about it.

3–4–5 △ The third of our special triangles is the 3–4–5 triangle that was discussed in Chapter 8. If a triangle is found to have sides of length 3, 4, and 5 units (any units—in., ft, cm), it must be a right triangle. The 3–4–5 triangle is the smallest right triangle with sides whose lengths are whole numbers.

Construction workers or surveyors can use this triangle to set up right angles.

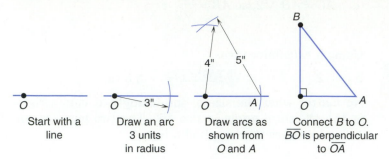

| Start with a line | Draw an arc 3 units in radius | Draw arcs as shown from O and A | Connect B to O. BO is perpendicular to OA |

Notice that in any 3–4–5 triangle the acute angles are approximately 37° and 53°. The smallest angle is always opposite the smallest side.

You will find these three triangles appearing often in triangle trigonometry. Now, for some practice in working with angles and triangles, turn to **15**.

___

15

Exercises 10-1   **Angles and Triangles**

A. Label the shaded angles as acute, obtuse, right, or straight.

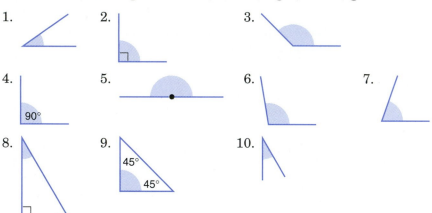

B. Write the following angles in degrees and minutes. (Round to the nearest minute.)

 1. $36\frac{1}{4}°$ 2. $73\frac{3}{5}°$ 3. 65.45°   4. 84.24°

 5. $17\frac{1}{2}°$ 6. $47\frac{3}{8}°$ 7. 16.11°   8. 165.37°

Write the following angles in decimal degrees. (Round to the nearest hundredth if necessary.)

 9. 27°30′   10. 80°15′   11. 154°39′   12. 131°6′

 13. 57°3′   14. 44°20′   15. 16°53′   16. 16′

Write the following angles in radians and round to the nearest hundredth.

 17. 24.75°   18. 185.8°   19. 1.054°   20. $14\frac{3}{4}°$

 21. $67\frac{3}{8}°$ 22. 9.65°   23. 80.60°   24. 216.425°

Write the following angles in degrees and round to the nearest hundredth.

 25. 0.10 rad   26. 0.84 rad   27. 2.1 rad   28. 3.5 rad

C. Solve the following problems involving triangles.

Place each of the following right triangles in standard position. Place an *H* near the hypotenuse, an *A* near the side adjacent to angle *a*, and a *B* near the side opposite angle *a*.

1. *Q*

R *C* *a*

2. *D* *E*

F *a*

3. *G* 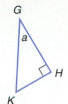 *a*

K *H*

4. 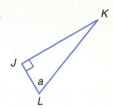 *K*

J *a* *L*

5. 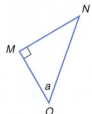 *N*

M *a* *O*

6. *P* 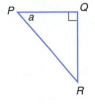 *Q*

a *R*

7. 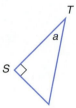 *T*

S *a* *P*

8. *X* 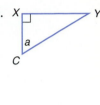 *Y*

a *C*

9. *D* 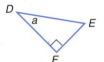 *E*

a *F*

10. 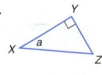 *Y*

X *a* *Z*

Find the lengths of the missing sides on the following triangles using the Pythagorean theorem. (Round to one decimal place.)

11.

24

10

12.

7

12

13.

25

7

14.

15

11

15.

15'

9'

16.

2"

3"

17.

11.3 11.2

18.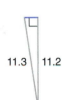

6.5 cm

5.6 cm

19.

4.8'

6.4'

20.

1.7"

2.4"

Each of the following right triangles is an example of one of the three "special triangles" described in this chapter. Find the quantities indicated without using the Pythagorean theorem. (Round the sides to the nearest tenth if necessary.)

21.

a

A

3'

3'

22.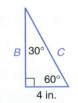

B 30° *C*

60°

4 in.

(a) *a* = _____

(b) *A* = _____

(a) *B* = _____

(b) *C* = _____

23.

    (a) $X =$ _____

    (b) $a =$ _____

    (c) $b =$ _____

24.

    (a) $E =$ _____

    (b) $F =$ _____

    (c) $t =$ _____

25.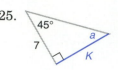

    (a) $K =$ _____

    (b) $a =$ _____

26.

    (a) $R =$ _____

    (b) $q =$ _____

27.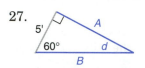

    (a) $A =$ _____

    (b) $B =$ _____

    (c) $d =$ _____

28.

    (a) $M =$ _____

    (b) $N =$ _____

    (c) $t =$ _____

29.

    (a) $Q =$ _____

    (b) $u =$ _____

30.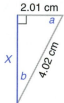

    (a) $X =$ _____

    (b) $a =$ _____

    (c) $b =$ _____

Calculate the arc length S and area A for each of the following sectors. (Round to the nearest whole number if necessary.)

31.	$a = 40°$	32.	$a = 120°$	33.	$a = 0.8$ rad	34.	$a = 2$ rad
	$r = 30$ cm		$r = 24$ ft		$r = 110$ ft		$r = 18$ m
35.	$a = 72°$	36.	$a = 3$ rad	37.	$a = 0.5$ rad	38.	$a = 180°$
	$r = 5$ in.		$r = 80$ cm		$r = 10$ m		$r = 30$ in.

D. Practical Problems

    1. **Sheet Metal Technology** A transition duct is constructed from a circular sector of radius 18 in. and central angle 115°. (a) What is the area of the metal used? (b) Calculate the arc length of the sector. (Round to the nearest tenth.)

2. **Landscaping** Calculate the area (in square yards) of a flower garden shaped like a circular sector with radius 60 yd and central angle 40°.

3. If the indicator on a flow rate meter moves 75° in 16.4 seconds, what is its average angular speed during this time? (Round to two decimal places.)

4. **Machine Technology** If a lathe makes 25 revolutions per second, calculate its average angular speed. (Round to the nearest whole number.)

5. A ferris wheel in an amusement park takes an average of 16 seconds for each revolution. What is its average angular speed? (Round to two decimal places.)

Check your answers on page 605, then continue in **16**.

10-2 TRIGONO-METRIC RATIOS

16 In Section 10-1 we showed that in *all* 45° right triangles, the adjacent side is equal in length to the opposite side, and the hypotenuse is roughly 1.4 times this length. Similarly, in *all* 30°–60° right triangles, the hypotenuse is twice the length of the side opposite the 30° angle, and the adjacent side is about 1.7 times this length.

The key to understanding trigonometry is to realize that in *every* right triangle there is a fixed relationship connecting the length of the adjacent side, opposite side, hypotenuse, and the angle that characterizes the triangle.

For example, in all right triangles that contain the angle 20°,

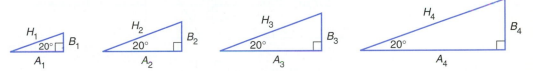

the ratio $\dfrac{\text{adjacent side}}{\text{hypotenuse}}$ will always have the same value. Therefore,

$$\frac{A}{H} = \frac{A_1}{H_1} = \frac{A_2}{H_2} = \frac{A_3}{H_3} = \ldots \text{ and so on}$$

This ratio will be approximately 0.9397.

Sine Ratio This kind of ratio of the side lengths of a right triangle is called a *trigonometric ratio*. It is possible to write down six of these ratios, but we will use only three here. Each ratio is given a special name.

For the triangle

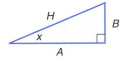

the ratio sin x is defined as

$$\sin x = \frac{\text{opposite side}}{\text{hypotenuse}} = \frac{B}{H}$$

(Pronounce "sin" as "sign" to rhyme with "dine" not with "sin." *Sin* is an abbreviation for *sine*. Sin x is read "sine of angle x.")

For example, in the triangle

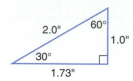

$$\sin 36° = \frac{47 \text{ ft}}{80 \text{ ft}} \approx 0.59$$

Use this triangle to find sin 60°. (Round to two decimal places.)

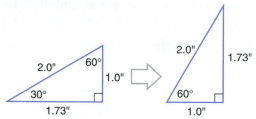

sin 60° = _____

Check your work in **17**.

---

17 Place the triangle in standard position with the 60° angle we are working with at the lower left.

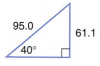

In standard position

$$\sin 60° = \frac{\text{opposite side}}{\text{hypotenuse}}$$

$$\sin 60° = \frac{1.73 \text{ in.}}{2.0 \text{ in.}}$$

$$\sin 60° \approx 0.87 \quad \text{approximately}$$

Every possible angle x will have some number sin x associated with it, and we can calculate the value of sin x from any right triangle that contains the angle x.

For example, sin 40° can be calculated from the following triangle.

$$\sin 40° = \frac{61.1}{95.0} \begin{array}{l} \Leftarrow \boxed{\text{Opposite side}} \\ \Leftarrow \boxed{\text{Hypotenuse}} \end{array}$$

$$\approx 0.643 \quad \text{rounded}$$

What is the value of sin 0°?

(*Hint:* Construct a right triangle with a very small angle and guess.)

Check your answer in **18**.

---

18   sin 0° = 0   In this triangle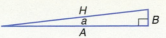

side *B* is much smaller than *H*; therefore, the number $\frac{B}{H}$ is very small.

As angle *a* decreases to zero, side *B* gets smaller and smaller, and sin *a* gets closer and closer to zero.

In a similar way you can show that sin 90° = 1. (Try it.)

Now, remember those three special triangles from Section 10-1.

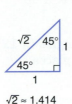 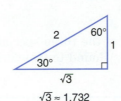 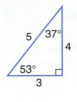

Use these drawings to calculate the following values of sin *x*. (Round to the nearest hundredth.)

x	30°	37°	45°	53°	60°
sin *x*					

Check your answers in **19**.

19

x	30°	37°	45°	53°	60°
sin *x*	0.50	0.60	0.71	0.80	0.87

Note ▶ Sin *x* is a trigonometric *ratio* or trigonometric *function* associated with the angle *x*. The value of sin *x* can never be greater than 1, since the legs of any right triangle can never be longer than the hypotenuse. ◀

Cosine Ratio A second useful trigonometric ratio is the *cosine* of angle *x*, abbreviated cos *x*. This ratio is defined as

$$\cos x = \frac{\text{adjacent side}}{\text{hypotenuse}} = \frac{A}{H}$$

For example, in the triangle

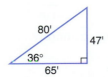

$$\cos 36° = \frac{65 \text{ ft}}{80 \text{ ft}}$$ ← Adjacent side ← Hypotenuse

cos 36° ≈ 0.81 approximately

Use the following triangle to find cos 60°.

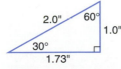

Check your work in **20**.

20

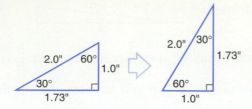

In standard position

$$\cos 60° = \frac{\text{adjacent side}}{\text{hypotenuse}}$$

$$\cos 60° = \frac{1.0 \text{ in.}}{2.0 \text{ in.}} = 0.50$$

Now use the drawing of the three special triangles in **12** on page 502 to help you to complete this table. (Round to two decimal places.)

x	0°	30°	37°	45°	53°	60°	90°
$\cos x$	1						0

We have entered the values for 0° and 90°.

Check your table against ours in **21**.

21

x	0°	30°	37°	45°	53°	60°	90°
$\cos x$	1	0.87	0.80	0.71	0.60	0.50	0

Tangent Ratio A third trigonometric ratio may also be defined. The ratio tangent x, abbreviated tan x, is defined as

$$\tan x = \frac{\text{opposite side}}{\text{adjacent side}} = \frac{B}{A}$$

For example, in the triangle

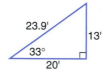

$$\tan 33° = \frac{13 \text{ ft}}{20 \text{ ft}} \quad \begin{array}{l} \text{Opposite side} \\ \text{Adjacent side} \end{array}$$

$$\tan 33° \approx 0.65 \text{ approximately}$$

Use the following triangle to find tan 30°.

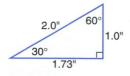

Check your work in **22**.

22

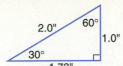

$$\tan 30° = \frac{\text{opposite side}}{\text{adjacent side}}$$

$$\tan 30° = \frac{1.0 \text{ in.}}{1.73 \text{ in.}} \approx 0.58$$

Complete this table for tan x using the three special triangles. (Round to two decimal places.)

x	0°	30°	37°	45°	53°	60°	90°
tan x							

(tan x for $x = 90°$ is very tricky.)

Check your table against ours in **23**.

23

x	0°	30°	37°	45°	53°	60°	90°
tan x	0	0.58	0.75	1.00	1.33	1.73	not defined

To find tan 90° draw a triangle with x very large.

As angle x gets larger and larger, the opposite side B gets longer and longer.

The ratio tan $x = \frac{B}{A}$ gets larger and larger. As x approaches 90°, tan x becomes infinitely large, and we have no number large enough to name it.

Finding Values of Trigonometric Functions

You have been asked to calculate these "trig" ratios to help you get a feel for how the ratios are defined and to encourage you to memorize a few values. In actual practical applications of trigonometry, the values of the ratios are found by using a calculator with built-in trigonometric function keys.

To find the value of a trigonometric ratio using a calculator, enter the value of the angle, then press the key for the given function. For example, to find sin 38°, enter this sequence:

38 (sin) → *0.6156615*   sin 38° ≈ 0.616

Careful ▶

Angles can be measured in units of radians and gradients as well as degrees, and most calculators are capable of accepting all three units of measure. Before entering an angle in degrees, make certain the calculator is in "degree mode" by pressing the (DRG) key until the "DEG" indication flashes on the display. On some calculators absence of any mode indication means that it is in degree mode. ◀

Now you try it. Find the value of each of the following to two decimal places.

(a) cos 26°	(b) tan 67.8°	(c) sin 8°
(d) tan 14.25°	(e) cos $47\frac{1}{2}°$	(f) sin 39.46°
(g) sin 138°	(h) cos 151.6°	(i) tan 141.12°

Check your answers in **24**.

---

24 (a) **26** $\boxed{\cos}$ → *0.8987940*   cos 26° ≈ 0.90

(b) **67.8** $\boxed{\tan}$ → *2.4504252*   tan 67.8° ≈ 2.45

(c) **8** $\boxed{\sin}$ → *0.1391731*   sin 8° ≈ 0.14

(d) **14.25** $\boxed{\tan}$ → *0.2539676*   tan 14.25° ≈ 0.25

(e) **47.5** $\boxed{\cos}$ → *0.6755902*   cos $47\frac{1}{2}°$ ≈ 0.68

(f) **39.46** $\boxed{\sin}$ → *0.6355394*   sin 39.46° ≈ 0.64

(g) **138** $\boxed{\sin}$ → *0.6691306*   sin 138° ≈ 0.67

(h) **151.6** $\boxed{\cos}$ → *−0.8796486*   cos 151.6° ≈ −0.88

(i) **141.12°** $\boxed{\tan}$ → *−0.8063221*   tan 141.12° ≈ −0.81

- Notice in problems (g), (h), and (i) that the trigonometric functions exist for angles greater than 90°. The values of the cosine and tangent functions are negative for angles between 90° and 180°.
- Notice in problem (g) that

$$\sin 138° = 0.6691306 \ldots$$
$$\text{and } \sin \quad 42° = 0.6691306 \ldots \quad 138° + 42° = 180°$$

Also   cos 151.6° = −0.8796486 . . .

and   cos   28.4 = 0.8796486 . . .   151.6° + 28.4° = 180°

This suggests that the following formulas are true:

$\sin A = \sin (180 - A)$

$\cos A = -\cos (180 - A)$

Angles in Degrees and Minutes

Finding the value of a trigonometric function with a calculator is easy when the angle is given as a decimal number of degrees. However, when the angle is given in degrees and minutes, it must first be converted to decimal form. Some scientific calculators have a single key that will perform this conversion. If your calculator does not have such a key, try it this way:

Find cos 59°37′.

First, write the given angle as a sum: $59°37′ = 59° + \dfrac{37′}{60′}$ ⬅ Divide minutes by 60′ to convert to degrees

Next, enter the sum on your calculator: **59** $\boxed{+}$ **37** $\boxed{÷}$ **60** $\boxed{=}$ → *59.616667*

Finally, enter the trigonometric function: $\boxed{\cos}$ → *0.5057828*

Therefore, cos 59°37′ ≈ 0.506

Try these problems for practice. Find each value to three decimal places.

(a)   sin 71°26′   (b)   tan 18°51′   (c)   cos 42°16′   (d)   cos 117°22′

Check your answers in **25**.

---

25 (a) **71** $\boxed{+}$ **26** $\boxed{÷}$ **60** $\boxed{=}$ $\boxed{\sin}$ → *0.9479538* ≈ 0.948

(b) 0.341

(c) 0.740

(d) −0.460

Finding the Angle

In some applications of trigonometry it may happen that we know the value of the trigonometric ratio number and we need to find the angle associated with it. This is called finding the **inverse** of the trigonometric function. To find the angle associated with any given value of a trigonometric function, enter the function value followed by the [INV] or [2ndF] key, and then press the key of the trigonometric function given.

For example, to find the angle whose sine is 0.728, enter the following:

.728 [INV] [sin] → *46.718988*

The angle is approximately 46.7°; therefore, sin 46.7° ≈ 0.728.*

Careful ▶

Be certain your calculator is in the degree mode before you enter this kind of calculation. ◀

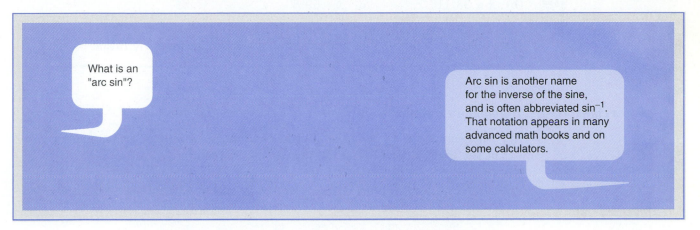

What is an "arc sin"?

Arc sin is another name for the inverse of the sine, and is often abbreviated $\sin^{-1}$. That notation appears in many advanced math books and on some calculators.

Use your calculator to find the angle x for the following trigonometric function values. (Round to the nearest tenth of a degree.)

(a) cos x = 0.589   (b) sin x = 0.248   (c) tan x = 1.75

Compare your answers with ours in **26**.

---

26   (a) $x \approx 53.9°$ (b) $x \approx 14.4°$ (c) $x \approx 60.3°$

To find the angle in degrees and minutes, follow the procedure described and add the following two steps:

1. Write the whole number portion of the angle displayed, and subtract this whole number from the quantity displayed.

2. Multiply by 60.

The fractional part of the answer, in minutes, will be displayed. For example, to find x in degrees and minutes for cos x = 0.296, follow this sequence:

.296 [INV] [cos] → *72.782489* [−] **72** [=] [×] **60** [=] → *46.949314*

Decimal degrees
Write 72°

0.782489°
converted
to minutes

Therefore, $x \approx 72°47'$.

*We have found the angle between 0 and 90° that corresponds to the given trigonometric function value, but there are many angles greater than 90° whose sine ratio has this same value. Since such values of the inverse are generally not very important in the trades, we will not discuss them in detail here.

Note ▶ Some calculators have a single key (usually labeled ⟨DMS⟩) that enables you to perform this calculation in one step. ◀

Now try these problems. Find x to the nearest minute.

(a) $\sin x = 0.818$ (b) $\tan x = 0.654$ (c) $\cos x = 0.513$

Check your results in **27**.

---

27   (a)   **.818** ⟨INV⟩ ⟨sin⟩ → `54.885084` ⟨−⟩ **54** ⟨=⟩ ⟨×⟩ **60** ⟨=⟩ → `53.105060`

        $x \approx 54°53'$

(b) $x \approx 33°11'$ (c) $x \approx 59°8'$

Now turn to **28** for a set of practice problems on trigonometric ratios.

28

Exercises 10-2   **Trigonometric Ratios**

A. For each of the following triangles *calculate* the indicated trigonometric ratios of the given angle. (Round to nearest hundredth.)

1.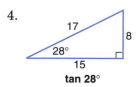

sin 33°

2.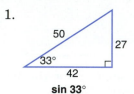

cos 42°

3.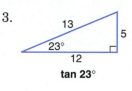

tan 23°

4.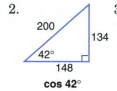

tan 28°

5.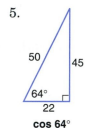

cos 64°

6.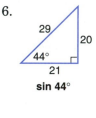

sin 44°

B. Find each of the following trig values. (Round to three decimal places.)

1. $\sin 27°$	2. $\cos 38°$	3. $\tan 12°$	4. $\sin 86°$
5. $\cos 79°$	6. $\tan 6°$	7. $\cos 87°$	8. $\cos 6°30'$
9. $\tan 50°20'$	10. $\sin 75°40'$	11. $\cos 41.25°$	12. $\cos 81°25'$
13. $\sin 50.4°$	14. $\tan 74.15°$	15. $\tan 81.06°$	16. $\sin 12.6°$
17. $\cos 98°$	18. $\sin 106°$	19. $\sin 144.2°$	20. $\cos 134.5°$

C. Find the acute angle x. Round to the nearest minute.

1. $\sin x = 0.974$	2. $\cos x = 0.719$
3. $\tan x = 2.05$	4. $\sin x = 0.077$
5. $\cos x = 0.262$	6. $\tan x = 0.404$
7. $\sin x = 0.168$	8. $\cos x = 0.346$

Find the acute angle x. Round to the nearest tenth of a degree.

9. $\tan x = 1.165$ 10. $\cos x = 0.662$

11. $\cos x = 0.437$ 12. $\tan x = 0.225$

13. $\tan x = 0.872$ 14. $\sin x = 0.472$

15. $\sin x = 0.605$ 16. $\cos x = 0.154$

Check your answers on page 606, then turn to **29** to learn how to use trigonometric ratios to solve problems.

10-3 SOLVING RIGHT TRIANGLES

29 In the first two sections of this chapter you learned about angles, triangles, and the trigonometric ratios that relate the angles to the side lengths in right triangles. Now it's time to look at a few of the possible applications of these trig ratios.

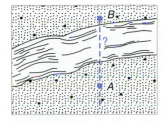

If you know the length of one of the three sides of a right triangle and the size of either of the two acute angles, it is always possible to determine the other side lengths and the third angle. For example, suppose you need to measure the distance between two points A and B on opposite sides of a river. You have a tape measure and a protractor but no way to stretch the tape measure across the river. How can you measure the distance AB?

Try it this way:

1. Choose a point C in line with points A and B. If you stand at C, A and B will appear to "line up."

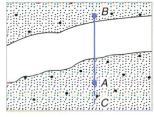

2. Construct a perpendicular to line BC at A and pick any point D on this perpendicular.

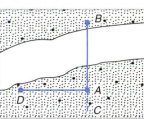

3. Find a point E where B and D appear to "line up." Draw DE and extend DE to F.

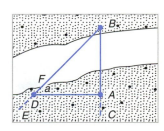

4. Triangle ABD is a right triangle. You can measure AD and angle a, and, using the trigonometric ratios, you can calculate AB. Suppose $AD = 100$ ft and $\measuredangle a = 49°$. Find AB.

Try it. Check your work in **30**.

30

B

AD is the adjacent side to angle *a*.

AB is the side opposite angle *a*.

?

49°

D   *A*

← 100 ft →

The trig ratio that relates the opposite to the adjacent side is the tangent.

$$\tan a = \frac{\text{opposite side}}{\text{adjacent side}}$$

$$\tan a = \frac{AB}{AD}$$

$$\tan 49° = \frac{AB}{100 \text{ ft}}$$

Multiply both sides of this equation by 100 to get

$$100 \tan 49° = AB$$

Use a calculator to find that tan 49° ≈ 1.150 approximately.

Therefore,

$$AB = 100 \times 1.150$$
$$AB \approx 115 \text{ ft} \quad \text{rounded to the nearest foot}$$

Using trig ratios, you have managed to calculate the distance from *A* to *B* without ever coming near point *B*.

Another Example: Find the length of side *A* in the triangle shown.

Step 1 Decide which trig ratio is the appropriate one to use. In this case, we are given the hypotenuse (21 cm), and we need to find the adjacent side (*A*). The trig ratio that relates the adjacent side and the hypotenuse is the cosine.

$$\cos a = \frac{\text{adjacent side}}{\text{hypotenuse}}$$

Step 2 Write the cosine of the given angle in terms of the sides of the given right triangle.

$$\cos 50° = \frac{A}{21 \text{ cm}}$$

Step 3 Solve for the unknown quantity. In this case multiply both sides of the equation by 21. Find cos 50° with a calculator.

$$A = 21 \cdot \cos 50°$$
$$A \approx 21 \cdot 0.64278 \ldots$$
$$A \approx 13.4985 \text{ cm} \quad \text{or} \quad 13.5 \text{ cm} \quad \text{rounded}$$

21 ⊗ **50** [cos] [=] → *13.498540*

Your turn. Find the length of the hypotenuse *H* in the triangle at the left. Round to the nearest tenth.

Work it out using the three steps shown above, then turn to **31** to check your answer.

_____

31 **Step 1** We are given the side opposite to the angle 71°, and we need to find the hypotenuse. The trig ratio relating the opposite side and the hypotenuse is the sine.

$$\sin a = \frac{\text{opposite side}}{\text{hypotenuse}}$$

Step 2 $\sin 71° = \dfrac{10 \text{ in.}}{H}$

Step 3   Solve for H. $H \cdot \sin 71° = 10$ in.

or $H = \dfrac{10 \text{ in.}}{\sin 71°} \approx \dfrac{10}{0.946}$

$H \approx 10.6$ in.

 10 ÷ 71 sin = → 10.576207

Here are a few problems for practice in solving right triangles. Round to the nearest tenth.

(a) Find H
12'
42°

(b) Find X
16 m   34°
X

(c) 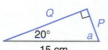 Find A
80° A
35"

(d) Find P, Q, and a
Q
20° P
a
15 cm

Check your work in **32**.

---

29   (a) $\sin 42° = \dfrac{12 \text{ ft}}{H}$

$H = \dfrac{12}{\sin 42°}$

$H \approx \dfrac{12}{0.669} \approx 17.9$ ft

(b) $\cos 34° = \dfrac{X}{16 \text{ m}}$

$X = 16 \cdot \cos 34°$

$X \approx (16)(0.829) \approx 13.3$ m

16
34°
X

(c) $\tan 80° = \dfrac{35 \text{ in.}}{A}$

$A = \dfrac{35 \text{ in.}}{\tan 80°} \approx 6.2$ in.

35"
80°

(d) $\sin 20° = \dfrac{P}{15 \text{ cm}}$

$P = 15 \cdot \sin 20° \approx 5.1$ cm

$\cos 20° = \dfrac{Q}{15 \text{ cm}}$

15
20° P
Q

$Q = 15 \cdot \cos 20° \approx 14.1$ cm

$a = 90° - 20° = 70°$

On problems (b), (c), and (d) redraw the triangle so that it is in standard position as shown.

Practical problems are usually a bit more difficult to set up, but they are solved with the same three-step approach. Try the following practical problems.

(a) Find the angle a in the casting shown. Round to the nearest minute.

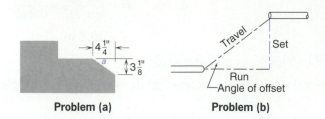

Problem (a)   **Problem (b)**

(b) In a pipe-fitting job, what is the length of set if the offset angle is $22\frac{1}{2}°$ and the travel is $32\frac{1}{8}$ in.? Round to the nearest tenth.

(c) Find the angle of taper, x. Round to the nearest minute.

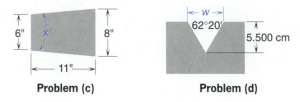

Problem (c)   **Problem (d)**

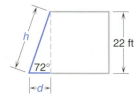

Problem (e)

(d) Find the width, w, of the V-slot shown. (Round to 0.001 cm.)

(e) Find the dimensions d and h on the metal plate shown. (Round to the nearest tenth.)

Our step-by-step solutions are given in **33**.

---

33   (a) In this problem the angle a is unknown. The information given allows us to calculate $\tan a$:

$$\tan a = \frac{\text{opposite side}}{\text{adjacent side}} = \frac{3\frac{1}{8} \text{ in.}}{4\frac{1}{4} \text{ in.}}$$

$$\tan a \approx 0.7353$$

$$a \approx 36°20'$$

 3.125 ÷ **4.25** = INV tan → **36.326826**

⊖ **36** = × **60** = → **19.609557**

(b) In this triangle the opposite side is unknown and the hypotenuse is given. Use the formula for sine:

$$\text{sine } 22\frac{1}{2}° = \frac{\text{opposite side}}{\text{hypotenuse}}$$

Then substitute, $0.383 = \dfrac{S}{32\frac{1}{8} \text{ in.}}$

And solve, $S = (32\frac{1}{8} \text{ in.})(0.383) \approx 12.3$ in.

(c) Notice that angle x, the *taper angle*, is twice as large as angle m. We can find angle m from triangle ABC.

$$\tan m = \frac{\text{opposite side}}{\text{adjacent side}} = \frac{AC}{BC}$$

$$\tan m = \frac{1 \text{ in.}}{11 \text{ in.}} \approx 0.090909 \ldots$$

$$m \approx 5°11.7'$$

$$x = 2m \approx 10°23'$$

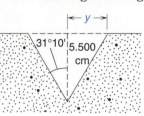

(d) The V-slot creates an isosceles triangle whose height is 5.500 cm. The height divides the triangle into two identical right triangles as shown in the figure.

To find y,

$$\tan 31°10' = \frac{y}{5.500}$$

or

$$y = (5.500)(\tan 31°10')$$

$$y \approx 3.3265 \text{ cm}$$

$$w = 2y \approx 6.653 \text{ cm}$$

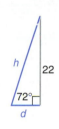

(e) $\sin 72° = \dfrac{22}{h}$

$$h = \frac{22}{\sin 72°} \approx 23.1 \text{ ft}$$

$$\tan 72° = \frac{22}{d}$$

$$d = \frac{22}{\tan 72°} \approx 7.1 \text{ ft}$$

Now turn to **34** for a set of problems on solving right triangles using trigonometric ratios.

<hr/>

34

Exercises 10-3 **Solving Right Triangles**

A. Find the missing quantities as indicated. (Round all angles to the nearest minute and sides to the nearest tenth.)

1.

2.

3.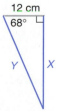

4.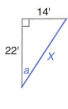

$X =$ _____ $X =$ _____ $X =$ _____ $\angle a =$ _____

$Y =$ _____ $Y =$ _____ $Y =$ _____ $X =$ _____

5.

$\angle a =$ _____

$X =$ _____

6.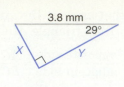

$X =$ _____

$Y =$ _____

7.

$X =$ _____

$Y =$ _____

8.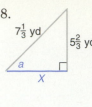

$\angle a =$ _____

$X =$ _____

Find the missing quantities as indicated. (Round all angles to the nearest tenth of a degree and sides to the nearest hundredth.)

9.

$\angle a =$ _____

$X =$ _____

10.

$X =$ _____

$Y =$ _____

11.

$X =$ _____

$Y =$ _____

12.

$\angle a =$ _____

$X =$ _____

13.

$\angle m =$ _____

$X =$ _____

14.

$X =$ _____

$Y =$ _____

15.

$X =$ _____

$Y =$ _____

16.

$\angle m =$ _____

$X =$ _____

B. Practical Problems

1. **Manufacturing**   The most efficient operating angle for a certain conveyor belt is 31°. If the parts must be moved a vertical distance of 16 ft, what length of conveyor is needed? (Round to the nearest tenth.)

2. **Metalworking**   Find the angle m in the casting shown. (Round to the nearest tenth of a degree.)

3. **Plumbing**   A pipe fitter must connect a pipeline to a tank as shown. The run from the pipeline to the tank is 62 ft, while the set is 38 ft.

    (a)   How long is the connection? (Round to the nearest inch.)

    (b)   Will the pipe fitter be able to use standard pipe fittings (i.e., $22\frac{1}{2}°$, 30°, 45°, 60°, or 90°)?

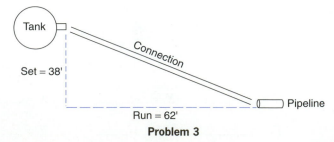

Problem 3

Problem 1

Problem 2

4. A helicopter, flying directly over a fishing boat at an altitude of 1200 ft, spots an ocean liner at a 15° angle of depression. How far from the boat is the liner? (Round to the nearest foot.)

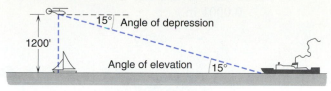

Problem 4

5. **Metalworking**   Find the angle of taper on the steel bar shown if it is equal to twice *m*. Round to the nearest minute.

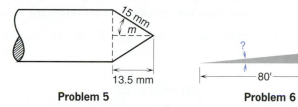

Problem 5   **Problem 6**

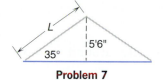

Problem 7

6. **Construction**   A road has a rise of 6 ft in 80 ft. What is the gradient angle of the road? (Round to the nearest tenth of a degree.)

7. **Carpentry**   Find the length of the rafter shown. (Round to the nearest inch.)

8. **Metalworking**   Three holes are drilled into a steel plate. Find the distances *A* and *B* as shown in the figure. (Round to the nearest hundredth of an inch.)

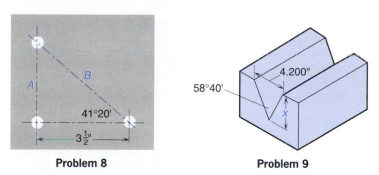

Problem 8   **Problem 9**

9. **Machine Technology**   Find the depth of cut *x* needed for the V-slot shown. (Assume that the V-slot is symmetric. Round to the nearest hundredth.)

10. **Masonry**   A 20-ft ladder leans against a building at an 80° angle with the ground. Will it reach a window 17 ft above the ground?

Problem 10

11. **Metalworking**   Find the distance *B* in the rivet shown. (Round to the nearest tenth.)

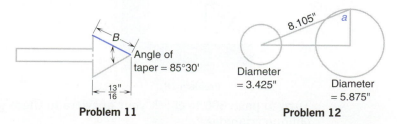

Problem 11   **Problem 12**

12. Find angle *a* in the figure. (Round to the nearest hundredth.)

13. **Construction** A road has a slope of 2°25′. Find the rise in 6500 ft of horizontal run. (Round to the nearest foot.)

14. Find the missing dimension *d* in the flathead screw shown. (Round to 0.0001 in.)

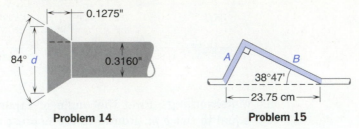

Problem 14   **Problem 15**

15. **Plumbing** What total length of conduit is needed for sections *A* and *B* of the pipe connection shown? (Round to one decimal place.)

16. **Machine Technology** Find the included angle *m* of the taper shown. (Round to the nearest minute.)

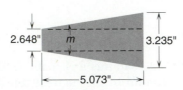

Problem 16

17. **Machine Technology** Ten holes are spaced equally around a 5-in.-diameter circle. Find the center-to-center distance *x* in the figure. Round to two decimal places.

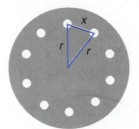

Problem 17

18. **Police Science** Detectives investigating a crime find a bullet hole in a wall at a height of 7 ft 6 in. from the floor. The bullet passed through the wall at an angle of 34°. If they assume that the gun was fired from a height of 4 ft above the floor, how far away from the wall was the gun when it was fired?

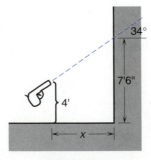

Problem 18

Turn to page 606 to check your answers to these problems. Then go to **35** to study oblique triangles.

Chap. 10 Triangle Trigonometry

10-4 OBLIQUE TRIANGLES (OPTIONAL)

35 We have already seen how the trigonometric functions can be used to solve right triangles. To solve for the missing parts of a triangle that does not contain a right angle, we need two new formulas, the *law of sines* and the *law of cosines*. In this section we will derive these formulas and show how they may be used to solve triangles.

Oblique Triangles Any triangle that is not a right triangle is called an **oblique triangle.** There are two types of oblique triangles: acute and obtuse. In an **acute triangle** all three angles are acute—each is less than 90°. In an **obtuse triangle** one angle is obtuse—that is, greater than 90°. Notice in these figures that the angles have been labeled A, B, and C, and the sides opposite these angles have been labeled a, b, and c, respectively.

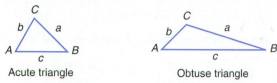

Acute triangle   Obtuse triangle

Law of Sines The law of sines states that the lengths of the sides of a triangle are proportional to the sines of the corresponding angles. To show this for an acute triangle, first construct a perpendicular from C to side AB. Then in triangle ACD,

$$\sin A = \frac{h}{b} \quad \text{or} \quad h = b \sin A$$

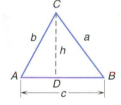

and in triangle BCD,

$$\sin B = \frac{h}{a} \quad \text{or} \quad h = a \sin B$$

Then $a \sin B = b \sin A$ or $\dfrac{a}{\sin A} = \dfrac{b}{\sin B}$

By repeating this process with a perpendicular from B to side AC, we can obtain a similar equation involving angles A and C and sides a and c. Combining these results gives the **law of sines.**

LAW OF SINES

$$\frac{a}{\sin A} = \frac{b}{\sin B} = \frac{c}{\sin C}$$

This law is sometimes written as

$$\frac{\sin A}{a} = \frac{\sin B}{b} = \frac{\sin C}{c}$$

For the triangle shown, calculate the ratios

$$\frac{a}{\sin A}, \frac{b}{\sin B} \text{ and } \frac{c}{\sin C}$$

Round to one decimal place.

Check your work in **36**.

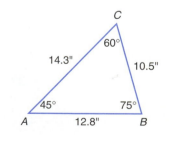

---

36 All three ratios are equal to 14.8, rounded to one decimal place. Using a calculator, we obtain

10.5 ÷ **45** [sin] [=] → *14.849242* ≈ 14.8 rounded

To *solve* a triangle means to calculate the values of all the unknown sides and angles from the information given. The law of sines enables us to solve any oblique triangle if the following information is given.

Case 1. Two angles and a side are known.

Case 2. Two sides and the angle opposite one of them are known.

Case 1

Example: For the triangle shown, $A = 47°$, $B = 38°$, and $a = 8.0$ in. Find all its unknown parts, in this case angle C and sides b and c.

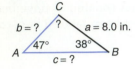

First, sketch the triangle as shown. Then since the angles of a triangle sum to 180°, find angle C by subtracting.

$$C = 180° - A - C = 180° - 47° - 38° = 95°$$

Next, apply the law of sines. Substitute A, B, and a to find b.

$$\frac{b}{\sin B} = \frac{a}{\sin A} \quad \text{gives} \quad b = \frac{a}{\sin A} \cdot \sin B = \frac{a \sin B}{\sin A}$$

$$= \frac{8.0 \sin 38°}{\sin 47°}$$

$$\approx \frac{8.0(0.616)}{0.731} \approx 6.7 \text{ in.}$$

Using a calculator, we have **8** ⊗ **38** (sin) ÷ **47** (sin) (=) → ⌐6.7344867⌐

Finally, apply the law of sines, again using A, C, and a to find c.

$$\frac{c}{\sin C} = \frac{a}{\sin A} \quad \text{gives} \quad c = \frac{a \sin C}{\sin A}$$

$$= \frac{8.0 \sin 95°}{\sin 47°} \approx 10.9 \text{ in.}$$

In any triangle the largest angle is opposite the longest side, and the smallest angle is opposite the shortest side. Always check to make certain that this is true for your solution.

Careful ▶ Never use the Pythagorean theorem with an oblique triangle. It is valid only for the sides of a right triangle. ◀

Your turn. Solve the triangle for which $c = 6.0$ ft, $A = 52°$, and $C = 98°$. Round sides to the nearest tenth. $B = ?$, $a = ?$, $b = ?$

Check your solution in **37**.

---

37 $B = 180° - 52° - 98° = 30°$

$$\frac{a}{\sin A} = \frac{c}{\sin C} \quad \text{or} \quad a = \frac{c}{\sin C} \cdot \sin A = \frac{c \sin A}{\sin C}$$

$$= \frac{6 \sin 52°}{\sin 98°} \approx 4.8 \text{ ft}$$

$$\frac{b}{\sin B} = \frac{c}{\sin C} \quad \text{or} \quad b = \frac{c \sin B}{\sin C}$$

$$= \frac{6 \sin 30°}{\sin 98°} \approx 3.0 \text{ ft}$$

Case 2 In case 2, when two sides and the angle opposite one of these sides are known, the numbers may not always lead to a solution. For example, if $A = 50°$, $a = 6$, and $b = 10$, we may make a sketch of the triangle like this:

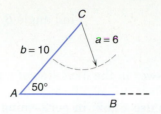

No matter how we draw side a it will not intersect line AB. There is no triangle with these measurements.

By looking at the given information before trying to solve the problem you can get an idea of what the solution triangle looks like. Let's examine all the possibilities.

- If $A = 90°$, then the triangle is a right triangle. Solve it using the methods of Section 10-3.
- If $A > 90°$ and $a > b$, then the triangle looks like this $\longrightarrow$
- If $A > 90°$ and $a \leq b$, then there will be no solution, like this

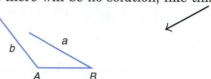

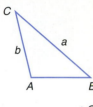

- If $A < 90°$ and $a = b \sin A$, then the triangle is a right triangle. You can solve it using the methods of Section 10-3. $\longrightarrow$

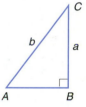

- If $A < 90°$ and $a \geq b$, then the triangle looks like this $\rightarrow$

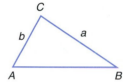

- If $A < 90°$ and $a < b$ and $a > b \sin A$, then the triangle looks either like this

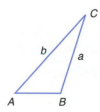

or like this

Because two different triangles may satisfy the given information, this is called the **ambiguous** situation.

- If $A < 90°$ and $a < b \sin A$, there will be no solution.

Now, let's look at some examples.

Turn to **38** for examples showing the solution of case 2 problems.

---

38 **Example** Given $A = 30.0°$ and $b = 10.0$ ft, solve the triangle where (a) $a = 12.0$ ft, (b) $a = 5.00$ ft, (c) $a = 4.50$ ft, (d) $a = 8.00$ ft.

(a) **First,** note that $a > b$ so the triangle will look like this:

Make a quick sketch of the triangle.

Second, solve for angle B using the law of sines.

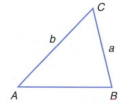

$$\frac{b}{\sin B} = \frac{a}{\sin A} \quad \text{or} \quad a \sin B = b \sin A, \text{ then}$$

$$\sin B = \frac{b \sin A}{a} = \frac{10 \, (\sin 30°)}{12} \qquad \text{Calculate } \sin B.$$

$$\sin B \approx 0.4166 \ldots \qquad \text{Find angle } B \text{ using inverse sine.}$$

$$B \approx 24.62° \approx 24.6°$$

Note ▶ When using an intermediate answer in the next step of a problem, always carry at least one additional decimal place to avoid rounding errors. In this case the rounded answer is 24.6°, but we use 24.62° in performing the next calculation. ◀

 Using a calculator, we obtain **10** ✕ **30** (sin) ÷ **12** (=) (INV) (sin) → | 24.624318 |

Next, calculate angle C. $C = 180° - A - B \approx 180° - 30.00° - 24.62°$

$$C \approx 125.38 \approx 125.4°$$

Finally, use the law of sines to find side c.

$$\frac{c}{\sin C} = \frac{a}{\sin A} \quad \text{gives} \quad c = \frac{a \sin C}{\sin A} \approx \frac{12 \, (\sin 125.38°)}{\sin 30°}$$

$$c \approx 19.6 \text{ ft}$$

(b) **First,** note that $a < b$. Then test to see if a is less than $b \sin A$. $b \sin A = 10 \sin 30° = 5$. Since $a = b \sin A$, the triangle is a right triangle and there is only one possible solution. We can show that the triangle is a right triangle by using the law of sines:

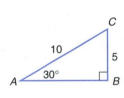

$$\frac{b}{\sin B} = \frac{a}{\sin A} \quad \text{gives} \quad \sin B = \frac{b \sin A}{a} = \frac{10 \sin 30°}{5}$$

$$\sin B = 1$$

$$B = 90°$$

Next, calculate angle C: $C = 180° - 30° - 90° = 60°$

Finally, calculate side c: $\cos A = \dfrac{c}{b}$

$$c = b \cos A = 10 \cos 30°$$

$$c \approx 8.66 \text{ ft}$$

(c) Note that $a < b$, then test to see if a is less than $b \sin A$. $b \sin A = 5$, and $4.5 < 5$; therefore, there is *no* triangle that satisfies the given conditions.

We can check this by using the law of sines:

$$\sin B = \frac{b \sin A}{a} = \frac{10 \sin 30°}{4.5} \approx 1.11$$

But the value of $\sin B$ can never be greater than 1; therefore, no such triangle exists. Using a calculator, we get

 10 ✕ **30** (sin) ÷ **4.5** (=) (INV) (sin) → ▨▨▨▨▨ *E* ▨

The error message tells that there is no solution.

(d) **First,** note that $a < b$. Because $8 > 5$, $a > b \sin A$, and we have the ambiguous case. There will be two possible triangles that satisfy these given conditions. Sketch them.

As given earlier, p. 512, $\sin A = \sin (180 - A)$. Therefore we know that $B_2 = 180 - B_1$.

and

Second, apply the law of sines to find angle B.

$$\frac{a}{\sin A} = \frac{b}{\sin B} \qquad \text{or} \qquad a \sin B = b \sin A$$

$$\sin B = \frac{b \sin A}{a} = \frac{10 \sin 30°}{8}$$

$$\sin B = 0.625$$

$$B \approx 38.68° \approx 38.7°$$

The two possible angles are

$$B_1 \approx 38.68° \qquad \text{and} \qquad B_2 \approx 180° - 38.68° \approx 141.32° \approx 141.3°$$

- For the acute value of angle B,

$$C_1 \approx 180° - 30.00° - 38.68° \approx 111.32° \approx 111.3°$$

and from the law of sines,

$$\frac{c_1}{\sin 111.32°} = \frac{8.00}{\sin 30°} \qquad \text{gives} \qquad c_1 \approx 14.9 \text{ ft}$$

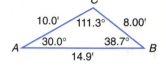

- For the obtuse value of angle B,

$$C_2 = 180° - 30.00° - 141.32° \approx 8.68° \approx 8.7°$$

and from the law of sines,

$$\frac{c_2}{\sin 8.68°} = \frac{8.00}{\sin 30°} \qquad \text{gives} \qquad c_2 \approx 2.41 \text{ ft}$$

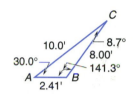

Summary of Procedures for Case 2

Step 1 Check the given angle. Is it $=$, $<$, or $> 90°$?

Step 2 If the angle is **equal to 90°**, the triangle is a right triangle.

Step 3 If the angle is **greater than 90°**, compare a and b. (a is the side opposite the given angle A.)

- If $a > b$,

Use the law of sines to find this solution.

- If $a = b$, or $a < b$, there is no solution. Your calculator will display an error message when you try to calculate angle B.

Step 4 If the angle is **less than 90°**, compare a and b.

- If $a = b$ or $a > b$,

Use the law of sines to find this solution.

- If $a < b$, compare a and $b \sin A$.
- If $a = b \sin A$, the triangle is a right triangle. Solve using either the law of sines or right triangle trigonometry.

- If $a > b \sin A$, there are two solutions.

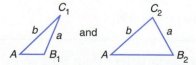

After finding B_1, subtract from $180°$ to find B_2, then complete both solutions.

- If $a < b \sin A$, there is no solution.

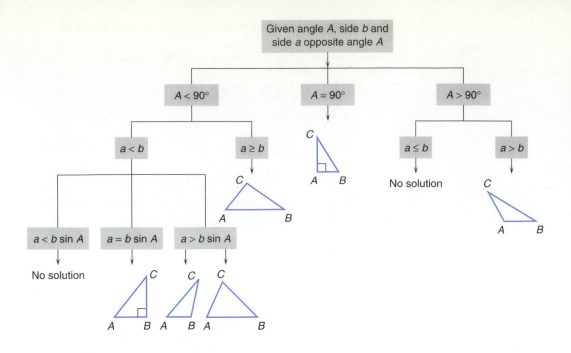

Example: If $A = 105.0°$ and $b = 125$ ft, solve the triangle where
(a) $a = 95.0$ ft   (b) $a = 156$ ft.

(a) The given angle is greater than 90°, and since $a < b$, from **step 3,** we know there will be no solution. Check this using the law of sines:

$$\frac{a}{\sin A} = \frac{b}{\sin B} \qquad \text{gives} \qquad \sin B = \frac{b \sin A}{a} = \frac{125 \sin 105°}{95}$$

$$\sin B \approx 1.27$$

But this is not an acceptable value; $\sin B$ cannot be greater than 1.

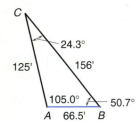

(b) In this case $a > b$, and we expect from **step 3** to find a unique solution. Using the law of sines, we have

$$\sin B = \frac{b \sin A}{a} = \frac{125 \sin 105°}{156}$$

$$\approx 0.7739\ldots$$

$$B \approx 50.71° \approx 50.7°$$

Then $C \approx 180° - 105.00° - 50.71° \approx 24.29° \approx 24.3°$

Use the law of sines to calculate side c:

$$\frac{c}{\sin C} = \frac{a}{\sin A} \qquad \text{gives} \qquad c = \frac{a \sin C}{\sin A} \approx \frac{156 \sin 24.29°}{\sin 105°}$$

$$c \approx 66.4 \text{ ft}$$

Use these steps and the law of sines to solve the triangle where $B = 36.0°$, $c = 16.4$ cm, and   (a) $b = 20.5$ cm,   (b) $b = 15.1$ cm. (Round angles to the nearest 0.1° and sides to three significant digits.)

Check your answers in **39**.

---

39 (a) The given angle is less than 90°, so by step 4 we need to compare the two sides. The side opposite the given angle is b, and b is greater than the other given side, c, so by step 4 we know there is only one solution. Sketch it.

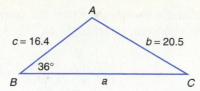

Now use the law of sines to solve for angle C.

$$\frac{c}{\sin C} = \frac{b}{\sin B} \quad \text{gives} \quad \sin C = \frac{c \sin B}{b} = \frac{16.4 \sin 36°}{20.5}$$

$$\sin C \approx 0.4702 \ldots$$

$$C \approx 28.049° \approx 28.0°$$

$$A \approx 180° - 28.05° - 36.00° \approx 115.95°$$

$$\frac{a}{\sin A} = \frac{b}{\sin B} \quad \text{gives} \quad a = \frac{b \sin A}{\sin B} \approx \frac{20.5 \sin 115.95°}{\sin 36°}$$

$$a \approx 31.4 \text{ cm}$$

(b) For this case $b < c$, so we need to compare b and $c \sin B$.

$$c \sin B = 16.4 \, (\sin 36°)$$

$$\approx 9.64$$

so $b > c \sin B$ and we know by step 4 that there are two solutions.

$$\sin C = \frac{c \sin B}{b} = \frac{16.4 \sin 36°}{15.1}$$

$$\sin C \approx 0.6383 \ldots$$

$$C \approx 39.67° \approx 39.7°$$

The two possible solutions are

$$C_1 \approx 39.67° \quad \text{and} \quad C_2 \approx 180° - 39.67° \approx 140.33° \quad \text{Sketch the triangles.}$$

$$\approx 140.3°$$

For the first value of angle C,

$$A_1 \approx 180° - 39.67° - 36.00° \approx 104.33° \approx 104.3°$$

$$\text{and} \quad \frac{a}{\sin A_1} = \frac{b}{\sin B} \quad \text{or} \quad a \approx \frac{15.1 \sin 104.33°}{\sin 36°} \approx 24.9 \text{ cm}$$

For the second value of C, $A_2 \approx 180° - 140.33° - 36.00° \approx 3.67°$ and

$$a \approx \frac{15.1 \sin 3.67°}{\sin 36°} \approx 1.64 \text{ cm}$$

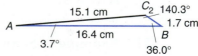

A second formula that is needed for solving oblique triangles is discussed in **40**.

40　In addition to the two cases already mentioned, it is also possible to solve any oblique triangle if the following information is given.

Case 3.　Two sides and the angle included by them.

Case 4.　All three sides.

But the law of sines is not sufficient for solving triangles given this information. We need an additional formula called the **law of cosines.**

Law of Cosines   To obtain this law, in the triangle shown construct the perpendicular from C to side AB. Using the Pythagorean theorem, in triangle BCD,

$$a^2 = (c - x)^2 + h^2$$

$$= c^2 - 2cx + x^2 + h^2$$

and in triangle ACD,

$$b^2 = x^2 + h^2$$

Therefore,

$$a^2 = c^2 - 2cx + b^2$$

$$= b^2 + c^2 - 2cx$$

In triangle ACD, $\cos A = \dfrac{x}{b}$ or $x = b \cos A$; therefore,

$$a^2 = b^2 + c^2 - 2bc \cos A$$

By rotating the triangle ABC, we can derive similar equations for b^2 and c^2.

LAW OF COSINES

$$a^2 = b^2 + c^2 - 2bc \cos A$$

$$b^2 = c^2 + a^2 - 2ca \cos B$$

$$c^2 = a^2 + b^2 - 2ab \cos C$$

Learning Hint ▶   Notice the similarity in these formulas. Rather than memorize all three formulas, you may find it easier to remember the following statement from which they may be written. ◀

LAW OF COSINES

The square of the length of any side of a triangle equals the sum of the squares of the lengths of the other two sides minus twice the product of the lengths of these two sides and the cosine of the angle between them.

The law of cosines may be used to solve any oblique triangle for which the information in case 3 and case 4 is given.

Now turn to **41** to work through some examples of the law of cosines being used to solve a case 4 problem.

---

41   **Example:**   Solve the triangle with sides $a = 106$ m, $b = 135$ m, and $c = 165$ m.

First, use the law of cosines to find the largest angle, the angle opposite the longest side. Side c is the longest side, so find C.

$$c^2 = a^2 + b^2 - 2ab \cos C$$

Solve for $\cos C$.

$$c^2 - a^2 - b^2 = -2ab \cos C$$

or

$$2ab \cos C = a^2 + b^2 - c^2$$

$$\cos C = \frac{a^2 + b^2 - c^2}{2ab} = \frac{106^2 + 135^2 - 165^2}{2(106)(135)}$$

Using a calculator, we get

$$106 \boxed{x^2} \boxed{+} 135 \boxed{x^2} \boxed{-} 165 \boxed{x^2} \boxed{=} \boxed{\div} \boxed{(} 2 \boxed{\times} 106 \boxed{\times} 135 \boxed{)} \boxed{=} \boxed{\text{INV}} \boxed{\text{COS}}$$

$\rightarrow \boxed{85.519076}$

$C \approx 85.5°$

Next, use the law of sines to find the smallest angle, the one opposite the shortest side.

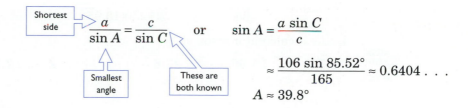

$$\frac{a}{\sin A} = \frac{c}{\sin C} \quad \text{or} \quad \sin A = \frac{a \sin C}{c}$$

$$\approx \frac{106 \sin 85.52°}{165} \approx 0.6404\ldots$$

$$A \approx 39.8°$$

Finally, since the angles of a triangle must sum to 180°, we can calculate the remaining angle by subtracting.

$$B \approx 180° - 85.5° - 39.8° \approx 54.7°$$

As a check on this process, note that the last angle calculated is indeed intermediate between the first two in size.

Careful ▶ To avoid difficulties that can arise in the second step of the last example, you must find the largest angle first. Remember, the largest angle is always opposite the longest side. ◀

Now you try it. Solve the triangle with sides $a = 9.5$ in., $b = 4.2$ in., and $c = 6.4$ in. Check your solution in **42**.

---

42 **First,** find the largest angle A using the law of cosines.

$$a^2 = b^2 + c^2 - 2bc \cos A$$

$$\cos A = \frac{b^2 + c^2 - a^2}{2bc} = \frac{4.2^2 + 6.4^2 - 9.5^2}{2(4.2)(6.4)}$$

$$A \approx 126.07° \approx 126.1°$$

Next, use the law of sines to find the smallest angle B.

$$\frac{b}{\sin B} = \frac{a}{\sin A} \quad \text{gives} \quad \sin B = \frac{b \sin A}{a} \approx \frac{4.2 \sin 126.07°}{9.5}$$

$$B \approx 20.9°$$

Finally, subtract to find angle C.

$$C \approx 180° - 126.1° - 20.9° \approx 33.0°$$

Next, let's solve a triangle given two sides and the included angle.

Solve the triangle with $a = 22.8$ cm, $b = 12.3$ cm, and $C = 42.0°$.

Turn to **43** for our solution showing the law of cosines applied to this problem.

---

43 **First,** make a sketch and notice that the given angle C is the angle formed by the given sides a and b. Use the form of the law of cosines that involves the given angle to find the third side c.

$$c^2 = a^2 + b^2 - 2ab \cos C$$

$$= 22.8^2 + 12.3^2 - 2(22.8)(12.3) \cos 42°$$

Using a calculator, we get

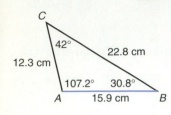

22.8 x^2 $+$ **12.3** x^2 $-$ **2** $\times$ **22.8** $\times$ **12.3** $\times$ **42** $\cos$ $=$ $\sqrt{\ }$ $\rightarrow$ $\boxed{15.947255}$

$c \approx 15.9$ cm

Next, since we now know all three sides and one angle, we can use the law of cosines to find one of the remaining angles.

$a^2 = b^2 + c^2 - 2bc \cos A$

or $\quad \cos A = \dfrac{b^2 + c^2 - a^2}{2bc} \approx \dfrac{12.3^2 + 15.95^2 - 22.8^2}{2(12.3)(15.95)}$

Using a calculator gives us

12.3 x^2 $+$ **15.95** x^2 $-$ **22.8** x^2 $=$ $\div$ $($ **2** $\times$ **12.3** $\times$ **15.95** $)$ $=$ INV $\cos$

$\rightarrow$ $\boxed{106.91292}$

$A \approx 106.9°$

Finally, find the remaining angle by subtracting.

$B \approx 180° - 106.9° - 42.0° \approx 31.1°$

Practice using the law of cosines by solving the triangle with $B = 34.4°$, $a = 145$ ft, and $c = 112$ ft. Check your solution in **44**.

---

44 Use the law of cosines to find the third side b.

$b^2 = a^2 + c^2 - 2ac \cos B$

$\quad = 145^2 + 112^2 - 2(145)(112) \cos 34.4°$

$b \approx 82.28$ ft ≈ 82.3 ft

Now find angle A. $\quad \cos A = \dfrac{b^2 + c^2 - a^2}{2bc} \approx \dfrac{82.28^2 + 112^2 - 145^2}{2(82.28)(112)}$

$A \approx 95.3°$

$C \approx 180° - 34.4° - 95.3° \approx 50.3°$

Now turn to **45** for a set of exercises on solving oblique triangles.

45

Exercises 10-4   **Oblique Triangles**

A. Solve each triangle.

1. $a = 6.5$ ft, $A = 43°$, $B = 62°$

2. $b = 17.2$ in., $C = 44.0°$, $B = 71.0°$

3. $b = 165$ m, $B = 31.0°$, $C = 110.0°$

4. $c = 2300$ yd, $C = 120°$, $B = 35°$

5. $a = 96$ in., $b = 58$ in., $B = 30°$

6. $b = 265$ ft, $c = 172$ ft, $C = 27.0°$

7. $a = 8.5$ m, $b = 6.2$ m, $C = 41°$

8. $b = 19.3$ m, $c = 28.7$ m, $A = 57.0°$

9. $a = 625$ ft, $c = 189$ ft, $B = 102.0°$

10. $b = 1150$ yd, $c = 3110$ yd, $A = 125.0°$

11. $a = 27.2$ in., $b = 33.4$ in., $c = 44.6$ in.

12. $a = 4.8$ cm, $b = 1.6$ cm, $c = 4.2$ cm

13. $a = 7.42$ m, $c = 5.96$ m, $B = 99.7°$

14. $b = 0.385$ in., $c = 0.612$ in., $A = 118.5°$

15. $a = 1.25$ cm, $b = 5.08$ cm, $c = 3.96$ cm

16. $a = 6.95$ ft, $b = 9.33$ ft, $c = 7.24$ ft

B. Practical Problems

1. **Landscape Design** To measure the height of a tree, Steve measures the angle of elevation of the treetop as 46°. He then moves 15 ft closer to the tree and from this new point measures the angle of elevation to be 59°. How tall is the tree?

2. **Construction Technology** In the channel shown, angle $B = 122.0°$, angle $A = 27.0°$, and side $BC = 32.0$ ft. Find the length of the slope AB.

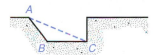

Problem 2

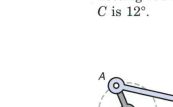

Problem 4

3. **Construction Technology** A triangular traffic island has sides 21.5, 46.2, and 37.1 ft. What are the angles at the corners?

4. **Surveying** The lot shown is split along a diagonal as indicated. What length of fencing is needed for the boundary line?

5. **Carpentry** Two sides of a sloped ceiling meet at an angle of 105.5°. If the distances along the sides to the opposite walls are 11.0 and 13.0 ft, what length of beam is needed to join the walls?

Problem 5

6. **Machine Technology** For the crankshaft shown, $AB = 4.2$ in., and the connecting rod $AC = 12.5$ in. Calculate the size of angle A when the angle at C is 12°.

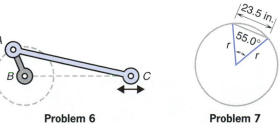

Problem 6   **Problem 7**

7. **Carpentry** If the chord of a circle is 23.5 in. long and subtends a central angle of 55.0°, what is the radius of the circle?

Turn to page 606 to check your answers to these problems. Then go to **46** for a set of problems covering the work of this chapter.

Triangle Trigonometry

46 Answers are given on page 606.

A. Which of the following are right triangles?

1.

2.

3.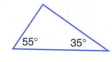

4.
4" 4"
4"

5.
1
1
√2

6.
4 5
3

7.
48°
32°

8.
23
24
7

9.
√3 2
1

10.
18 24
9

B. Find each of the following trig values using a calculator. (Round to three decimal places.)

1. cos 67°

2. tan 81°

3. sin 4°

4. cos 63°10′

5. tan 35.75°

6. sin 29.2°

7. sin 107°

8. cos 123°

Use a calculator to find the acute value of x to the nearest minute.

9. sin $x = 0.242$

10. tan $x = 1.54$

11. sin $x = 0.927$

12. cos $x = 0.309$

13. tan $x = 0.194$

14. cos $x = 0.549$

15. tan $x = 0.823$

16. sin $x = 0.672$

17. cos $x = 0.118$

Find the acute value of x to the nearest tenth of a degree.

18. tan $x = 0.506$

19. cos $x = 0.723$

20. sin $x = 0.488$

21. sin $x = 0.154$

22. cos $x = 0.273$

23. tan $x = 2.041$

24. tan $x = 0.338$

25. cos $x = 0.608$

26. sin $x = 0.772$

Express in radians. (Round to three decimal places.)

27. 35°

28. 21.4°

29. 74°30′

30. 112.2°

Name

Date

Course/Section

Express in degrees. (Round to the nearest tenth.)

31. 0.45 rad   **32.** 1.7 rad   **33.** 2.3 rad   **34.** 0.84 rad

C. Find the missing dimensions in the right triangles as indicated. (Round all distances to one decimal place, and all angles to the nearest degree.)

1.   **2.**   **3.**   **4.**

$X = \underline{\hspace{1.5cm}}$ $X = \underline{\hspace{1.5cm}}$ $\angle a = \underline{\hspace{1.5cm}}$ $X = \underline{\hspace{1.5cm}}$

$Y = \underline{\hspace{1.5cm}}$ $Y = \underline{\hspace{1.5cm}}$ $X = \underline{\hspace{1.5cm}}$ $Y = \underline{\hspace{1.5cm}}$

Round all distances to the nearest tenth and all angles to one decimal place.

5.   **6.** 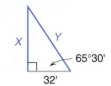   **7.**   **8.**

$\angle a = \underline{\hspace{1.5cm}}$ $X = \underline{\hspace{1.5cm}}$ $\angle t = \underline{\hspace{1.5cm}}$ $X = \underline{\hspace{1.5cm}}$

$X = \underline{\hspace{1.5cm}}$ $Y = \underline{\hspace{1.5cm}}$ $X = \underline{\hspace{1.5cm}}$ $Y = \underline{\hspace{1.5cm}}$

D. Solve the following oblique triangles.

1. $a = 17.9$ in., $A = 65.0°$, $B = 39.0°$

2. $a = 721$ ft, $b = 444$ ft, $c = 293$ ft

3. $a = 260$ yd, $c = 340$ yd, $A = 37°$

4. $b = 87.5$ in., $c = 23.4$ in., $A = 118.5°$

5. $a = 51.4$ m, $b = 43.1$ m, $A = 64.3°$

6. $a = 166$ ft, $c = 259$ ft, $B = 47.0°$

7. $a = 1160$ m, $c = 2470$ m, $C = 116.2°$

8. $a = 7.6$ in., $b = 4.8$ in., $B = 30°$

E. Practical Problems

Round all angles to the nearest minute, and all distances to one decimal place, unless told otherwise.

1. **Machine Technology**   What height of gauge blocks is required to set an angle of 7°15′ for 10-in. plots?

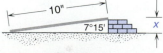

Problem 1

20 mph

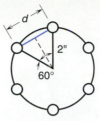

120 mph v a

Problem 2

2. The destination of an airplane is due north, and the plane flies at 120 mph. There is a crosswind of 20 mph from the west. What heading angle a should the plane take? What is its relative ground speed v?

3. **Woodworking** Six holes are spaced evenly around a 4-in.-diameter circle. Find the distance, d, between the holes.

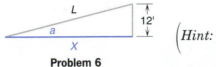

d

2"

60°

Problem 3

(*Hint:* The dashed line creates two identical right triangles.)

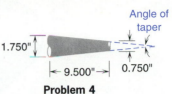

Angle of taper

1.750"

9.500"

0.750"

Problem 4

4. **Machine Technology** Find the angle of taper of the shaft in the figure.

5. A TV technician installs a 50-ft antenna on a flat roof. Safety regulations require a minimum angle of 30° between mast and guy wires.
   (a) Find the minimum value of X.
   (b) Find the length of the guy wires.

50' 30°

X

Problem 5

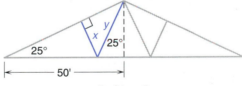

L

12'

a

X

$\left(Hint: \dfrac{15}{100} = \dfrac{12}{x}\right)$

Problem 6

6. **Construction** A bridge approach is 12 ft high. The maximum slope allowed is 15%—that is, $\frac{15}{100}$.
   (a) What is the length of the approach?
   (b) What is the angle a of the approach?

7. **Machine Technology** The *helix angle* of a screw is the angle at which the thread is cut. The *lead* is the distance advanced by one turn of the screw. The circumference refers to the circumference of the screw as given by $C = \pi d$. The following formula applies:

   $$\text{Tangent of helix angle} = \frac{\text{lead}}{\text{circumference}}$$

   (a) Find the helix angle for a 2-in.-diameter screw if the lead is $\frac{1}{8}$ in.
   (b) Find the lead of a 3-in.-diameter screw if the helix angle is $2°$.

Lead Helix angle

Circumference

Problem 7

8. **Construction** Find the lengths x and y of the beams shown in the bridge truss.

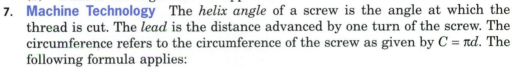

y

x

25° 25°

50'

Problem 8

9. **Machine Technology** Find the width X of the V-thread shown. (Round to the nearest thousandth.)

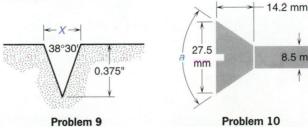

X

38°30'

0.375"

Problem 9

14.2 mm

27.5 mm

8.5 mm

a

Problem 10

10. **Machine Technology** Find the head angle a of the screw shown.

11. **Machine Technology** A machinist makes a cut 13.8 cm long in a piece of metal. Then, another cut is made that is 18.6 cm long and at an angle of 62°0′ with the first cut. How long a cut must be made to join the two endpoints?

12. **Drafting** Determine the center-to-center measurement x in the adjustment bracket shown in the figure. Round to the nearest hundredth.

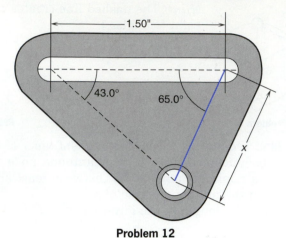

Problem 12

Advanced Algebra

Objective	Sample Problems		Where to go for Help	
			Page	Frame

When you finish this unit you will be able to:

1. Solve a system of two linear equations in two variables.

(a) $5x + 2y = 20$ $x = $ _____   542   **1**

     $3x - 2y = 12$ $y = $ _____

(b) $x = 3y$ $x = $ _____

     $2y - 4x = 30$ $y = $ _____

(c) $3x - 5y = 7$ $x = $ _____

     $6x - 10y = 14$ $y = $ _____

(d) $2y = 5x - 7$ $x = $ _____

     $10x - 4y = 9$ $y = $ _____

2. Solve word problems involving systems of equations in two variables.

(a) The sum of two numbers is 35. Twice the smaller decreased by the larger is 10. Find them. _____   554   **14**

(b) The perimeter of a rectangular vent is 28 in. The length of the vent is 2 in. greater than its width. Find the dimensions. _____

(c) A mixture of 800 screws costs $60. If the mixture consists of one type costing 5 cents each and another type costing 9 cents each, how many of each kind are there? _____

Name _____

Date _____

Course/Section _____

3. Solve quadratic equations.

(a) $x^2 = 16$ $x =$ _____   560   **20**

(b) $x^2 - 7x = 0$ $x =$ _____

(c) $x^2 - 5x = 14$ $x =$ _____

(d) $3x^2 + 2x - 16 = 0$ $x =$ _____

(e) $2x^2 + 3x + 11 = 0$ $x =$ _____

4. Solve word problems involving quadratic equations.

(a) Find the side length of a square opening whose area is 36 cm². _____   568   **29**

(b) Find the radius of a circular pipe whose cross-sectional area is 220 sq in. _____

(c) In the formula $P = RI^2$, find I (in amperes) if $P = 2500$ watts and $R = 15$ ohms. _____

(d) The length of a rectangular vent is 3 in. longer than its width. Find the dimensions if the cross-sectional area is 61.75 sq in. _____

(Answers to these preview problems are given on page 579.)

If you are certain that you can work *all* these problems correctly, turn to page 575 for a set of practice problems. If you cannot work one or more of the preview problems, turn to the page indicated to the right of the problem. Those who wish to master this material with the greatest success should turn to frame **1** and begin work there.

Advanced Algebra

Reprinted by permission: Tribune Media Services

11-1 SYSTEMS OF EQUATIONS

1 This chapter is designed for students who are interested in highly technical occupations. We explain how to solve systems of linear equations and how to solve quadratic equations. Before beginning, you may want to return to Chapter 7 to review the basic algebraic operations explained there. When you have had the review you need, return here and continue.

Before you can solve a system of two equations in two unknowns, you must be able to solve a single linear equation in one unknown. Let's review what you learned in Chapter 7 about solving linear equations.

A *solution* to an equation such as $3x - 4 = 1$ is a number that we may use to replace x in order to make the equation a true statement. To find such a number, we *solve* the equation by changing it to an *equivalent* equation with only x on the left. The process goes like this:

Solve: $3x - 4 = 11$

Step 1 Add 4 to both sides of the equation

$$3x \underbrace{- 4 + 4}_{0} = \underbrace{11 + 4}_{15}$$

$$3x = 15$$

Step 2 Divide both sides of the equation by 3.

$$\frac{3x}{3} = \frac{15}{3}$$

$x = 5$ This is the solution and you can check it by substituting 5 for x in the original equation.

Check $3(5) - 4 = 11$
$15 - 4 = 11$ which is true

Solve each of the following equations and check your answer.

(a) $\dfrac{3x}{2} = 9$ (b) $17 - x = 12$

(c) $2x + 7 = 3$ (d) $3(2x + 5) = 4x + 17$

The correct answers are given in **2**.

---

2 (a) 6 (b) 5 (c) −2 (d) 1

A *system of equations* is a set of equations in two or more variables that may have a common solution. For example, the pair of equations

$2x + y = 11$
$4y - x = 8$

has the common solution $x = 4$, $y = 3$. This pair of numbers will make each equation a true statement. If we substitute 4 for x and 3 for y, the first equation becomes

$2(4) + 3 = 8 + 3$ or 11

and the second equation becomes

$4(3) - 4 = 12 - 4$ or 8

The pair of values $x = 4$ and $y = 3$ satisfies *both* equations. This solution is often written $(4, 3)$, where the x-value is listed first and the y-value is listed second.

Note ▶ Each of these equations by itself has an infinite number of solutions. For example, the pairs $(0, 11)$, $(1, 9)$, $(2, 7)$, and so on, all satisfy $2x + y = 11$, and the pairs $(0, 2)$, $\left(1, \frac{9}{4}\right)$, $\left(2, \frac{5}{2}\right)$, and so on, all satisfy $4y - x = 8$. However, $(4, 3)$ is the only pair that satisfies both equations. ◀

By substituting, show that the numbers $x = 2$, $y = -5$ give the solution to the pair of equations

$5x - y = 15$
$x + 2y = -8$

Check your work in **3**.

---

3 The first equation is

$5(2) - (-5) = 15$
$10 + 5 = 15$, which is correct.

The second equation is

$(2) + 2(-5) = -8$
$2 - 10 = -8$, which is also correct.

Solution by Substitution In this chapter we will show you two different methods of solving a system of two linear equations in two unknowns. The first method is called the method of *substitution*. For example, to solve the pair of equations

$$y = 3 - x$$
$$3x + y = 11$$

follow these steps.

Step 1 *Solve* the first equation for x or y.

The first equation is already solved for y:

$$y = 3 - x$$

Step 2 *Substitute* this expression for y in the second equation.

$3x + y = 11$ becomes

$3x + \boxed{(3 - x)} = 11$

Substitute this for y

Step 3 *Solve* the resulting equation.

$3x + (3 - x) = 11$ becomes
$2x + 3 = 11$ Subtract 3 from each side.
$2x = 8$ Divide each side by 2.
$x = 4$

Step 4 *Substitute* this value of x into the first equation and find a value for y.

$y = 3 - x$
$y = 3 - (4)$
$y = -1$

x value

The solution is $x = 4$, $y = -1$ or $(4, -1)$.

y value

Step 5 *Check* your solution by substituting it back into the second equation.

$3x + y = 11$ becomes
$3(4) + (-1) = 11$
$12 - 1 = 11$ which is correct

Try it. Use this substitution procedure to solve the system of equations

$x - 2y = 3$
$2x - 3y = 7$

Check your work in **4**.

---

4 **Step 1** *Solve* the first equation for x by adding $2y$ to both sides of the equation.

$x - 2y + \boxed{2y} = 3 + \boxed{2y}$
$x = 3 + 2y$

Step 2 *Substitute* this expression for x in the second equation.

$2x - 3y = 7$ becomes

$2 \boxed{(3 + 2y)} - 3y = 7$

Step 3 *Solve:*

$6 + 4y - 3y = 7$ Simplify by combining the y-terms.
$6 + y = 7$ Subtract 6 from each side.
$y = 1$

Step 4 *Substitute* this value of y in the first equation to find x.

$$x - 2y = 3 \quad \text{becomes}$$
$$x - 2(1) = 3$$
$$x - 2 = 3 \quad \text{Add 2 to each side.}$$
$$x = 5$$

The solution is $x = 5$, $y = 1$ or $(5, 1)$.

Step 5 *Check* the solution in the second equation.

$$2x - 3y = 7 \quad \text{becomes}$$
$$2(5) - 3(1) = 7$$
$$10 - 3 = 7 \quad \text{which is correct.}$$

Of course, it does not matter which variable, x or y, we solve for in step 1, or which equation we use in step 4. For example, in the pair of equations

$$2x + 3y = 22$$

$$x - y = 1$$

the simplest procedure is to solve the *second* equation for x to get

$x = 1 + y$ and substitute this expression for x into the first equation.

Solve this set of equations.

Check your work in **5**.

---

5 $2x + 3y = 22$
$x - y = 1$

Step 1 From the second equation, $x = 1 + y$.

Step 2 When we substitute into the first equation,

$$2x + 3y = 22 \quad \text{becomes}$$

$$2 \boxed{(1 + y)} + 3y = 22$$

Step 3 *Solve:*

$$2 + 2y + 3y = 22 \quad \text{Combine terms.}$$
$$2 + 5y = 22 \quad \text{Subtract 2 from each side.}$$
$$5y = 20 \quad \text{Divide each side by 5.}$$
$$y = 4$$

Step 4 *Substitute* 4 for y in the second equation.

$$x - y = 1 \quad \text{becomes}$$
$$x - (4) = 1$$

or

$$x = 5 \quad \text{The solution is } x = 5, y = 4.$$

Step 5 *Check* the solution by substituting these values into the first equation.

$$2x + 3y = 22 \quad \text{becomes}$$
$$2(5) + 3(4) = 22$$
$$10 + 12 = 22 \quad \text{which is correct.}$$

Solve the following systems of equations by using the substitution method.

(a) $x = 1 + y$
    $2y + x = 7$

(b) $3x + y = 1$
    $y + 5x = 9$

(c) $x - 3y = 4$
    $3y + 2x = -1$

(d) $y + 2x = 1$
    $3y + 5x = 1$

(e) $x - y = 2$
    $y + x = 1$

(f) $y = 4x$
    $2y - 6x = 0$

Check your answers in **6**.

---

6 (a) $x = 3, y = 2$ (b) $x = 4, y = -11$ (c) $x = 1, y = -1$

(d) $x = 2, y = -3$ (e) $x = 1\frac{1}{2}, y = -\frac{1}{2}$ (f) $x = 0, y = 0$

Dependent and Inconsistent Systems

So far we have looked only at systems of equations with a single solution—one pair of numbers. Such a system of equations is called a *consistent* and *independent* system. However, it is possible for a system of equations to have no solution at all or to have very many solutions.

For example, the system of equations

$$y + 3x = 5$$
$$2y + 6x = 10$$

has *no unique* solution. If we solve for y in the first equation

$$y = 5 - 3x$$

and substitute this expression into the second equation

$$2y + 6x = 10$$

or $\quad 2\,(5 - 3x)\, + 6x = 10$

This resulting equation simplifies to

$$10 = 10$$

There is no way of solving to get a unique value of x or y.

A system of equations that does not have a single unique number-pair solution but has an unlimited number of solutions is said to be *dependent*. The two equations are essentially the same. For the system shown in this example, the second equation is simply twice the first equation. There are infinitely many pairs of numbers that will satisfy this pair of equations. For example,

$$x = 0, y = 5$$
$$x = 1, y = 2$$
$$x = 2, y = -1$$
$$x = 3, y = -4$$

and so on.

If a system of equations is such that our efforts to solve it produce a false statement, the equations are said to be *inconsistent*. The system of equations has no solution. For example, the pair of equations

$$y - 1 = 2x$$
$$2y - 4x = 7 \quad \text{is inconsistent.}$$

If we solve the first equation for y

$$y = 2x + 1$$

and substitute this expression into the second equation,

$$2y - 4x = 7 \quad \text{becomes}$$

$$2\,(2x + 1)\, - 4x = 7$$

$$4x + 2 - 4x = 7$$

or $\quad\quad\quad 2 = 7 \quad \text{which is false.}$

All the variables have dropped out of the equation, and we are left with an incorrect statement. The original pair of equations is said to be inconsistent, and it has no solution.

Try solving the following systems of equations.

(a) $2x - y = 5$ 　　　　(b) $3x - y = 5$
　　 $2y - 4x = 3$ 　　　　　　 $6x - 10 = 2y$

Check your work in **7**.

---

7 (a) Solve the first equation for y.

$y = 2x - 5$

Substitute this into the second equation.

$2y - 4x = 3$ 　　becomes

$2\,(2x - 5) - 4x = 3$

$4x - 10 - 4x = 3$

$-10 = 3$ 　　This is impossible. There is no solution for this system of equations. The equations are inconsistent.

(b) Solve the first equation for y.

$y = 3x - 5$

Substitute this into the second equation.

$6x - 10 = 2y$ 　　becomes

$6x - 10 = 2\,(3x - 5)$

$6x - 10 = 6x - 10$

$0 = 0$ 　　This is true, but all of the variables have dropped out and we cannot get a single unique solution. The equations are dependent.

Solution by Elimination The second method for solving a system of equations is called the *method of elimination*. When it is difficult or "messy" to solve one of the two equations for either x or y, the method of elimination may be the simplest way to solve the system of equations. For example, in the system of equations

$2x + 3y = 7$
$4x - 3y = 5$

neither equation can be solved for x or y without introducing fractions that are difficult to work with. But we can simply add these equations, and the y-terms will be eliminated.

$2x + 3y = 7$
$\underline{4x - 3y = 5}$ 　　Add like terms.
$6x + 0\ \ = 12$
$6x = 12$
$x = 2$

Now substitute this value of x back into either one of the original equations to obtain a value for y.

The first equation

$2x + 3y = 7$ 　　becomes
$2(2) + 3y = 7$

$$4 + 3y = 7 \qquad \text{Subtract 4 from each side.}$$
$$3y = 3 \qquad \text{Divide each side by 3.}$$
$$y = 1$$

The solution is $x = 2$, $y = 1$, or $(2, 1)$.

Check the solution by substituting it back into the second equation.

$$4x - 3y = 5$$
$$4(2) - 3(1) = 5$$
$$8 - 3 = 5 \qquad \text{which is correct.}$$

Try it. Solve the following system of equations by adding.

$$2x - y = 3$$
$$y + x = 9$$

Check your work in **8**.

---

8 $2x - y = 3$
 $y + x = 9$

becomes

$2x - y = 3$
$x + y = 9$ $\qquad$ by rearranging the terms in the
$\qquad\qquad\qquad\quad$ second equation

Add to get

$$3x + 0 = 12$$
$$3x = 12$$
$$x = 4$$

Substitute 4 for x in the first equation.

$2x - y = 3$
$2(4) - y = 3$ $\qquad$ Simplify
$8 - y = 3$ $\qquad$ Subtract 8 from each side.
$-y = 3 - 8$ $\qquad$ Combine terms.
$-y = -5$ $\qquad$ Multiply both sides by -1.
$y = 5$

The solution is $x = 4$, $y = 5$, or $(4, 5)$. Check it.

Careful ▶ It is important to rearrange the terms in the equations so that the x and y terms appear in the same order in both equations.

$$2\ x\ -\ y\ =\ 3$$
$$x\ +\ y\ =\ 9$$

constant terms

y column

x column ◀

Solve the following systems of equations by this process of elimination.

(a) $\quad x + 5y = 17$ $\qquad$ (b) $\quad x + y = 16$ $\qquad$ (c) $\quad 3x - y = -5$ $\qquad$ (d) $\quad \frac{1}{2}x + 2y = 10$
$ -x + 3y = 7$ $\qquad x - y = 4$ $\qquad y - 5x = 9$ $\qquad y + 1 = \frac{1}{2}x$

Our step-by-step solutions are given in **9**.

---

9 (a) **Solve:**

$$x + 5y = 17$$
$$\underline{-x + 3y = 7}$$
$$0 + 8y = 24 \qquad \text{adding like terms}$$
$$8y = 24$$
$$y = 3$$

Substitute 3 for y in the first equation.

$$x + 5(3) = 17$$
$$x + \quad 15 = 17$$
$$x = 2 \qquad \text{This solution is } x = 2, y = 3, \text{ or } (2, 3).$$

Check:
$$-x + 3y = 7$$
$$-(2) + 3(3) = 7$$
$$-2 + 9 = 7 \qquad \text{which is correct.}$$

(b) **Solve:**

$$x + y = 16$$
$$\underline{x - y = 4}$$
$$2x + 0 = 20 \qquad \text{adding like terms}$$
$$2x = 20$$
$$x = 10$$

Substitute 10 for x in the first equation.

$$(10) + y = 16$$
$$y = 6 \qquad \text{The solution is } x = 10, y = 6, \text{ or } (10, 6).$$

Check:
$$x - y = 4$$
$$(10) - (6) = 4$$
$$10 - 6 = 4 \qquad \text{which is correct.}$$

(c) **Solve:**

$$3x - y = -5$$
$$y - 5x = 9$$

Rearrange the order of the terms in the second equation.

$$3x - y = -5$$
$$\underline{-5x + y = \quad 9}$$
$$-2x + 0 = \quad 4 \qquad \text{adding like terms}$$
$$-2x = \quad 4$$
$$x = -2$$

Substitute -2 for x in the first equation.

$$3(-2) - y = -5$$
$$-6 - y = -5$$
$$-y = -5 + 6 = 1$$
$$y = -1 \qquad \text{The solution is } x = -2, y = -1, \text{ or } (-2, -1).$$

Check:
$$y - 5x = 9$$
$$(-1) - 5(-2) = 9$$
$$-1 + 10 = 9 \qquad \text{which is correct.}$$

(d) **Solve:**

$$\tfrac{1}{2}x + 2y = 10$$
$$y + 1 = \tfrac{1}{2}x$$

Rearrange terms in the second equation so that they are in the same order as in the first equation. Subtract 1 from both sides, then subtract $\tfrac{1}{2}x$ from both sides.

$$\tfrac{1}{2}x + 2y = 10$$
$$\underline{-\tfrac{1}{2}x + \ y = -1}$$
$$0 + 3y = \ 9 \qquad \text{adding like terms}$$
$$3y = \ 9$$
$$y = \ 3$$

Substitute 3 for y in the first equation.

$$\tfrac{1}{2}x + 2(3) = 10$$
$$\tfrac{1}{2}x + 6 = 10$$
$$\tfrac{1}{2}x = 4$$
$$x = 8 \qquad \text{The solution is } x = 8, y = 3, \text{ or } (8, 3).$$

Check: $\quad y + 1 = \tfrac{1}{2}x$

$$(3) + 1 = \tfrac{1}{2}(8)$$
$$3 + 1 = 4 \qquad \text{which is correct.}$$

Multiplication with the Elimination Method

With some systems of equations neither x nor y can be eliminated by simply adding like terms. For example, in the system

$$3x + y = 17$$
$$x + y = \ 7$$

adding like terms will not eliminate either variable. To solve this system of equations, multiply all terms of the second equation by -1 so that

$$\boxed{x + y = 7} \quad \text{becomes} \Longrightarrow \quad \boxed{-x - y = -7}$$

and the system of equations becomes

$$3x + y = 17$$
$$-x - y = -7$$

The system of equations may now be solved by adding the like terms as before.

$$3x + y = 17$$
$$\underline{-x - y = -7}$$
$$2x + 0 = 10 \qquad \text{adding like terms}$$
$$2x = 10$$
$$x = \ 5$$

Substitute 5 for x in the first equation.

$$3(5) + y = 17$$
$$15 + y = 17$$
$$y = 17 - 15 = 2 \qquad \text{The solution is } x = 5, y = 2, \text{ or } (5, 2).$$

Check: $x + y = 7$
$(5) + (2) = 7$
$5 + 2 = 7$ which is correct.

Careful ▶ When you multiply an equation by some number, be careful to multiply *all* terms on *both* sides by that number. It is very easy to forget to multiply on the right side. ◀

Use this "multiply and add" procedure to solve the following system of equations.

$2x + 7y = 29$
$2x + y = 11$

Check your solution in **10**.

---

10 $2x + 7y = 29$
$2x + y = 11$

Multiply all terms in the second equation by –1.

$2x + 7y = 29$
$\underline{-2x - y = -11}$
$0 + 6y = 18$ adding like terms
$6y = 18$
$y = 3$

Substitute 3 for y in the first equation.

$2x + 7(3) = 29$
$2x + 21 = 29$
$2x = 8$
$x = 4$ The solution is $x = 4$, $y = 3$, or $(4, 3)$.

Check the solution by substituting in the original equations.

$2x + y = 11$ $2x + 7y = 29$
$2(4) + (3) = 11$ $2(4) + 7(3) = 29$
$8 + 3 = 11$ $8 + 21 = 29$

Solving by the multiply and add procedure may involve multiplying by constants other than –1, of course. For example, use this method to solve the following system of equations.

$7x + 4y = 25$
$3x - 2y = 7$

Look for our solution in **11**.

---

11 $7x + 4y = 25$
$3x - 2y = 7$

First, look at these equations carefully. Notice that the y-terms can be eliminated easily if we multiply all terms in the second equation by 2.

The y column $\boxed{\begin{array}{c} +4y \\ -2y \end{array}}$ becomes ⟹ $\boxed{\begin{array}{c} +4y \\ -4y \end{array}}$ when we multiply by 2.

sum = 0

The second equation becomes $2(3x) - 2(2y) = 2(7)$

or $6x - 4y = 14$

and the system of equations is converted to the equivalent system

$$7x + 4y = 25$$
$$6x - 4y = 14$$

$$13x + 0 = 39 \quad \text{adding like terms}$$
$$13x = 39$$
$$x = 3$$

Substitute 3 for x in the first equation.

$$7(3) + 4y = 25$$
$$21 + 4y = 25$$
$$4y = 4$$
$$y = 1 \quad \text{The solution is } x = 3, y = 1, \text{ or } (3, 1).$$

Check the solution by substituting it back into the original equations.

Try these problems to make certain you understand this procedure.

(a) $3x - 2y = 14$ (b) $6x + 5y = 14$
         $5x - 2y = 22$ $-2x + 3y = -14$

(c) $5y - x = 1$ (d) $-x - 2y = 1$
         $2x + 3y = 11$ $19 - 2x = -3y$

Our solutions are given in **12**.

---

12   (a)   **Solve:**

$$3x - 2y = 14$$
$$5x - 2y = 22$$

Multiply each term in the first equation by -1.

$$(-1)\,(3x) - (-1)\,(2y) = (-1)\,(14)$$

$$-3x + 2y = -14$$

The system of equations is therefore

$$-3x + 2y = -14$$
$$5x - 2y = 22$$

$$2x + 0 = 8 \quad \text{adding like terms}$$
$$x = 4$$

Substitute 4 for x in the first equation.

$$3(4) - 2y = 14$$
$$12 - 2y = 14$$
$$-2y = 2$$
$$-y = 1$$
$$y = -1 \quad \text{The solution is } x = 4, y = -1, \text{ or } (4, -1).$$

Check: $3x - 2y = 14$ $5x - 2y = 22$
             $3(4) - 2(-1) = 14$ $5(4) - 2(-1) = 22$
                 $12 + 2 = 14$ $20 + 2 = 22$
                     $14 = 14$ $22 = 22$

(b)   **Solve:**

$$6x + 5y = 14$$
$$-2x + 3y = -14$$

Multiply each term in the second equation by 3.

$$(3)(-2x) + (3)(3y) = (3)(-14)$$

$$-6x + 9y = -42$$

The system of equations is now

$$\begin{array}{r} 6x + 5y = 14 \\ -6x + 9y = -42 \\ \hline \end{array}$$

$$0 + 14y = -28 \qquad \text{adding like terms}$$
$$y = -2$$

Substitute −2 for y in the first equation.

$$6x + 5(-2) = 14$$
$$6x - 10 = 14$$
$$6x = 24$$
$$x = 4 \qquad \text{The solution is } x = 4, y = -2, \text{ or } (4, -2).$$

Be certain to check your solution.

(c) **Solve:**

$$5y - x = 1$$
$$2x + 3y = 11$$

Rearrange to put the terms in the first equation in the same order as they are in the second equation.

$$-x + 5y = 1$$
$$2x + 3y = 11$$

Multiply each term in the first equation by 2.

$$\begin{array}{r} -2x + 10y = 2 \\ 2x + 3y = 11 \\ \hline \end{array}$$

$$0 + 13y = 13 \qquad \text{adding like terms}$$
$$13y = 13$$
$$y = 1$$

Substitute 1 for y in the first equation.

$$5(1) - x = 1$$
$$5 - x = 1$$
$$-x = 1 - 5 = -4$$
$$x = 4 \qquad \text{The solution is } x = 4, y = 1, \text{ or } (4, 1).$$

Check it.

(d) **Solve:**

$$-x - 2y = 1$$
$$19 - 2x = -3y$$

Rearrange the terms in the second equation in the same order as they are in the first equation.

$$-x - 2y = 1$$
$$-2x + 3y = -19$$

Multiply each term in the first equation by −2.

$$(-2)(-x) - (-2)(2y) = (-2)(1)$$

$$2x + 4y = -2$$

The system of equations is now

$$2x + 4y = -2$$
$$-2x + 3y = -19$$

$$0 + 7y = -21 \qquad \text{adding like terms}$$
$$7y = -21$$
$$y = -3$$

Substitute -3 for y in the first equation of the original problem.

$$-x - 2(-3) = 1$$
$$-x + 6 = 1$$
$$-x = 1 - 6 = -5$$
$$x = 5 \qquad \text{The solution is } x = 5, y = -3, \text{ or } (5, -3).$$

Check your solution.

Multiplying Both Equations

If you examine the system of equations

$$3x + 2y = 7$$
$$4x - 3y = -2$$

you will find that there is no single integer we can use as a multiplier that will allow us to eliminate one of the variables when the equations are added. Instead we must convert each equation separately to an equivalent equation, so that when the new equations are added one of the variables is eliminated. For example, with the system of equations given, if we wish to eliminate the y variable, we must multiply the first equation by 3 and the second equation by 2.

First equation: $\boxed{3x + 2y = 7}$ $\boxed{\text{Multiply by 3}}$ $\Rightarrow$ $\boxed{9x + 6y = 21}$

Second equation: $\boxed{4x - 3y = -2}$ $\boxed{\text{Multiply by 2}}$ $\Rightarrow$ $\boxed{8x - 6y = -4}$

The new system of equations is

$$9x + 6y = 21$$
$$8x - 6y = -4$$

$$17x + 0 = 17 \qquad \text{adding like terms}$$
$$x = 1$$

Substitute 1 for x in the original first equation.

$$3x + 2y = 7$$
$$3(1) + 2y = 7$$
$$3 + 2y = 7$$
$$2y = 4$$
$$y = 2 \qquad \text{The solution is } x = 1, y = 2, \text{ or } (1, 2).$$

Check this solution by substituting it back into the original pair of equations.

Check:
$$3x + 2y = 7 \qquad\qquad 4x - 3y = -2$$
$$3(1) + 2(2) = 7 \qquad\qquad 4(1) - 3(2) = -2$$
$$3 + 4 = 7 \qquad\qquad 4 - 6 = -2$$
$$7 = 7 \qquad\qquad -2 = -2$$

Use this same procedure to solve the following system of equations:

$$2x - 5y = 9$$
$$3x + 4y = 2$$

Check your work in **13**.

13 We can eliminate x from the two equations as follows:

First equation: $\boxed{2x - 5y = 9}$ $\boxed{\text{Multiply by } -3}$ ⇨ $\boxed{-6x + 15y = -27}$

Second equation: $\boxed{3x + 4y = 2}$ $\boxed{\text{Multiply by } 2}$ ⇨ $\boxed{6x + 8y = 4}$

The new system of equations is:

$$-6x + 15y = -27$$
$$\underline{6x + 8y = \quad 4}$$
$$0 + 23y = -23 \qquad \text{adding like terms}$$
$$y = -1$$

Substitute -1 for y in the first original equation.

$$2x - 5y = 9$$
$$2x - 5(-1) = 9$$
$$2x + 5 = 9$$
$$2x = 4$$
$$x = 2 \qquad \text{The solution is } x = 2, y = -1, \text{ or } (2, -1).$$

Note ▶ Of course, we could have chosen to eliminate the y-variable and we would have arrived at the same solution. Try it. ◀

When you are ready to continue, practice your new skills by solving the following systems of equations.

(a) $\quad 2x + 2y = \quad 4$
$\qquad 5x + 7y = 18$

(b) $\quad 5x + 2y = 11$
$\qquad 6x - 3y = 24$

(c) $\quad 3x + 2y = 10$
$\qquad 2x = 5y - 25$

(d) $\quad -7x - 13 = 2y$
$\qquad 3y + 4x = 0$

When you have solved these systems of equations, check your answers in **14**.

---

14 (a) $\quad x = -2, y = 4$

(b) $\quad x = 3, y = -2$

(c) $\quad x = 0, y = 5$

(d) $\quad x = -3, y = 4$

In (a) multiply the first equation by -5 and the second equation by 2.

In (b) multiply the first equation by 3 and the second equation by 2.

In (c) rearrange the terms of the second equation, then multiply the first equation by -2 and the second equation by 3.

In (d) rearrange the terms of the first equation to agree with the second equation, then multiply the first equation by 3 and the second equation by 2.

Word Problems In many practical situations not only must you be able to solve a system of equations you must also be able to write the equations in the first place. You must be able to set up and solve word problems. In Chapter 7 we listed some "signal words" and showed how to translate English sentences and phrases to mathematical equations and expressions. If you need to review the material on word problems in Chapter 7, turn to page 340 now; otherwise, continue here.

Translate the following sentence into *two* equations.

The sum of two numbers is 26 and their difference is 2. (Let x and y represent the two numbers.)

Check your work in **15**.

---

15 The sum of two numbers is 26

$$x + y = 26$$

. . . their difference is 2.

$$x - y = 2$$

The two equations are

$x + y = 26$
$x - y = 2$

Solve this pair of equations using the elimination method. Check your answer in **16**.

---

16 The solution is $x = 14$, $y = 12$. Check the solution by seeing if it fits the original problem. The sum of these numbers is 26 ($14 + 12 = 26$) and their difference is 2 ($14 - 12 = 2$).

Ready for another word problem? Translate and solve this one:

> The difference of two numbers is 14, and the larger number is three more than twice the smaller number.

Our step-by-step solution is given in **17**.

---

17 The first phrase in the sentence should be translated as

"The difference of two numbers is 14 . . ."   The larger number is L; the
$$L - S = 14$$ smaller number is S; the
difference must be $L - S$.

and the second phrase should be translated as

". . . the larger number is three more than twice the smaller . . ."

$$L = 3 + 2S$$

The system of equations is

$L - S = 14$
$L = 3 + 2S$

To solve this system of equations, substitute the value of L from the second equation into the first equation. Then the first equation becomes

$(3 + 2S) - S = 14$

or $3 + S = 14$
                 $S = 11$

Now substitute this value of S into the first equation to find L.

$L - (11) = 14$
          $L = 25$ The solution is $L = 25$, $S = 11$.

Check the solution by substituting it back into both of the original equations. Never neglect to check your answer.

Translating word problems into systems of equations is a very valuable and very practical algebra skill. Translate each of the following problems into a system of equations, then solve.

(a) The total value of an order of nuts and bolts is $1.40. The nuts cost 5 cents each and the bolts cost 10 cents each. If the number of bolts is four more than twice the number of nuts, how many of each are there? (*Hint:* Keep all money values in cents to avoid decimals.)

(b) The perimeter of a rectangular lot is 350 ft. The length of the lot is 10 ft more than twice the width. Find the dimensions of the lot.

(c) A materials yard wishes to make a 900-cu ft mixture of two different types of rock. One type of rock costs $3.40 per cubic foot and the other costs $4.60 per cubic foot. If the cost of the mixture is to run $3330, how many cubic feet of each should go into the mixture?

(d) A lab technician wishes to mix a 5% salt solution and a 15% salt solution to obtain 4 liters of a 12% salt solution. How many liters of each solution must be added?

Check your work in **18**.

---

18 (a) In problems of this kind it is often very helpful to first set up a table:

Item	Number of Items	Cost per Item	Total Cost
Nuts	N	5	$5N$
Bolts	B	10	$10B$

We can write the first equation as

"The total value of an order of nuts and bolts is 140 cents."

The second equation would be

". . . the number of bolts is four more than twice the number of nuts . . ."

$$B = 4 + 2N$$

The system of equations to be solved is

$$5N + 10B = 140$$
$$B = 4 + 2N$$

Use substitution. Since B is equal to $4 + 2N$, replace B in the first equation with $4 + 2N$.

$$5N + 10\,(4 + 2N) = 140$$
$$5N + 40 + 20N = 140$$
$$25N + 40 = 140 \qquad \text{collecting like terms}$$
$$25N = 100$$
$$N = 4$$

When we replace N with 4 in the second equation,

$B = 4 + 2(4)$
$B = 4 + 8$
$B = 12$

There are 4 nuts and 12 bolts.

Check:

$$5N + 10B = 140 \qquad B = 4 + 2N$$
$$5(4) + 10(12) = 140 \qquad 12 = 4 + 2(4)$$
$$20 + 120 = 140 \qquad 12 = 4 + 8$$
$$140 = 140 \qquad 12 = 12$$

(b) Recalling the formula for the perimeter of a rectangle, we have

"The perimeter of a rectangular lot is 350 ft."

$$2L + 2W \qquad = 350$$

The second sentence gives us

"The length of the lot is 10 ft more than twice the width."

$$L \qquad = 10 \qquad + \qquad 2W$$

The system of equations to be solved is

$2L + 2W = 350$
$L = 10 + 2W$

Using substitution, we replace L with $10 + 2W$ in the first equation.

$$2\,(10 + 2W) + 2W = 350$$
$$20 + 4W + 2W = 350$$
$$20 + 6W = 350$$
$$6W = 330$$
$$W = 55 \text{ ft}$$

Now substitute 55 for W in the second equation:

$L = 10 + 2(55)$
$L = 10 + 110$
$L = 120 \text{ ft}$

The lot is 120 ft long and 55 ft wide.

Check:

$$2L + 2W = 350 \qquad L = 10 + 2W$$
$$2(120) + 2(55) = 350 \qquad 120 = 10 + 2(55)$$
$$240 + 110 = 350 \qquad 120 = 10 + 110$$
$$350 = 350 \qquad 120 = 120$$

(c) First set up the following table:

Item	Amount (cu ft)	Cost per cu ft	Total Cost
Cheaper rock	x	\$3.40	\$3.40x
More expensive rock	y	\$4.60	\$4.60y

The first equation comes from the statement:

". . . a 900 cubic foot mixture of two different types of rock."

$$900 = x + y$$

Consulting the table, we write the second equation as follows:

". . . the cost of the mixture is to run $3330."

$$3.40x + 4.60y = 3330$$

Multiply this last equation by 10 to get the system of equations

$$x + y = 900$$
$$34x + 46y = 33{,}300$$

Multiply the first equation by −34 and add it to the second equation:

$$
\begin{aligned}
-34x + (-34y) &= -30{,}600 \\
\underline{34x + 46y} &= \underline{33{,}300} \\
12y &= 2{,}700 \\
y &= 225 \text{ cu ft}
\end{aligned}
$$

Replacing y with 225 in the first equation, we have:

$$
\begin{aligned}
x + (225) &= 900 \\
x &= 675 \text{ cu ft}
\end{aligned}
$$

There should be 225 cu ft of the $4.60 per cu ft mixture, and 675 cu ft of the $3.40 per cu ft mixture.

Check:

$$
\begin{aligned}
x + y &= 900 & 3.40x + 4.60y &= 3330 \\
675 + 225 &= 900 & 3.40(675) + 4.60(225) &= 3330 \\
900 &= 900 & 2295 + 1035 &= 3330 \\
& & 3330 &= 3330
\end{aligned}
$$

(d)

Solution	Amount (liters)	Salt Fraction	Total Salt
5%	A	0.05	0.05A
15%	B	0.15	0.15B
12%	4	0.12	0.12(4)

Since the final solution is to contain 4 liters, we have

$$A + B = 4$$

The second equation represents the total amount of salt:

$$0.05A + 0.15B = (0.12)4$$

Multiplying this last equation by 100 to eliminate the decimals, we have the system

$$A + B = 4$$
$$5A + 15B = 48$$

To solve this system, multiply each term in the first equation by −5 and add to get

$$-5A + (-5B) = -20$$
$$\underline{5A + 15B = 48}$$
$$10B = 28$$
$$B = 2.8 \text{ liters}$$

Substituting back into the first equation, we have

$$A + 2.8 = 4$$
$$A = 1.2 \text{ liters}$$

Check:
$$A + B = 4 \qquad 5A + 15B = 48$$
$$1.2 + 2.8 = 4 \qquad 5(1.2) + 15(2.8) = 48$$
$$4 = 4 \qquad 6 + 42 = 48$$
$$48 = 48$$

Now turn to **19** for a set of exercises on systems of equations.

19

Exercises 11-1 **Systems of Equations**

A. Solve each of the following systems of equations using the method of substitution. If the system is inconsistent or dependent, say so.

1. $y = 10 - x$
   $2x - y = -4$

2. $3x - y = 5$
   $2x + y = 15$

3. $2x - y = 3$
   $x - 2y = -6$

4. $2y - 4x = -3$
   $y = 2x + 4$

5. $3x + 5y = 26$
   $x + 2y = 10$

6. $x = 10y + 1$
   $y = 10x + 1$

7. $3x + 4y = 5$
   $x - 2y = -5$

8. $2x - y = 4$
   $4x - 3y = 11$

B. Solve each of the following systems of equations. If the system is inconsistent or dependent, say so.

1. $x + y = 5$
   $x - y = 13$

2. $2x + 2y = 10$
   $3x - 2y = 10$

3. $2y = 3x + 5$
   $2y = 3x - 7$

4. $2y = 2x + 2$
   $4y = 5 + 4x$

5. $x = 3y + 7$
   $x + y = -5$

6. $3x - 2y = -11$
   $x + y = -2$

7. $5x - 4y = 1$
   $3x - 6y = 6$

8. $y = 3x - 5$
   $6x - 3y = 3$

9. $y - 2x = -8$
   $x - \frac{1}{2}y = 4$

10. $x + y = a$
    $x - y = b$

11. $3x + 2y = 13$
    $5x + 3y = 20$

12. $3x + 4y = 11$
    $2x + 2y = 7$

C. Practical Problems

Translate each problem statement into a system of equations and solve.

1. The sum of two numbers is 39 and their difference is 7. What are the numbers?

2. The sum of two numbers is 14. The larger is two more than three times the smaller. What are the numbers?

3. Separate a collection of 20 objects into two parts so that twice the larger amount equals three times the smaller amount.

4. The average of two numbers is 25 and their difference is 8. What are the numbers?

5. Four bleebs and three freems cost $11. Three bleebs and four freems cost $10. What does a bleeb cost?

6. **Carpentry**   The perimeter of a rectangular window is 14 ft, and its length is 2 ft less than twice the width. What are the dimensions of the window?

7. Harold exchanged a $1 bill for change and received his change in nickels and dimes, with seven more dimes than nickels. How many of each coin did he receive?

8. If four times the larger of two numbers is added to three times the smaller, the result is 26. If three times the larger number is decreased by twice the smaller, the result is 11. Find the numbers.

9. Mr. Brown bought five cans of peas and four cans of corn, but he forgot what each cost. He knows that the total cost was $5.60 and he recalls that a can of peas cost 5 cents less than a can of corn. How much did each can cost?

10. **Sheet Metal Technology**   The length of a piece of sheet metal is twice the width. The difference in length and width is 20 in. What are the dimensions?

11. **Electrical Technology**   A 30-in. piece of wire is to be cut into two parts, one part being four times the length of the other. Find the length of each.

12. **Painting and Decorating**   A painter wishes to mix paint worth $15 per gallon with paint worth $18 per gallon to make a 12-gallon mixture worth $16.75 per gallon. How many gallons of each should he mix?

13. A chemist wishes to mix a 10% salt solution with a 2% salt solution to obtain 6 liters of a 4% salt solution. How many liters of each should be added?

14. The perimeter of a rectangular field is 520 ft. The length of the field is 20 ft more than three times the width. Find the dimensions.

15. **Landscaping**   A 24-ton mixture of crushed rock is needed in a construction job; it will cost $800. If the mixture is composed of rock costing $30 per ton and $40 per ton, how many tons of each should be added?

When you have completed these problems check your answers on page 607. Then turn to **20** to learn about quadratic equations.

11-2 QUADRATIC EQUATIONS

20 Thus far in your study of algebra you have worked only with linear equations. In a linear equation the variable appears only to the first power. For example, $3x + 5 = 2$ is a linear equation. The variable appears as x or x^1. No powers of x such as x^2, x^3, or x^4 appear in the equations.

An equation in which the variable appears to the second power, but to no higher power, is called a *quadratic equation*.

Which of the following are quadratic equations?

(a) $x^2 = 49$

(b) $2x - 1 = 4$

(c) $3x - 2y = 19$

(d) $5x^2 - 8x + 3 = 0$

(e) $x^3 + 3x^2 + 3x + 1 = 0$

Check your answers in **21**.

---

21 Equations (a) and (d) are quadratic equations. Equation (b) is a linear equation in one variable. Equation (c) is a linear equation in two variables, x and y. Equation (e) is a cubic or third-order equation. Because (e) contains an x^3 term, it is not a quadratic.

Standard Form   Every quadratic equation can be put into a *standard quadratic form*.

$$ax^2 + bx + c = 0 \qquad \text{where } a \text{ cannot equal zero}$$

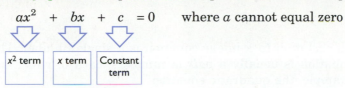

| x^2 term | x term | Constant term |

Every quadratic equation must have an x^2 term, but the x term and the constant term may be missing. For example,

$2x^2 + x - 5 = 0$ is a quadratic equation in standard form: the x^2 term is first, the x term second, and the constant term last on the left.

$x^2 + 4 = 0$ is also a quadratic equation in standard form. The x term is missing, but the other terms are in the proper order. We could rewrite this equation as

$x^2 + 0 \cdot x + 4 = 0$

Which of the following quadratic equations are written in standard form?

(a) $7x^2 - 3x + 6 = 0$ (b) $8x - 3x^2 - 2 = 0$

(c) $2x^2 - 5x = 0$ (d) $x^2 = 25$

(e) $x^2 - 5 = 0$ (f) $4x^2 - 5x = 6$

Check your answer in **22**.

---

22 Equations (a), (c), and (e) are in standard form.

To solve a quadratic equation, the first step is usually to rewrite it in standard form. For example, the equation

$x^2 = 25x$	becomes	$x^2 - 25x = 0$	in standard form
$8x - 3x^2 - 2 = 0$	becomes	$-3x^2 + 8x - 2 = 0$	in standard form
	or	$3x^2 - 8x + 2 = 0$	if we multiply all terms by -1
$4x^2 - 5x = 6$	becomes	$4x^2 - 5x - 6 = 0$	in standard form

In each case we add or subtract a term on both sides of the equation until all terms are on the left, then rearrange terms until the x^2 is first on the left, the x term next, and the constant term third.

Try it. Rearrange the following quadratic equations in standard form.

(a) $5x - 19 + 3x^2 = 0$ (b) $7x^2 = 12 - 6x$

(c) $9 = 3x - x^2$ (d) $2x - x^2 = 0$

(e) $5x = 7x^2 - 12$ (f) $x^2 - 6x + 9 = 49$

(g) $3x + 1 = x^2 - 5$ (h) $1 - x^2 + x = 3x + 4$

Check your answers in **23**.

---

23 (a) $3x^2 + 5x - 19 = 0$ (b) $7x^2 + 6x - 12 = 0$

(c) $x^2 - 3x + 9 = 0$ (d) $x^2 - 2x = 0$

(e) $7x^2 - 5x - 12 = 0$ (f) $x^2 - 6x - 40 = 0$

(g) $x^2 - 3x - 6 = 0$ (h) $x^2 + 2x + 3 = 0$

Solutions to Quadratic Equations The solution to a linear equation is a single number. The solution to a quadratic equation is usually a *pair* of numbers, each of which satisfies the equation. For example, the quadratic equation

$$x^2 - 5x + 6 = 0$$

has the solutions

$$x = 2 \quad \text{or} \quad x = 3$$

To see that either 2 or 3 is a solution, substitute each into the equation.

For $x = 2$

$$(2)^2 - 5(2) + 6 = 0$$
$$4 - 10 + 6 = 0$$
$$10 - 10 = 0$$

For $x = 3$

$$(3)^2 - 5(3) + 6 = 0$$
$$9 - 15 + 6 = 0$$
$$15 - 15 = 0$$

Show that $x = 5$ or $x = 3$ gives a solution of the quadratic equation

$$x^2 - 8x + 15 = 0$$

Check your work in **24**.

---

24 $x^2 - 8x + 15 = 0$

For $x = 5$

$$(5)^2 - 8(5) + 15 = 0$$
$$25 - 40 + 15 = 0$$
$$40 - 40 = 0$$

For $x = 3$

$$(3)^2 - 8(3) + 15 = 0$$
$$9 - 24 + 15 = 0$$
$$24 - 24 = 0$$

Solving $x^2 - a = 0$ The easiest kind of quadratic equation to solve is one in which the linear term is missing. For example, to solve the quadratic equation

$$x^2 - 25 = 0$$

simply rewrite it as

$$x^2 = 25$$

and take the square root of both sides of the equation.

$$\sqrt{x^2} = \sqrt{25}$$

or $x = \pm \sqrt{25}$

or $x = 5 \quad$ or $\quad x = -5$

Both 5 and -5 satisfy the original equation.

For $x = 5$

$$x^2 - 25 = 0$$
$$(5)^2 - 25 = 0$$
$$25 - 25 = 0$$

and

For $x = -5$

$$x^2 - 25 = 0$$
$$(-5)^2 - 25 = 0$$
$$25 - 25 = 0$$

Every positive number has two square roots, one positive and the other negative. Both of them may be important in solving a quadratic equation.

Solve each of the following quadratic equations and check *both* solutions. (Round to two decimal places if necessary.)

(a) $x^2 - 36 = 0$ (b) $x^2 = 8$ (c) $x^2 - 64 = 0$

(d) $4x^2 = 81$ (e) $3x^2 = 27$ (f) $x^2 - 1 = 2$

Our solutions are in **25**.

---

25 (a) $x^2 = 36$
$x = \pm \sqrt{36}$
$x = 6$ or $x = -6$

(b) $x^2 = 8$
$x = \pm \sqrt{8}$
$x \approx 2.83$ or $x \approx -2.83$ rounded

(c) $x^2 = 64$
$x = \pm \sqrt{64}$
$x = 8$ or $x = -8$

(d) $4x^2 = 81$
$x^2 = \dfrac{81}{4}$
$x = \pm \sqrt{\dfrac{81}{4}}$
$x = \dfrac{9}{2}$ or $x = -\dfrac{9}{2}$

(e) $3x^2 = 27$
$x^2 = 9$
$x = \pm\sqrt{9}$
$x = 3$ or $x = -3$

(f) $x^2 = 3$
$x = \pm \sqrt{3}$
$x \approx 1.73$ or $x \approx -1.73$ rounded

Notice in each case that first we rewrite the equation so that x^2 appears alone on the left and a number appears alone on the right. Second, take the square root of both sides. The equation will have two solutions.

You should also notice that an equation such as

$x^2 = -4$

has no solution. There is no number x whose square is a negative number.

Solve the following quadratic equations. (Round to two decimal places if necessary.)

(a) $x^2 - 3.5 = 0$ (b) $x^2 = 18$ (c) $6 - x^2 = 0$

(d) $9x^2 = 49$ (e) $7x^2 = 80$ (f) $\dfrac{3x^2}{5} = 33.3$

Check your answers in **26**.

---

26 (a) $x^2 = 3.5$
$x = \pm \sqrt{3.5}$
$x \approx 1.87$ or $x \approx -1.87$

(b) $x^2 = 18$
$x = \pm \sqrt{18}$
$x \approx 4.24$ or $x \approx -4.24$

(c) $6 - x^2 = 0$
$x^2 = 6$
$x = \pm \sqrt{6}$
$x \approx 2.45$ or ≈ -2.45

(d) $9x^2 = 49$
$x^2 = \dfrac{49}{9}$
$x = \pm \sqrt{\dfrac{49}{9}}$
$x = \dfrac{7}{3}$ or $x = \dfrac{-7}{3}$

(e) $7x^2 = 80$

$$x^2 = \frac{80}{7}$$

$$x = \pm \sqrt{\frac{80}{7}}$$

$$x \approx \pm \sqrt{11.4286}$$
$$x \approx 3.38 \quad \text{or} \quad x \approx -3.38$$

(f) $\dfrac{3x^2}{5} = 33.3$

$$x^2 = \frac{(33.3)(5)}{3} = 55.5$$

$$x = \pm \sqrt{55.5}$$
$$x \approx 7.45 \quad \text{or} \quad x \approx -7.45$$

 Do problem (f) this way on a calculator:

33.3 ⊠ **5** ⊡ **3** ⊜ ☑ → *7.4498322*

Quadratic Formula In general, a quadratic equation will contain all three terms, an x^2 term, an x term, and a constant term. The solution of any quadratic equation

$$ax^2 + bx + c = 0$$

is

$$x = \frac{-b + \sqrt{b^2 - 4ac}}{2a} \quad \text{or} \quad x = \frac{-b - \sqrt{b^2 - 4ac}}{2a}$$

or

THE QUADRATIC FORMULA

$$x = \frac{-b \pm \sqrt{b^2 - 4ac}}{2a}$$

Careful ▶ Note that the entire expression in the numerator is divided by $2a$. ◀

For example, to solve the quadratic equation

$$2x^2 + 5x - 3 = 0$$

follow these steps.

Step 1 *Identify* the coefficients a, b, and c for the quadratic equation.

$$2x^2 \quad + \quad 5x \quad - \quad 3 \quad = 0$$

$a = 2$ $b = 5$ $c = -3$

Step 2 *Substitute* these values of a, b, and c into the quadratic formula.

$$x = \frac{-(5) \pm \sqrt{(5)^2 - 4(2)(-3)}}{2(2)}$$

The $\pm$ sign means that there are two solutions, one to be calculated using the $+$ sign and the other solution calculated using the $-$ sign.

Step 3 *Simplify* these equations for x.

$$x = \frac{-5 \pm \sqrt{25 + 24}}{4}$$

$$x = \frac{-5 \pm \sqrt{49}}{4}$$

$$x = \frac{-5 \pm 7}{4}$$

$$x = \frac{-5 + 7}{4} \quad \text{or} \quad x = \frac{-5 - 7}{4}$$

$$x = \frac{2}{4} \quad \text{or} \quad \frac{-12}{4}$$

$$x = \frac{1}{2} \quad \text{or} \quad x = -3$$

The solution is $x = \frac{1}{2}$ or $x = -3$.

Step 4 *Check* the solution numbers by substituting them into the original equation.

Check: $2x^2 + 5x - 3 = 0$

For $x = \frac{1}{2}$ For $x = -3$

$$2\left(\frac{1}{2}\right)^2 + 5\left(\frac{1}{2}\right) - 3 = 0 \qquad 2(-3)^2 + 5(-3) - 3 = 0$$

$$2\left(\frac{1}{4}\right) + 5\left(\frac{1}{2}\right) - 3 = 0 \qquad \begin{aligned} 2(9) + 5(-3) - 3 &= 0 \\ 18 - 15 - 3 &= 0 \\ 18 - 18 &= 0 \end{aligned}$$

$$\frac{1}{2} + 2\frac{1}{2} - 3 = 0$$

$$3 - 3 = 0$$

Your turn. Use the quadratic formula to solve $x^2 + 4x - 5 = 0$.

Check your work in **27**.

---

27 **Step 1** $x^2 \quad + \quad 4x \quad - \quad 5 \quad = 0.$

$\boxed{a = 1} \quad \boxed{b = 4} \quad \boxed{c = -5}$

Step 2 $x = \dfrac{-(4) \pm \sqrt{(4)^2 - 4(1)(-5)}}{2(1)}$

Step 3 Simplify

$$x = \frac{-4 \pm \sqrt{36}}{2}$$

$$x = \frac{-4 \pm 6}{2}$$

$$x = \frac{-4 + 6}{2} \quad \text{or} \quad x = \frac{-4 - 6}{2}$$

$x = 1 \quad$ or $\quad x = -5 \quad$ The solution is $x = 1$ or $x = -5$.

Step 4 **Check:** $x^2 + 4x - 5 = 0$

For $x = 1$ For $x = -5$

$$\begin{aligned} (1)^2 + 4(1) - 5 &= 0 \\ 1 + 4 - 5 &= 0 \\ 5 - 5 &= 0 \end{aligned} \qquad \begin{aligned} (-5)^2 + 4(-5) - 5 &= 0 \\ 25 - 20 - 5 &= 0 \\ 25 - 25 &= 0 \end{aligned}$$

Here is one that is a bit tougher. Solve $3x^2 - 7x = 5$.

Check your work in **28**.

---

28 **Step 1** Rewrite the equation in standard form.

$$3x^2 - 7x - 5 = 0$$

$a = 3$ $b = -7$ $c = -5$

$-(-7) = +7$

Step 2 $x = \dfrac{-(-7) \pm \sqrt{(-7)^2 - 4(3)(-5)}}{2(3)}$

$(-7)^2 = (-7) \cdot (-7) = 49$

Step 3 Simplify

$$x = \frac{7 \pm \sqrt{109}}{6}$$

$$x = \frac{7 + \sqrt{109}}{6} \quad \text{or} \quad x = \frac{7 - \sqrt{109}}{6}$$

$$x \approx \frac{7 + 10.44}{6} \quad \text{or} \quad x \approx \frac{7 - 10.44}{6} \quad \text{rounded to two decimal places}$$

$$x \approx \frac{17.44}{6} \quad \text{or} \quad x \approx \frac{-3.44}{6}$$

$$x \approx 2.91 \quad \text{or} \quad x \approx -0.57 \qquad \begin{array}{l}\text{The solution is}\\ x \approx 2.91 \quad \text{or} \quad x \approx -0.57.\end{array}$$

Step 4 Check both answers by substituting them back into the original quadratic equation.

Use the following calculator sequence to find this solution.

 7 $\boxed{x^2}$ $\boxed{-}$ 4 $\boxed{\times}$ 3 $\boxed{\times}$ 5 $\boxed{+/-}$ $\boxed{=}$ $\boxed{\sqrt{}}$ $\boxed{\text{STO}}$ $\boxed{+}$ 7 $\boxed{=}$ $\boxed{\div}$ 6 $\boxed{=}$ → *2.9067178* ≈ 2.91

7 $\boxed{-}$ $\boxed{\text{RCL}}$ $\boxed{=}$ $\boxed{\div}$ 6 $\boxed{=}$ → *-0.5733844* ≈ -0.57

Use the quadratic formula to solve each of the following equations. (Round to two decimal places if necessary.)

(a) $6x^2 - 13x + 2 = 0$ (b) $3x^2 - 13x = 0$

(c) $2x^2 - 5x + 17 = 0$ (d) $8x^2 = 19 - 5x$

Check your work in **29**.

---

29 (a) $6x^2 - 13x + 2 = 0$

$a = 6$ $b = -13$ $c = 2$

$$x = \frac{-(-13) \pm \sqrt{(-13)^2 - 4(6)(2)}}{2(6)}$$

$$x = \frac{13 \pm \sqrt{121}}{12} = \frac{13 \pm 11}{12}$$

The solution is $x = \dfrac{13 + 11}{12}$ or $x = \dfrac{13 - 11}{12}$

$$x = 2 \qquad \text{or} \qquad x = \dfrac{1}{6}$$

The solution is $x = 2$ or $x = \frac{1}{6}$.

Check it.

(b) $3x^2 - 13x = 0$

$$3x^2 \quad - \quad 13x \quad + \quad 0 \quad = 0$$

⬇   ⬇   ⬇

$\boxed{a = 3}$ $\boxed{b = -13}$ $\boxed{c = 0}$

$$x = \dfrac{-(-13) \pm \sqrt{(-13)^2 - 4(3)(0)}}{2(3)}$$

$$x = \dfrac{13 \pm \sqrt{169}}{6}$$

The solution is $x = \dfrac{13 + 13}{6}$ or $x = \dfrac{13 - 13}{6}$

$$x = \dfrac{13}{3} \qquad \text{or} \qquad x = 0$$

The solution is $x = 0$ or $x = \frac{13}{3}$.

(c)

$$2x^2 \quad - \quad 5x \quad + \quad 17 \quad = 0$$

⬇   ⬇   ⬇

$\boxed{a = 2}$ $\boxed{b = -5}$ $\boxed{c = 17}$

$$x = \dfrac{-(-5) \pm \sqrt{(-5)^2 - 4(2)(17)}}{2(2)}$$

$$x = \dfrac{5 \pm \sqrt{-111}}{4}$$

But the square root of a negative number is not acceptable if our answer must be a real number. This quadratic equation has no real number solution.

(d) $8x^2 = 19 - 5x$

or

$$8x^2 \quad + \quad 5x \quad - \quad 19 \quad = 0$$

⬇   ⬇   ⬇

$\boxed{a = 8}$ $\boxed{b = 5}$ $\boxed{c = -19}$

$$x = \dfrac{-(5) \pm \sqrt{(5)^2 - 4(8)(-19)}}{2(8)}$$

$$x = \dfrac{-5 \pm \sqrt{633}}{16}$$

$$x \approx \dfrac{-5 \pm 25.159}{16}$$ Find the square root to three decimal places to ensure an accurate answer.

The solution is $x \approx -1.88$ or $x \approx 1.26$ rounded

When you check a solution that includes a rounded value, the check may not give an exact fit. The differences should be very small if you have the correct solution.

Word Problems and Quadratic Equations

Example: In a dc circuit the power P dissipated in the circuit is given by the equation

$$P = RI^2$$

where R is the circuit resistance in ohms and I is the current in amperes.

What current will produce 1440 watts of power in a 10-ohm resistor?

Substituting into the equation yields

$$1440 = 10I^2$$

or

$$I^2 = 144$$

$$I = \pm \sqrt{144}$$

The solution is $I = 12$ amperes or $I = -12$ amperes.

Another Example: One side of a rectangular opening for a heating pipe is 3 in. longer than the other side. If the total cross-sectional area is 70 sq in., find the dimensions of the cross section.

Let L = length, W = width.

Then $L = 3 + W$

and the area is

$$\text{area} = LW$$

or

$$70 = LW$$

Substituting $L = \boxed{3 + W}$ into the area equation, we have

$$70 = \boxed{(3 + W)}\, W$$

or

$$70 = 3W + W^2$$

or

$$W^2 + 3W - 70 = 0 \qquad \text{in standard quadratic form.}$$

$$\boxed{a = 1} \qquad \boxed{b = 3} \qquad \boxed{c = -70}$$

To find the solution, substitute a, b, and c into the quadratic formula.

$$W = \frac{-(3) \pm \sqrt{(3)^2 - 4(1)(-70)}}{2(1)}$$

or

$$W = \frac{-3 \pm \sqrt{289}}{2}$$

$$W = \frac{-3 \pm 17}{2}$$

The solution is

$$W = -10 \qquad \text{or} \qquad W = 7$$

Only the positive value makes sense. The answer is $W = 7$ in.

Substituting back into the equation $L = 3 + W$, we find that $L = 10$ in.

Check to see that the area is indeed 70 sq in.

Solve the following problems. (Round each answer to two decimal places if necessary.)

(a) Find the side length of a square whose area is 200 sq m.

(b) The cross-sectional area of a rectangular duct must be 144 sq in. If one side must be twice as long as the other, find the length of each side.

(c) Find the diameter of a circular pipe whose cross-sectional area is 3.00 sq in.

(d) The SAE horsepower rating of an engine is given by $HP = \dfrac{D^2 N}{2.5}$, where D is the bore of the cylinder in inches, and N is the number of cylinders. What must the bore be for an eight-cylinder engine to have a horsepower rating of 300?

(e) One side of a rectangular plate is 6 in. longer than the other. The total area is 216 sq in. How long is each side?

(f) A man has a long strip of sheet steel 12 ft wide. He wishes to make an open-topped water channel with a rectangular cross section. If the cross-sectional area must be 16 sq ft, what should the dimensions of the channel be? (*Hint:* $H + H + W = 12$ in.)

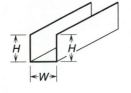

Our worked solutions are given in **30**.

---

30 (a) $A = s^2$

$200 = s^2$

$s = \pm\sqrt{200}$

$s \approx 14.14$ m, rounded   The negative solution is not possible.

(b) $L \cdot W = A$

and $L = 2W$

Therefore, $(2W)\,W = A$
or $2W^2 = A$
                $2W^2 = 144$
                $W^2 = 72$
                $W = +\sqrt{72}$
                $W \approx 8.485 \approx 8.49$ in.   rounded   for the positive root

Then $L = 2W$

or $L \approx 16.98$ in.   rounded

(c) $A = \dfrac{\pi D^2}{4}$

$3.00 = \dfrac{\pi D^2}{4}$

or $D^2 \approx \dfrac{4(3.00)}{3.1416}$

$D^2 \approx 3.8197$

$D \approx +\sqrt{3.8197}$

$D \approx 1.95$ in.   rounded

(d) $HP = \dfrac{D^2 N}{2.5}$

$300 = \dfrac{D^2 \cdot 8}{2.5}$

or $D^2 = \dfrac{(300)(2.5)}{8}$

$D^2 = 93.75$

$D = +\sqrt{93.75}$

or $D \approx 9.68$ in. rounded

(e) "One side . . . must be 6 in. longer than the other."

$L = 6 + W$

For a rectangle, area $= LW$.

Then,

$216 = LW$

or $216 = (6 + W)W$

$216 = 6W + W^2$

$W^2 + 6W - 216 = 0$ in standard quadratic form

$a = 1$ $b = 6$ $c = -216$

When we substitute into the quadratic formula,

$W = \dfrac{-(6) \pm \sqrt{(6)^2 - 4(1)(-216)}}{2(1)}$

$W = \dfrac{-6 \pm \sqrt{900}}{2}$

$W = \dfrac{-6 \pm 30}{2}$

The solution is

$W = -18$ or $W = 12$

Only the positive value is a reasonable solution. The answer is $W = 12$ in.

Substituting into the first equation yields

$L = 6 + W$

$L = 6 + (12) = 18$ in.

(f)

$H + H + W = 12$

$2H + W = 12$

or $W = 12 - 2H$

Area $= HW$

or $16 = HW$ Substituting $W = 12 - 2H$ into the area equation.

$16 = H(12 - 2H)$

$16 = 12H - 2H^2$ or $8 = 6H - H^2$ Divide each term by 2 to get a simpler equation.

$H^2 - 6H + 8 = 0$ in standard quadratic form

$a = 1$ $b = -6$ $c = 8$

Substituting a, b, and c into the quadratic formula yields

$$H = \frac{-(-6) \pm \sqrt{(-6)^2 - 4(1)(8)}}{(2)1}$$

$$H = \frac{6 \pm \sqrt{4}}{2}$$

$$H = \frac{6 \pm 2}{2}$$

The solution is $H = 4$ ft   or $H = 2$ ft

For $H = 4$ ft, $W = 4$ ft, since $16 = HW$
For $H = 2$ ft, $W = 8$ ft

The channel can be either 4 ft by 4 ft or 2 ft by 8 ft. Both dimensions give a cross-sectional area of 16 sq ft.

Now turn to **31** for a set of exercises on solving quadratic equations.

31

Exercises 11-2   **Quadratic Equations**

A. Which of the following are quadratic equations?

  1. $5x - 13 = 23$ 2. $2x + 5 = 3x^2$

  3. $2x^3 - 6x^2 - 5x + 3 = 0$ 4. $x^2 = 0$

  5. $8x^2 - 9x = 0$

Which of the following quadratic equations are in standard form? For those that are not, rearrange them into standard form.

  6. $7x^2 - 5 + 3x = 0$ 7. $14 = 7x - 3x^2$

  8. $13x^2 - 3x + 5 = 0$ 9. $23x - x^2 = 5x$

10. $4x^2 - 7x + 3 = 0$

B. Solve each of these quadratic equations. (Round to two decimal places if necessary.)

  1. $x^2 = 25$ 2. $3x^2 - 27 = 0$

  3. $5x^2 = 22x$ 4. $2x^2 - 7x + 3 = 0$

  5. $4x^2 = 81$ 6. $6x^2 - 13x - 63 = 0$

  7. $15x = 12 - x^2$ 8. $4x^2 - 39x = -27$

  9. $0.4x^2 + 0.6x - 0.8 = 0$ 10. $0.001 = x^2 + 0.03x$

11. $x^2 - x - 13 = 0$ 12. $2x - x^2 + 11 = 0$

13. $7x^2 - 2x = 1$ 14. $3 + 4x - 5x^2 = 0$

15. $2x^2 + 6x + 3 = 0$ 16. $5x^2 - 7x + 1 = 0$

17. $3x^2 = 8x - 2$

18. $x = 1 - 7x^2$

19. $0.2x^2 - 0.9x + 0.6 = 0$

20. $1.2x^2 + 2.5x = 1.8$

C. Practical Problems (Round to the nearest hundredth.)

1. The area of a square is 625 mm^2. Find its side length.

2. **Manufacturing** The capacity in gallons of a cylindrical tank can be found using the formula

$$C = \frac{0.7854D^2L}{231}$$

where C = capacity in gallons
D = diameter of tank in inches
L = length of tank in inches

(a) Find the diameter of a 42-in.-long tank that has a capacity of 30 gallons.

(b) Find the diameter of a 60-in.-long tank that has a capacity of 50 gallons.

3. **Plumbing** The length of a rectangular pipe is three times longer than its width. Find the dimensions that give a cross-sectional area of 75 sq in.

4. **Sheet Metal Technology** Find the radius of a circular vent that has a cross-sectional area of 250 cm^2.

5. **Sheet Metal Technology** The length of a rectangular piece of sheet metal must be 5 in. longer than the width. Find the exact dimensions that will provide an area of 374 sq in.

6. **Auto Mechanics** If the horsepower rating of an engine is given by

$$HP = \frac{D^2N}{2.5}$$

Find D (the bore of the cylinder in inches) if

(a) N (the number of cylinders) is 4 and the horsepower rating is 80.

(b) N is 6 and HP = 200.

7. **Plumbing** Find the diameter of a circular pipe whose cross-sectional area is 40 sq in.

8. **Electrical Technology** If $P = RI^2$ for a direct current circuit, find I (current in amperes) if

(a) The power (P) is 405 watts and the resistance (R) is 5 ohms.

(b) The power is 800 watts and the resistance is 15 ohms.

9. **Auto Mechanics** Total piston displacement is given by the following formula:

P.D. $= 0.7854D^2LN$

where P.D. = piston displacement, in cubic inches
D = diameter of bore of cylinder
L = length of stroke, in inches
N = number of cylinders

Find the diameter if

(a) P.D. = 400 cu in., $L = 4.5$ in., $N = 8$

(b) P.D. = 392.7 cu in., $L = 4$ in., $N = 6$

10. **Sheet Metal Technology**   An open-topped channel must be made out of a 20-ft-wide piece of sheet steel. What dimensions will result in a cross-sectional area of 48 sq ft?

Check your answers on page 608. Then turn to **32** for a problem set covering both systems of equations and quadratic equations.

Advanced
Algebra

32 Answers are given on page 608.

A. Solve and check each of the following systems of equations. If the system is inconsistent or dependent, say so.

1. $x + 4y = 27$
$x + 2y = 21$

2. $3x + 2y = 17$
$x = 5 - 2y$

3. $5x + 2y = 20$
$3x - 2y = 4$

4. $x = 10 - y$
$2x + 3y = 23$

5. $3x + 4y = 45$
$x - \frac{1}{3}y = 5$

6. $2x - 3y = 11$
$4x - 6y = 22$

7. $2x + 3y = 5$
$3x + 2y = 5$

8. $5x = 1 - 3y$
$4x + 2y = -8$

9. $3x - 2y = 10$
$4x + 5y = 12$

B. Solve and check each of the following quadratic equations.

1. $x^2 = 9$

2. $x^2 - 3x - 28 = 0$

3. $3x^2 + 5x + 1 = 0$

4. $4x^2 = 3x + 2$

5. $x^2 = 6x$

6. $2x^2 - 7x - 15 = 0$

7. $x^2 - 4x + 4 = 9$

8. $x^2 - x - 30 = 0$

9. $\frac{5x^2}{3} = 60$

10. $7x + 8 = 5x^2$

C. Practical Problems

For each of the following, set up either a system of equations in two variables or a quadratic equation and solve. (Round to two decimal places if necessary.)

1. The sum of two numbers is 38. Their difference is 14. Find them.

2. The area of a square is 196 sq in. Find its side length.

3. The difference of two numbers is 21. If twice the larger is subtracted from five times the smaller, the result is 33. Find the numbers.

4. **Carpentry** The perimeter of a rectangular door is 22 ft. Its length is 2 ft more than twice its width. Find the dimensions of the door.

5. **Heating and Air Conditioning** One side of a rectangular heating pipe is four times as long as the other. The cross-sectional area is 125 cm^2. Find the dimensions of the pipe.

Name

Date

Course/Section

6. **Carpentry**   A mixture of 650 nails costs $43.50. If some of the nails cost 5 cents apiece, and the rest cost 7 cents apiece, how many of each are there?

7. **Electrical Technology**   In the formula $P = RI^2$ find the current I in amperes if
   (a)   The power P is 1352 watts and the resistance R is 8 ohms.
   (b)   The power P is 1500 watts and the resistance is 10 ohms.

8. **Sheet Metal Technology**   The length of a rectangular piece of sheet steel is 2 in. longer than the width. Find the exact dimensions if the area of the sheet is 168 sq in.

9. **Electrical Technology**   A 42-in. piece of wire is to be cut into two parts. If one part is 2 in. less than three times the length of the other, find the length of each piece.

10. **Landscaping**   The perimeter of a rectangular field is 750 ft. If the length is four times the width, find the dimensions of the field.

11. **Sheet Metal Technology**   Find the diameter of a circular vent with a cross-sectional area of 200 sq in.

12. **Painting and Decorating**   A painter mixes paint worth $8 per gallon with paint worth $11 per gallon. He wishes to make 15 gallons of a mixture worth $10 per gallon. How many gallons of each kind of paint must be included in the mixture?

13. **Auto Mechanics**   In the formula $\text{HP} = \dfrac{D^2 N}{2.5}$ find the diameter of the cylinder bore D if
   (a)   The number of cylinders N is 6 and the horsepower rating HP is 60.
   (b)   The number of cylinders is 8 and the horsepower rating is 125.

14. **Plumbing**   The length of a rectangular pipe is 3 in. less than twice the width. Find the dimensions if the cross-sectional area is 20 sq in.

15. **Landscaping**   Fifty tons of a mixture of decorative rock cost $3150. If the mixture consists of rock costing $60 per ton and a more expensive rock costing $75 per ton, how many tons of each are used to make the mixture?

Answers to Previews

Answers to Preview 1

1. (a) 125   (b) 8607   (c) 37   (d) 2068   (e) 77

2. (a) 2368   (b) 74,115   (c) 640,140   (d) 334 remainder 2
   (e) 203

3. 445 lb

4. (a) 33   (b) 33   (c) 33   (d) 48

Answers to Preview 2

1. (a) $7\frac{3}{4}$ (b) $\frac{31}{8}$ (c) $\frac{20}{64}$ (d) $\frac{56}{32}$ (e) $\frac{5}{32}$ (f) $1\frac{7}{8}$

2. (a) $\frac{35}{256}$ (b) 3   (c) $\frac{9}{10}$ (d) $1\frac{1}{2}$ (e) $2\frac{3}{10}$ (f) 8

3. (a) $\frac{5}{8}$ (b) $1\frac{15}{16}$ (c) $\frac{11}{20}$ (d) $2\frac{11}{16}$

Answers to Preview 3

1. (a) 5.916   (b) 2.791   (c) 22.97   (d) 2.18225
   (e) 3256.25   (f) 225   (g) 3.045

2. 4.3

3. (a) 0.1875   (b) 106.027   (c) twenty-six and thirty-five thousandths
   (d) 3.452   (e) 9.225   (f) 0.305

4. (a) $311.48   (b) 0.34 lb

Answers to Preview 4

1. (a) 5:2 or 2:5   (b) $8\frac{2}{3}$ to 1

2. (a) $x = 10$ (b) $y = 14.3$

3. 64 ounces

4. (a) 25%   (b) 46%   (c) 500%   (d) 7.5%

5. (a) 0.35   (b) 0.0025   (c) 1.12

6. (a) 225   (b) 54   (c) 6.25% or $6\frac{1}{4}$%   (d) 75
   (e) 0.90%   (f) 10.4 lb   (g) $55.09   (h) 80%
   (i) 11.4%

Answers to Preview 5

1. (a) 140 mi   (b) 230 min or 3 hr 50 min
   (c) 3 hr

2. (a) 25,200 min   (b) 1.5 bbl   (c) 4.4 cu yd
   (d) 960 oz   (e) $283\frac{1}{2}$ in.   (f) 206 oz

3. (a) 99 lb   (b) 66 ft   (c) 7.6 cm
   (d) 4.4°C   (e) 150 cm, 1.5 m

Answers to Preview 6

1. (a) −14   (b)
   (c) −250

2. (a) −8   (b) $11\frac{3}{4}$ (c) −7.5   (d) −3

3. (a) −3   (b) −8   (c) 10.9   (d) $-1\frac{7}{8}$

4. (a) −72   (b) 6   (c) 31   (d) $-\frac{3}{20}$ (e) 1500

5. (a) 64   (b) 4.2025

6. (a) 11   (b) 500   (c) 22

7. (a) 13   (b) 3.81

Answers to Preview 7

1. (a) 7   (b) 5   (c) $A = 12$ (d) $T = 4$ (e) $P = 120$

2. (a) $6ax^2$ (b) $-3x - y$ (c) $4x - 2$ (d) $3x^2 + 7x + 12$

3. (a) $x = 5$ (b) $x = 9$ (c) $x = 12$ (d) $N = 2S - 52$
   (e) $A = \dfrac{8M + BL}{L}$

4. (a) $4A$ (b) i^2R (c) $E = \dfrac{100u}{I}$ (d) $R = \dfrac{12L}{D^2}$
   (e) $24.07   (f) 10.37 hr, or 11 hr, rounded

5. (a) $6y^2$ (b) $-12x^5y^4$ (c) $3xy - 6x^2$ (d) $-5x^5$ (e) $\dfrac{b^2}{3a^3}$
   (f) $4m^2 - 3m + 5$

6. (a) 18 ft   (b) 12 machines

7. (a) 1.84×10^{-4} (b) 2.13×10^5 (c) 1.44×10^{-3}
   (d) 6.5×10^{-8}

Answers to Preview 8

1. 54°

2. (a) 80°, 80°, 100°, 80°, 100°   (b) 20°

3. (a) Triangle   (b) Parallelogram   (c) Trapezoid
   (d) Square   (e) Hexagon

4. 1.5 in.

5. (a) 14 sq in.   (b) 24 cm²   (c) 0.42 sq in., 2.4 in.
   (d) 22.9 sq in., 17.0 in.   (e) 19.3 sq in.

Answers to Preview 9

1. (a) Cube (b) Frustum of a cone (c) Rectangular prism

2. (a) 12.6 cu ft (b) 38.8 cm^3, 55.4 cm^2 (c) 3.7 cu in.
   (d) 80.5 sq in., 75.4 cu in. (e) 180 cm^2, 336 cm^3

3. See frames 24 and 28.

Answers to Preview 10

1. (a) Yes (b) Yes (c) Yes (d) Yes (e) No

2. (a) 0.438 (b) 0.105 (c) 0.943

3. (a) 14°0′ (b) 56°43′ (c) 75.3° (d) 48.6°

4. (a) 0.80 (b) 0.15 (c) 1.40
   (d) 4.2 in., 12.6 sq in. (e) 1.7 rad/sec

5. (a) $X \approx 10.5$, $Y \approx 13.4$ (b) $\angle m = 44°25′$, $X \approx 3.7$ in.
   (c) ≈ 39.8 in. (d) 5°43′

6. (a) $A = 82°$, $B = 66°$, $C = 32°$ (b) $C = 33°$, $a = 7.6$ ft, $c = 4.4$ ft

Answers to Preview 11

1. (a) (4, 0) (b) (−9, −3) (c) Dependent (d) Inconsistent, no solution

2. (a) 20 and 15 (b) 8 in. and 6 in. (c) 300 of the 5¢ screws, and 500 of the 9¢ screws

3. (a) $x = 4$ or $x = -4$ (b) $x = 0$ or $x = 7$

   (c) $x = 7$ or $x = -2$ (d) $x = -2\frac{2}{3}$ or $x = 2$ (e) No solution

4. (a) 6 cm (b) ≈ 8.4 in. (c) 12.9 amp (d) 6.5 in. by 9.5 in.

Answers

Chapter 1

Practice, page 6

A.
10	11	11	12	16	7	9	15	12	13
10	13	15	8	14	9	13	18	9	11
16	8	10	17	9	14	12	7	10	11
13	13	12	16	9	11	10	10	14	17
11	14	11	10	11	14	13	12	13	15

B.
11	12	14	17	18	21	11	20	18	15
15	14	15	18	22	17	14	14	18	16
12	19	8	10	22	12	19	15	14	16

Exercises 1–1, page 12

A.
1. 70	2. 65	3. 80	4. 103	5. 123
6. 124	7. 132	8. 136	9. 393	10. 1003
11. 1390	12. 831	13. 1009	14. 806	15. 861
16. 5525	17. 9461	18. 9302	19. 11,428	20. 15,715
21. 25,717	22. 47,111	23. 11,071	24. 14,711	25. 175,728

B.
1. 1042	2. 5211	3. 2442	4. 6441
5. 7083	6. 16,275	7. 6352	8. 7655
9. 6514	10. 9851	11. 64	12. 141
13. 55	14. 148	15. 357	

C.
1. Three hundred fifty-seven
2. Two thousand three hundred four
3. Seventeen thousand ninety-two
4. Two hundred seven thousand six hundred thirty
5. Two million thirty-four
6. Ten thousand seven
7. Seven hundred forty thousand one hundred six
8. Five million fifty-five thousand five hundred fifty
9. One hundred eighteen million one hundred eighty thousand eighteen
10. Six thousand seven hundred nine

11. 3006	12. 17,024	13. 11,100
14. 3,002,017	15. 4,040,006	16. 720,000,010
17. 360	18. 4400	19. 4000
20. 5000	21. 230,000	22. 226,000

D.
1. 4861 ft	2. 11,365 fbm	3. 1636 screws
4. $5148	5. 1129 minutes	6. 6670 shingles

7. (a) 3607 watts   (b) 1997 watts   (c) 850 watts   8. 111 qt
9. 3114   10. 97 in.

E. 1. (a) *Daily Totals*   (b) *Machine Totals*   (c) yes
             1. 1919   A. 4074
             2. 2125   B. 4462
             3. 1958   C. 2185
             4. 1594   D. 5665
             5. 2192   E. 2883
             6. 1953
             7. 1974
             8. 1988
             9. 1667
            10. 1899

    2. (a) $307,225   (b) $732,813   (c) $2,298,502   (d) $7156
    3. (a) $8200   (b) $7735   (c) SELL
    4. (a) #12 BHD: 11,453   (b) A3:3530
             #Tx: 258   A4:8412
             410 AAC: 12,715   B1:4294
             110 ACSR: 8792   B5:5482
             6B: 7425   B6:5073
                                          C4:6073
                                          C5:7779

Exercises 1–2, page 20

A. 1. 6   2. 7   3. 2   4. 8   5. 4   6. 9
    7. 3   8. 0   9. 3   10. 3   11. 8   12. 8
   13. 9   14. 9   15. 9   16. 9   17. 3   18. 6
   19. 8   20. 5   21. 7   22. 18   23. 7   24. 8
   25. 0   26. 5   27. 8   28. 9   29. 6   30. 8
   31. 6   32. 9   33. 5   34. 7   35. 4   36. 4

B. 1. 13   2. 29   3. 12   4. 19   5. 15
    6. 36   7. 38   8. 29   9. 46   10. 22
   11. 25   12. 38   13. 189   14. 85   15. 281
   16. 254   17. 408   18. 154   19. 273   20. 715
   21. 574   22. 29   23. 2809   24. 5698   25. 12,518
   26. 56,042   27. 4741   28. 9614   29. 47,593   30. 22,422

C. 1. $247   2. 2353 sq ft   3. 1758 ft   4. 29 cm
    5. $330,535 6. $3144   7. 3 drums, by 44 liters
    8. (a) 9 in.   (b) 4 in.   (c) 4 in.   (d) 5 in.
    9. 174 gal   10. 5840 lb   11. 7750   12. $A = 32$ in.; $B = 32$ in.

D. 1.

Truck No.	1	2	3	4	5	6	7	8	9	10
Mileage	1675	1167	1737	1316	1360	299	1099	135	1461	2081

    Total mileage 12,330

    2. First sum: 1,083,676,269
        Second sum: 1,083,676,269
        They are the same.
    3. $21,286
    4. $116,805

5. (a) $2065

(b)

Deposits	Withdrawals	Balance
		$6375
	$ 379	5996
$1683		7679
474		8153
487		8640
	2373	6267
	1990	4277
	308	3969
	1090	2879
	814	2065

Practice, page 25

A. 12   32   63   36   12   18   0   24   14   8
    48   16   45   30   10   9   72   35   18   4
    28   15   36   49   8   40   42   54   64   24
    20   0   25   27   81   6   1   48   16   63

B. 16   30   9   35   18   20   28   48   12   63
    32   0   18   24   9   25   24   45   10   72
    15   49   40   54   36   8   42   64   0   4
    25   27   7   56   36   12   81   0   2   56

Exercises 1–3, page 30

A. 1. 42   2. 56   3. 48   4. 72   5. 63
    6. 87   7. 54   8. 28   9. 45   10. 84
    11. 296   12. 168   13. 576   14. 423   15. 320
    16. 156   17. 290   18. 564   19. 416   20. 644
    21. 792   22. 1088   23. 1404   24. 153   25. 282
    26. 308   27. 720   28. 1728   29. 5040   30. 7138
    31. 1938   32. 1650   33. 4484   34. 928   35. 3822
    36. 8930

B. 1. 37,515   2. 375,750   3. 297,591   4. 38,023
    5. 378,012   6. 41,064   7. 30,780   8. 1,368,810
    9. 397,584   10. 60,241   11. 7281   12. 4263
    13. 25,000   14. 325,200   15. 3,532,536 16. 8954
    17. 10,112   18. 31,868   19. 89,577   20. 113,680

C. 1. $1840   2. 2100 ft   3. 1300 ft   4. $64,698
    5. 8000 envelopes   6. 225 in.   7. 2430   8. 132 in.
    9. 1200   10. 295,188 lb   11. 7650   12. 44,280
    13. 23,040

D. 1. $355 left   2. (a) Nets $36,500   (b) Nets $36,400
                             (c) Nets $31,200   (d) Nets $671,088.63
                        Therefore, (d) gives you the most money.
    3. (a) 111,111,111; 222,222,222; 333,333,333
        (b) 111,111; 222,222; 333,333
        (c) 1; 121; 12,321; 1,234,321; 123,454,321
        (d) 42; 4422; 444,222; 44,442,222; 4,444,422,222

4.   34,600 lb
5.   Alpha   Beta   Gamma   Delta   Tau
     $3510 $5695 $4640 $9065 $7020

Factors and Primes, page 38

A.   1.   1, 2, 4   2.   1, 2, 5, 10   3.   1, 3, 9   4.   Prime
     5.   1, 2, 3, 4, 6, 8, 12, 24   6.   1, 2, 3, 6, 9, 18   7.   1, 3, 5, 15
     8.   Prime   9.   1, 2, 7, 14   10.   Prime   11.   1, 3, 7, 21
     12.   1, 2, 4, 8, 16, 32   13.   1, 2, 4, 5, 10, 20   14.   Prime
     15.   Prime   16.   1, 2, 13, 26   17.   1, 2, 4, 11, 22, 44
     18.   1, 2, 3, 5, 6, 9, 10, 15, 18, 30, 45, 90
     19.   1, 3, 13, 39   20.   Prime

B.   1.   2   2.   2, 5   3.   3   4.   Prime   5.   2, 3
     6.   2, 3   7.   3, 5   8.   Prime   9.   2, 7   10.   Prime
     11.   3, 7   12.   2   13.   2, 5   14.   Prime   15.   Prime
     16.   2, 13   17.   2, 11   18.   2, 3, 5   19.   3, 13   20.   Prime

Exercises 1–4, page 40

A.   1.   9   2.   11 rem. 4   3.   Not defined
     4.   7 rem. 2   5.   10 rem. 1   6.   1
     7.   8   8.   4   9.   6
     10.   35   11.   23 rem. 6   12.   57
     13.   51 rem. 4   14.   1103 rem. 1   15.   21
     16.   52   17.   37   18.   50 rem. 1
     19.   23   20.   20 rem. 2   21.   39
     22.   25   23.   9 rem. 1   24.   53
     25.   22   26.   63 rem. 23   27.   8 rem. 35

B.   1.   120   2.   9 rem. 6   3.   56 rem. 8
     4.   95 rem. 6   5.   96   6.   142 rem. 6
     7.   222 rem. 2   8.   32   9.   305 rem. 5
     10.   84 rem. 41   11.   119   12.   3001
     13.   501   14.   8001 rem. 3   15.   604
     16.   20,720   17.   200   18.   50 rem. 4
     19.   108 rem. 4   20.   2009 rem. 2   21.   600
     22.   61   23.   102 rem. 98   24.   81
     25.   100 rem. 11   26.   19 rem. 66   27.   17 rem. 123

C.   1.   27 in.   2.   6   3.   13 hr   4. $W = 12$ in., $H = 8$ in.
     5.   28   6.   50   7.   7 in.   8.   373 HP
     9. $42   10.   11 hr   11.   60   12.   13 cu ft
     13.   27 reams   14.   36 days

D.   1.   (a)   21,021,731   (b)   449   (c)   93   (d)   27,270
     2. $9380
     3.   19.148255 or 20 rivets to be sure
     4.   123,091 lb   5.   51 hours 40 minutes, or 52 hr, rounded
     6.   (a)   16,094   (b)   201,023   (c)   2283   (d)   357

Exercises 1–5, page 43

A.   1.   50   2.   14   3.   36   4.   24   5.   57   6.   14
     7.   4   8.   12   9.   42   10.   1   11.   6   12.   99
     13.   61   14.   23   15.   112   16.   2   17.   31   18.   26
     19.   17   20.   49   21.   84   22.   27   23.   4   24.   32
     25.   6   26.   8   27.   36   28.   26   29.   2   30.   4
     31.   8   32.   3   33.   13   34.   3   35.   7   36.   8
     37.   6   38.   3   39.   5   40.   43

B. 1. $3 \times \$8 + 5 \times \$10 = \$74$ 2. $\$168 + \$7 \times 8 = \$224$
   3. $12 \times \$8 - \$7 \times 5 = \$61$ 4. $520 - 48 \times 5 + 300 = 580$

C. 1. 8347 2. 403 3. 7386 4. 1262
   5. 5 6. 1,003,632 7. 1359 8. 1265
   9. 1691 10. 12,998 11. 1458 12. 165
   13. 13,920 14. 30 15. 63 16. 44

Problem Set 1, page 47

A. 1. 93 2. 83 3. 528 4. 860
   5. 934 6. 2980 7. 15 8. 26
   9. 649 10. 196 11. 195 12. 2615
   13. 1504 14. 3423 15. 1407 16. 3690
   17. 13,041 18. 290,764 19. 230,384 20. 1,575,056
   21. 37 22. 213 23. 57 24. 62
   25. 9 26. 43 27. 18 28. 69
   29. 6 30. 9 31. 115 32. 40
   33. 7 34. 627 35. 1245 36. 833

B. 1. 43 ft 2. 64 rods 3. 1892 sq ft 4. 445 lb
   5. 6 6. 24 hr 7. 207 lb 8. $494
   9. $539 10. 252 ft 11. 733,967 mi 12. $105,359
   13. 650 gpm 14. (a) 513 (b) 5068 ft
   15. 2839 lb 16. 263° 17. 24 hr 18. $34.91
   19. 87,780 cu in. 20. 646 21. 193 rpm
   22. 101 lb 23. 6 ft

Chapter 2 *Exercises 2–1, page 63*

A. 1. $\frac{7}{3}$ 2. $\frac{15}{2}$ 3. $\frac{67}{8}$ 4. $\frac{17}{16}$ 5. $\frac{23}{8}$
   6. $\frac{2}{1}$ 7. $\frac{8}{3}$ 8. $\frac{259}{64}$ 9. $\frac{29}{6}$ 10. $\frac{29}{16}$

B. 1. $8\frac{1}{2}$ 2. $1\frac{3}{5}$ 3. $1\frac{3}{8}$ 4. $2\frac{8}{16}$ or $2\frac{1}{2}$ 5. $1\frac{1}{2}$
   6. $3\frac{2}{3}$ 7. $16\frac{4}{6}$ or $16\frac{2}{3}$ 8. $1\frac{1}{3}$ 9. $2\frac{16}{32}$ or $2\frac{1}{2}$ 10. $2\frac{1}{2}$

C. 1. $\frac{3}{4}$ 2. $\frac{2}{3}$ 3. $\frac{3}{8}$ 4. $\frac{9}{2}$ 5. $\frac{2}{5}$
   6. $\frac{7}{6}$ 7. $\frac{4}{5}$ 8. $\frac{5}{2}$ 9. $4\frac{1}{4}$ 10. $\frac{17}{16}$
   11. $\frac{21}{32}$ 12. $\frac{2}{7}$ 13. $\frac{5}{12}$ 14. $\frac{5}{2}$ 15. $\frac{19}{12}$

D. 1. 14 2. 12 3. 8 4. 24 5. 20 6. 92
   7. 36 8. 34 9. 5 10. 24 11. 42 12. 34

E. 1. $\frac{3}{5}$ 2. $\frac{13}{8}$ 3. $1\frac{1}{2}$ 4. $\frac{13}{16}$ 5. $\frac{7}{8}$ 6. $2\frac{1}{2}$
   7. $\frac{6}{4}$ 8. $\frac{25}{60}$ 9. $\frac{13}{5}$ 10. $2\frac{7}{4}$ 11. $\frac{5}{12}$ 12. $1\frac{1}{5}$

F. 1. $15\frac{3}{4}$ in. 2. $\frac{3}{4}$ 3. $\frac{19}{6}, \frac{25}{8}$ 4. $\frac{13}{64}$-in. fastener
   5. No. 6. $2\frac{2}{3}$ in. 7. $\frac{3}{5}$ 8. $1\frac{1}{2}$ 9. $\frac{1}{5}$

Exercises 2–2, page 68

A. 1. $\frac{1}{8}$ 2. $\frac{4}{15}$ 3. $\frac{2}{15}$ 4. 3 5. $2\frac{2}{3}$
   6. $\frac{11}{45}$ 7. $1\frac{1}{9}$ 8. $\frac{13}{16}$ 9. $2\frac{1}{2}$ 10. 14
   11. 3 12. 8 13. $3\frac{1}{4}$ 14. $1\frac{1}{21}$ 15. 69
   16. $35\frac{3}{4}$ 17. 74 18. $9\frac{7}{8}$ 19. $10\frac{3}{8}$ 20. $21\frac{1}{3}$
   21. $\frac{1}{8}$ 22. $\frac{3}{10}$ 23. $\frac{1}{15}$ 24. $1\frac{1}{3}$ 25. 2

B. 1. $\frac{1}{6}$ 2. $\frac{3}{32}$ 3. $\frac{1}{2}$ 4. $\frac{7}{16}$ 5. $\frac{3}{4}$

6. $\frac{15}{16}$ 7. $1\frac{5}{16}$ 8. $2\frac{7}{9}$ 9. 1 10. $\frac{7}{10}$

11. $1\frac{1}{20}$ 12. $2\frac{2}{5}$ 13. $2\frac{5}{8}$ 14. $\frac{5}{14}$ 15. 1 16. $1\frac{1}{2}$

C. 1. $137\frac{3}{4}$ in. 2. $99\frac{3}{4}$ in. 3. $11\frac{2}{3}$ ft

4. $4\frac{5}{16}$ in. 5. 14 ft $1\frac{1}{2}$ in. 6. $110\frac{1}{4}$ in.

7. $319\frac{1}{5}$ mi 8. $36\frac{3}{4}$ in. 9. $356\frac{1}{2}$ lb

10. 118 ft 11. $\frac{3}{4}$ in. 12. 210 in. or 17 ft 6 in.

13. $9\frac{3}{4}$ in. 14. $431\frac{41}{64}$ cu in. 15. $348\frac{3}{4}$ min

16. 1001 cu in. 17. $10\frac{2}{3}$ hr 18. 126 in.

19. $5\frac{2}{5}$ in. 20. 390 lb 21. $22\frac{1}{2}$ picas

22. 24 MGD 23. $31\frac{1}{8}$ 24. 45°, 72°, 60°

25. $3\frac{1}{4}$ in. 26. $2\frac{1}{2}$ in. 27. $\frac{15}{32}$ in.

Exercises 2–3, page 74

A. 1. $1\frac{2}{3}$ 2. 9 3. $\frac{5}{16}$ 4. 32 5. $\frac{1}{2}$ 6. 1

7. $\frac{1}{4}$ 8. $2\frac{2}{5}$ 9. 9 10. 4 11. $1\frac{1}{3}$ 12. $1\frac{3}{4}$

13. $1\frac{1}{5}$ 14. $8\frac{1}{3}$ 15. 16 16. $\frac{1}{9}$ 17. 18 18. $\frac{6}{7}$

19. $7\frac{1}{2}$ 20. $\frac{3}{5}$ 21. $\frac{5}{6}$ 22. $\frac{7}{8}$ 23. $\frac{1}{8}$ 24. $1\frac{1}{3}$

B. 1. 8 ft 2. 14 ft 3. 48 4. 12 5. 84

6. $40\frac{1}{2}$ ft 7. 18 8. 8 9. 210 10. 29 ft by 34 ft

11. 7 sheets 12. 108 rev 13. 45 threads 14. 14 ft by $18\frac{1}{2}$ ft

Exercises 2–4, page 84

A. 1. $\frac{1}{4}$ 2. $1\frac{1}{3}$ 3. $\frac{3}{4}$ 4. $\frac{5}{6}$ 5. $\frac{1}{2}$ 6. $\frac{5}{8}$

7. $\frac{2}{5}$ 8. $\frac{1}{4}$ 9. $\frac{15}{16}$ 10. $1\frac{3}{8}$ 11. $1\frac{1}{2}$ 12. $2\frac{1}{4}$

13. $\frac{3}{4}$ 14. $\frac{13}{16}$ 15. $\frac{17}{24}$ 16. $\frac{29}{48}$ 17. $\frac{1}{8}$ 18. $\frac{7}{32}$

19. $\frac{7}{16}$ 20. $\frac{13}{32}$ 21. $\frac{29}{40}$ 22. $1\frac{7}{15}$ 23. $\frac{19}{40}$ 24. $\frac{7}{36}$

25. $\frac{5}{8}$ 26. $\frac{11}{48}$ 27. $1\frac{3}{4}$ 28. $3\frac{3}{16}$ 29. $4\frac{1}{8}$ 30. $3\frac{7}{20}$

31. $3\frac{8}{15}$ 32. $2\frac{1}{8}$ 33. $2\frac{3}{8}$ 34. $3\frac{2}{3}$ 35. $1\frac{7}{60}$ 36. $2\frac{14}{15}$

B. 1. $5\frac{1}{8}$ 2. $1\frac{13}{16}$ 3. $2\frac{13}{16}$ 4. $1\frac{17}{60}$ 5. $\frac{7}{8}$ 6. $20\frac{3}{8}$

7. $2\frac{1}{8}$ 8. $3\frac{3}{8}$ 9. $2\frac{11}{16}$ 10. $5\frac{5}{12}$ 11. $3\frac{9}{10}$ 12. $\frac{15}{56}$

C. 1. $10\frac{13}{20}$ min 2. $\frac{13}{16}$ in. 3. $\frac{843}{1000}$ 4. $\frac{3}{32}$ in.

5. $1\frac{9}{16}$ in. 6. $\frac{15}{16}$ in. 7. $25\frac{1}{4}$ ci 8. $2\frac{1}{6}$ ft

9. $1\frac{1}{8}$ in. 10. $7\frac{31}{32}$ in. 11. $23\frac{5}{16}$ in. 12. $1\frac{5}{8}$ in.

13. $2\frac{13}{16}$ in. 14. No 15. $\frac{3}{8}$ in. 16. $23\frac{3}{4}$ in.

17. $7\frac{3}{4}$ in. by $6\frac{1}{2}$ in. 18. 5 in. 19. $\frac{9}{16}$ in. 20. $1\frac{3}{32}$ in.

Problem Set 2, page 87

A. 1. $\frac{9}{8}$ 2. $\frac{21}{5}$ 3. $\frac{5}{3}$ 4. $\frac{35}{16}$ 5. $\frac{99}{32}$ 6. $\frac{33}{16}$

7. $\frac{13}{8}$ 8. $\frac{55}{16}$ 9. $2\frac{1}{2}$ 10. $9\frac{1}{2}$ 11. $8\frac{1}{3}$ 12. $1\frac{1}{8}$

13. $1\frac{9}{16}$ 14. $1\frac{5}{16}$ 15. $8\frac{3}{4}$ 16. $2\frac{1}{3}$ 17. $\frac{3}{16}$ 18. $\frac{1}{4}$

19. $\frac{3}{8}$ 20. $\frac{3}{4}$ 21. $\frac{1}{6}$ 22. $1\frac{4}{7}$ 23. $1\frac{4}{5}$ 24. $3\frac{2}{5}$

25. 9 26. 28 27. 44 28. 44 29. 68 30. 18

31. 15 32. 26 33. $\frac{7}{16}$ 34. $\frac{2}{3}$ 35. $\frac{7}{8}$ 36. $1\frac{1}{4}$

37. $\frac{3}{5}$ 38. $\frac{2}{10}$ 39. $\frac{7}{4}$ 40. $\frac{1}{9}$

B. 1. $\frac{3}{32}$ 2. $\frac{1}{2}$ 3. $\frac{7}{12}$ 4. $\frac{5}{256}$ 5. $1\frac{1}{4}$ 6. $\frac{49}{80}$

7. $\frac{5}{64}$ 8. $5\frac{1}{4}$ 9. $7\frac{1}{2}$ 10. $\frac{2}{3}$ 11. 27 12. 34

13. $11\frac{2}{3}$ 14. $7\frac{1}{2}$ 15. 2 16. $\frac{4}{5}$ 17. 32 18. $10\frac{2}{3}$

19. $\frac{1}{6}$ 20. $\frac{3}{4}$ 21. $\frac{7}{10}$ 22. $\frac{5}{6}$ 23. $2\frac{4}{9}$ 24. $1\frac{13}{15}$

C. 1. $1\frac{1}{4}$ 2. $1\frac{1}{4}$ 3. $\frac{7}{32}$ 4. $1\frac{5}{8}$ 5. $1\frac{13}{30}$ 6. $\frac{29}{40}$

7. $\frac{3}{8}$ 8. $\frac{3}{8}$ 9. $\frac{7}{16}$ 10. $\frac{19}{30}$ 11. $\frac{23}{40}$ 12. $1\frac{13}{32}$

13. $3\frac{3}{8}$ 14. $2\frac{7}{16}$ 15. $4\frac{1}{2}$ 16. $1\frac{1}{8}$ 17. $1\frac{19}{24}$ 18. $1\frac{5}{12}$

19. $1\frac{1}{30}$ 20. $3\frac{19}{20}$ 21. $1\frac{1}{6}$ 22. $1\frac{1}{3}$ 23. $\frac{2}{5}$ 24. $3\frac{1}{3}$

D. 1. $37\frac{1}{8}$ in. 2. 22 3. $25\frac{5}{8}$ in; $23\frac{1}{8}$ in. 4. $7\frac{9}{16}$ in.

5. $\frac{25}{32}$ in. 6. Yes 7. $4\frac{3}{5}$ cu ft 8. $15\frac{15}{16}$ in.

9. $92\frac{13}{16}$ in. 10. 13 ft 11. $\frac{15}{16}$ in.

12. A: $2\frac{11}{16}$ in. B: $2\frac{5}{32}$ in. C: $6\frac{9}{32}$ in. D: $8\frac{7}{16}$ in.

13. $4\frac{3}{32}$ in. 14. $21\frac{3}{8}$ in. 15. $244\frac{1}{2}$ in. 16. $859\frac{3}{8}$ in. or 71 ft $7\frac{3}{8}$ in.

17. $1\frac{1}{16}$ in. and $\frac{11}{16}$ in. 18. $\frac{3}{5}$ in. and $\frac{9}{10}$ in. 19. $5\frac{3}{16}$ in.

20. $22\frac{5}{16}$ in.; $22\frac{7}{16}$ in. 21. $26\frac{2}{3}$ min 22. $\frac{15}{32}$ in.

Chapter 3 *Exercises 3–1, page 100*

A. 1. Seventy-two hundredths
2. Eight and seven tenths
3. Twelve and thirty-six hundredths
4. Five hundredths
5. Three and seventy-two thousandths
6. Fourteen and ninety-one thousandths
7. Three and twenty-four ten-thousandths
8. Six and eighty-three ten-thousandths
9. 0.004 10. 3.4 11. 6.7
12. 0.005 13. 12.8 14. 3.021
15. 10.032 16. 40.7 17. 0.0116
18. 0.0047 19. 2.0374 20. 10.0222

B. 1. 21.01 2. 78.17 3. $15.02 4. $151.11
5. 1.617 6. 5.916 7. 828.6 8. 238.16
9. 63.7305 10. 462.04 11. 6.97 12. 1.04
13. $15.36 14. $6.52 15. 42.33 16. 36.18
17. $22.02 18. $24.39 19. 113.96 20. 13.22
21. 45.195 22. 245.11 23. $27.51 24. 151.402
25. 95.888 26. 39.707 27. 15.16 28. 86.07
29. 8.618 30. 18.6373 31. 31.23 32. 292.19
33. 17.608 34. 29.395 35. 0.0776 36. 0.04262
37. 24.22 38. 0.612 39. 1.748 40. 1.833

C. 1. 0.473 in. 2. 11.85 lb
3. (a) 0.013 in. (b) smaller; 0.021 in. (c) #14
4. $1426.75 5. A: 2.246 in. B: 0.455 in. C: 4.21 in.
6. 32.275 in. 7. 2.267 in. 8. $767.05 9. No 10. 68 ft
11. 3.4 hr 12. 0.013 cm 13. 3.37 in. 14. 19.3 cm

D. 1. $308.24   2. $19,759.97
    3. (a) 0.7399   (b) 4240.775   (c) 510.436   (d) 7.4262

Exercises 3–2, page 115

A. 1. 0.00001   2. 21.5   3. 4   4. 0.09
    5. 0.84   6. 0.00006   7. 0.00003   8. 2.18225
    9. 0.07   10. 0.03   11. 2.16   12. 3.6225
   13. 6.03   14. 120   15. 20   16. 130
   17. 0.045   18. 126   19. 60   20. 10,000
   21. 400   22. 0.037   23. 6.6   24. 3256.25
   25. 605   26. 13.915   27. 0.00378 28. 0.29
   29. 0.048   30. 3200   31. 45000   32. 0.000365
   33. 0.00008   34. 0.001128 35. 0.000364 36. 0.000048
   37. 0.000096   38. 55.53   39. 1.705

B. 1. 3.33   2. 0.83   3. 10.53   4. 37.04
    5. 0.12   6. 2.62   7. 33.86   8. 4.96
    9. 33.3   10. 0.2   11. 0.3   12. 11.1
   13. 0.2   14. 0.336   15. 0.143   16. 0.224
   17. 65   18. 13.268   19. 2.999   20. 1109.001

C. 1. $126   2. $15.60   3. 18.8 lb, or 19 rounded   4. 7.8 hr
    5.

	W	C
A	16.65 lb	$16.32
B	23.46	20.88
C	8.55	8.98
D	2.775	5.97

$T = \$52.15$

    6. 40.035 lb   7. 48 lb   8. 8.36 lb
    9. 153.37 ft   10. 6255 lb   11. 0.075 volts
   12. $1642.08   13. 335   14. 1600   15. 27 gal

D. 1. 8.00000007
    2. (a) 0.01234567   (b) 0.00112233   (c) 0.000111222
    3. 4.2435 in.   4. 0.0659 mm   5. $580
    6. (a) 0.031 lb   (b) 10¢   7. 26.2 psi

Exercises 3–3, page 123

A. 1. 0.25   2. 0.67   3. 0.75   4. 0.4   5. 0.8
    6. 0.83   7. 0.29   8. 0.57   9. 0.86   10. 0.38
   11. 0.75   12. 0.1   13. 0.3   14. 0.17   15. 0.42
   16. 0.19   17. 0.38   18. 0.56   19. 0.81   20. 0.15
   21. 0.22   22. 0.65   23. 0.46   24. 0.61   25. 0.03
   26. 0.213   27. 0.019   28. 0.34

B. 1. 4.385   2. 1.77   3. 1.5   4. 0.7681
    5. 7.88   6. 2.33   7. 1.43   8. 2.98
    9. 3.64   10. 10.908   11. 7.65   12. 1.07

C. 1. 1.375 g; 5.2 tablets
    2. (a) 420 sq ft   (b) 3712.5 sq ft of 4 in. and 4851 sq ft of 6 in.
        (c) $25,707.12
    3. 2.3 squares   4. 2 in.
    5. (a) $64.94   (b) $385.94   (c) $316.74
        (d) $295.83; Total: $1063.45
    6. 523 full lots   7. 0.396 in.

8. 3.4775 in.   9. (a) 120.2 ft   (b) 12 ft $6\frac{5}{8}$ in.

10. $19\frac{3}{8}$ in.

D. 1.

Gauge No.	Thickness (in.)	Gauge No.	Thickness (in.)
7–0	0.5	14	0.078
6–0	0.469	15	0.070
5–0	0.438	16	0.063
4–0	0.406	17	0.056
3–0	0.375	18	0.05
2–0	0.344	19	0.044
0	0.313	20	0.038
1	0.281	21	0.034
2	0.266	22	0.031
3	0.25	23	0.028
4	0.234	24	0.025
5	0.219	25	0.022
6	0.203	26	0.019
7	0.188	27	0.017
8	0.172	28	0.016
9	0.156	29	0.014
10	0.141	30	0.013
11	0.125	31	0.011
12	0.109	32	0.010
13	0.094		

2. $6866.54; $16,618.88; $10,902.33; $14,774.40; Total: $49,162.15

3. $438.48

Problem Set 3, page 127

A. 1. Ninety-one hundredths
2. Eighty-four hundredths
3. Twenty-three and one hundred sixty-four thousandths
4. Sixty-three and two hundred nineteen thousandths
5. Nine and three tenths
6. Three and forty-five hundredths
7. Ten and six hundredths
8. Fifteen and thirty-seven thousandths

9. 0.07   10. 0.018   11. 200.8
12. 16.17   13. 63.063   14. 110.021
15. 5.0063   16. 11.0218

B. 1. 23.19   2. 174.96   3. $19.29   4. 26.06   5. 1.94
6. $3.95   7. 88.26   8. 12.55   9. 277.104   10. 5.311
11. 239.01   12. 253.01   13. 83.88   14. 3.985   15. 33.672
16. 4.71   17. 4.28   18. 0.34   19. 1.92   20. 6.77

C. 1. 0.00008   2. 0.003   3. 0.84   4. 37.68   5. 0.108
6. 2.12   7. 19.866   8. 1.24   9. 61.7   10. 0.26
11. 3.8556   12. 28.585   13. 0.006   14. 5.26   15. 18
16. 0.2   17. 4.34   18. 4.36   19. 0.23   20. 1.50
21. 2.78   22. 0.22   23. 0.526   24. 0.503   25. 214.634
26. 0.028   27. 1.5   28. 23.5   29. 4.6   30. 0.9

D. 1. 0.0625   2. 0.875   3. 0.15625   4. 0.4$\overline{375}$
5. 1.375   6. 1.1875   7. 2.34375   8. 1.1$\overline{6}$
9. 2.$\overline{6}$ 10. 1.03125   11. 2.3125
12. 19.28   13. 30.67   14. 1.92   15. 78.57
16. 0.96   17. 0.13

E.   1.   291   2.   2.75 in.   3.   9.772 in.   4. $322.13
     5.   0.00857 in.   6.   1.132 ohms   7.   489.4 lb   8.   0.02374 in.
     9.   (a)   0.1875 in.   (b)   0.15625 in.   (c)   0.375 in.
          (d)   0.203125 in.
    10.   (a) $\frac{9}{64}$ in.   (b) $\frac{6}{64}$ in.   (c) $\frac{19}{64}$ in.   (d) $\frac{12}{64}$ in.
    11.   (a)   0.0089 in.   (b)   0.0027 in.   (c)   0.55215 in.   (d)   0.306175 in.
    12. $1408.74   13.   17.5 ft   14.   0.0423 in.
    15.   Max: 2.53125 in. Min: 2.46875 in.   16. $423.13   17. $229.90
    18. $23.85   19.   56.241 cu ft   20.   27 in.   21.   19
    22.   (a) $1859.38   (b) $201.25   (c) $804.88   (d) $554.84
          (e) $298.41
    23.   11.4 hr   24.   1.4045 in.   25.   52 lines   26.   359 gal

Chapter 4   *Exercises 4–1, page 146*

A.   (In each case only the missing answer is given.)
     1.   (a)   7:1   (b)   12:7   (c)   6   (d)   6
          (e)   45   (f)   9   (g)   16   (h)   50
          (i)   3:2   (j)   2:5
     2.   (a)   8:3   (b)   5:4   (c)   16 in.   (d)   6 cm
          (e)   40 cm   (f) $2\frac{1}{2}$:1 or 5:2   (g)   0.75 or 3:4   (h)   6.5 cm
          (i)   5.4 cm   (j)   17.8 cm
     3.   (a)   2:3   (b)   8 ft   (c)   28 ft   (d)   4:7
          (e) $\frac{3}{5}$ or 3:5   (f)   4 ft   (g)   20 ft   (h)   5 ft 1 in.

B.   1. $x = 12$ 2. $R = 86.4$ 3. $y = 100$ 4. $H = 60$
     5. $P = \frac{7}{6}$ 6. $x = 102$ 7. $A = 3.25$ 8. $M = 7.2$
     9. $T = 1.8$ 10. $N = 0.42$ 11. $x = 3$ 12. $A = 1.3$
    13. $x = 5.04$ 14. $R = 6\frac{1}{2}$ 15. $L = 5$ ft   16. $x = 58.9$ cm
    17. $x = 4$ cm   18. $W = \frac{1}{8}$ in. or 0.125 in.

C.   1.   5 cu in.   2.   7 gal   3.   21 pins
     4.   25.5 lb   5.   5.25 cu yd   6.   300 lb
     7.   288.75 gal   8.   224 ounces   9.   124.8 lb
    10.   12 ounces   11. $512   12.   72 parts
    13.   375 ft   14. $20\frac{5}{8}$ lb or 20.625 lb

Exercises 4–2, page 155

A.   1.   32%   2.   100%   3.   50%   4.   210%
     5.   25%   6.   375%   7.   4000%   8.   67.5%
     9.   200%   10.   7.5%   11.   50%   12. $16\frac{2}{3}$%
    13.   33.5%   14.   0.1%   15.   0.5%   16.   30%
    17.   150%   18.   7.5%   19.   330%   20.   20%

B.   1.   0.06   2.   0.45   3.   0.01   4.   0.33
     5.   0.71   6.   4.56   7.   0.005   8.   0.0005
     9.   0.0625   10.   0.0875   11.   0.3   12.   0.021
    13.   8   14.   0.08   15.   0.0025   16. $0.16\overline{3}$

Exercises 4–3, page 162

A.   1.   80%   2.   64%   3.   15   4.   100%
     5.   54   6.   12.5%   7.   150   8.   80
     9. $66\frac{2}{3}$%   10.   500%   11.   100   12.   52%
    13. $21.25   14.   600%   15.   1.5   16. $23\frac{1}{3}$
    17.   43.75%   18. $133\frac{1}{3}$%   19.   2.4   20.   375
    21.   40   22.   3.25   23.   5000   24.   17.6

B. 1. 225 2. 25% 3. 20¢ 4. 180%
    5. 160 6. $2.72 7. 150% 8. 2%
    9. 17.5 10. 400% 11. 427 12. 22.1
    13. 2% 14. 3.625 15. 460 16. 1400
    17. 50 18. 2.475

C. 1. 88% 2. $48,200 3. 5.5% 4. $36,900
    5. $117.60 6. $130 7. $1.84 8. $51.50
    9. 9.6 in. by 12 in. 10. 67% 11. 1447 ft

Exercises 4–4, page 179

1. 64% 2. 3.78 kW 3. 63.6%
4. 5035 to 5565, 2695 to 2805, 6120 to 7480, 4536 to 6804
    tolerances: 265, 55, 680, 1134
5. 67.5% 6. 550 ohms 7. 5%
8. $504 9. 12.5%
10. $16\frac{2}{3}\%$ ($\frac{2}{3}$ lb) tin; 2% (0.08 lb) zinc; $81\frac{1}{3}\%$ (3.25 lb) copper
11. 0.9% 12. 1660 lb 13. 81.25 hp 14. Yes (1910)
15. 0.13% 16. $31.96
17. (a) 0.03% (b) 0.44% (c) ±0.007 in.
18. $9.69 19. 37.5% 20. $82.13 21. $16.69 22. $725
23. $1312.50 24. (a) $10,625 (b) $460.42
25. $350.64 26. 21.1 mpg 27. B, by $1.74 28. $170.64
29. 37.5% 30. 8% 31. 15.6 MGD 32. 39% 33. 150 lb
34.

Measurement	Tolerance	Maximum	Minimum	Percent Tolerance
1.58 in.	±0.002 in.	1.582 in.	1.578 in.	0.13%
	±0.005 in.	1.585 in.	1.575 in.	0.32%
	±0.002 in.	1.582 in.	1.578 in.	0.15%
0.647 in.	±0.004 in.	0.651 in.	0.643 in.	0.62%
	±0.001 in.	0.648 in.	0.646 in.	0.15%
	±0.001 in.	0.648 in.	0.646 in.	0.20%
165.00 mm	±0.15 mm	165.15 mm	164.85 mm	0.09%
	±0.50 mm	165.50 mm	164.50 mm	0.30%
	±0.08 mm	165.08 mm	164.92 mm	0.05%
35.40 mm	±0.01 mm	35.41 mm	35.39 mm	0.03%
	±0.07 mm	35.47 mm	35.33 mm	0.20%
	±0.02 mm	35.42 mm	35.38 mm	0.05%
	±0.03 mm	35.43 mm	35.37 mm	0.08%
	±0.04 mm	35.44 mm	35.56 mm	0.10%

35. 28% 36. $1389.98 37. $228.31
38. $588,571.43 39. Approx. 2.8%

Problem Set 4, page 183

A. 1. (a) 30 in. (b) 5 to 2 (c) 25 in.
    2. (a) 1 to 3 (b) 65 (c) 42

B. 1. $x = 35$ 2. $x = 22.5$ 3. $x = 20$
   4. $x = 12$ 5. $x = 14.3$ 6. $x = 8.928$

C. 1. 72%   2. 6%   3. 60%   4. 35.8%   5. 130%
   6. 303%   7. 400%   8. 70%   9. $16\frac{2}{3}$%   10. 260%

D. 1. 0.04   2. 0.37   3. 0.11   4. 0.94   5. 0.0125
   6. 0.0009   7. 0.002   8. 0.017   9. 0.03875   10. 0.0802
   11. 1.15   12. 2.1

E. 1. 60%   2. $6   3. 5.6   4. 200%   5. 42
   6. 120%   7. $28.97   8. 125   9. 8000   10. 0.099

F. 1. 29.92 in.   2. 0.9 lb of carbon in cast iron; 0.016 lb in wrought iron
   3. (a) $4267.50   (b) $299.50   (c) $429   4. $172.58
   5. (a) $9.52   (b) $11.77   (c) 73¢   (d) $35.62   (e) $472.71
   6. $427.27   7. 1748 ft   8. 4%
   9. (a) 0.06%   (b) 2.82%   (c) ±0.009 in.   (d) 0.03%
      (e) ±0.62 in.
   10. 22%   11. $7216.25   12. 14.71 cm
   13. ±135 ohms; 4365 to 4635 ohms
   14. 68.6%   15. 0.9%   16. $5.06   17. 72%
   18. (a) $7267.50   (b) $729.65   19. 9.12 HP
   20. 22.5%   21. 3.6%
   22. 1950 bricks   23. (a) 66 fps   (b) 15 mph
   24. (a) 0.07 in. per foot   (b) $1\frac{3}{4}$ in. per 25 feet
   25. $5\frac{1}{4}$ in.   26. 8.8 lb
   27. 150 seconds or $2\frac{1}{2}$ minutes
   28. $1184 per month   29. 450 parts   30. 74

Chapter 5   *Exercises 5–1, page 202*

A. 1. 11 in.   2. 50 lb   3. 9.0 in.   4. 38.6 psi
   5. 0.334 in.   6. 8.93 ft   7. 47.5 mph   8. 2.8 gal
   9. 0.08 in.   10. 25.4 psi

B. 1. 40 sq ft   2. 5.3 sq in.   3. 7.2 sq ft   4. 25 miles
   5. 1.9 sq ft   6. 380 sq in.   7. 4.6 cu ft   8. 8.6 ft
   9. 6.2 mph   10. $0.43/lb   11. 1.6 hr   12. 1.9 ft

C. 1. (a) 1.88 in.   (b) 4.05 in.   (c) 3.13 sec   (d) 0.06 in.
      (e) 0.09 in.   (f) 1.19 lb   (g) 2.27 in.   (h) 0.59 in.
      (i) 0.38 lb

   2. (a) $\frac{15}{16}$ in.   (b) $2\frac{9}{16}$ in.   (c) $1\frac{13}{16}$ in.   (d) $3\frac{11}{16}$ in.
      (e) $\frac{13}{16}$ in.   (f) $\frac{5}{16}$ in.   (g) $1\frac{15}{16}$ in.   (h) $1\frac{9}{16}$ in.
      (i) $\frac{13}{16}$ in.

   3. (a) $1\frac{29}{32}$ in.   (b) $\frac{27}{32}$ in.   (c) $2\frac{11}{32}$ in.   (d) $\frac{21}{32}$ in.
      (e) $2\frac{3}{32}$ in.   (f) $\frac{9}{32}$ in.   (g) $\frac{19}{32}$ in.   (h) $\frac{22}{32}$ in.
      (i) $1\frac{17}{32}$ in.

4. (a) $\frac{15}{64}$; error of 0.0006   (b) $\frac{33}{64}$; error of 0.0006

    (c) $1\frac{51}{64}$; error of 0.0031   (d) $2\frac{27}{64}$; error of 0.0019

    (e) $3\frac{11}{64}$; error of 0.0031   (f) $2\frac{55}{64}$; error of 0.0006

    (g) $1\frac{60}{64}$; error of 0.0025   (h) $\frac{41}{64}$; error of 0.0044

    (i) $\frac{31}{64}$; error of 0.0044

D. 1. Yes   2. $\frac{13}{32}$; error 0.0034   3. $\frac{30}{64}$ in.
   4. $\frac{24}{32}$ in.   5. 4.28 in.   6. 0.02325 in.   7. $3\frac{5}{16}$ in.
   8. 17.5 ft   9. 354 mi   10. 23 ft.

Box, page 209

1. (a) 2 bf   (b) 16 bf   (c) 8 bf   (d) $10\frac{2}{3}$ bf   (e) 7 bf   (f) 3 bf
2. 1280 bf   3. 4410 bf

Exercises 5–2, page 211

A. 1. 51 in.   2. 99 ft   3. 17,952 ft   4. 272
   5. 4 bbl   6. 14,960 yd   7. 3.1 atm   8. 120 oz
   9. 582.8 gal   10. $2\frac{2}{3}$ ft   11. $\frac{1}{2}$ mi   12. 3.5 rods
   13. 8.7 yd   14. 14.4 lb   15. 1180 cu in.   16. 26,300 cu in.
   17. 0.80 mi   18. 6.4 gal   19. 83 in.   20. 206 oz
   21. 20 lb 6 oz   22. 9 yd 2 ft

B. 1. 864   2. 108,900   3. 72.6   4. 42.8
   5. 0.563   6. 5.42   7. 0.03   8. 7.1
   9. $0.\overline{3}$ or $\frac{1}{3}$ 10. 500   11. 12,800   12. 95
   13. 2700   14. 380   15. 0.30   16. 220

C. 1. (a) 43,560   (b) 325,829   (c) 10,344   (d) 1613
   2. (a) $\frac{1}{12}$ (b) 144   3. 4
   4. (a) $12\frac{1}{2}$ gal   (b) \$20.37   (c) 2000 sq ft   (d) 2.3 gal
     (e) \$56.18
   5. 32.6 ± 0.6 in.
   6. (a) 5400 in. × 900 in. × 540 in. or 450 ft × 75 ft × 45 ft
     (b) 117 in. or 9 ft 9 in.
   7. 141 sq yd   8. 0.1 lb/cu in.   9. 10.2 rps   10. 28 pieces
   11. 1.93 cfs   12. 270 min   13. $466\frac{1}{2}$ in.   14. 18,000 oz

Exercises 5–3, page 224

A. 1. (c)   2. (b)   3. (a)   4. (a)   5. (b)   6. (c)   7. (a)
   8. (c)   9. (a)   10. (b)   11. (c)   12. (c)   13. (a)   14. (c)
   15. (a)   16. (a)   17. (a)   18. (b)   19. (a)   20. (b)

B. 1. (b)   2. (c)   3. (a)   4. (a)   5. (b)   6. (c)
   7. (b)   8. (b)   9. (b)   10. (c)   11. (a)   12. (a)

C. 1. 7.6   2. 130   3. 3.2   4. 6.25   5. 23
   6. 68   7. 69.0   8. 1.5   9. 19   10. 295
   11. 11.9   12. 2.6   13. 6.1   14. 98   15. 5.0
   16. 110   17. 50   18. 89   19. 19   20. 6.6
   21. 26.5 m^2 22. 8390 cm^3
   23. 3.59 cu yd   24. 7.16 sq in.

D.

in.	mm
0.030	0.762
0.035	0.889
0.040	1.016
0.045	1.143

in.	mm
$\frac{1}{16}$	1.588
$\frac{5}{64}$	1.984
$\frac{3}{32}$	2.381
$\frac{1}{8}$	3.175

in.	mm
$\frac{5}{32}$	3.969
$\frac{3}{16}$	4.763
$\frac{3}{8}$	9.525
$\frac{11}{64}$	4.366

2. (a) 2.2 lb (b) 28.3 kg (c) 1.0 kg
3. 11.9 mph; 19.1 kmh
4. (a) 103,000 (b) 689,000 (c) 2000 (d) 120,000
5. (a) 0.6093 (b) 564 (c) 0.89 (d) 0.057 (e) 1.02
6. 57.15 mm by 69.85 mm or 57 by 70, rounded
7. 12 km/liter 8. $1.39 9. 9.3 m^2
10. (a) 2.4 m × 1.2 m × 1.2 m (b) 3.6 m^3
11. (a) 1.6093 (b) 76.2 (c) 21.59 cm × 27.94 cm (d) 2.54
(e) 804.65 (f) 0.0648 (g) 28.35; 453.6 (h) 0.0648
(i) 7.57 (j) 37.85 (k) 7568 12. 7 liters

Exercises 5–4, page 245

A. 1. (a) $\frac{5}{8}$ in. (b) $1\frac{7}{8}$ in. (c) $2\frac{1}{2}$ in. (d) 3 in.

(e) $\frac{1}{4}$ in. (f) $1\frac{1}{16}$ in. (g) $2\frac{5}{8}$ in. (h) $3\frac{1}{2}$ in.

2. (a) $\frac{5}{16}$ in. (b) $\frac{3}{4}$ in. (c) $1\frac{1}{32}$ in. (d) $1\frac{17}{32}$ in.

(e) $\frac{1}{8}$ in. (f) $\frac{17}{32}$ in. (g) $\frac{53}{64}$ in. (h) $1\frac{7}{16}$ in.

3. (a) $\frac{2}{10}$ in. = 0.2 in. (b) $\frac{5}{10}$ in. = 0.5 in. (c) $1\frac{3}{10}$ in. = 1.3 in.

(d) $1\frac{6}{10}$ in. = 1.6 in. (e) $\frac{25}{100}$ in. = 0.25 in. (f) $\frac{72}{100}$ in. = 0.72 in.

(g) $1\frac{49}{100}$ in. = 1.49 in. (h) $1\frac{75}{100}$ in. = 1.75 in.

B. 1. 0.650 in. 2. 0.705 in. 3. 0.287 in. 4. 0.061 in.
5. 0.850 in. 6. 0.441 in. 7. 0.4068 in. 8. 0.5997 in.
9. 0.2581 in. 10. 0.7796 in. 11. 0.0888 in. 12. 0.8451 in.
13. 11.57 mm 14. 3.41 mm 15. 22.58 mm 16. 18.21 mm
17. 18.94 mm 18. 13.27 mm

C. 1. 3.256 in. 2. 0.836 in. 3. 2.078 in. 4. 0.609 in.
5. 2.908 in. 6. 1.682 in. 7. 2.040 in. 8. 0.984 in.
9. 0.826 in. 10. 2.656 in.

D. 1. 62°21′ 2. 19°38′ 3. 34°56′ 4. 65°10′
5. 35°34′ 6. 63°48′ 7. 20°26′ 8. 44°15′

E. 1. 2.000 in. + 0.200 in. + 0.050 in. + 0.057 in. + 0.0503 in.
2. 1.000 in. + 0.500 in. + 0.200 in. + 0.140 in. + 0.056 in. + 0.0507 in.
3. 0.200 in. + 0.120 in. + 0.053 in. + 0.0502 in.
4. 2.000 in. + 1.000 in. + 0.500 in. + 0.200 in. + 0.050 in. + 0.058 in. + 0.0507 in.
5. 1.000 in. + 0.500 in. + 0.070 in. + 0.058 in. + 0.0509 in.
6. 2.000 in. + 0.500 in. + 0.100 in. + 0.050 in. + 0.058 in. + 0.0508 in.
7. 0.050 in. + 0.053 in. + 0.0509 in.
8. 2.000 in. + 0.500 in. + 0.300 in. + 0.120 in. + 0.057 in. + 0.0506 in.
9. 2.000 in. + 0.140 in. + 0.0505 in.
10. 1.000 in. + 0.400 in. + 0.056 in. + 0.0506 in.

Problem Set 5, page 249

A. 1. 3.37 sec   2. 6.8 sq ft   3. 0.74 in.   4. 3.8 cm^2
    5. 16$\frac{7}{8}$ in.   6. 3 lb 14 oz   7. 21.1 mi/gal   8. 9.6 cm
    9. 17.65 psi   10. 3.2 ft   11. 3.869 in.   12. 403.9 lb
    13. 21 ft 3 in.   14. 12$\frac{25}{64}$ in.

B. 1. 15.24   2. 73.33   3. ≈72°   4. 2.75
    5. 45,792   6. 5°C   7. 31.53   8. 8.41
    9. 23.33   10. 5735.47   11. 60.22   12. 14.88
    13. 11.48   14. 1.82   15. 979.2   16. 11.59
    17. 103.63 cm   18. 114.3 cm/sec   19. 52.82 mph   20. 12.71 kg

C. 1. (b)   2. (b)   3. (a)   4. (a)   5. (b)   6. (c)

D. 1. (a) $\frac{5}{10}$ in. = 0.5 in.   (b) 1$\frac{3}{10}$ in. = 1.3 in.   (c) $\frac{17}{100}$ in. = 0.17 in.
        (d) 1$\frac{35}{100}$ in. = 1.35 in.   (e) $\frac{3}{4}$ in.   (f) 3$\frac{3}{4}$ in.   (g) $\frac{5}{16}$ in.
        (h) 1$\frac{19}{32}$ in.   (i) $\frac{3}{8}$ in.   (j) 1$\frac{27}{64}$ in.
    2. (a) 0.807   (b) 0.550   (c) 12.13 mm   (d) 19.07 mm
        (e) 0.5469 in.   (f) 0.8179 in.
    3. (a) 2.156 in.   (b) 3.030 in.   (c) 0.612 in.   (d) 1.925 in.
        (e) 1.706 in.   (f) 1.038 in.

E. 1. 22$\frac{3}{64}$ in.   2. 53 mph   3. 6.1 mph
    4. 0.0505 in., 0.057 in., 0.130 in., 0.200 in., 2.000 in.
    5. $\frac{15}{32}$ in.   6. 500°F   7. 936 sq in.   8. 18.9 gallons
    9. 293°C   10. 97.6 cu in.   11. 7.623 in.   12. (a) 3.85 m^3/hr
        (b) 7.62 m   (c) 6.35 cm   13. 0.28 lb/cu in.
    14. 28.3 liter   15. 5.08 cm × 10.16 cm or 5 cm by 10 cm   16. 1.5 m

Chapter 6 *Exercises 6–1, page 260*

A. 1. −4   2. −17   3. 18   4. −15
    5. −10   6. −31   7. −65   8. 23
    9. −47   10. −3   11. 0   12. −1.4
    13. −2.8   14. −3$\frac{1}{4}$ 15. −11.4   16. −0.1

B. 1.
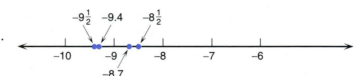

2.
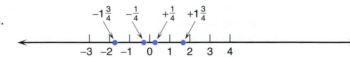

3.
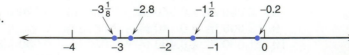

4.

$$-3.9 \qquad -1.4 \qquad +1\tfrac{1}{4} \qquad +4.1$$

(number line with points at −3.9, −1.4, +1¼, +4.1; marks from −4 to 5)

C.
1. −6	2. −300	3. +12,000	4. +240
5. −9	6. −7	7. −80	8. +15
9. −5	10. −10	11. +6000	12. −30

D.
1. 5	2. −4	3. −15	4. −25
5. 17	6. −5	7. 46	8. 31
9. −16	10. −16	11. 9	12. −27
13. 29	14. −65	15. −30	16. −80
17. −0.9	18. −2.2	19. −28.09	20. −133.84
21. $-1\tfrac{1}{4}$	22. $2\tfrac{1}{2}$	23. −14	24. $-15\tfrac{1}{3}$
25. $-\tfrac{3}{4}$	26. $3\tfrac{1}{8}$	27. $1\tfrac{9}{16}$	28. $-22\tfrac{1}{10}$
29. −10	30. 3	31. −16	32. 0
33. 2	34. −3	35. 18	36. −15
37. −230	38. 10	39. 7	40. 23.69
41. 3599	42. −4.89	43. −26,620	44. −0.0655

E.
1. +22 (22 above quota)   2. 6 yd; $\tfrac{6}{7}$ yards per carry
3. −9° (9 degrees below zero)   4. $884
5. −$136,000 (a loss of $136,000)   6. $-2\tfrac{15}{16}$ in.

Exercises 6–2, page 266

A.
1. −2	2. −13	3. −16	4. −24
5. 9	6. 21	7. −6	8. −12
9. 8	10. 14	11. 11	12. 3
13. −18	14. −35	15. −31	16. −29
17. 8	18. 27	19. 26	20. 45
21. 9	22. −11	23. −19	24. −26
25. −17	26. −55	27. 8	28. 14
29. −87	30. −112	31. −67	32. −27
33. −10	34. 4	35. −42	36. 25
37. 6	38. −13	39. −1	40. −9
41. $-\tfrac{5}{8}$	42. $5\tfrac{3}{4}$	43. $-6\tfrac{1}{2}$	44. $-\tfrac{1}{6}$
45. 9.2	46. −1.65	47. −63.5	48. −0.0725
49. 41.02	50. −6.14	51. −0.85	52. −1.575

B.
1. 14,698 ft   2. 263°
3. 35°F   4. 4423 ft
5. $1817   6. −9 (9 under par)
7. lost $5211   8. −0.625

Exercises 6–3, page 270

A.
1. −63	2. −40	3. 77	4. 36
5. −4	6. −4	7. 12	8. 4
9. −4	10. 9	11. $12\tfrac{3}{8}$	12. $-\tfrac{1}{2}$
13. 5.1	14. $-\tfrac{1}{4}$ (−0.25)	15. $\tfrac{1}{5}$ (0.2)	16. 52
17. −90	18. −54	19. 84	20. −20
21. −28	22. 20	23. 5	24. −3.45
25. −8	26. 8	27. $\tfrac{1}{6}$	28. −15
29. $-2\tfrac{11}{20}$	30. $1\tfrac{1}{9}$	31. 0.4	32. $11\tfrac{3}{4}$
33. $-19\tfrac{1}{2}$	34. 0.077	35. −4.48	36. −0.06
37. −90	38. 224	39. −20	40. 700
41. −0.4	42. 50		

B. 1. −19.3   2. 57   3. 0.0019
    4. 160,000   5. −3.16   6. 3.87
    7. 0.23   8. −0.030   9. 46
    10. 215   11. −70,000,000   12. −8,500,000

C. 1. −8⅓°C   2. −$23.6 million   3. (a) −120;   (b) 40
    4. −11.2 mm   5. −6000 ft/min   6. 17.6°F

Exercises 6–4, page 280

A. 1. 16   2. 9   3. 64   4. 125
    5. 1000   6. 49   7. 256   8. 36
    9. 512   10. 81   11. 625   12. 100,000
    13. −8   14. −243   15. 729   16. 1
    17. 5   18. 1   19. 32   20. 64
    21. 196   22. 441   23. 3375   24. 256
    25. 108   26. 576   27. 1125   28. 2744
    29. 2700   30. 9216   31. 18   32. 6
    33. 216   34. 49   35. 19   36. 2

B. 1. 9   2. 12   3. 6   4. 4
    5. 5   6. 3   7. 16   8. 20
    9. 15   10. 7   11. 18   12. 11
    13. 2.12   14. 22.36   15. 3.52   16. 26.46
    17. 14.49   18. 17.92   19. 28.46   20. 9.62
    21. 31.62   22. 44.72   23. 158.11   24. 50
    25. 12.25   26. 17.32   27. 54.77   28. 173.21
    29. 1.12   30. 1.00   31. 13.42   32. 8.47
    33. 51.26   34. 7.35   35. 4.80   36. 11.45

C. 1. 13.6 ft   2. 35.6 ft   3. 89,000 psi
    4. 19,000 ohms   5. 96 fps   6. 43 mph

Problem Set 6, page 283

A. 1. −13, −8, −2, 4, 7
    2. −0.96, −0.6, −0.55, −0.138, 0, 0.4
    3. −180, −160, −150, −140, −120, 0
    4. $-\frac{3}{4}, -\frac{5}{8}, -\frac{1}{2}, \frac{3}{8}, \frac{1}{2}, \frac{3}{4}$

B. 1. −10   2. −26   3. −6   4. 0
    5. −15   6. −50   7. 25   8. 8
    9. 2.3   10. $-2\frac{1}{2}$ 11. −$9.15   12. $-9\frac{7}{8}$
    13. −16   14. −27.35

C. 1. −60   2. 35   3. 12   4. −40
    5. $7\frac{1}{32}$ 6. $-\frac{3}{4}$ 7. −0.4   8. −36
    9. −120   10. 54   11. −16   12. 0.318

D. 1. −43.75   2. 15   3. −4   4. $-1\frac{1}{32}$
    5. 64   6. 243   7. 289   8. 12,167
    9. 0.25   10. 1.728   11. 0.0004   12. 0.000027
    13. 0.000000001   14. 4.0401   15. 16.1604
    16. −27   17. 16   18. 37   19. 28
    20. 2.1   21. 0.4   22. 2.3   23. 16.29
    24. 34.201   25. 121.68   26. 2.2   27. 432
    28. 22   29. 10   30. 20.52   31. 0.5
    32. 6.7   33. 8.94   34. 10.30   35. 17.61
    36. 1.34   37. 2.05   38. 1.74   39. 1.04

40. 0.28   41. 2.76   42. 0.21   43. 0.78
44. 7.71

E.   1. +7 amp   2. 15 ft   3. $-22°F$
    4. $-\$848$ 5. 0.17   6. 539 gal/min
    7. 700   8. \$61,780   9. 3200 nanosec
   10. -300 volts   11. (a) $-20°C$ (b) $-4°C$
   12. (a) $-5.8°F$ (b) $10.4°F$

Chapter 7   *Exercises 7–1, page 297*

A.   1. $A = 6$ 2. $D = 7$ 3. $T = 13$ 4. $H = 9$
   5. $K = 8$ 6. $Q = 38$ 7. $F = 19$ 8. $W = 6$
   9. $L = 1$ 10. $A = 8$ 11. $B = 13$ 12. $F = 9$

B.   1. $A = 11$ 2. $V = 250$ sq ft   3. $I = 15$ 4. $H = 10$ cm^2
   5. $T = 23$ m^2 6. $V = 221$ cu in.   7. $P = 40$ 8. $W = 40$ lb
   9. $V = 121$ m^3 10. $V = 14$ cu ft

C.   1. $P = 39$ in.   2. $i = 12$ amp   3. $P = 96$ watts
   4. $A = 1260$ cm^2 5. $F = 104°$ 6. $V = 140$ cu in.
   7. $A = \$1450$ 8. $R = 120$ ohms   9. $T = 19$ cm^2
   10. (a) 0.3125 in.   (b) 0.1185 in.   (c) 0.2625 cm   (d) $\frac{5}{32}$ in.
   11. $L = 18\frac{1}{4}$ in.   12. $L = 16.5$ in.   13. $L = 5700$ in.
   14. $L = 120$ in.   15. $X = 1000$ ft   16. $C = 179$ ft/min
   17. $P = 26.4$ kW   18. $V = 600$ mph   19. $L = 920$ cm
   20. $F = 1250$ lb   21. 36 lines   22. 3 pipes
   23. $V = 4.2850$ cu in.   24. $T = 39.3$ min
   25. (a) 0.7860 sq in.   (b) 0.6013 sq in.
       (c) 13.3803 sq in.   (d) 3.3345 cm^2
   26. 79.84 ft   27. 6.5 cu ft
   28. (a) $-8.5°C$ (b) $-18.1°C$ or $-0.6°F$

Exercises 7–2, page 306

A.   1. $9y$ 2. $9x^2y$ 3. $6E$
   4. $-4ax$ 5. $7B$ 6. 0
   7. $-2x^2$ 8. $10x + 16y$ 9. $2R^2 + 5R$
   10. $1.45A - 0.8A^2$ 11. $\frac{1}{8}x$ 12. $-2x$
   13. $1\frac{1}{2} - 3.1W$ 14. $q - 2\frac{1}{2}p$ 15. $3x + 6xy$
   16. $4ab$ 17. $5x^2 + x^2y + 3x$ 18. $4.6p + 0.3pq$

B.   1. $3x^2 + 2x - 5$ 2. $6 - 3a + 8b$
   3. $4m^2 + 10m$ 4. $11x - 5x^2$
   5. $2 - x - 5y$ 6. $7x - 4 - 2y$
   7. $3a - 8 + 6b$ 8. $5 - w + 6z$
   9. $-10x^2 - 3x$ 10. $8m - 6n$
   11. $23 - 3x$ 12. $-3 - 5y$
   13. $x + 8y$ 14. $8 - 9m$
   15. $-14 - 7w + 3z$ 16. $-14x - 4y + 8$
   17. $9x - 12y$ 18. $20a + 24b$
   19. $-56m - 48$ 20. $-12x + 6$
   21. $-9x - 15$ 22. $-2y - 16$
   23. $23m - 42$ 24. $-19w + 15$
   25. $3 - 8x - 12y$ 26. $6 - 6a + 10b$
   27. $28 - 6w$ 28. $-75 - 55a$
   29. $6x + 8y - 24x^2 + 20y^2$ 30. $-20a + 24b + 14ab - 28b^2$
   31. $4x + 10y$ 32. $18w - 31z$
   33. $3x + 9y$ 34. $33x^2 + 6x$
   35. $-2x^2 - 42x + 62$ 36. $-10a - 18b + 44ab$

A. 1. $x = 9$ 2. $A = 17$ 3. $x = -24$ 4. $z = 47$
    5. $a = 8\frac{1}{2}$ 6. $x = -62$ 7. $y = 24.66$ 8. $R = 7.9$
    9. $x = -13$ 10. $y = -13$ 11. $a = 0.006$ 12. $x = 21$
    13. $z = 0.65$ 14. $N = \frac{3}{4}$ 15. $m = -12$ 16. $x = -27$
    17. $y = 6.5$ 18. $a = -36$ 19. $T = -18$ 20. $y = 36$
    21. $Q = 0.7$ 22. $Z = 10$ 23. $x = 25$ 24. $y = 24$
    25. $K = 4.24$ 26. $z = -0.3$

B. 1. 384 volts   2. 5.64 hr or   3. 28.8 parts per   4. 12.8 ft
                         5 hr 38 min   million
    5. 1.715 m   6. 18 psi   7. 21 windings   8. 2900 rpm

A. 1. $x = 10$ 2. $x = -1$ 3. $x = 35$ 4. $y = -2.5$
    5. $m = 2.5$ 6. $x = \frac{1}{6}$ 7. $n = 10$ 8. $a = 5$
    9. $z = -6$ 10. $q = 10$ 11. $x = 48$ 12. $m = -5.5$
    13. $n = -9$ 14. $y = 25$ 15. $x = \frac{1}{4}$ 16. $x = 10$
    17. $a = 7$ 18. $x = 14$ 19. $z = -5\frac{1}{2}$ 20. $x = -3$
    21. $x = \frac{2}{5}$ 22. $x = -24$ 23. $P = 3$ 24. $x = 8$
    25. $x = 10$ 26. $x = 6$ 27. $x = 5\frac{1}{3}$ 28. $x = -7$

B. 1. 4.5 hr   2. 22,000 ft   3. 32.5 years   4. 10%
    5. 0.125 in.   6. 0.25 in.   7. 31.75 hr   8. 18 in.

A. 1. $x = 9$ 2. $y = -17$ 3. $n = 4\frac{5}{6}$ 4. $a = 1\frac{1}{6}$
    5. $x = -7$ 6. $m = 11$ 7. $y = 7\frac{1}{2}$ 8. $z = -2$
    9. $x = 1$ 10. $w = -\frac{1}{2}$ 11. $c = 5$ 12. $y = 2\frac{1}{3}$
    13. $x = -7$ 14. $a = -\frac{3}{4}$ 15. $t = \frac{1}{5}$ 16. $y = 1\frac{1}{2}$
    17. $x = -2$ 18. $n = -3$ 19. $y = 2$ 20. $p = 4$
    21. $x = 2$ 22. $y = 6$ 23. $t = -3\frac{3}{4}$ 24. $A = \frac{8}{9}$
    25. $x = 4\frac{2}{7}$ 26. $x = -1\frac{2}{3}$ 27. $x = 3$ 28. $x = 1$

B. 1. $L = \dfrac{S}{W}$ 2. $B = \dfrac{2A}{H}$ 3. $I = \dfrac{V}{R}$

    4. $D = 2H + R$

    5. $T = \dfrac{2S - WA}{W}$ 6. $H = \dfrac{V}{\pi R^2}$ 7. $B = \dfrac{P - 2A}{2}$

    8. $R = 2H - 0.1$ 9. $P = \dfrac{T(R + 2)}{R}$ 10. $V = IR - E$

C. 1. (a) $a = \dfrac{360L}{2\pi R}$ (b) $R = \dfrac{360L}{2\pi a}$ (c) $L = 5.23$ in.

    2. (a) $a = \dfrac{360A}{\pi R^2}$ (b) $A = 57$ sq in.

    3. (a) $R_1 = \dfrac{V - R_2 i}{i}$ (b) $R_2 = 50$ ohms

    4. (a) $N = \dfrac{3.78P}{t^2 d}$ (b) $P = 0.05$ HP

    5. (a) $V_1 = \dfrac{V_2 P_2}{P_1}$ (b) $V_2 = \dfrac{V_1 P_1}{P_2}$ (c) $P_1 = \dfrac{V_2 P_2}{V_1}$

        (d) $P_2 = \dfrac{V_1 P_1}{V_2}$ (e) $P_1 = 300$ psi

6. (a) $L = \dfrac{6V}{\pi T^2}$ (b) $T^2 = \dfrac{6V}{\pi L}$

7. (a) $D = \dfrac{CA + 12C}{A}$ (b) 0.14 gram

8. $H = PF - T$

9. (a) $i_L = \dfrac{i_s T_p}{T_s}$ (b) $i_s = \dfrac{i_L T_s}{T_p}$ (c) $i_L = 22.5$ amp

10. (a) $P = IV$ (b) $I = \dfrac{P}{V}$ (c) $V = 375$ volts

11. $L = 4138$ microhenrys
12. (a) 1.11 in.   (b) 0.989 in.   (c) 0.741 in.
13. 7732 parts   14. 3.125 years   15. 110.7°

Exercises 7–6, page 348

A. 1. $H = 1.4\,W$ 2. $W_1 + W_2 = 167$ 3. $V = \tfrac{1}{4}h\pi d^2$

4. $K = 0.454\,P$ 5. $V = AL$ 6. $V = \tfrac{1}{3}\pi h r^2$

7. $D = \dfrac{N}{P}$ 8. $T = \dfrac{L}{FR}$ 9. $W = 0.785\,hDd^2$

10. $V = iR$

B. 1. $133\tfrac{1}{3}$ gal and $266\tfrac{2}{3}$ gal
2. 36 lb and 48 lb
3. $13\tfrac{5}{7}$ ft, $6\tfrac{6}{7}$ ft, $3\tfrac{3}{7}$ ft
4. 10 V, 12 V, 16 V
5. 22, 22, 22, 20, 18, 16, 14, 12, 10 blocks
6. $6\tfrac{3}{4}$ cu ft of cement, $13\tfrac{1}{2}$ cu ft of water, $13\tfrac{1}{2}$ cu ft of aggregate, $20\tfrac{1}{4}$ cu ft of sand
7. Jo: \$48,800; Ellen: \$36,600
8. Pavement: \$320,000; Base material: \$160,000 Sidewalk: \$80,000
9. 21 (with some left over)
10. 64 ft   11. $6\tfrac{2}{3}$ hr   12. 36 avocados   13. 5000 parts
14. \$35.16   15. \$50.02   16. 523 pieces

Exercises 7–7, page 360

A. 1. (a) $\tfrac{5}{8}$ in.   (b) 31.2 cm   (c) 8 ft   (d) 3 m
2. (a) $2\tfrac{4}{5}$ in.   (b) $12\tfrac{1}{2}$ ft   (c) 1.4 m   (d) 42 cm
3. (a) 100   (b) 9   (c) 25   (d) 1500
4. (a) 150   (b) 15   (c) 8   (d) 700

B. 1. 24 in.   2. (a) 43.5 ft   (b) 48 to 1
3. 139 HP   4. 216 ft   5. 109.1 lb
6. 120 min   7. 980 rpm   8. 144 lb
9. 23 lb   10. 2500 rpm   11. 500 rpm
12. 21.6 watts   13. 6 hr   14. 33 turns
15. 1470 ft   16. 3.8 hr, rounded   17. 1.4 in.
18. 12.5 cu yd cement, 37.5 cu yd sand, 75 cu yd gravel
19. 4 in.   20. $9\tfrac{3}{8}$ in. by $4\tfrac{3}{4}$ in.   21. 22.4 psi
22. 4.766 ohms   23. \$377.54
24. $A = 5.95$ cm, $B = 5.10$ cm, $C = 4.25$ cm
$D = 3.40$ cm, $E = 2.55$ cm, $F = 1.70$ cm
$G = 0.85$ cm
25. 14.4 hr, rounded   26. 86.1 psi

A. 1. $20x$ 2. a^5 3. $6R^2$ 4. $9x^2$
    5. $-8x^3y^4$ 6. $8x^3$ 7. $0.6a^2$ 8. $2x^4A^3$
    9. $\frac{1}{16}Q^4$ 10. $1.2p^4q^4$ 11. $24M^6$ 12. $6x^2-2x$
    13. $-2+4y$ 14. a^3b-ab^3 15. p^6 16. $6x^2$
    17. $5x^7$ 18. $8x^6$ 19. $10y^7$ 20. $4a^4b^3$
    21. x^4y^4 22. x^2+2x 23. $3p^3$ 24. x^2y+xy^2
    25. 5^{10} 26. 3^9 27. 10^{-3} 28. 2^3
    29. $-6x^4+15x^5$ 30. $24x^7+18x^5$
    31. $6a^3b^2-10a^2b^3+14ab^4$ 32. $-20x^3y^3+30x^2y^4-40xy^5$

B. 1. 4^2 2. 10^6 3. x^3 4. y^4

    5. $\dfrac{1}{10^2}$ 6. $\dfrac{1}{m^5}$ 7. $2a^4$ 8. $-3m^6$

    9. $-\dfrac{2}{3y}$ 10. $\dfrac{4}{t^2}$ 11. $3ab^3$ 12. $-4x^2y^4$

    13. $-\dfrac{3}{m^4n}$ 14. $\dfrac{3c^4}{4d^2}$ 15. $-\dfrac{3}{5b^3}$ 16. $-\dfrac{10}{xy^3}$

    17. $3x^2-4$ 18. $3y^2+2y$ 19. $-2a^5+a^3-3a$
    20. $-3m^4+2m+1$

A. 1. 5×10^3 2. 4.5×10^2 3. 9×10^1
    4. 4.07×10^4 5. 3×10^{-3} 6. 7.1×10^{-2}
    7. 4×10^{-4} 8. 5.9×10^{-3} 9. 6.77×10^6
    10. 3.82×10^4 11. 2.92×10^{-2} 12. 9.901×10^{-3}
    13. 1.001×10^3 14. 2×10^{-3} 15. 1.07×10^{-4}
    16. 8.1×10^5 17. 3.14×10^1 18. 6×10^{-1}
    19. 1.25×10^2 20. 7.4×10^{-1} 21. 2.9×10^7 lb/sq in.
    22. 2.4×10^{-4} cal/cm · sec   23. 9.55×10^7 watts
    24. 6.588×10^8 mph

B. 1. 200,000   2. 7,000,000   3. 0.00009   4. 0.0003
    5. 0.0017   6. 0.037   7. 51,000   8. 870
    9. 40,500   10. 0.00000701   11. 0.003205   12. 1007
    13. 2,450,000   14. 0.000319   15. 647,000   16. 0.000000826

C. 1. 8×10^7 2. 8.9×10^{-9} 3. 7.8×10^2
    4. 9.9×10^2 5. 2.8×10^{-9} 6. 2.5×10^{-8}
    7. 5×10^6 8. 1.5×10^6 9. 1.4×10^2
    10. 1.5×10^{-2} 11. 6.3×10^{-2} 12. 9.5×10^{-2}
    13. 2.3×10^{10} 14. 6.5×10^{-1} 15. 8×10^{-11}
    16. 5.9×10^{-3} 17. 2.8×10^{-2} 18. 1.2×10^{-6}
    19. 5.9×10^{-12} 20. 2.0×10^{15} 21. 6.6×10^{13}
    22. 6.5×10^{-10} 23. 4.0×10^2 24. 4.1×10^{-4}

D. 1. 13.3 cal/sec   2. 2×10^{-5} 3. 1.53×10^{-4} μF
    4. 3.98×10^{-19} Joules

A. 1. $10x$ 2. $12y$ 3. $12xy$ 4. $15xy^3$
    5. $1\frac{11}{24}x$ 6. 0 7. $0.4G$ 8. $16y-8y^2$
    9. $21x^2$ 10. $8m^3$ 11. $-10x^2y^2z$ 12. $27x^3$
    13. $12x-21$ 14. $6a^3b-10ab^3$ 15. $9x+11$ 16. $5y-7$

    17. $9x-3y$ 18. $-5x^2+9$ 19. 6^4 20. $\dfrac{1}{x^5}$

    21. $-3m^8$ 22. $-\dfrac{3}{5y^2}$ 23. $\dfrac{4a}{b}$ 24. $2x^2-x+3$

B. 1. $L = 13$ 2. $M = 8$ 3. $I = 144$ 4. $I = 183{,}333\frac{1}{3}$
   5. $V = 42\frac{21}{64}$ 6. $N = -95$ 7. $L = 115.2$ 8. $f = \frac{1}{48}$
   9. $t = 4$ 10. $V = 270.64$

C. 1. $x = 12$ 2. $m = 0.35$ 3. $e = 44$ 4. $a = 0.9$
   5. $x = -5$ 6. $x = 10\frac{1}{2}$ 7. $x = 4$ 8. $m = -6$
   9. $y = 32$ 10. $M = 60$ 11. $B = 3.5$ 12. $x = 6.23$
   13. $y = 40$ 14. $E = 48$ 15. $F = 0.72$ 16. $x = 3$
   17. $x = -6$ 18. $f = -1.15$ 19. $x = -4$ 20. $y = 2$
   21. $g = 10$ 22. $m = 1.5$ 23. $x = 16$ 24. $x = -4$
   25. $x = 3$ 26. $x = 6$ 27. $b = \dfrac{A}{H}$ 28. $P = R - S$

   29. $L = \dfrac{P - 2W}{2}$ 30. $F = \dfrac{w}{P}$

   31. $g = \dfrac{2S + 8}{t}$ 32. $A = \dfrac{\pi R^2 H - V}{B}$

D. 1. 7.5×10^3 2. 1.28×10^4 3. 4.1×10^{-2}
   4. 2.36×10^{-4} 5. 5.72×10^{-3} 6. 4.82×10^{-6}
   7. 4.47×10^5 8. 2.127×10^6 9. 8.02×10^7
   10. 4.671×10^4 11. 7.05×10^{-6} 12. 1.006×10^{-3}
   13. 930,000   14. 6020   15. 0.00029
   16. 0.00000305   17. 0.0000005146   18. 0.006203
   19. 90,710   20. 4,006,000   21. 5.7×10^{10}
   22. 7.5×10^{12} 23. 3.0×10^{-6} 24. 2.1×10^{-7}
   25. 4.2×10^{-1} 26. 5.6×10^{-1} 27. 1.2×10^2
   28. 8.1×10^{-3} 29. 2.8×10^{-2} 30. 2.2×10^3
   31. 2.4×10^{-8} 32. 1.6×10^8

E. 1. (a) 31.25%   (b) $L = \dfrac{100(H - X)}{CG}$; $L = 166.7$ in.

   (c) $H = \dfrac{L \cdot CG + 100x}{100}$; $H = 140$ in.

   (d) $x = \dfrac{100H - L \cdot CG}{100}$; $x = 156$ in.

   2. (a) 12:5 or 2.4 to 1   (b) 3600 rpm

   3. (a) $L = \dfrac{FDU}{A}$ (b) 42 foot candles   (c) $A = \dfrac{FDU}{L}$

   (d) 66.7 sq ft
   4. $3\frac{3}{4}$ hr   5. $8W + 2W = 80$; $W = 8$ cm, $L = 32$ cm
   6. (a) 9 ft   (b) 30 ft   (c) $\frac{1}{8}$ ft   (d) $7\frac{1}{2}$ ft
   7. 6 sacks   8. $L = 225$ tons
   9. 60 teeth   10. 960 lb   11. 360 turns
   12. $2x + x = 0.048$; $x = 0.016$ in., $2x = 0.032$ in.   13. 1.4 ampere
   14. 12,632 copies   15. Sal: $8,763.64
                  Gloria: $13,145.45
                  Martin: $26,290.91
   16. 9.7×10^6 m/sec   17. about 16 pipes
   18. $7\frac{9}{16}$ in.
   19. $673.98   20. 546 rpm   21. 999,600 BTU   22. 146 employees
   23. 1.984 in.   24. 83.96 psi   25. 0.617   26. 0.1139 in. per in.
   27. 1,770,000

Exercises 8–1, page 392

A. 1. (a) $\angle B$, $\angle x$, $\angle ABC$ (b) $\angle O$, $\angle POQ$ (c) $\angle a$
    (d) $\angle T$ (e) $\angle 2$ (f) $\angle s$, $\angle RST$
    2. (a) $\angle AOB$, $\angle AOC$ (b) $\angle HOG$, $\angle HOF$ (c) $\angle AOF$, $\angle AOG$
    (d) $\angle AOE$, $\angle HOE$
    (e) $\angle AOB = 25°$, $\angle GOF = 39°$, $\angle HOC \approx 132°$
        $\angle FOH = 75°$, $\angle BOF \approx 83°$, $\angle COG \approx 98°$
    4. (a) 145°   (b) 49°   (c) 30°   (d) 72°   (e) 77° approx.
    (f) 90°   (g) 45°   (h) 45°   (i) 151°

B. 1. (a) $a = 55°$, $b = 125°$, $c = 55°$ (b) $d = 164°$, $e = 16°$, $f = 164°$
    (c) $p = 110°$ (d) $x = 75°$, $t = 105°$, $k = 115°$
    (e) $s = 60°$, $t = 120°$ (f) $p = 30°$, $q = 30°$, $w = 90°$
    2. (a) $A = 68°$ (b) $E = 137°$
    (c) $G = 42°$ (d) 133°
    (e) $a = 70°$ (f) 53°
    (g) 63°   (h) 44°45′   (i) 60°2′
    3. (a) $w = 80°$, $x = 100°$, $y = 100°$, $z = 100°$
    (b) $a = 60°$, $b = 60°$, $c = 120°$, $d = 120°$, $e = 120°$, $f = 120°$, $g = 60°$
    (c) $n = 100°$, $q = 80°$, $m = 100°$, $p = 100°$, $h = 80°$, $k = 100°$
    (d) $z = 140°$, $s = 40°$, $y = 140°$, $t = 140°$, $u = 40°$, $w = 140°$, $x = 40°$

C. 1. $\angle ACB \approx 63°$ 2. 45° and 135°   3. $B \approx 125°$, $A \approx 27\frac{1}{2}°$, $C \approx 27\frac{1}{2}°$
    4. $a \approx 35°$ 5. $2''$ 6. 45°   7. 29°5′36″
    8. 60°   9. $a = 112°$, $b = 68°$ 10. $22\frac{1}{2}°$
    11. $\angle 2 = 106°$ 12. $\angle y = 48°$

Exercises 8–2, page 407

A. 1. 20.0 m   2. 206 ft   3. 115 in.   4. 42 ft
    5. 48 in.   6. 104.8 m   7. 39 cm   8. 0.252 in.
    9. 22.81 in.   10. 23.8 ft   11. 64 ft   12. 26.4 in.

B.
		Perimeter	*Area*		*Perimeter*	*Area*
	1.	20 in.	25 sq in.	2.	22.7 yd	29.6 sq yd
	3.	70.6 in.	285 sq in.	4.	98 ft	420 sq ft
	5.	56 m	160 m^2	6.	17.3 yd	18.8 sq yd
	7.	124 m	468 m^2	8.	42.5 ft	109.4 sq ft
	9.	86 ft	432 sq ft	10.	23 m	33.1 m^2
	11.	114.9 m	722.2 m^2	12.	847.5 ft	39496.7 sq ft

C. 1. 960   2. $35\frac{7}{8}$ sq ft   3. $25\frac{1}{3}$ sq yd, $657.40
    4. 1840 sq   5. 11 bundles   6. 4385 sq in.
    7. 640 sq in.   8. 450 blocks
    9. 10 ft 8 in.   10. 56.25 lb
    11. $6\frac{2}{3}$ in.   12. 19.5 lb
    13. Cut the 6-in. card along the 38-in. side; 18 cards, 86 sq in. of waste
    14. 3500 BTU

Exercises 8–3, page 424

A. 1. 10.6 in.   2. 13.4 ft   3. 20 cm   4. 28.6 yd
    5. 46.8 mm   6. 39 in.   7. 27.7 in.   8. 4.5 ft
    9. 28 ft   10. 4.4 ft   11. 21.2 in.   12. 28 in.
    13. 6 in.   14. 8 cm   15. 17.4 ft   16. 5.4 in.
    17. 9.369 cm   18. 8.288 in.   19. 0.333 in.   20. 21.35 in.
    21. 118 ft   22. 43.91 m

B. 1. 173.2 mm^2 2. 439.1 cm^2 3. 336 sq ft   4. 53.2 sq ft
    5. 36 sq yd   6. 84.9 cm^2 7. 259.8 sq in.   8. 162 sq in.

9.	520 sq ft	10.	77 sq in.	11.	5.1 cm^2	12.	161 m^2
13.	433 sq ft	14.	234 sq in.	15.	354 sq ft	16.	2250 m^2
17.	364.8 sq ft	18.	120 sq in.	19.	39 sq in.	20.	19.4 sq in.
21.	0.586 sq in.	22.	0.039 cm^2	23.	7.94 m^2	24.	0.117 sq in.

C. 1. 7 gal 2. 268 sq in. 3. 5 squares 4. 74 ft
  5. 15.2 ft 6. 11.3 in. 7. 174.5 sq ft
  8. 318 sq ft 9. $y = 36$ ft 5 in., $x = 64$ ft 11 in.

Box, page 434

1. 4 in. 2. 8.5 in. 3. 5 in. 4. 10.4 in. 5. 5.1 in.

Exercises 8–4, page 435

A.   *Circumference*   *Area*
  1. 88.0 in.   616 sq in.
  2. 66.0 ft   346 sq ft
  3. 19.2 m   29.2 m^2
  4. 44.0 cm   154 cm^2
  5. 75.4 ft   452 sq ft
  6. 1.57 m   0.196 m^2
  7. 3.77 cm   1.13 cm^2
  8. 2.51 in.   0.503 sq in.

B. 1. 17.1 sq in. 2. 238.8 cm^2 3. 5.2 sq in. 4. 5.8 sq ft
  5. 15.1 sq in. 6. 15.9 m^2 7. 26.7 cm^2 8. 247.8 sq ft
  9. 0.7 sq in. 10. 135.3 cm^2 11. 246.1 sq in. 12. 8037.5 sq ft
  13. 25.1 in. 14. 55.1 in. 15. 52.3 cm

C. 1. 30.1 in. 2. (a) 908 sq ft (b) 226 sq ft 3. 0.6 in.
  4. 0.05 sq in. 5. 163 sq in. 6. 316 sq in. 7. 103.7 sq in.
  8. $1\frac{3}{4}$ in. 9. 314 10. 4.24 cm 11. $25.71
  12. 99.52 m 13. 174 in. 14. 10 ft 4.5 in. 15. 2.4 in.
  16. 67.0 sq yd 17. 2.6 cm
  18. 182 mm 19. 2.79 in.
  20. About $1\frac{1}{2}$ turns 21. 0.021 fps
  22. 393 lb

Problem Set 8, page 439

A. 1. $\angle$ 1, acute 2. $\angle$ ABC or $\angle$ B, obtuse 3. $\angle$ m, acute
  4. $\angle$ E, right 5. $\angle$ PQR or $\angle$ Q, obtuse 6. $\angle$ G, right
  7. 65° approx. 8. 50° approx. 9. 68° approx.
  10. 90° approx. 11. 180° approx. 12. 153° approx.
  13. 60° approx. 14. 70° approx. 15. 102° approx.
  17. (a) 55° (b) $a = 95°, b = 85°, c = 95°$
    (c) $a = 130°, b = 50°, c = 130°, d = 50°, e = 130°, f = 50°, g = 130°$
    (d) 131° (e) 100° (f) All equal 60° (g) 58°
    (h) 80° (i) 35°

B.   *Perimeter* *Area*   *Perimeter* *Area*
  1. 12 in. 6 sq in. 2. 30 ft 48 sq ft
  3. 31.4 cm 78.5 cm^2 4. 3 in. $\frac{9}{16}$ sq in.
  5. 12 ft 10.4 sq ft 6. 74.4 in. 289.8 sq in.
  7. 17 yd 8 sq yd 8. 100 mm 561 mm^2
  9. 36 in. 53.7 sq in. 10. 42 ft 102 sq ft
  11. 12 in. 6.9 sq in. 12. 94.2 cm 706.9 cm^2
  13. 17.2 ft 12.0 sq ft 14. 74 in. 145 sq in.
  15. 51.7 ft 169.3 sq ft 16. 70 m 284 m^2
  17. 28.3 in. 51.1 sq in. 18. 7.5 in. 2.3 sq in.

19.	5.77 mm	20.	15.59 in.
21.	3.46 in.	22.	23.56 ft
23.	$\frac{3}{4}$ in.	24.	11.70 in.

C.

1.	$88,825	2.	7.07 in.	3.	45°	4.	640 sq ft
5.	251 ft	6.	105°	7.	$1020.24	8.	32.5 in.
9.	18 in.	10.	36 ft 5 in.	11.	36	12.	34.7 in.
13.	110°	14.	15 oz	15.	94 ft	16.	0.707 in.
17.	$170.82	18.	153 sq ft	19.	69.1 cm	20.	68.0 in.
21.	$954.56	22.	118°	23.	27.3 lb	24.	3.005 in.
25.	12.17 in.						

Chapter 9

Exercises 9–1, page 457

A.

1.	729 cu in.	2.	931.5 cu ft	3.	13,800 cm^3	4.	35 cu in.
5.	132.4 mm^3	6.	2.8 cu in.	7.	144 cm^3	8.	108 cm^3

B.

	Lateral Surface Area	Volume		Lateral Surface Area	Volume
1.	280 sq ft	480 cu ft	2.	453.6 cm^2	519.6 cm^3
3.	784 mm^2	2744 mm^3	4.	608 sq ft	1120 cu ft
5.	48.6 sq ft.	46.7 cu ft	6.	72 sq ft	69.3 cu ft
7.	55.2 m^2	31.2 m^3	8.	360 sq in.	480 cu in.

C.

	Total Surface Area	Volume		Total Surface Area	Volume
1.	150 sq yd	125 cu yd	2.	311.3 m^2	357.9 m^3
3.	162.0 sq in.	105.2 cu in.	4.	11.8 sq in.	2.3 cu in.
5.	172.8 sq ft	144 cu ft	6.	5290 cm^2	23,940 cm^3
7.	484 sq in.	616.7 cu in.	8.	396 sq yd	288 cu yd

D.

1.	333.3 cu ft	2.	27.2 lb	3.	47,124 gal	4.	8 gal
5.	450 trips	6.	322 bushels	7.	2012 lb	8.	71.3 cu yd
9.	384 sq in.	10.	66.9 cu yd	11.	16.9 lb	12.	63.9 lb
13.	3.8 ft	14.	5670 lb				
15.	238 trips	16.	143,000 gal	17.	8.3 cu yd	18.	8100 lb
19.	3.5 cu yd	20.	2.4 gal	21.	7000 lb		

Exercises 9–2, page 468

A.

	Lateral Surface Area	Volume		Lateral Surface Area	Volume
1.	7.07 sq in.	1.77 cu in.	2.	4070 sq ft	24,400 cu ft
3.	1490 mm^2	6350 mm^3	4.	245 sq in.	367 cu in.
5.	301 sq yd	603 cu yd	6.	427 sq in.	1000 cu in.
7.	1580 sq ft	7070 cu ft	8.	647 sq ft	1880 cu ft
9.	747 sq in.	2610 cu in.	10.	188 sq ft	462 cu ft

Total area: 3. 1950 mm^2 8. 961 sq ft   9. 1110 sq in.

B.

1.	240 m^3	2.	1100 cu in.	3.	49 cu in.	4.	4.7 cu ft

C.

1.	104.7 m^3	2.	12.3 lb	3.	0.24 ft
4.	746.1 cm^2	5.	6283.2 liters	6.	2.6 lb
7.	253.2 bu	8.	57.4 in.	9.	No. (4512 gal)
10.	4071.5 sq in.	11.	1073.6 cm^2	12.	49.6 lb
13.	7.5 gal	14.	581.7 gal	15.	2.0 lb
16.	8.9 cu yd	17.	0.43 lb	18.	11 gal
19.	28.1 in.	20.	53,300 gal		

Problem Set 9, page 487

A.

	Lateral Surface Area	Volume		Lateral Surface Area	Volume
1.	226.2 sq in.	452.4 cu in.	2.	240 sq in.	504 cu in.

604

	Lateral Surface Area	Volume		Lateral Surface Area	Volume
3.	43.2 sq ft	26.7 cu ft	4.	816.9 sq ft	1638 cu ft
5.	259.2 mm^2	443.5 mm^3	6.	630 sq in.	1909.5 cu in.
7.	377.0 sq ft	980.2 cu ft	8.	300 sq ft	912 cu ft
9.	384 sq ft	512 cu ft	10.	107.8 sq in.	56.6 cu in.
11.	1767.2 cm^2	5301.5 cm^3	12.	256.2 sq in.	259.8 cu in.
13.	450 sq ft	396 cu ft	14.	879.6 sq in.	1570.8 cu in.
15.	1256.6 cm^2	4188.8 cm^3	16.	648 sq in.	1008 cu in.
17.	530.9 sq in.	1150.3 cu in.	18.	1381.6 sq in.	3479.1 cu in.
19.	1263.6 sq in.	2520 cu in.	20.	35.3 sq in.	14.7 cu in.
21.	329.2 sq in.	312 cu in.			

C. 1. 1125 boxes 2. 1204 cu yd 3. 2992 lb 4. 423.0 gal
5. 40 qt 6. 84 trips 7. 1.5 cu ft 8. 8 ft
9. 294 lb 10. 22,000 sq in. 11. 14 cu yd 12. 738 sq in.
13. 2.8 gal 14. 18.5 lb 15. 6.3 ft

Chapter 10 *Exercises 10–1, page 504*

A. 1. Acute 2. Right 3. Obtuse 4. Right
5. Straight 6. Obtuse 7. Acute 8. Acute
9. Right 10. Acute

B. 1. 36°15′ 2. 73°36′ 3. 65°27′ 4. 84°14′
5. 17°30′ 6. 47°23′ 7. 16°7′ 8. 165°22′
9. 27.5° 10. 80.25° 11. 154.65° 12. 131.1°
13. 57.05° 14. 44.33° 15. 16.88° 16. 0.27°
17. 0.43 18. 3.24 19. 0.02 20. 0.26
21. 1.18 22. 0.17 23. 1.41 24. 3.78
25. 5.73° 26. 48.13° 27. 120.32° 28. 200.54°

C. 1. 2. 3. 4.

5. 6. 7. 8.

9. 10.

11. 26 12. 9.7 13. 24 14. 18.6
15. 12 ft 16. 3.6 in. 17. 1.5 18. 3.3 cm
19. 8.0 ft 20. 1.7 in. 21. (a) 45° (b) 4.2 ft
22. (a) 6.9 in. (b) 8 in. 23. (a) 4 (b) 53.1° (c) 36.9°
24. (a) 3 cm (b) 5.2 cm (c) 60° 25. (a) 7 (b) 45°
26. (a) 5 in. (b) 53.1° 27. (a) 8.7 ft (b) 10 ft (c) 30°
28. (a) 0.4 in. (b) 0.3 in. (c) 90° 29. (a) 10 in. (b) 45°
30. (a) 3.48 cm (b) 60° (c) 30°
31. $S = 21$ cm, $A = 314$ cm^2 32. $S = 50$ ft, $A = 603$ sq ft
33. $S = 88$ ft, $A = 4840$ sq ft 34. $S = 36$ m, $A = 324$ m^2

35. $S = 6$ in., $A = 16$ sq in.   36. $S = 240$ cm, $A = 9600$ cm^2
37. $S = 5$ m, $A = 25$ m^2 38. $S = 94$ in., $A = 1414$ sq in.

D. 1. (a) 325.2 sq in.   (b) 36.1 in.   2. 1256.6 sq yd
    3. 0.08 rad/sec   4. 157 rad/sec   5. 0.39 rad/sec

Exercises 10–2, page 514

A. 1. 0.54   2. 0.74   3. 0.42   4. 0.53   5. 0.44   6. 0.69

B. 1. 0.454   2. 0.788   3. 0.213   4. 0.998
    5. 0.191   6. 0.105   7. 0.052   8. 0.994
    9. 1.206   10. 0.969   11. 0.752   12. 0.149
    13. 0.771   14. 3.522   15. 6.357   16. 0.218
    17. −0.139   18. 0.961   19. 0.585   20. −0.701

C. 1. 76°54′   2. 44°2′   3. 64°0′   4. 4°25′
    5. 74°49′   6. 22°0′   7. 9°40′   8. 69°45′
    9. 49.4°   10. 48.5°   11. 64.1°   12. 12.7°
    13. 41.1°   14. 28.2°   15. 37.2°   16. 81.1°

Exercises 10–3, page 519

A. 1. $X = 11.0$ in., $Y = 8.6$ in.   2. $X = 4.9$ ft, $Y = 7.7$ ft
    3. $X = 29.7$ cm, $Y = 32.0$ cm   4. $a = 32°28′, X = 26.1$ ft
    5. $a = 60°, X = 1.3$ in.   6. $X = 1.8$ mm, $Y = 3.3$ mm
    7. $X = 44.4$ in., $Y = 14.5$ in.   8. $a = 50°36′, X = 4.7$ yd
    9. $a = 36.0°, X = 1.54$ in.   10. $X = 33.50$ cm, $Y = 22.85$ cm
    11. $X = 1.16$ in., $Y = 1.47$ in.   12. $a = 61.3°, X = 1.47$ in.
    13. $m = 41.6°, X = 27.67$ cm   14. $X = 28.39$ in., $Y = 15.75$ in.
    15. $X = 1.27$ in, $Y = 1.42$ in.   16. $m = 16.5°, X = 17.97$ cm

B. 1. 31.1 ft   2. 43.8°   3. (a) 72 ft 9 in.   (b) No.   4. 4478 ft
    5. 51°41′   6. 4.3°   7. 115 in.   8. $A = 3.08$ in., $B = 4.66$ in.
    9. 3.74 in.   10. Yes   11. 1.1 in.   12. 68.75°
    13. 274 ft   14. 0.5456 in.   15. 33.4 cm   16. 6°37′
    17. 1.55 in.   18. 5 ft 2 in.

Exercises 10–4, page 532

A. 1. $C = 75°, c = 9.2$ ft, $b = 8.4$ ft
    2. $c = 12.6$ in., $A = 65.0°, a = 16.5$ in.
    3. $A = 39.0°, a = 202$ m, $c = 301$ m
    4. $b = 1500$ yd, $a = 1100$ yd, $A = 25°$
    5. $A = 55.9°, C = 94.1°, c = 116$ in. or $A = 124.1°, C = 25.9°, c = 50.6$ in.
    6. $A = 108.6°, B = 44.4°, a = 359$ ft or $A = 17.4°, B = 135.6°, a = 113$ ft
    7. $c = 5.6$ m, $A = 92°, B = 47°$ 8. $a = 24.3$ m, $B = 41.7°, C = 81.3°$
    9. $b = 690$ ft, $A = 62.4°, C = 15.6°$
    10. $a = 3890$ yd, $B = 14.0°, C = 41.0°$
    11. $C = 94.2°, A = 37.5°, B = 48.3°$ 12. $A = 102°, B = 19°, C = 59°$
    13. $b = 10.3$ m, $A = 45.4°, C = 34.9°$
    14. $a = 0.865$ in., $B = 23.0°, C = 38.5°$
    15. $B = 149.9°, A = 7.1°, C = 23.0°$
    16. $A = 47.6°, B = 82.2°, C = 50.2°$

B. 1. 41 ft   2. 36.3 ft   3. 27.2°, 52.1°, 100.7°
    4. 182 ft   5. 19.1 ft   6. 26° and 130°
    7. 25.4 in.

Problem Set 10, page 535

A. 1. Yes   2. Yes   3. No   4. No   5. Yes
    6. Yes   7. No   8. No   9. Yes   10. No

B. 1. 0.391   2. 6.314   3. 0.070   4. 0.451
    5. 0.720   6. 0.488   7. 0.956   8. −0.545
    9. 14°   10. 57°   11. 67°58′   12. 72°
    13. 10°59′ 14. 56°42′ 15. 39°27′ 16. 42°13′
    17. 83°13′ 18. 26.8°   19. 43.7°   20. 29.2°
    21. 8.9°   22. 74.2°   23. 63.9°   24. 18.7°
    25. 52.6°   26. 50.5°   27. 0.611   28. 0.374
    29. 1.300   30. 1.958   31. 25.8°   32. 97.4°
    33. 131.8° 34. 48.1°

C. 1. $X = 37.1, Y = 32.5$ 2. $X = 20.2, Y = 16.4$
    3. $a = 46°, X = 4.9$ 4. $X = 5.3, Y = 8.4$
    5. $a = 54.2°, X = 5.6$ 6. $X = 70.2$ ft, $Y = 77.2$ ft
    7. $t = 39.3°, X = 9.8$ in.   8. $X = 11.7$ ft, $Y = 10.9$ ft

D. 1. $C = 76.0°, c = 19.2$ in., $b = 12.4$ in.
    2. $A = 155.5°, B = 14.8°, C = 9.7°$
    3. $C = 52°, B = 91°, b = 430$ yd or $C = 128°, B = 15°, b = 110$ yd
    4. $a = 101$ in., $B = 49.7°, C = 11.8°$
    5. $B = 49.1°, C = 66.6°, c = 52.4$ m
    6. $b = 190$ ft, $A = 39.8°, C = 93.2°$
    7. $A = 24.9°, B = 38.9°, b = 1730$ m
    8. $A = 52°, C = 98°, c = 9.5$ in. or $A = 128°, C = 22°, c = 3.6$ in.

E. 1. 1.3 in.   2. $a = 9°36′, V = 118.3$ mph   3. 2 in.
    4. 6°2′   5. (a) $X = 28.9$ ft   (b) 57.7 ft
    6. (a) 80.9 ft   (b) 8°32′   7. (a) 1°8′   (b) 0.3 in.
    8. $x = 16.5$ ft, $y = 25.7$ ft   9. 0.262 in.   10. 67°34′   11. 17.2 cm
    12. 1.08 in.

Chapter 11 *Exercises 11–1, page 559*

A. 1. $x = 2, y = 8$ 2. $x = 4, y = 7$ 3. $x = 4, y = 5$
    4. Inconsistent   5. $x = 2, y = 4$ 6. $x = -\frac{1}{9}, y = -\frac{1}{9}$
    7. $x = -1, y = 2$ 8. $x = \frac{1}{2}, y = -3$

B. 1. $x = 9, y = -4$ 2. $x = 4, y = 1$ 3. Inconsistent
    4. Inconsistent   5. $x = -2, y = -3$ 6. $x = -3, y = 1$
    7. $x = -1, y = -1\frac{1}{2}$ 8. $x = 4, y = 7$ 9. Dependent
    10. $x = \dfrac{a + b}{2}, y = \dfrac{a - b}{2}$ 11. $x = 1, y = 5$ 12. $x = 3, y = \frac{1}{2}$

C. 1. $x + y = 39$ Solution: $x = 23$
        $x - y = 7$ $y = 16$
    2. $x + y = 14$ Solution: $x = 11$
        $x = 2 + 3y$ $y = 3$
    3. $x + y = 20$ Solution: $x = 12$
        $2x = 3y$ $y = 8$
    4. $\frac{1}{2}(x + y) = 25$ Solution: $x = 29$
        $x - y = 8$ $y = 21$
    5. $4b + 3f = 11$ Solution: $b = \$2$ A bleeb costs $2.00.
        $3b + 4f = 10$ $f = \$1$
    6. $2L + 2W = 14$ Solution: $L = 4$ ft
        $L = 2W - 2$ $W = 3$ ft
    7. $10d + 5n = 100$ Solution: $n = 2$
        $d = n + 7$ $d = 9$
    8. $4L + 3S = 26$ Solution: $L = 5$
        $3L - 2S = 11$ $S = 2$
    9. $5p + 4c = 560$ Solution: $c = 65¢$
        $p = c - 5$ $p = 60¢$

10. $L = 2W$ Solution: $L = 40$ in.
$L - W = 20$ $W = 20$ in.

11. $x + y = 30$ Solution: $x = 24$ in.
$x = 4y$ $y = 6$ in.

12. $x + y = 12$ Solution: $x = 5$ gal (of \$15 paint)
$15x + 18y = 201$ $y = 7$ gal (of \$18 paint)

13. $x + y = 6$ Solution: $x = 1.5$ liter (10%)
$0.1x + 0.02y = 0.24$ $y = 4.5$ liter (2%)

14. $2L + 2W = 520$ Solution: $L = 200$ ft
$L = 20 + 3W$ $W = 60$ ft

15. $x + y = 24$ Solution: $x = 16$ tons (\$30 rock)
$30x + 40y = 800$ $y = 8$ ton (\$40 rock)

Exercises 11–2, page 571

A.   1. No   2. Yes   3. No
     4. Yes   5. Yes   6. $7x^2 + 3x - 5 = 0$
     7. $3x^2 - 7x + 14 = 0$ 8. OK   9. $x^2 - 18x = 0$
   10. OK

B.   1. $x = 5$ or $x = -5$ 2. $x = 3$ or $x = -3$
     3. $x = 0$ or $x = 4\frac{2}{5}$ 4. $x = \frac{1}{2}$ or $x = 3$
     5. $x = 4\frac{1}{2}$ or $x = -4\frac{1}{2}$ 6. $x = 4\frac{1}{2}$ or $x = -2\frac{1}{3}$
     7. $x \approx 0.76$ or $x \approx -15.76$ 8. $x = 9$ or $x = \frac{3}{4}$
     9. $x \approx -2.35$ or $x \approx 0.85$ 10. $x = 0.02$ or $x = -0.05$
   11. $x \approx 4.14$ or $x \approx -3.14$ 12. $x \approx 4.46$ or $x \approx -2.46$
   13. $x \approx 0.55$ or $x \approx -0.26$ 14. $x \approx 1.27$ or $x \approx -0.47$
   15. $x \approx -2.37$ or $x \approx -0.63$ 16. $x \approx 1.24$ or $x \approx 0.16$
   17. $x \approx 2.39$ or $x \approx 0.28$ 18. $x \approx -0.46$ or $x \approx 0.31$
   19. $x \approx 3.69$ or $x \approx 0.81$ 20. $x \approx 0.57$ or $x \approx -2.65$

C.   1. 25 mm   2. (a) 14.49 in.   (b) 15.66 in.
     3. 5 in. by 15 in.   4. 8.92 cm   5. 17 in. by 22 in.
     6. (a) 7.07 in.   (b) 9.13 in.   7. 7.14 in.
     8. (a) 9 amp   (b) 7.30 amp
     9. (a) 3.76 in.   (b) 4.56 in.   10. 6 ft by 8 ft or 4 ft by 12 ft

Problem Set 11, page 575

A.   1. $x = 15, y = 3$ 2. $x = 6, y = -\frac{1}{2}$ 3. $x = 3, y = 2\frac{1}{2}$
     4. $x = 7, y = 3$ 5. $x = 7, y = 6$ 6. Dependent
     7. $x = 1, y = 1$ 8. $x = -13, y = 22$ 9. $x = \frac{74}{23}, y = -\frac{4}{23}$

B.   1. $x = 3$ or $x = -3$ 2. $x = 7$ or $x = -4$
     3. $x \approx -1.43$ or $x \approx -0.23$ 4. $x \approx 1.18$ or $x \approx -0.43$
     5. $x = 0$ or $x = 6$ 6. $x = 5$ or $x = -1\frac{1}{2}$
     7. $x = 5$ or $x = -1$ 8. $x = 6$ or $x = -5$
     9. $x = 6$ or $x = -6$ 10. $x \approx 2.15$ or $x \approx -0.75$

C.   1. $x + y = 38$ 2. $s^2 = 196$ 3. $x - y = 21$
        $x - y = 14$ $s = 14$ in. $5y - 2x = 33$
        $x = 26$ $x = 46$
        $y = 12$ $y = 25$
     4. $2L + 2W = 22$ 5. $L = 4W$ 6. $x + y = 650$
        $L = 2 + 2W$ $4W^2 = 125$ $5x + 7y = 4350$
        $W = 3$ ft $W \approx 5.59$ cm $x = 100$ (at 5¢ each)
        $L = 8$ ft $L \approx 22.36$ cm $y = 550$ (at 7¢ each)
     7. (a) $I = 13$ amp   (b) $I = 12.25$ amp

8. $L = 2 + W$
    $W(2 + W) = 168$
    $W = 12$ in.
    $L = 14$ in.

9. $x + y = 42$
    $x = 3y - 2$
    $y = 11$ in.
    $x = 31$ in.

10. $2L + 2W = 750$
    $L = 4W$
    $W = 75$ ft
    $L = 300$ ft

11. $d = 15.96$ in.

12. $x + y = 15$
    $8x + 11y = 150$
    $x = 5$ gal of \$8 paint
    $y = 10$ gal of \$11 paint

13. (a) $D = 5$ in.
    (b) $D = 6.25$ in.

14. $L = 2W - 3$
    $W(2W - 3) = 20$
    $W = 4$ in.
    $L = 5$ in.

15. $x + y = 50$
    $60x + 75y = 3150$
    $x = 40$ tons (of \$60 rock)
    $y = 10$ tons (of \$75 rock)

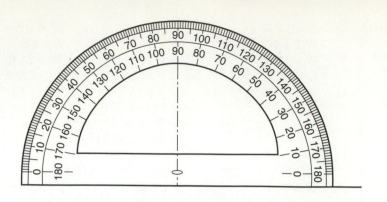

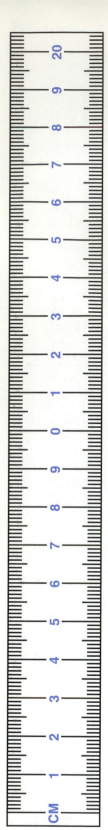

×	0	1	2	3	4	5	6	7	8	9	10
0	0	0	0	0	0	0	0	0	0	0	0
1	0	1	2	3	4	5	6	7	8	9	10
2	0	2	4	6	8	10	12	14	16	18	20
3	0	3	6	9	12	15	18	21	24	27	30
4	0	4	8	12	16	20	24	28	32	36	40
5	0	5	10	15	20	25	30	35	40	45	50
6	0	6	12	18	24	30	36	42	48	54	60
7	0	7	14	21	28	35	42	49	56	63	70
8	0	8	16	24	32	40	48	56	64	72	80
9	0	9	18	27	36	45	54	63	72	81	90
10	0	10	20	30	40	50	60	70	80	90	100

TABLE OF SQUARE ROOTS

Number	Square root	Number	Square root	Number	Square root	Number	Square root
1	1.0000	51	7.1414	101	10.0499	151	12.2882
2	1.4142	52	7.2111	102	10.0995	152	12.3288
3	1.7321	53	7.2801	103	10.1489	153	12.3693
4	2.0000	54	7.3485	104	10.1980	154	12.4097
5	2.2361	55	7.4162	105	10.2470	155	12.4499
6	2.4495	56	7.4833	106	10.2956	156	12.4900
7	2.6458	57	7.5498	107	10.3441	157	12.5300
8	2.8284	58	7.6158	108	10.3923	158	12.5698
9	3.0000	59	7.6811	109	10.4403	159	12.6095
10	3.1623	60	7.7460	110	10.4481	160	12.6491
11	3.3166	61	7.8102	111	10.5357	161	12.6886
12	3.4641	62	7.8740	112	10.5830	162	12.7279
13	3.6056	63	7.9373	113	10.6301	163	12.7671
14	3.7417	64	8.0000	114	10.6771	164	12.8062
15	3.8730	65	8.0623	115	10.7238	165	12.8452
16	4.0000	66	8.1240	116	10.7703	166	12.8841
17	4.1231	67	8.1854	117	10.8167	167	12.9228
18	4.2426	68	8.2462	118	10.8628	168	12.9615
19	4.3589	69	8.3066	119	10.9087	169	13.0000
20	4.4721	70	8.3666	120	10.9545	170	13.0384
21	4.5826	71	8.4261	121	11.0000	171	13.0767
22	4.6904	72	8.4853	122	11.0454	172	13.1149
23	4.7958	73	8.5440	123	11.0905	173	13.1529
24	4.8990	74	8.6023	124	11.1355	174	13.1909
25	5.0000	75	8.6603	125	11.1803	175	13.2288
26	5.0990	76	8.7178	126	11.2250	176	13.2665
27	5.1962	77	8.7750	127	11.2694	177	13.3041
28	5.2915	78	8.8318	128	11.3137	178	13.3417
29	5.3852	79	8.8882	129	11.3578	179	13.3791
30	5.4772	80	8.9443	130	11.4018	180	13.4164
31	5.5678	81	9.0000	131	11.4455	181	13.4536
32	5.6569	82	9.0554	132	11.4891	182	13.4907
33	5.7446	83	9.1104	133	11.5326	183	13.5277
34	5.8310	84	9.1652	134	11.5758	184	13.5647
35	5.9161	85	9.2195	135	11.6190	185	13.6015
36	6.0000	86	9.2736	136	11.6619	186	13.6382
37	6.0828	87	9.3274	137	11.7047	187	13.6748
38	6.1644	88	9.3808	138	11.7473	188	13.7113
39	6.2450	89	9.4340	139	11.7898	189	13.7477
40	6.3246	90	9.4868	140	11.8322	190	13.7840
41	6.4031	91	9.5394	141	11.8743	191	13.8203
42	6.4807	92	9.5917	142	11.9164	192	13.8564
43	6.5574	93	9.6437	143	11.9583	193	13.8924
44	6.6332	94	9.6954	144	12.0000	194	13.9284
45	6.7082	95	9.7468	145	12.0416	195	13.9642
46	6.7823	96	9.7980	146	12.0830	196	14.0000
47	6.8557	97	9.8489	147	12.1244	197	14.0357
48	6.9282	98	9.8995	148	12.1655	198	14.0712
49	7.0000	99	9.9499	149	12.2066	199	14.1067
50	7.0711	100	10.0000	150	12.2474	200	14.1421

TABLE OF TRIGONOMETRIC FUNCTIONS

Angle	Sine	Cosine	Tangent	Angle	Sine	Cosine	Tangent
0°	0.0000	1.0000	0.0000	45°	0.7071	0.7071	1.000
1	0.0175	0.9998	0.0175	46	0.7193	0.6947	1.036
2	0.0349	0.9994	0.0349	47	0.7314	0.6820	1.072
3	0.0523	0.9986	0.0524	48	0.7431	0.6691	1.111
4	0.0698	0.9976	0.0699	49	0.7547	0.6561	1.150
5	0.0872	0.9962	0.0875	50	0.7660	0.6428	1.192
6	0.1045	0.9945	0.1051	51	0.7771	0.6293	1.235
7	0.1219	0.9925	0.1228	52	0.7880	0.6157	1.280
8	0.1392	0.9903	0.1405	53	0.7986	0.6018	1.327
9	0.1564	0.9877	0.1584	54	0.8090	0.5878	1.376
10	0.1736	0.9848	0.1763	55	0.8192	0.5736	1.428
11	0.1908	0.9816	0.1944	56	0.8290	0.5592	1.483
12	0.2079	0.9781	0.2126	57	0.8387	0.5446	1.540
13	0.2250	0.9744	0.2309	58	0.8480	0.5299	1.600
14	0.2419	0.9703	0.2493	59	0.8572	0.5150	1.664
15	0.2588	0.9659	0.2679	60	0.8660	0.5000	1.732
16	0.2756	0.9613	0.2867	61	0.8746	0.4848	1.804
17	0.2924	0.9563	0.3057	62	0.8829	0.4695	1.881
18	0.3090	0.9511	0.3249	63	0.8910	0.4540	1.963
19	0.3256	0.9455	0.3443	64	0.8988	0.4384	2.050
20	0.3420	0.9397	0.3640	65	0.9063	0.4226	2.145
21	0.3584	0.9336	0.3839	66	0.9135	0.4067	2.246
22	0.3746	0.9272	0.4040	67	0.9205	0.3907	2.356
23	0.3907	0.9205	0.4245	68	0.9272	0.3746	2.475
24	0.4067	0.9135	0.4452	69	0.9336	0.3584	2.605
25	0.4226	0.9063	0.4663	70	0.9397	0.3420	2.747
26	0.4384	0.8988	0.4877	71	0.9455	0.3256	2.904
27	0.4540	0.8910	0.5095	72	0.9511	0.3090	3.078
28	0.4695	0.8829	0.5317	73	0.9563	0.2924	3.271
29	0.4848	0.8746	0.5543	74	0.9613	0.2756	3.487
30	0.5000	0.8660	0.5774	75	0.9659	0.2588	3.732
31	0.5150	0.8572	0.6009	76	0.9703	0.2419	4.011
32	0.5299	0.8480	0.6249	77	0.9744	0.2250	4.331
33	0.5446	0.8387	0.6494	78	0.9781	0.2079	4.705
34	0.5592	0.8290	0.6745	79	0.9816	0.1908	5.145
35	0.5736	0.8192	0.7002	80	0.9848	0.1736	5.671
36	0.5878	0.8090	0.7265	81	0.9877	0.1564	6.314
37	0.6018	0.7986	0.7536	82	0.9903	0.1392	7.115
38	0.6157	0.7880	0.7813	83	0.9925	0.1219	8.144
39	0.6293	0.7771	0.8098	84	0.9945	0.1045	9.514
40	0.6428	0.7660	0.8391	85	0.9962	0.0872	11.43
41	0.6561	0.7547	0.8693	86	0.9976	0.0698	14.30
42	0.6691	0.7431	0.9004	87	0.9986	0.0523	19.08
43	0.6820	0.7314	0.9325	88	0.9994	0.0349	28.64
44	0.6947	0.7193	0.9657	89	0.9998	0.0175	57.29
45	0.7071	0.7071	1.0000	90	1.0000	0.0000	∞

Index

Diameter
of a circle, 427
of a sphere, 466
Difference, 16
Digits, 4
decimal, 95
Dividend, 33
Division
of algebraic expressions, 292, 364–67
of decimal numbers, 107–10
of fractions, 70–73
of measurement numbers, 193–95, 198–99
in scientific notation, 371–74
shortcuts, 39–41, 109
of signed numbers, 266–67
of whole numbers, 33–39
by zero, 33
Divisor, 33
Dual dimensioning, 216

Efficiency, 173
Equations, 140–44, 307–37
containing square roots, 335–36
equivalent, 309
quadratic, 560–73
solution of, 307–24, 326, 541
system of, 541–60
inconsistent, 545
Equilateral triangles, 409, 438
Equivalent equations, 309
Equivalent fractions, 59
Estimating, 8–10
Expanded form
of decimals, 94
of a whole number, 4
Exponents, 364–74
negative, 367
Expressions, algebraic, 291, 293
Exterior angles, 392

Factors, 23, 37, 272, 293, 364
prime, 36, 38
Formulas
evaluating, 293–97
solving, 333–40
Fractions, 55–82
addition of, 75–81
comparing, 62–63
decimal, 117–22
decimal equivalents of, 121
division of, 74–77
equivalent, 59
improper, 57
like, 76
lowest terms, 61
multiplication of, 68–70
proper, 57
subtraction of, 82–86
Frustum
of a cone, 464
lateral surface area of, 465
of a pyramid, 455
volume of, 456

Gauge blocks, 244–45
Geometric constructions, 471–85
bisecting an angle, 472–73
bisecting a line segment, 471–72
constructing a circle, 482

constructing an equilateral triangle, 479–80
dividing a line segment, 481
duplicating an angle, 477–78
duplicating a triangle, 479
finding the center of a circle, 482
parallel to a line, 480
perpendicular at an end point, 475–76
perpendicular from a line, 475
perpendicular to a line, 474
tangent to a circle, 483
Geometry
plane, 385–435
solid, 447–71

Hero's formula, 417
Hexagon, 418–21, 438
area of, 418–19, 438
dimensions of, 419, 438
Hypotenuse, of a right triangle, 410

Improper fractions, 57
Inconsistent system of equations, 545
Integers, 256
Interest, 169–70
Inverse trigonometric function, 513

Lateral surface area, 449–57, 462–64
LCD, 77–81
Least Common Denominator, 77–81
Legs, of a triangle, 500
Length
of an arc, 498
units, 213–16
List price, 164
Literal expression, 291
Loans, 174
Lowest terms, of fraction, 61
Lumber measure, 209

Mass, metric units of, 221
Mean, 114
Measurement, 189–224
length, 210–11, 213–16
Measurement numbers
addition, 191–93, 197
division, 193–95, 198–99
multiplication, 193–95, 198–99
rounding, 110–13, 193, 196–99, 201
subtraction, 191–93, 197
units, 189, 203–10
Meters, 241–43
range, 241
Metric units, 211–24
area, 208–10
length, 213–16
mass, 221
speed, 207, 218
temperature, 222–24
volume, 210, 219–20
weight, 221
Micrometers, 232–35
vernier, 235
Minuend, 16
Mixed number, 57
Multiplication
of algebraic expressions, 291–92, 364–67
of decimal numbers, 104–07
of fractions, 65–68
of measurement numbers, 193

in scientific notation, 371–74
shortcuts, 27, 109
of signed numbers, 263–66
of whole numbers, 23–30
by zero, 26

Negative numbers, 255–70
adding, 257–60
dividing, 269–70
multiplying, 268–69
subtracting, 263–66
Numbers
decimal, 93–122
expanded form, 4, 94
measurement, 185–224
mixed, 57
negative, 255–70
prime, 38
whole, 3
Numerator, of fractions, 56

Obtuse angle, 495
Opposites, 263
Order of operations, 41–44, 275–76, 295

Parallel lines, 391–92
Parallelogram, 398–405
area, 403–405, 438
perimeter, 397, 438
Parentheses, 292, 304–05, 326–30
Pentagon, 418, 421
Percent, 148–79
applications, 164–79
change, 176–79
commission, 170–72
credit cards, 171
discount, 164–68
efficiency, 173
interest, 169
sales tax, 168
tolerance, 173–74
changing decimal numbers to, 150
changing to decimal numbers, 152–54
changing fractions to, 151
equivalents, 154
problems, 155–62
Percentage, 156, 158
Perimeter, 397
Pi, 427
Place value, 4, 93–95
Polygons, 396–407, 421
diagonal of, 397
irregular, 421–23
parallelogram, 398–405
area of, 403–05, 438
perimeter of, 397, 438
perimeter of, 397
quadrilateral, 398–407
rectangle, 398–401
area of, 399–400
regular, 396–404, 418
sides of, 397
square, 397
area of, 401–404
perimeter of, 404–05
side of, 398
trapezoid, 398, 405–7
area, 405–06

triangle, 391, 409–17
area, 414–15
equilateral, 410
isosceles, 410
right, 410, 499–532
scalene, 410
Precision, 190, 196
Prime factors, 38
Prime numbers, 38
Principal, 174
Prism, 447–53
altitude, 448
edge of, 448
face of, 448
rectangular, 448
right, 448
surface area of, 449–53
trapezoidal, 449
triangular, 449
vertex of, 448
volume of, 449–53
Product, 23
Proper fraction, 56
Proportion, 141–45, 350–60
direct, 355–57
inverse, 355–57
Protractor, 240–41, 387, 610
bevel, 240
vernier, 241
Pyramid, 453
altitude of, 453
frustum of, 455–57
lateral surface area of, 454–55
right, 453
slant height of, 453
volume of, 454–55
Pythagorean theorem, 411–14, 501–02

Quadratic equations, 560–72
applications, 568–71
formula for solving, 564–68
solution of, 562–68
standard form of, 561
Quadratic formula, 564–68
Quadrilateral, 398–407
Quotient, 33

Radian, 497–98
Radius of circle, 427
Rate of interest, 169–70
Ratio, 137–40, 350
compression, 138–39
gear, 137–38, 351, 357
and proportion, 141, 350–60
pulley, 138, 351, 357
scale, 351
trigonometric, 507–14
Rectangle, 398–401
area of, 399–400
perimeter of, 397, 400
Rectangular solid, 447–71
Regular polygon, 396–404
Repeating decimal, 119–22
Right angle, 387
Right triangle, 410, 499–532
solution of, 515–19
Rings, 433–35, 438
Rounding, 111–13, 193, 203–10

Index of Applications